ANNALS OF
THE NEW YORK ACADEMY
OF SCIENCES

Volume 400

EDITORIAL STAFF

Executive Editor
BILL BOLAND

Managing Editor
JOYCE HITCHCOCK

Associate Editor
LINDA MEHTA

The New York Academy of Sciences
2 East 63rd Street
New York, New York 10021

NEUROTENSIN, A BRAIN AND GASTROINTESTINAL PEPTIDE

NEUROTENSIN, A BRAIN AND GASTROINTESTINAL PEPTIDE

Edited by Charles B. Nemeroff and Arthur J. Prange, Jr.

The New York Academy of Sciences
New York, New York
1982

Library of Congress Cataloging in Publication Data
Main entry under title:
Neurotensin, a brain and gastrointestinal peptide.
    (Annals of the New York Academy of Sciences; 400)
    Papers presented at a conference held March 24-26,
1982, by the New York Academy of Sciences.
    Bibliography: p.
    Includes index.
    1. Neurotensin—Congresses. I. Nemeroff, Charles B.
II. Prange, Arthur J., 1926-   . III. New York
Academy of Sciencs. IV. Series. [DNLM: 1. Neurotensin
—Congresses. 2. Central nervous system—Physiology—
Congresses. 3. Gastrointestinal system—Physiology—
Congresses. W1 AN626YL v. 400 / WL 104 N4945 1982]
Q11.N5 vol. 400   500s [599'.0192456]   82-24692

[QP572.N47]

PCP
Printed in the United States of America
ISBN 0-89766-190-7 (cloth)
ISBN 0-89766-191-5 (paper)

ANNALS OF THE NEW YORK ACADEMY OF SCIENCES

Volume 400

December 28, 1982

NEUROTENSIN, A BRAIN AND GASTROINTESTINAL PEPTIDE*

Editors and Conference Chairmen
CHARLES B. NEMEROFF AND ARTHUR J. PRANGE, JR.

*This volume is the result of a conference entitled Neurotensin, a Brain and Gastrointestinal Peptide held on March 24-26, 1982, by the New York Academy of Sciences.

Financial assistance was received from:

- Burroughs Wellcome Company
- Department of the Army—U.S. Army Medical Research and Development Command
- Department of the Army—U.S. Army Research Office
- E. I. du Pont de Nemours & Company
- Endo Laboratories, Inc.
- Hoffmann-La Roche, Inc.
- ICI Americas, Inc.
- International Society of Neuroendocrinology
- Lederle Laboratories
- McNeil Pharmaceutical
- Merck Sharp & Dohme Research Laboratories
- Norwich-Eaton Pharmaceuticals
- Office of Naval Research
- Smith Kline & French Laboratories
- The Upjohn Company
- Warner-Lambert Company
- Wyeth Laboratories

PREFACE

Charles B. Nemeroff and Arthur J. Prange, Jr.

Biological Sciences Research Center and Department of Psychiatry
University of North Carolina School of Medicine
Chapel Hill, North Carolina 27514

This volume contains the proceedings of the first conference on neurotensin, a peptide discovered almost ten years ago and now known to be present within the central nervous system and the gastrointestinal tract of a variety of mammals. The conference was organized along two major themes: (1) the localization and effects of neurotensin in the periphery, including its endocrine and cardiovascular actions, and (2) the localization, behavioral and neurochemical effects of neurotensin in the central nervous system.

We were fortunate in assembling many of the leading investigators in the various sub-disciplines of neurotensin research. Thus, R. E. Carraway and S. E. Leeman, the co-discoverers of the tridecapeptide, described respectively the original isolation of the peptide and current problems in the assay of neurotensin and other oligopeptides. Use of immunohistochemical and radioimmunoassay techniques to describe the distribution of the peptide in the brain and gut received much attention, and to a lesser extent the characterization and distribution of putative neurotensin receptors was discussed. An entire session was devoted to endocrine effects of the peptide. Within the three sessions on neurotensin and the central nervous system, its inactivation, electrophysiological effects, antinociceptive and hypothermic properties and interaction with centrally active drugs were scrutinized. Moreover, its distribution in post-mortem human brain tissue in psychiatric and neurologic disease was described. Two poster sessions served to amplify the themes mentioned above and to introduce novel areas of investigation.

The goals for the conference which we had in mind are implicit from the above. In simplest terms we sought to provide a forum in which most of the leading neurotensin workers could present their work and, in turn, be enriched by the presentations of the others. For all participants, the whole, we hoped, would be greater than the sum of the parts. In such a context, at least the outline for a comprehensive view of the role of neurotensin might appear. From our subsequent discussions with the conference contributors and participants, we have concluded that much of the task was successfully completed. This is evident in the discussions which followed each presentation.

We would like to acknowledge the excellent advice and aid provided by the staff of the New York Academy of Sciences, including the Conference Committee and its advisory subcommittee. In particular, Mrs. Ellen Marks and Ms. Renée Wilkinson were supportive and efficient and made our organizational duties a pleasure. Finally without the financial support of the sponsors, this conference certainly would not have taken place; for this we are indeed grateful.

NEUROTENSIN: DISCOVERY, ISOLATION, CHARACTERIZATION, SYNTHESIS AND POSSIBLE PHYSIOLOGICAL ROLES*

Susan E. Leeman and Robert E. Carraway

Department of Physiology
University of Massachusetts Medical School
Worcester, Massachusetts 01605

DISCOVERY

The discovery of the biologic activity that led to the isolation of the peptide neurotensin (NT) occurred while one of us (Leeman) was screening column eluates from an ion-exchange column for sialagogic activity in the course of work directed toward the isolation of substance P. A characteristic vasodilation of the exposed cutaneous regions of the rat, particularly noticeable around the face and ears, occurred after intravenous injection of the material designated in FIGURE 1. It was this simple vasoactive response that was used to monitor purification procedures through the purification of this new peptide.

ISOLATION

Although NT was discovered in and first isolated from bovine hypothalamic extracts,[1] the same peptide has been obtained in pure form from extracts of calf[2,3] as well as human small intestine.[4] The purification scheme employed initially (TABLE 1) effected a more than 10^7-fold enrichment of activity based upon the initial content of protein, with an overall yield near 20%. The procedure has been modified somewhat over the years, the major changes being the introduction of an affinity chromatography step (utilizing antibody to NT) and the use of high-performance liquid chromatography (HPLC). Whereas the earlier studies required large quantities of tissue (*ca* 45 kg), more recent work has led to the isolation of chicken NT from only 8 kg of intestine[5] and human NT from about 60 g of scraped intestinal mucosa.[4] This trend reflects the advances in chromatographic procedures and peptide analyses that have come about this past decade.[6,7]

STRUCTURE

Sequence studies have been performed on NT isolated (a) from bovine hypothalami using bioassay, (b) from bovine hypothalami using radioimmunoassay (RIA), and (c) from bovine intestine using RIA. The structure deduced from the information obtained on each of these preparations was identical. The results of a typical analysis are shown in FIGURE 2, which summarizes information obtained while sequencing papain-generated fragments of NT. The sites of cleavage by

*This work was supported by National Institutes of Health Grants AM 28565 and AM 29876.

1

Annals New York Academy of Sciences

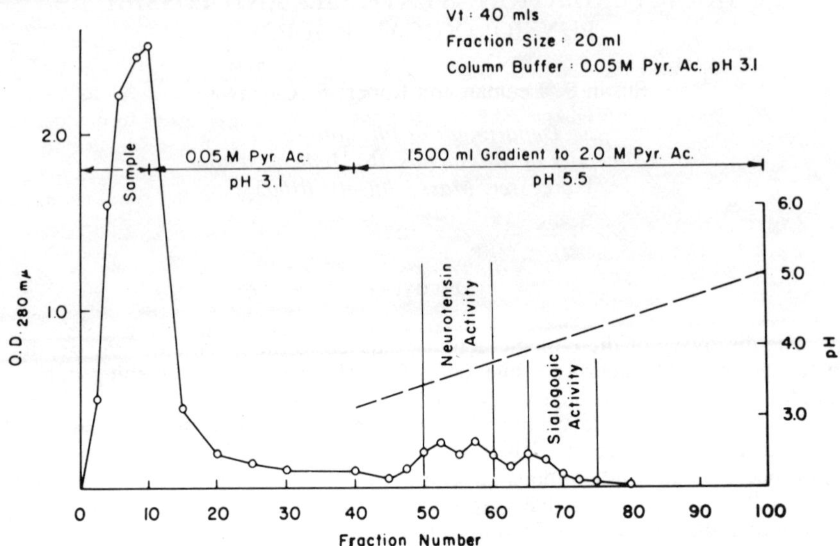

FIGURE 1. Ion-exchange chromatography of a bovine hypothalamic extract on sulfoethyl Sephadex C-25. Neurotensin and substance P (sialogogic activity) were detected using bioassays and protein concentration was monitored at 280 mμ. Pyr Ac = pyridine acetate.

other endopeptides that have proved useful in these analyses are also indicated. The results of total enzymic digestions indicated that all amino acids were in the L-configuration, that there were no non-amino-acid substituents attached to the molecule, and that the aspartic acid was in the amidated form. Also consistent with this structure were the results of charge measurements for each of the peptides at pH 1.9 and 6.5 by a technique involving paper electrophoresis. The presence of an NH_2-terminal pyrrolidone carboxylic acid residue (<Glu) was established by treatment of NT and its fragments with L-pyroglutamyl peptide hydrolase. Kinetic experiments using carboxypeptidase A and B served to establish COOH-terminal

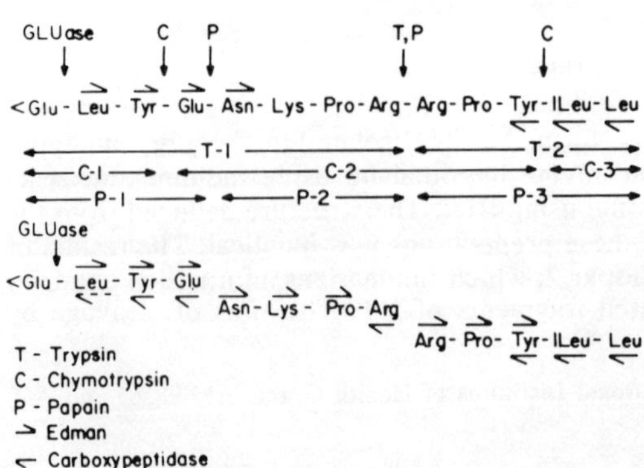

FIGURE 2. Complete amino acid sequence of NT and the alignment of the fragments obtained by enzymic cleavage. (<GLUase) pyrrolidonecarboxylyl peptidase; (T-1,2) tryptic peptides: (C-1-3) chymotryptic peptides; (P-1-3) papain peptides. From reference 8 with permission.

sequences for each peptide, while the Edman degradation proved useful for NH$_2$-terminal sequences.

The amino acid sequence of NT appears at this time to be unique and not contained within any other known peptide or protein.[8] There are, however, similarities between NT and other biologically active, small peptides. NT, as well as xenopsin, thyrotropin-releasing hormone, and luteinizing-hormone-releasing hormone (LRH), has an NH$_2$-terminal < Glu residue, possibly in order to protect the peptide from NH$_2$-terminal degradation. Using the methods described by Dayhoff[9] to assess sequence similarity, we obtained results suggesting a distant relatedness between vasopressin, LRH, and NT. The COOH-terminal sequence of NT (-Arg-Pro-Tyr-Ile-Leu-OH) is also similar to that of angiotensin I

TABLE 1

SUMMARY OF PURIFICATION OF NEUROTENSIN FROM
45 KG OF BOVINE HYPOTHALAMI

Purification Step	Total Protein (mg)	Total Sialogogic Doses Recovered	Total Neurotensin Doses Recovered	Neurotensin Yield	Neurotensin Specific Activity (doses/mg)
80% Acetone extraction	270,000*	40,000	15,000‡	100	0.06
First gel chromatography on Sephadex G-25	10,000*	33,000			
Second gel chromatography on Sephadex G-25	1,500*	30,000	11,000	75	7.0
First chromatography on SE-Sephadex	250*	20,000	7,500	50	30.0
Second chromatography on SE-Sephadex	10*		6,000	40	600.0
Paper electrophoresis pH 3.5	0.3†		3,100	21	10,300.0

*Protein is expressed as absorbance units at 280 nm.

†Protein was calculated from quantitative amino acid analyses.

‡Calculated by assuming a 75% yield of neurotensin through the two initial gel chromatography stages.

(-His-Pro-Phe-His-Leu-OH). These relationships are interesting to note and it is tempting to suggest that they reflect some common ancestry for these different peptides. Xenopsin (< Glu-Gly-Lys-Arg-Pro-Trp-Ile-Leu-OH), which is a peptide of nonmammalian origin (*Xenopus laevis* skin[10]), bears a striking resemblance to NT and also shares a number of its biological properties. This peptide is likely to be a member of the NT family and recent work in our laboratory suggests that xenopsin-like substances, apart from NT, exist in mammalian tissue.[11]

SYNTHESIS

The first synthesis of NT was performed using solid-phase methodology.[12] The synthetic scheme employed put to use the tendency of NH$_2$-terminal glutaminyl peptides to rearrange to their pyroglutamyl derivatives. Thus, the peptide was synthesized as GLn[1]-NT which when treated with acid, cyclized to the < Glu form. Because Gln[1]-NT and NT are quite easily separated, this approach permitted us to obtain both substances in pure form. This was useful since it allowed us to ask whether Gln[1]-NT would be cyclized to the < Glu form by the treatments used to purify NT from tissues. We established that there was, indeed, quantitative conversion when synthetic Gln[1]-NT was added to an acid–acetone extract of hypothalamic tissue. Thus, it is possible that native NT exists *in situ* in the Gln[1] form; however, because this reaction is catalyzed by heat and acid, devising an extraction procedure to test this possibility would be difficult. Chang *et al.*[13] have also suggested that the Glu[4] residue in NT may have resulted from deamidation of a Gln residue. This possibility is not very likely since substance P was isolated from the same tissue preparations with its Gln[5] and Gln[6] residues intact.[14]

Considerable evidence supports the identity of synthetic NT to the extracted native peptide. Initially, they were shown to be indistinguishable on the basis of multiple biologic and chromatographic criteria, in addition to giving similar peptide maps and sequence results.[12] More recent studies employing RIA have demonstrated that native and synthetic NT cross-react equally with multiple region-specific antisera towards both ends of the molecule,[15, 16] and that they co-elute during high-resolution HPLC.[4]

STRUCTURE–FUNCTION STUDIES

Quite a bit of information is now available regarding structure–activity relationships for NT. The earliest studies concerned the biologic capabilities of COOH-terminal partial sequences of NT that became available during solid-phase synthesis of NT.[17] The results indicated that the major participants in the hypotensive, hyperglycemic, and gut-contracting effects of NT resided within its COOH-terminal five to six residues. Although the COOH-terminal pentapeptide possessed only about 1% potency relative to NT in a number of bioassays, it displayed full intrinsic biologic activity, indicating that it contained all of the function groups necessary to fully activate these biologic receptors. In contrast, all NH_2-terminal partial sequences tested, including NT[1−10] and NT[13]-amide were essentially ineffective. These findings led us to propose the model of a NT–receptor interaction shown in FIGURE 3.

All structure–activity studies performed since confirm the importance of COOH-terminal structures as determinants of specific binding and biologic action. Studies with partial peptides and analogs of NT have now been performed in regard to other biologic effects as well as binding to receptor preparations[18] and these are discussed and presented in more detail by Kitabgi *et al.* in this volume.[19]

Using RIA, information has been gathered that suggests that the COOH-terminal region of NT is highly conserved in evolution, while the NH_2-terminus is allowed to vary (See Carraway, this volume[20]). Although only a few relatives of mammalian NT have been isolated so far, all of these share COOH-terminal structures with NT. Xenopsin, for example, which is an octopeptide isolated from the skin of the frog, *Xenopus laevis*, bears a striking resemblance to the COOH-terminal six residues in NT, and although it lacks a segment corresponding to the NH_2-terminal five residues, it exhibits many of the biologic actions of NT. Thus, the available information is consistent with the proposed functional importance of COOH-terminal structures in NT, and furthermore suggests that NT and its counterparts in lower animals participate in important biologic processes that are essential to life.

DISTRIBUTION

Using synthetic NT conjugated by various means to carrier substances such as succinylated hemocyanin, antisera that recognize NT have been obtained. When characterized by RIA, several of these antisera have been shown to display quite different specificities for NT.[16, 21] Both heterogeneous and region-specific antisera have been characterized and shown to be useful in the analysis of NT-related substances (see Carraway in this volume[20]). The relative reactivities of test material with the different antisera can be used as an index of similarity to NT. Unequal measurements have been shown to indicate the presence of immunoreactive substances which differ from NT, while equal measurements have been con-

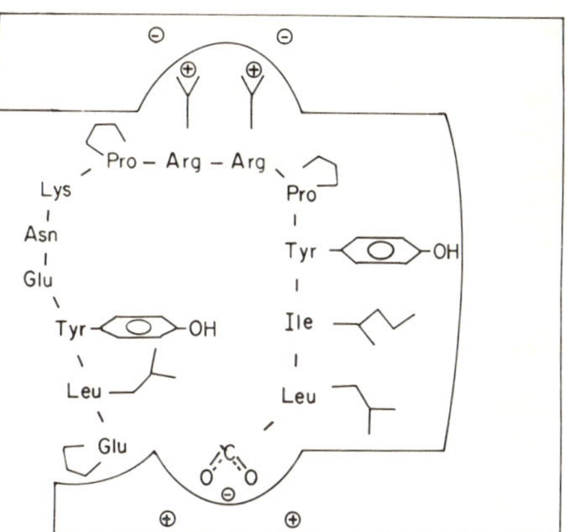

FIGURE 3. Proposed model for neurotensin receptor interaction. Strong ionic interactions involving the side chains of Arg[8], Arg[9] and the COOH-terminal carboxyl with oppositely charged groups on the receptor appear to be important for binding. Biologic efficacy derives to a large extent from hydrophobic interactions involving the side chains of -Pro-Tyr-Ile-Leu. From reference 17 with permission.

sidered strong evidence for the presence of NT[16]. Using this approach, we have determined the distribution and character of radioimmunoassayable NT (R-NT) through various organs of the rat,[14] chicken,[22] and human.[4] Concentrations of R-NT through the central nervous system (CNS) and gastrointestinal tissues of the rat are given in TABLE 2. These values, obtained using a region specific antiserum (HC-8), were in agreement with those determined using a heterogeneous antiserum (PGL-6), suggesting that they reflect the distribution of NT itself. Indeed, when careful chromatographic and immunochemical analyses were performed, these measurements were shown to be attributable to material co-eluting with synthetic NT and reacting equally well with both NH_2- and COOH-terminal directed antisera.[23]

Recently, we have studied the distribution of NT-related material in fresh postmortem muscosal scrapings of human small intestine.[4] FIGURE 4 shows that the concentration of R-NT increases from duodenum to jejunum and ileum, as has been shown for all other mammals examined to date. In the jejunoileal region, equal measurements were obtained with the three region-specific antisera utilized, while measurements in the duodenum were 2–50 times higher with the COOH-terminal directed, PGL-4 than with the NH_2-terminal directed TG-1. This disparity

TABLE 2

DISTRIBUTION OF RADIOIMMUNOASSAYABLE NT (R-NT) IN THE BRAIN
AND GASTROINTESTINAL TISSUES OF THE RAT

Tissue	R-NT* (pmol/g)
Hypothalamus	60.0 ± 3.1
Thalamus	16.1 ± 1.3
Spinal cord	13.7 ± 0.7
Brainstem	12.9 ± 0.7
Cerebral cortex	2.0 ± 0.1
Cerebellum	0.8 ± 0.1
Anterior pituitary	24.0 ± 3.4
Posterior pituitary	31.1 ± 4.1
Esophagus	0.1 ± 0.7
Stomach	0.6 ± 0.3
Duodenum (first 12–15 cm)	1.4 ± 0.6
Postduodenum (next 12–15 cm)	5.6 ± 2.4
Jejunum	50 ± 15
Ileum	48 ± 12
Large intestine	7.3 ± 2.1
Pancreas	0.15 ± 0.05

*Given are levels of R-NT/g wet weight of tissue as measured using antiserum HC-8. Mean ± SEM for 3–20 samples.

is probably due to the presence of cross-reacting NT-related peptides also observed in rat and bovine stomach and duodenum.[15]

The relative distribution of NT through rat CNS determined by RIA is in qualitative agreement with that observed by immunohistochemical techniques.[18] High concentrations of R-NT were found in the hypothalamus, thalamus, amygdaloid nucleus, and the substantia gelatinosa of the spinal cord, areas where high densities of NT-positive cell bodies and nerve terminals were also observed. Immunochemical staining of the hypothalamic region has demonstrated the presence of multiple clusters of cell bodies and terminals associated with the arcuate nucleus, the paraventricular nucleus, and the bed nucleus of the stria terminalis.[24] Fibers were also noted projecting to the hypophysial portal system in the median eminence. These findings suggest that NT may affect neuroendocrine mechanisms

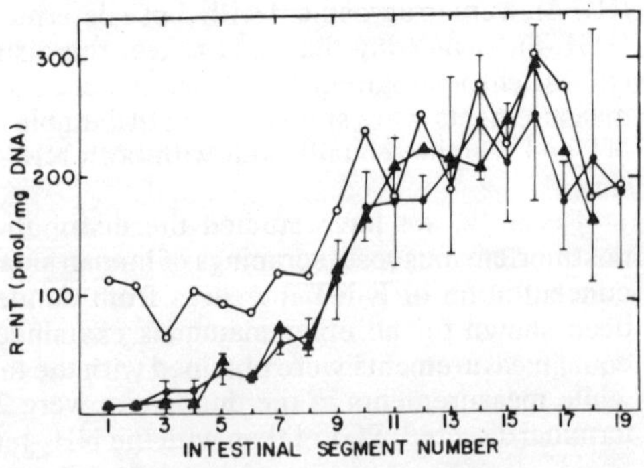

FIGURE 4. Average measurements of radioimmunoassayable NT (R-NT) at points sampled along the small intestine of four subjects, determined with antisera HC-8 (●), TG-1 (▲), and PGL-4 (○). Error bars indicate standard deviation of HC-8 measurements. From reference 4 with permission.

at several levels, including actions on the pituitary gland. For further details refer to Khan *et al.*, this volume.[25]

In the gastrointestinal tract, R-NT has been found to be localized to the mucosal region of the jejunoileum (in all mammals studied), where it resides within an endocrine-like cell called the "N-cell." FIGURE 5 shows several N-cells in the guinea pig ileum visualized by the peroxidase–antiperoxidase method. The N-cell has often been seen to make contact with the intestinal lumen via a narrow apical process and is therefore classified as an "open"-type endocrine cell. (For more details refer to Sundler *et al.* this volume[26]). The unique appearance of the villus-containing surface of the cell leads one to conjecture that it may possess a specialized receptor to monitor luminal contents and to signal for the release of NT from its storage granules upon arrival of the appropriate stimulus. The results of

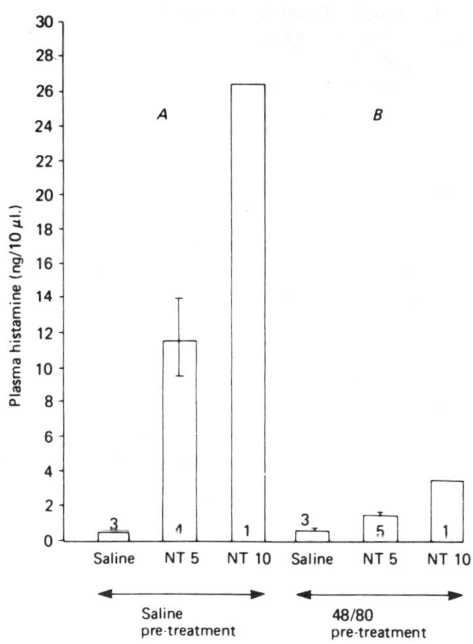

FIGURE 5. Plasma histamine levels in response to neurotensin in animals pretreated 24 hr previously with saline (a) or compound 48/80, 1 mg/kg (B). Neurotensin (5 or 10 nmol/kg) or saline (0.3 ml) were given intravenously. Mean ± S.E. of mean of n experiments, quoted within the column. From reference 28 with permission.

a number of studies presented at this symposium argue that intraluminal lipid is a potent stimulus for the release of NT into the circulation (refer to Rosell *et al.* this volume[27]), and one wonders if the luminal surface of the NT cell monitors lipid concentrations.

BIOLOGIC PROPERTIES

Some of the pharmacologic capabilities of NT are summaried in TABLE 3. Effects have been noted for both *in vivo* and *in vitro* preparations; however, as yet, it is not known whether endogenous NT participates physiologically in any of these actions. The number and diversity of these biologic activities is impressive and makes one wonder how all of these effects might be promoted by a single messenger in a controlled manner. One possibility is that NT operates as a local regulator, affecting only the cells immediately surrounding its site of release and

TABLE 3

SOME BIOLOGIC ACTIVITIES OF NT

In vitro:
   Contraction of guinea pig ileum, rat fundus strip and guinea pig atrium
   Relaxation of rat duodenum
In vivo (peripheral injection):
   Vasodilation, hypotension and cyanosis
   Increased vascular permeability
   Increased secretion of ACTH, LH, FSH, GH and prolactin
   Hyperglycemia involving glucagon, insulin and histamine
   Inhibition of gastric acid and pepsin secretion
   Delayed vasoconstriction in adipose tissue
   Inhibition of gastric motility
In vivo (central injection):
   Hypothermia
   Enhanced phenobarbital action
   Analgesic effect

being degraded rapidly as it diffuses away (as is the case for acetylcholine, for example). Consistent with this is the rapid degradation observed after intravenous injection of NT into rats ($t_{1/2}$ ca 0.5 min). Another suggestion is that some of these abilities for NT represent cross-reactions of NT with receptors meant for other substances, relatives of NT which are the actual physiologic mediators of these actions (analogous to the relationships between oxytocin and vasopressin, for example). The COOH-terminal variants of NT (which cross-react with antisera towards this region of NT) might be considered candidates for these biologically active members of the NT family. These substances are discussed in detail by Carraway, this volume.[20]

NT is a potent hypotensive agent that also increases vascular permeability and brings about a stasis of blood in peripheral tissues. These were the properties that led to its discovery, since they produce a visible vasodilitation in the skin followed by a cyanosis lasting several minutes. Recent studies suggest that these responses may be secondary to an action of NT on mast cells.[28] NT releases histamine from isolated mast cells and when injected intravenously into rats causes a pronounced (> 30-fold) elevation in plasma histamine levels within 1 minute (FIG. 5). Prior depletion of mast cells by prolonged treatment with compound 48/80 blocks the effect of NT on histamine levels, as does prior injection of disodium cromoglycate. Furthermore, the increase in vascular permeability observed after intradermal injection of NT into blued rats (Evans blue dye) is blocked by prior administration of the antihistamine, diphenhydramine. As discussed in detail by Rioux *et al.* in this volume,[29] histamine also appears to be involved in the induction of hypotension by NT and has been implicated as a mediator of the hyperglycemic action of NT.[30]

Recent studies by Cochrane *et al.*, this volume,[31] demonstrate that the concentration of histamine in perfusates of minced rat skin is elevated in a dose-dependent manner by NT and that the minimum effective concentration of NT in the medium is less than $10^{-11}M$. Thus, although isolated peritoneal mast cells require rather high levels of NT to stimulate histamine secretion ($> 10^{-9}M$), mast cells within intact segments of skin respond at doses near the estimated circulating levels of NT (ca $10^{-11}M$).

It should also be noted that isolated rat mast cells have been shown to bind

[125]I-labeled NT in a stereospecific fashion.[32] Although the affinity appears to be rather low (kd $10^{-7}M$), binding was highly dependent upon concentrations of Na, K, Ca, and Mg, and thus the conditions chosen may not have been ideal. The major determinants of binding were also found within the COOH-terminal half of NT as has been shown for a number of biologic actions of NT. These findings suggest that one primary action of NT is on the mast cell and that various products of mast cell secretion, including histamine, can then go on to induce hypotension, hyperglycemia, increased vascular permeability, and stasis. Although relatively high levels of NT are necessary to observe some of these actions, it may be that NT exerts local effects on the microcirculation by affecting mast cells immediately surrounding its site of release. Perhaps NT participates in vascular adjustments involved in digestion or even in inflammation through its actions on mast cells.

PLASMA LEVELS OF R-NT

Immunochemical and chromatographic analyses on extracts of rat, bovine and human plasma has led to the conclusion that NT is present in the circulation.[33] FIGURE 6 shows the profile of immunoreactivity observed when an acid–acetone extract of bovine plasma was subjected to chromatography on Sephadex G-25 and the eluates were examined in the RIA using both NH$_2$-terminal- and COOH-terminal-directed antisera. Only the material eluting in region I behaved like NT, giving equal measurements with the two antisera, and when this was further analyzed by ion exchange chromatography and HPLC, it eluted again in the region of NT. The plasma level of the component that was indistinguishable from NT was estimated to be 15–25 fmol/ml, which was about 30–50% of the measurements obtained on unfractionated extracts of plasma using antiserum HC-8. Although we can conclude that NT enters the circulation, it is not clear whether it is delivered as a hormone to affect distant targets or is simply enroute to degradation, having acted in a local, paracrine fashion.

In order to determine whether the NT-related peptides found in region II (FIG. 6) represented breakdown products of NT, rats were infused with NT for 30 min at 0.1 nmol/kg/min and then plasma was obtained for analysis. Chromatography on Sephadex G-25 revealed the generation of NH$_2$-terminal metabolites that eluted just behind the peak in region I (FIG. 6). No immunoreactivity was seen to

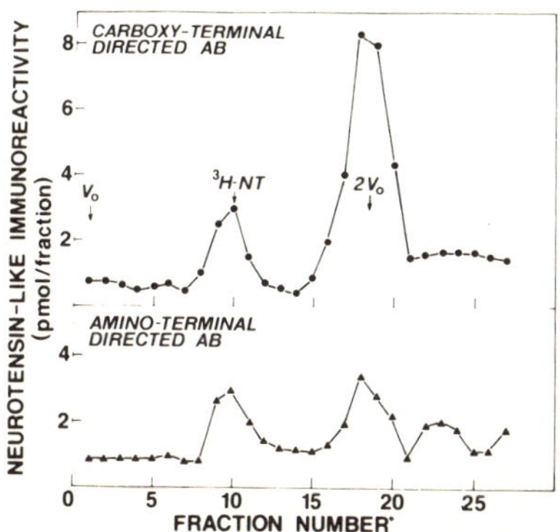

FIGURE 6. Gel chromatography of bovine plasma extracts on Sephadex G-25. The sample consisted of an acid/acetone extract of 1.2 liters bovine plasma dissolved in 200 ml column buffer. The column dimensions were 14 × 123 cm with a bed volume of approximately 18.5 liters. The column buffer was 0.2 M acetic acid. After the sample was applied, 7 liters of effluent were discarded and then fractions of 500 ml were collected. Multiples of the void volume (V_0) and the elution position for ^{3}H-labeled NT are indicated. Aliquots of each fraction were placed into the RIA for NT using the NH$_2$-terminal antiserum (TG-1) and COOH-terminal antiserum (HC-8). Recovery 66.

elute in region II, indicating that this material was likely to represent variants of NT from another source.

Meal-induced Release of Human R-NT

Interested by the reports of Rosell and co-workers[34, 35] that plasma levels of R-NT were elevated in humans following ingestion of a high-fat meal, Hammer *et al.*[36] in our group performed a careful analysis of the immunochemical and chromatographic behavior of the NT-like components of human plasma obtained before and after a defined meal (37 g protein, 59 g carbohydrate and 76 g fat, totaling 1050 calories). Measurements on unfractionated acid–acetone extracts indicated that NH_2-terminal immunoreactivity was elevated after the meal, while COOH-terminal immunoreactivity was not significantly changed (TABLE 4). Analysis by HPLC revealed that the elevated NH_2-terminal components co-chromatographed with synthetic preparations of NT^{1-8} and NT^{1-11}, fragments which have been identified as breakdown products of NT injected into rats (see Carraway, this volume[20]). These results indicate that there was a postprandial rise in the levels of substances similar to NT^{1-8} and NT^{1-11} in peripheral human plasma and suggest that the meal was a stimulus for release of NT. The short half-life of NT ($t_{1/2} = 0.5$ min in rat) and the finding that elevations were seen only for putative degradation products of NT lead one to conjecture that the released NT probably acted in a paracrine fashion. Rosell and co-workers (this volume) have suggested that NT, released from the small intestine into the circulation by ingestion of fat, acts as an enterogastrone hormone. If this is the case, then it would seem likely, from the results of Hammer and co-workers, that NT from the small intestine does not act directly on the stomach since its peripheral levels are not elevated. A recent study by Hirst *et al.*[37] indicates that COOH-terminal but not NH_2-terminal partial sequences of NT inhibit gastric acid and pepsin secretion in the cat. It is not very probable then, that the NH_2-terminal metabolites of NT, elevated in humans after a meal, promote enterogastrone-like effects. If intestinal NT plays a role in the normal control of gastric function, it seems reasonable to suggest that such actions

TABLE 4

LEVELS OF PLASMA R-NT MEASURED IN FIVE VOLUNTEERS AFTER
INGESTION OF A HIGH-FAT MEAL*

Time After Meal (min)	Plasma R-NT (fmol/ml)†	
	COOH-terminal (HC-8)	NH-terminal (TG-1)
Basal	19.7 ± 3.8	3.5 ± 2.5
40	25.8 ± 1.0	22.8 ± 4.9
90	16.3 ± 2.7	39.2 ± 6.3
180	14.2 ± 1.8	33.8 ± 4.9

*Blood was collected before and at the times indicated after a test meal and plasma from each sample was extracted, lyophilized and dissolved in a volume of assay buffer equal to one-fifth the original volume of plasma. Aliquots of 200 μl were assayed in duplicate with each antiserum.

†Given are levels of R-NT (mean ± SEM) measured using the antiserum indicated.

are mediated by effects on the enteric nervous system or by release of other substances that circulate to affect the stomach.

LIPID STIMULATION OF NEUROTENSIN RELEASE FROM RAT SMALL INTESTINE

Ferris *et al.*[38] have recently shown that perfusion of the rat small intestine *in vivo* with a micellar solution (2.4 mM Na-tauro-deoxycholate, 0.6 μM oleic acid, and 0.3 μM monolein in 0.9% saline) caused a significant increase in the concentration of R-NT in hepatic portal vein plasma (FIG. 7), while perfusion with amino acids, glucose, hyperosmotic saline, acidified saline, bile salt, and diluted rat bile had no effect. After HPLC of the immunoreactive material, an increase in the peak

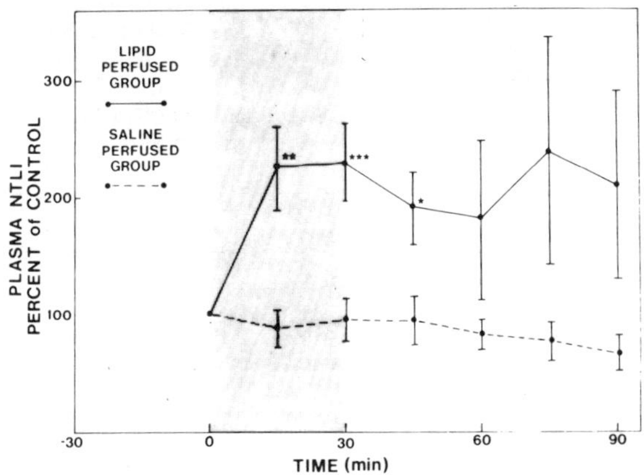

FIGURE 7. Measurements of immunoreactive neurotensin (NTLI) in superior mesenteric vein plasma during saline and lipid perfusion of the small intestine. The levels of NTLI were calculated as percent changes with each animal serving as its own control. The mean and SEM (vertical bars) of these values are presented. The levels in samples obtained at −15 and 0 minutes were averaged and used as the control values. The NTLI (mean ± SEM) in the control period for the saline perfused rats (n = 4) was 47 ± 5 fmol/ml and for the lipid perfused rat (n = 10) was 53 ± 9 fmol/mol. The shaded area denotes the duration of the lipid perfusion in the experimental group. *p < 0.025, **p < 0.01, ***p < 0.005. From reference 38 with permission.

of R-NT, which chromatographed with synthetic NT, was demonstrable (FIG. 8). That the intestine was the source of this material is supported by the fact that R-NT was higher in hepatic portal plasma than in arterial plasma, sampled concomitantly (Ferris, C. *et al.*, this volume[39]). These results demonstrate that intraintestinal lipid is an effective and specific stimulus for the release of NT from the small intestine. Even in the portal circulation, however, the increase in the level of R-NT was only modest (from *ca* 9 to 18 fmol/ml plasma) and this increase was, apparently, not communicated to the peripheral circulation within the time course of this experiment (15 min). Recent work indicates that the concentration of NT[1−8] is also elevated in portal plasma during lipid stimulation suggesting that NT is degraded in close proximity to its site of release, the intestine. These results are compatible with either an endocrine or a paracrine role for NT, but do suggest that if intestinal NT has a hormonal role, the target organ is nearby, possibly the liver, heart, or the intestine itself.

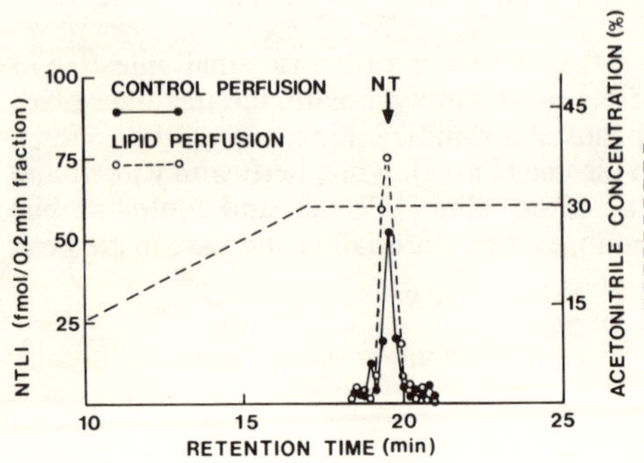

FIGURE 8. High-pressure liquid chromatographic analysis of immunoreactive NT (NTLI) in plasma collected during control and lipid perfusion of the small intestine. Samples from superior mesenteric vein plasma were pooled, extracted, lyophilized, and dissolved in 1.8 ml distilled water. Samples were injected onto a 3.9×300 mm μBondapak C-18 column running at 2 ml/min in 0.01 M KH_2PO_4, pH 4.6. Five minutes after sample application, a twelve-minute-long gradient was run to 30% acetonitrile in 0.01 M KH_2PO_4. The arrow denotes the retention of a synthetic preparation of neurotensin (NT). The chromatograms were obtained using pooled extracts of 17 ml of plasma from nine animals during control perfusion and 16 ml of plasma obtained during lipid perfusion. From reference 38 with permission.

NEUROTENSIN AND THE CENTRAL REGULATION OF LUTEINIZING HORMONE SECRETION

Neurons immunoreactive to gonadotropin-releasing hormone have been demonstrated in the medial preoptic-septal region of the rat hypothalamus, an area known to be essential for the surge of luteinizing (LH) secretion that preceeds ovulation. The presence of immunoreactive neurotensin (144 $\pm$ 20 fmol/mg protein) in this region prompted us to evaluate whether stereotaxic microinjection of neurotensin into this area would affect peripheral plasma LH levels. The rat model chosen for these initial studies was the long-term ovariectomized female rat anesthetized with Dial-urethane, an anesthetic that depresses the pulsatile release of gonadotropins. The responsiveness of this model was first assessed by the stereotaxic microinjection of norepinephrine (50 ng/nl) into this area, which was found to be effective in stimulating the secretion of LH (FIG. 9). The stereotaxic microinjection of neurotensin (40 ng/50 nl) into the rostal portion of the medial preoptic nucleus of the test rats also caused a significant elevation of plasma LH levels (FIG. 9). In contrast, saline injections were without effect. This data as well as the presence of high-affinity binding sites for neurotensin (Ferris *et al.*, this volume[37]) support a role for neurotensin in the central regulation of LH release.

CONCLUDING REMARKS

Eight years have elapsed since the isolation of neurotensin was first reported and in this time an impressive amount of information has accumulated concerning the distribution, localization and biologic capabilities of this "messenger"-peptide. NT has been implicated as a mediator of events in the CNS, as a signal for mast cell secretion, as a regulator of pancreatic and pituitary function, and as an enterogastrone. These are effects demonstrated for exogenously administered

NT; now the issue at hand is whether endogenous NT actually participates in any of these action. Our ability to address this question will be enhanced by the development of specific antagonists to NT that might be employed to interfere with the actions of endogenous peptide.

Although the presence of NT within the circulation has been demonstrated, it is, as yet, unclear whether this peptide operates hormonally or in a paracrine fashion. Its broad distribution and very short half-life in blood suggest that its effects might be confined to areas near to its site of release. The source(s) and the target(s) for circulating NT also have to be defined. These important questions challenge us to develop new approaches to study hormonal communication and to distinguish it from paracrine influences.

Although the degradation of NT has been studied, little is known concerning

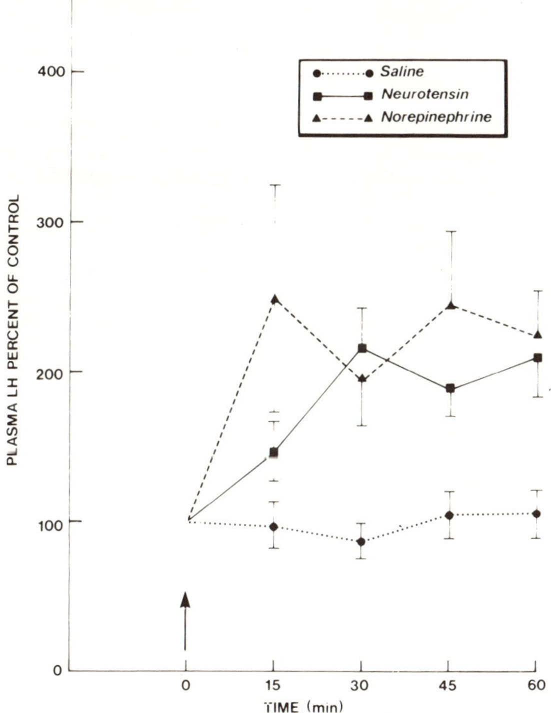

FIGURE 9. Stereotaxic microinjections into the rostral medial preoptic nucleus of ovariectomized rats. Blood samples were drawn from the femoral artery at 15 min intervals one hour before and following the microinjection (arrow) of saline (n = 7), NT (n = 9) or NE (n = 4). Plasma LH levels were calculated as percent changes with each animal serving as its own control. The mean and SEM (vertical bars) of these values are presented. The mean of the LH levels preceding microinjection were used as control values. The values obtained at 15, 30, 45, and 60 min in both the NT and NE microinjected groups were significantly greater than their controls (p < .05). Injection site: incisor bar = 5.0 mm above the intraural line, (AP = 8.2 mm from vertical zero plane, L = 2.0 mm from the midsagittal suture and angles medially at 8° from the perpendicular, V = 7.0 mm below dura. Coordinates taken from the stereotaxis atlas of Pelligrino et al., 1979.

its biosynthesis. One might predict that NT is synthesized as part of a larger molecule that undergoes processing within its storage vesicles as they mature. If so, then one might conceive of utilizing specific inhibitors to interfere with processing, storage, or degradation. Another virgin area concerns binding or transport proteins. Are there neurophysin-like substances specific for NT? If NT is co-stored and co-secreted with a binding protein that delivers it safely to its receptor then the molecule is (a) less susceptible to degradation and filtration, and (b) possibly in higher concentrations in plasma than presently thought. This is an important question that bears on the disparity between the circulating levels of NT ($ca\ 10^{-11}M$) and the concentrations required for many of its biologic effects *in vitro* ($ca\ 10^{-9}M$).

Although much has been accomplished over the years, providing an impetus for a meeting of this sort, there are many important questions yet to be addressed.

REFERENCES

1. CARRAWAY, R. & S. E. LEEMAN. 1973. The isolation of a new hypotensive peptide, neurotensin, from bovine hypothalami. J. Biol. Chem. **248**(19): 6854–6861.
2. KITABGI, P., R. CARRAWAY & S. E. LEEMAN. 1976. Isolation of a tridecapeptide from bovine intestine tissue and its partial characterization as neurotensin. J. Biol. Chem. **251**(22): 7053–7058.
3. CARRAWAY, R., P. KITABGI & S. E. LEEMAN. 1978. J. Biol. Chem. **253**: 7996–7998.
4. HAMMER, R. A., S. E. LEEMAN, R. CARRAWAY & R. H. WILLIAMS. 1980. Isolation of human intestinal neurotensin. J. Biol. Chem. **255**: 2476–2480.
5. CARRAWAY, R. & Y. M. BHATNAGER. 1980. Isolation, structure, and biological activity of chicken intestinal neurotensin. Peptides **1**: 167–174.
6. DAVID, T. P., H. SCHOEMAKER, A. CHEN & H. I. YAMAMURA. 1982. Life Sci. **30**: 971–987.
7. MOORE, G. J. Life Sci. **30**: 995–1002.
8. CARRAWAY, R. & S. E. LEEMAN. 1975. The amino acid sequence of a hypothalamic peptide neurotensin. J. Biol. Chem. **250**: 1907–1911.
9. BARKER, W. C. & M. O. DAYHOFF. 1972. *In* Atlas of Protein Sequence and Structure. M. D. Dayhoff, Ed.: Vol. **5**: 101–110. National Biomedical Research Foundation. Silver Spring, Maryland.
10. ARAKI, K., S. TACHIBANA, M. UCHIYAMA, T. NAKAJIMA & T. YASUHARA. 1973. Isolation and structure of a new active peptide "Xenopsin" on the smooth muscle, especially on a strip of fundus from a rat stomach, from the skin of *Xenopus laevis*. Chem. Pharm. Bull. (Tokyo) **21**: 2801–2804.
11. CARRAWAY, R. E. Unpublished results.
12. CARRAWAY, R. & S. E. LEEMAN. The synthesis of neurotensin. J. Biol. Chem. **250**: 1912–1918.
13. CHANG, D., J. HUMPHRIES, K. FOLKERS, R. E. CARRAWAY, S. E. LEEMAN & C. Y. BOWERS. 1976. Synthesis and activities of neurotensin and its amidated analogs and possible natural occurrence of (Gln⁴)-neurotensin. Proc. Nat. Acad. Sci. USA **73**: 3883.
14. CARRAWAY, R. E. & S. E. LEEMAN. 1979. The amino acid sequence of bovine hypothalamic substance P. J. Biol. Chem. **254**: 2944–2945.
15. CARRAWAY, R. & S. E. LEEMAN. 1976. Characterization of radioimmunoassayable neurotensin in the rat: Its differential distribution in the central nervous system, small intestine, and stomach. J. Biol. Chem. **251**(22): 7045–7052.
16. CARRAWAY, R. E. 1978. Neurotensin and related substances. *In* Methods of Hormone Radioimmunoassay, Second Ed. B. M. Jaffe & H. R. Behrman, Eds. Academic Press. New York. pp. 139–169.
17. CARRAWAY, R. & S. E. LEEMAN. 1975. Structural requirements for the biological activity of neurotensin, a new vasoactive peptide. *In* Peptides: Chemistry, Structure and Biology. R. Walter & J. Meienhofer, Eds. Ann Arbor Sciences. Ann Arbor, Michigan. pp. 679–685.

18. FERNSTROM, M. H., R. CARRAWAY & S. E. LEEMAN. 1980. Neurotensin. *In* Frontiers in Neuroendocrinology. L. Martini, Ed.: Vol. **6:** 103–127. Raven Press. New York.
19. KITABGI, P. 1982. Effects of neurotensin on intestinal smooth muscle: Applications to the study of structure-activity relationships. Ann. N.Y. Acad. Sci. **400:** 37–55. This volume.
20. CARRAWAY, R. E. 1982. A critical analysis of three approaches to radioimmunoassay of peptides: Applications to the study of the neurotensin family. Ann. N.Y. Acad. Sci. **400:** 17–36. This volume.
21. CARRAWAY, R. & S. E. LEEMAN. 1976. Radioimmunoassay for neurotensin, a hypothalamic peptide. J. Biol. Chem. **251**(22): 7035–7044.
22. CARRAWAY, R. & Y. M. BHATNAGER. 1980. Immunochemical characterization of neurotensin-like peptides in chicken. Peptides **1:** 159–165.
23. LEEMAN, S. E., E. A. MROZ & R. E. CARRAWAY. 1977. Substance P and neurotensin. *In* Peptides in Neurobiology. H. Gainer, Ed. New York. pp. 99–144.
24. KHAN, D., G. M. ABRAMS, E. A. ZIMMERMAN, R. E. CARRAWAY & S. E. LEEMAN. 1980. Neurotensin neurons in the rat hypothalamus: An immunocytochemical study. Endocrinology **107:** 47–53.
25. KAHN, D., A. HOU-YU & E. A. ZIMMERMAN. 1982. Localization of neurotensin in the hypothalamus. Ann. N.Y. Acad. Sci. **400:** 117–131. This volume.
26. SUNDLER, F., R. HÅKANSON, S. LEANDER & R. UDDMAN. 1982. Light and electron microscopic localization of neurotensin in the gastrointestinal tract. Ann. N.Y. Acad. Sci. **400:** 94–104. This volume.
27. ROSELL, S. 1982. The role of neurotensin in the regulation of fat metabolism. Ann. N.Y. Acad. Sci. **400:** 183–197. This volume.
28. CARRAWAY, R., D. E. COCHRANE, J. B. LANSMAN, S. E. LEEMAN, B. M. PATERSON & H. J. WELCH. 1982. Neurotensin stimulates exocytotic histamine secretion from rat mast cells and elevates plasma histamine levels. J. Physiol. **323:** 403–414.
29. RIOUX, F., R. QUIRION, M. KEROUAC & S. ST. PIERRE. 1982. Mechanisms of the cardiovascular effects of neurotensin. Ann. N.Y. Acad. Sci. **400:** 56–74. This volume.
30. CARRAWAY, R. E. L. M. DEMERS, & S. E. LEEMAN. 1976. Hyperclycemic effect of neurotensin, a hypothalamic peptide. Endocrinology **99:** 1452–1462.
31. COCHRANE, D. E., C. EMIGH, G. LEVINE, R. E. CARRAWAY & S. E. LEEMAN. 1982. Neurotensin alters cutaneous vascular permeability and stimulates histamine release from isolated slices of skin. Ann. N.Y. Acad. Sci. **400:** 396–397. This volume.
32. LAZARUS, L. H., M. H. PERRIN, M. R. BROWN & J. E. RIVIER. 1977. Mast cell binding of neurotensin II. Molecular conformation of NT involved in the stereospecific binding to mast cell receptor sites. J. Biol. Chem. **252:** 7180–7183.
33. CARRAWAY, R., R. A. HAMMER & S. E. LEEMAN. 1980. Neurotensin in plasma: Immunochemical and chromatographic character of acid/acetone soluble material. Endocrinology **107:** 400–406.
34. ROSELL, S. & A. ROKAEUS, 1979. The effect of ingestion of amino acids, glucose, and fat on circulating neurotensin-like immunoreactivity (NTLI) in man. Acta Physiol. Scand. **107:** 263–267.
35. ROSELL, S., A. ROKAEUS, M. L. MASHFORD, K. THOR, D. CHANG & K. FOLKERS. 1980. Neurotensin as a hormone in man. *In* Neuropeptides and Neural Transmission. C. A. Marsan & W. Z. Traczyk, Eds. Raven Press. New York. pp. 181–188.
36. HAMMER, R. A., R. E. CARRAWAY & S. E. LEEMAN. 1982. Elevation of plasma neurotensin-like immunoreactivity after a meal: Characterization of the elevated components. J. Clin. Invest. **70:** 74–81.
37. HIRST, B. H., B. SHAW & L. WILSON. 1982. Neurotensin inhibition of gastric exocrine secretions in the cat. Regulatory Peptides **3:** 289–301.
38. FERRIS, C. F., R. A. HAMMER & S. E. LEEMAN. 1981. Elevation of plasma neurotensin during lipid perfusion of rat small intestine. Peptides **2:** 263–266.
39. FERRIS, C. F., J. X. PAN, E. A. SINGER, N. BOYD, R. E. CARRAWAY & S. E. LEEMAN. 1982. Evaluation of neurotensin in the central regulation of luteinizing hormone release. Ann. N.Y. Acad. Sci. **400:** 379–380. This volume.

Discussion of the Paper

S. Bloom (*Royal Postgraduate Medical School, London, England*): Have you had an opportunity to look any further at immunoreactive peptides that resemble the COOH-terminal of neurotensin in the stomach?

S. E. Leeman (*University of Massachusetts, Worcester*): We have continued with some of that work, but if I may call on Dr. Carraway, I think that he can answer that and then, perhaps, Dr. Hammer would like to comment.

R. E. Carraway (*University of Massachusetts, Worcester*): Yes, we have established that there are quite a number of peptides in gastric extracts that have COOH-terminal homologies with neurotensin. We have not isolated very many of these. We have isolated some and established that they do share sequence similarities with neurotensin.

R. A. Hammer (*Northwestern University, Evanston, IL*): There are several substances that are COOH-terminal cross-reactive that appear in plasma and in mesenteric vein plasma after lipid stimulation.

J. M. Polak (*Royal Postgraduate Medical School, London, England*): I would just like to ask Dr. Leeman and her colleagues whether the amount of COOH-terminal reactive material is found throughout the evolutionary tree? And, whether there is any specific localization? I am aware of your published work on the characterization. I wonder if you have been able to do further morphological localization?

Carraway: Although we have very little information concerning their localization, work by George Forssman and coworkers has established that a variant of neurotensin with COOH-terminal homology appears to be present in endocrine-like cells in intestinal mucosa. These are sometimes in the same cells as the counterpart of neurotensin, which has also been localized, and sometimes in a distinct second population. We believe that it is the counterpart possibly to some of the variants that we have seen in gastric extracts in mammals.

Polak: This is very exciting, Dr. Carraway, because we have looked at reptiles and amphibians and we have localized a neurotensin-like immunoreactive material that we cannot characterize as specifically COOH-terminal, especially in the upper intestine and the stomach.

A CRITICAL ANALYSIS OF THREE APPROACHES TO RADIOIMMUNOASSAY OF PEPTIDES: APPLICATIONS TO THE STUDY OF THE NEUROTENSIN FAMILY*

Robert E. Carraway

Department of Physiology
University of Massachusetts Medical School
Worcester, Massachusetts 01605

INTRODUCTION

Radioimmunoassay (RIA) has revolutionized endocrinology and come to be one of the most common and necessary tools for the characterization and measurement of the various signal peptides employed in cellular communication.[1] Over the more than twenty years since its introduction by Yalow and Berson,[2] numerous variations on this technique have been devised, including such exciting developments as solid-phase RIA,[3] the immunoradiometric assay,[4] the two-site sandwich,[5] and the enzyme-linked immunoadsorbent assay.[6] In addition, the applications of RIA have been broadened considerably, so that it should no longer be viewed as simply a method for measuring a given substance in the midst of millions of others. Perhaps even more importantly, it is also a versatile tool that can be applied to probe the primary structures of peptides[7, 8] and even possibly, the secondary structures of proteins.[9, 10] In order to achieve these latter goals, however, one must employ an approach to RIA that differs fundamentally from that used presently by most researchers.

In this paper the advantages, problems, and limitations of three different approaches to the RIA of messenger peptides are discussed. These basic designs include (a) the heterogeneous antiserum approach, (b) the multiple region-specific antiserum method, and (c) the two-site sandwich technique. Emphasis is placed on the overall performance of each method as applied to (a) the measurement of plasma levels and (b) the characterization of related peptides. Little time is spent discussing the general problems and pitfalls of RIA, which have been very aptly described elsewhere.[11] Some of the concepts presented here are illustrated using data gathered in our laboratory for neurotensin (NT) and related peptides.

APPROACHES TO RIA OF PEPTIDES

TABLE 1 lists some characteristics of three different approaches to the RIA of peptides. Each has its advantages and disadvantages when applied to the quantification and characterization of peptides that exist in multiple variant forms and that can be degraded to yield NH_2- and COOH-terminal fragments. These are discussed below in order of appearance in the table.

*This work was supported by National Institute of Health Grants AM 28565 and AM 28557.

17

TABLE 1

SOME PROPERTIES OF THREE APPROACHES TO THE RADIOIMMUNOASSAY (RIA) OF PEPTIDES

Property*	Highly Heterogeneous Antiserum RIA	Multiple Region-specific Antisera RIA	Two-site Sandwich RIA
(a) Ease of development	Ideal antiserum difficult to get	Many options	Difficult
(b) Identity criteria	1. Parallelism	1. Parallelism 2. Equipotency	1. Parallelism 2. Need two sites
(c) Ability to detect single change	Poor	Excellent	Good
(d) Plasma Assay	Good	Fair	Excellent
(e) Characterization of fragments or variants	No	Excellent	No

*Refer to the text for detailed discussion of each property listed.

Heterogeneous Antiserum Method

The highly heterogeneous antiserum approach (TABLE 1, column 1) is by far the most widely used today in spite of its inherent lack of specificity and its limited applicability.

(a) Although the specificity of this system absolutely requires the use of an antiserum that senses all regions of the peptide equally well, this kind of antiserum is extremely difficult to obtain from a single animal. This is because in practice, one usually finds that certain sites on peptides are more highly antigenic than others and these dominate the antibody response.

(b) This system offers only one criterion of identity, parallelism, and without every region being represented in the antiserum this is weakened considerably.

(c) Even with an ideal antiserum, the ability of this system to detect a single change in the peptide is also quite poor. This is because the antibody power is spread over the entire surface of the molecule and only a small percentage of the total, especially for a large peptide, is directed toward any one site.[†]

(d) There are also problems when this system is applied to measurement of plasma levels of the peptide since plasma offers the special situation of containing high levels of multiple breakdown products as well as multiple variants of the peptide. The heterogeneous antiserum, having antibodies toward all parts of the molecule, will respond somewhat to each fragment and variant and since these affects are additive, the sum total could be a large measurement.

(e) Another disappointment with this approach is that it yields no information concerning the nature of other substances that crossreact in the assay; they are sometimes recognized as being dissimilar to the standard but how they differ is not determined.

As we have said, it is very difficult to obtain an ideal heterogeneous antiserum. In fact, all of the antisera we have ever raised towards peptides of 10 or more residues have been found to be largely towards one or two highly antigenic regions within these structures. This has also apparently been the experience of several

[†]A general property of antibodies is that their binding sites can accommodate no more than approximately 6 amino acids. See reference 12 for more discussion.

other researchers.[13] One method of circumventing this problem might be to prepare heterogeneous antisera by mixing well-characterized, region-specific antibodies. In this way, one could ensure that the prepared antiserum was directed uniformly over the surface of the peptide, sensing most elements of the structure. For best results, it would seem necessary to choose antibodies of similar affinity and to mix them in proportion to their titers.

In order to test the limits of the heterogeneous antiserum approach, we have prepared an "ideal" heterogeneous antiserum towards NT and examined its properties. This prepared antiserum was a mixture of three region-specific antisera, the average binding sites of which are shown diagrammatically at the lower left in Figure 1. Each of these antisera was directed exclusively towards the indicated regions, cross-reacting 20–100% with the appropriate partial peptide.[12]

Dose–response curves for NT and other substances obtained using the prepared mixture of antibodies are shown in Figure 1. As expected, fragments of NT cross-reacted only partially in the assay and gave rise to dose–response curves that did not parallel that of NT. However, when a mixture of the NH_2- and COOH-terminal fragments was tested, a parallel dose–response curve was obtained (Fig. 1, e.g., NT^{1-12}, NT^{2-13} and mixture). This illustrates one of the major problems encountered when this method is applied to measure the levels of NT in plasma, that is, the high levels of NH_2- and COOH-terminal fragments and variants of NT lead to additive effects in the assay that can even satisfy the one criterion of identity, parallelism.

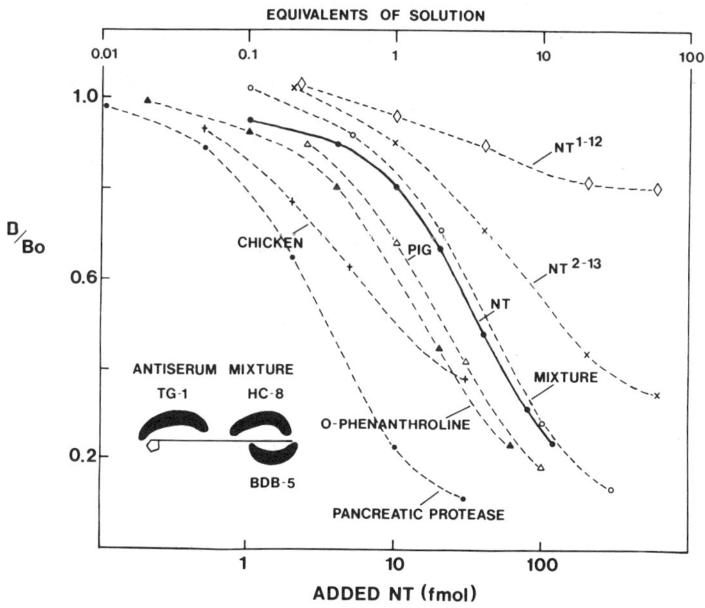

Figure 1. Comparison of the inhibition curves produced by various substances to that of synthetic bovine NT. The antiserum used was a mixture of TG-1 (1:150,000) HC-8 (1:300,000) and BDB-5 (1:300,000) given as final concentrations. One equivalent of the solutions contained the following amounts of material: NT^{1-12} (13 fmol), NT^{2-13} (9 fmol), mixture (NT^{1-12} and NT^{2-13} as above), O-phenantholine (200 µg), pancreatic protease (50 ng), porcine NT (10 fmol measured with HC-8) and chicken NT (100 fmol measured with HC-8).

There can be many false positives in RIA, since any substance (e.g., denaturants, acids, enzymes) that disrupts the antigen–antibody interaction will register in the assay. A surprise to us was the finding that many such artifacts also generate dose–response curves that parallel that of the standard (FIG. 1, e.g., o-phenanthroline, pancreatic protease). These results illustrate the flimsiness of the parallelism criterion and underline the need for more critical attention to its application.

Finally, we applied this ideal antiserum to the characterization of partially purified NT-related material obtained from pigs and chickens. While the preparation from pig satisfied the parallelism criterion, that from chicken gave a nonparallel curve (FIG. 1.). Although these results suggested that chicken NT differed from the bovine molecule, they could not indicate where the difference(s) lay.

Multiple Region-Specific Antiserum Method

This approach, which is not so commonly used, involves the comparison of results obtained in separate RIA's employing different region-specific antisera. It has a number of assets (TABLE 1, column 2).

(a) Region-specific antisera are quite easily obtained. In TABLE 2 are listed several means by which one can direct specificity using conventional immunization techniques (methods a, b), select particular antibodies from mixed populations (methods c, d), and generate monoclonal antibodies toward single determinants (method e). Together these methods provide the investigator with the ability to tailor antibody preparations to display the desired specificity for binding.

(b) An important attribute of this approach is that it affords two criteria of identity: parallelism and equal immunological potency. Therefore, if an unknown is to be considered identical to the standard, it must not only display parallelism but also give rise to equal measurements in all of the assays. As you will see later, this added criterion is a very sensitive and discerning measure.

(c) Another important property of this approach is that it concentrates one's antibody power to a small region of the peptide (as opposed to the first method which spreads it out). In effect, this allows one to detect a single change in the structure of the peptide much more effectively since instead of a nonparallel curve, one obtains nearly an all-or-nothing response. This is due to the fact that each antibody preparation is selected to bind exclusively to a 4–8 amino acid segment of the peptide and since this is about the size of an antibody binding site, a structural change within this region affects nearly all antibodies in the population.

TABLE 2

METHODS FOR DIRECTING THE SPECIFICITY OF ANTIBODY PREPARATIONS

(a) Orientation of Peptide-hapten on Carrier—The antibodies are usually directed towards regions most distal to the site of conjugation.

(b) Use Fragment as Hapten—Sometimes antibodies raised toward fragments recognize the parent peptide.

(c) Adsorption of Unwanted Antibodies—One can add fragments in excess or perform affinity purification.

(d) Use Labeled Fragment as Tracer—Only those antibodies that can bind the labeled fragment are expressed.

(e) Monoclonal Preparations—These homogeneous preparations usually bind a single determinant.

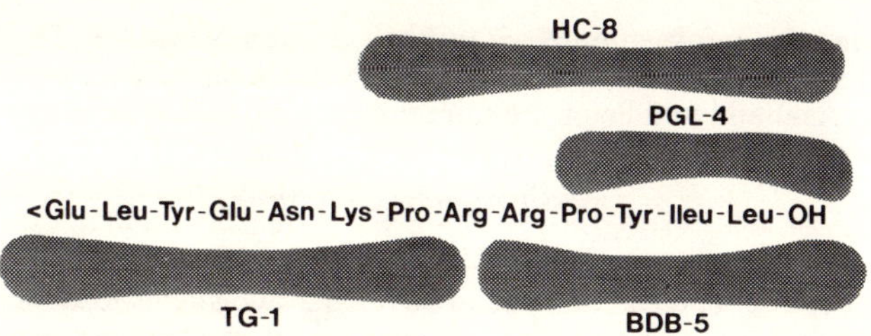

FIGURE 2. Diagrammatic representation of the average binding sites for several region-specific antisera towards NT.

(d) When applied to assay plasma, this method succeeds in being specific only when one of the region-specific antisera is directed towards a sequence that is present only in the peptide of interest and not in circulating breakdown products or variants of the peptide.

(e) This approach offers an extremely powerful tool for the rapid characterization of precursors, variants and metabolites of the original peptide.

Several well-characterized, region-specific antisera towards NT have been generated in our laboratory and applied successfully to measuring and characterizing multiple NT-related peptides.[14] The specificities of some of the antisera that we have available are illustrated diagramatically in FIGURE 2. The NH_2-terminal directed TG-1 was shown to cross-react about 20% with NT^{1-8}, about 100% with NT^{1-12} and $< 0.01\%$ with any COOH-terminal partial peptides.[15] The other antisera shown were all found to react exclusively with COOH-terminal portions of NT. While antiserum PGL-4 cross-reacted about 100% with NT^{10-13}, antisera BDB-5 and HC-8 required 6 and 8 COOH-terminal residues for full recognition.[16] Thus, each of these antibody populations is directed against a discrete region of the molecule that is not much larger than the size of an antibody-binding site.

Some results that were obtained when these different antisera were applied to analyze several NT-related substances are summarized in TABLE 3. The results are presented as ratios of measurements obtained in each assay; the ratio for NT is 1.0 since it was the standard employed. Single changes in the COOH-terminal region of NT (NT^{1-12} and NT^{1-13} amide) can be seen to greatly affect recognition by the COOH-terminal but not the NH_2-terminal antisera (TG-1/BDB-5 ratios near 1000). In contrast, single changes in the NH_2-terminal region of NT (NT^{2-13} and Gln^4NT) selectively diminished NH_2-terminal reactivity (TG-1/BDB-5 ratios near 1/1000). These results illustrate the tremendous sensitivity of this method and the ease of interpretation. Reagents and enzymes can give false positivies in this system; however, since they are not likely to generate identical measurements in all assays, they can usually be distinguished from NT. Carboxypeptidase, for example, yields a TG-1/BDB-5 ratio of less than 0.001 (TABLE 3).

The strength of the equipotency criterion is nicely illustrated by comparing the results obtained for the partially purified preparations of NT from pigs and chickens. Porcine NT gave rise to parallel curves and equal measurements in all the assays (TABLE 3; ratios 1.0) arguing that it did not differ from the bovine structure. In contrast, the preparation from chicken was not recognized by the NH_2-terminal antiserum, TG-1; and reacted better with the COOH-terminal antisera PGL-4 and BDB-5, than with the middle and COOH-terminal directed HC-8 (e.g., HC-8/BDB-5, 0.8). These immunochemical findings argue that bovine and chicken NT differ primarily at their NH_2-terminal and middle regions. It is noteworthy that after obtaining this information,[15] we isolated chicken NT and established that it was identical to the bovine substance except for three amino acid substitutions in the NH_2-terminal and middle of the molecule.[17]

Two-Site Sandwich Assay

Once one has available two region-specific antisera that bind to different portions of the peptide, one can attempt to develop a two-site procedure. The basic idea here is to construct an assay system that will require the presence of both an NH_2-terminal and a COOH-terminal site within the peptide; thus, eliminating the problem of cross-reacting NH_2- and COOH-terminal fragments. Although there are a number of possible designs, the reaction scheme initially proposed by Addison and Hales is one of the simplest.[4] This involves (a) anchoring an NH_2-terminal

TABLE 3

RATIOS OF MEASUREMENTS OBTAINED USING VARIOUS REGION-SPECIFIC
ANTISERA TOWARDS NEUROTENSIN (NT)

Substance*	Ratio of Activities Measured with Antisera†		
	TG-1/BDB-5	HC-8/BDB-5	HC-8/PGL-4
NT-$(NH_2)^{13}$	>400		
NT^{1-12}	>1000		
NT^{2-13}	<0.001	1.0	1.0
Gln^4-NT	<0.003	1.0	1.0
O-Phenanthroline	2		
Pancreatic protease	4		
Carboxypeptidase A	>1000		
Pig-NT	1.0	1.0	1.0
Chicken-NT	<0.001	0.8	0.8

*The solutions used were the same as described in FIGURE 1.

†Given are the ratios of the measurements obtained in the RIA using the antisera indicated. Measurements were taken at the point of 50% inhibition when possible. Antiserum specificities are given in FIGURE 2.

antiserum (in excess) to a solid matrix, such as a polyethelene tube, (b) incubating with sample or standards, (c) washing and incubating with labeled COOH-terminal antiserum, and (d) washing and counting the tube. The counts associated with the tube are found to be proportional to the amount of peptide present initially.

Although the concept of the two-site assay was introduced more than ten years ago, it has received surprisingly little attention. While there has been some effort to apply this method to measure rather large substances such as growth hormone, parathyroid hormone, and ferritin,[18] a two-site procedure for a small peptide such as NT has not yet been reported. In spite of its wonderful features, the two-site assay is a complex system and there are a number of difficulties encountered in its development. Refer to TABLE 1, column 3, for the following discussion.

(a) Firstly, it is required that one of the antibodies be affinity purified and iodinated to high specific activity without loss of binding ability. Secondly, the other antiserum (bound to the tube in excess) must be free of peptides. Since antisera toward secreted peptides are preloaded with peptide from the immunized animals, an unloading step may be necessary. Thirdly, the two-antibody preparations must be able to bind simultaneously to the same molecule without steric hindrance.

(b) There is only one measurement that can be used as an identity criterion; however, both ends of the molecule are required for binding.

(c) The ability to detect a single change is good within the two binding sites of the antisera used, but the middle of the molecule is not sensed.

(d) The two-site assay is an ideal method for measuring plasma levels since neither NH_2- nor COOH-terminal fragments should register. In addition, this procedure should be more sensitive than conventional RIA. A major advantage is the fact that this assay cannot display false positives due to reagents or enzymes that are washed away before incubation with labeled antibody. Only false negatives are possible.

(e) This assay system is not useful for the characterization of related peptides.

We are presently working toward obtaining a two-site assay for NT since this would permit direct analysis of plasma, obviating the need for chromatographic removal of the NH_2-terminal fragments and COOH-terminal variants of NT.[19] Using affinity columns prepared from CNBr-activated Sepharose, we have succeeded in obtaining purified antibodies for iodination.[15] We have also succeeded in binding purified antibodies in an irreversible and highly reproducible manner to polystyrene tubes using the technique of Kevin Catt and Tragear.[20, 21] The crucial question now is whether two antibodies can bind simultaneously with high affinity to a peptide as small as NT.

APPLICATIONS OF REGION-SPECIFIC RIA TO THE NT SYSTEM

In order to illustrate the usefulness of multiple region-specific antisera in addressing the biochemistry and physiology of messenger-peptide systems, we have selected a number of topics that have been studied in our laboratory using this technique.

Phylogenetic Studies

The application of region-specific RIA to the detection and characterization of immunoreactive NT (iNT) in lower animals has led to the conclusion that the biologically active COOH-terminal region of NT-like peptides has been highly conserved during evolution.[22, 23] The NH_2-terminal-directed antiserum (TG-1) recognized only mammalian forms of NT, failing to react even with avian, reptilian, and amphibian preparations (FIG. 3). In contrast, antisera towards the COOH-terminal region of NT (HC-8 and PGL-4) recognized material in extracts from representatives of all animal classes including Porifera and Protozoa. The distribution of some of these NT-related peptides through digestive tissues of both vertebrates and invertebrates has been examined successfully by RIA and immunocytochemistry using these COOH-terminal directed antisera.[24, 25] In addition, partially purified preparations of iNT from some animals, including the lobster, have been shown to exhibit biologic activities in the rat that are characteristic for NT.[22, 23] These findings are consistent with the idea that NT and related peptides participate in important processes that are basic to animal life and that their functioning depends highly upon elements located in their COOH-terminal regions.

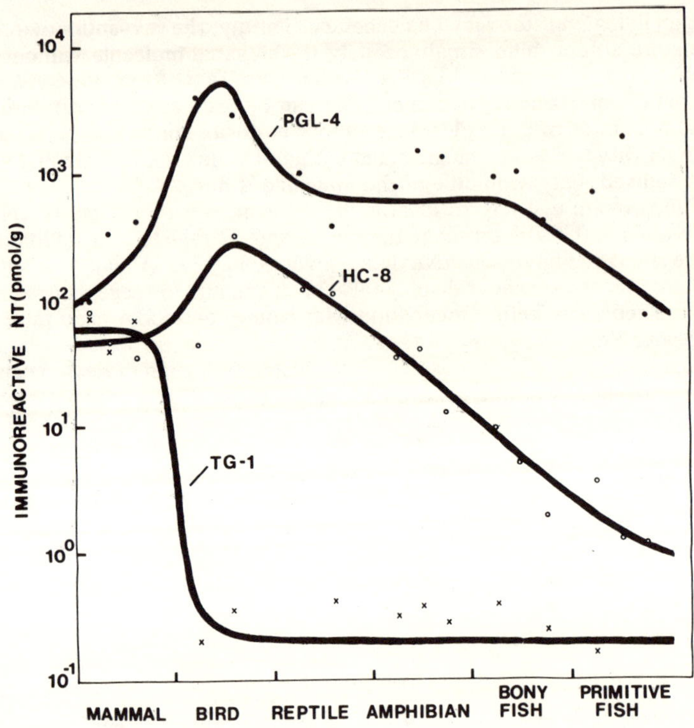

FIGURE 3. Concentration of iNT in intestinal tissues of representative vertebrates as measured using three region-specific antisera. Plotted is the average concentration measured in acetic acid extracts of small intestinal tissues as a function of animal class. (From Carraway, R., S. E. Ruane & H. R. Kim. 1982. Peptides 3: 115–123. Used with permission.)

Immunochemistry on Extracts of Mammalian Tissues

The results for mammals in FIGURE 3 show that the NH_2-terminal (TG-1) and COOH-terminal (HC-8) antisera gave roughly the same measurements on unfractionated extracts of tissues. In order to carefully analyze the character of the multiple peptides present, the extracts were subject to chromatography on Sephadex G-25 (FIG. 4). A major peak of HC-8 immunoreactivity was found to elute in the region of synthetic NT and to give equal measurements with the different antisera used (TG-1, PGL-4, and HC-8). This constitutes strong evidence that the structure of NT is very similar, if not identical, in dogs, rats, pigs, humans, and cattle. Consistent with these findings, we have recently demonstrated that human intestinal NT has the same amino acid composition as the bovine molecule.[26]

In addition to the material behaving like NT, other substances reacting preferentially with antiserum PGL-4 were seen to elute later from the Sephadex column (FIG. 4). The immunochemical behavior of these peptides suggests that they represent variants of NT that differ in their NH_2-terminal regions but share a common COOH-terminal sequence.[27] Work in the rat has indicated that NT and its COOH-terminal variants appear to be distributed differently through tissues, the ratio varying from about 2 in the jejunoileum to about 500 in the stomach.

Although NT has been shown to be capable of influencing multiple gastric functions, genuine NT appears to be absent from the mammalian stomach and present
in only low levels in peripheral plasma.[19] Since the COOH-terminal region of NT
appears to be its biologically active core, it seems reasonable to suggest that the
COOH-terminal variants of NT within the stomach are the actual physiological
mediators of these effects. COOH-terminal fragments of NT have recently been
shown to inhibit gastric acid and pepsin secretion in the cat.[28]

Immunochemistry on Extracts of Lower Animals

The results for crude extracts of lower vertebrates (FIG. 3) as well as invertebrates indicated that recognition was obtained only with antisera directed toward

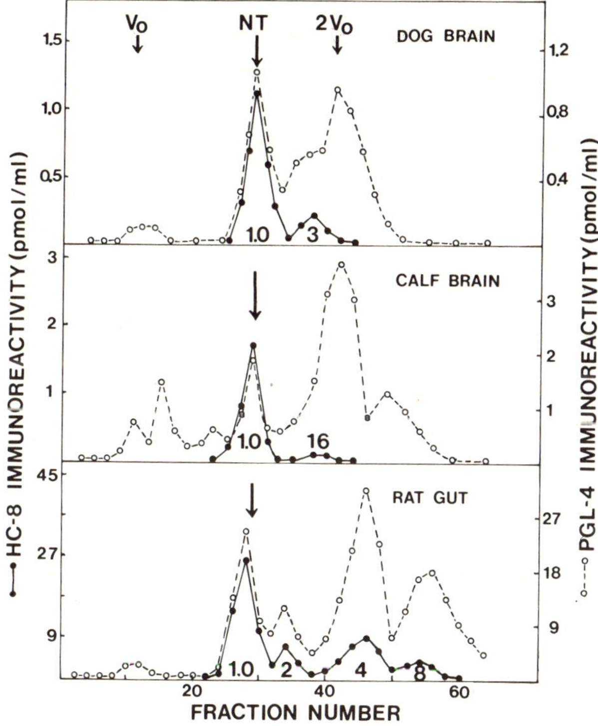

FIGURE 4. Gel chromatography of extracts of mammalian whole brain and small intestine
on Sephadex G-25. Column 1.6 × 90 cm; buffer 0.2 *M* acetic acid. After the sample was
applied, 50 ml of effluent was discarded and 2.4 ml fractions were collected. Aliquots were
assayed using antiserum PGL-4 (O) and HC-8 (●). The elution position for synthetic NT and
multiples of the void volume (V_0) are shown with arrows. The number under each peak
denotes the approximate ratio of the measurements obtained using antiserum PGL-4 to those
obtained with antiserum HC-8. Antiserum TG-1 (data not shown) gave a major peak of
immunoreactivity in each case coincident upon and about equal to that seen with antiserum
HC-8, peaking at tube 29. (From Carraway, R., S. E. Ruane & H. R. Kim. 1982. Peptides
3: 115–123. Used with permission.)

the COOH-terminal half of NT.[22, 23] When the extracts were chromatographed on Sephadex G-25, again multiple peaks of iNT were observed (e.g., Fig. 5). Some peaks reacted with both COOH-terminal antisera, while others reacted only with antiserum PGL-4 and the PGL-4/HC-8 ratios tended to be higher for the more primative animals. For example, the peak of material that behaved most like NT immunochemically and biologically gave ratios of 1.0 for mammals, 1.5–1.6 for birds and reptiles, 8–35 for fish and 5–200 for invertebrates. Assuming that these peaks represent counterparts to NT in the animals tested, these results suggest

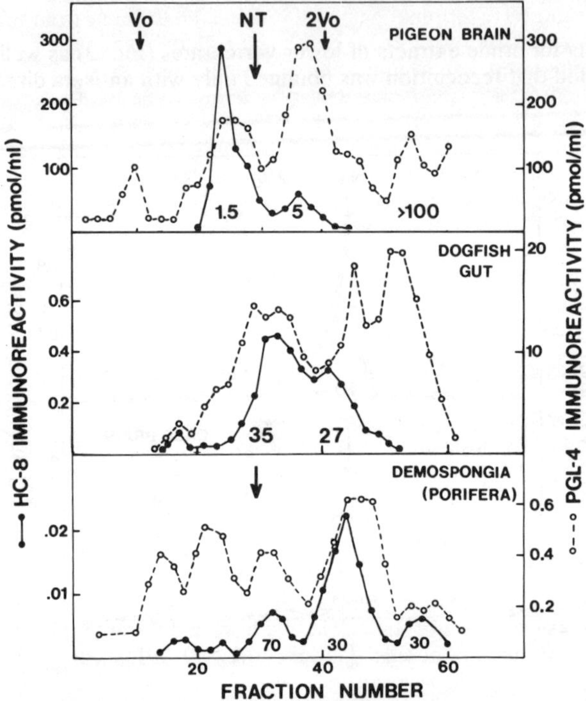

FIGURE 5. Gel chromatography of extracts of lower vertebrate and invertebrate tissues on Sephadex G-25. Column and conditions were as in FIGURE 4. The number under each peak denotes the approximate ratio of the measurements obtained using antiserum PGL-4 (O) to those obtained with antiserum HC-8 (●). Antiserum TG-1 (not shown) gave no recognition.

preferential conservation of the COOH-terminal 4 residues in NT (as compared to the middle 4 residues) during evolution. Considering that the NH_2-terminal antiserum (TG-1) did not react at all with NT from lower animals, these results indicate that nature has permitted the structure of NT to vary most at its NH_2-terminus, to a lesser extent in the middle, and least at its COOH-terminus.

NT-Related Peptides in Chicken

The high concentrations of NT-related material in birds (5–10 times higher than in mammals) permitted the recent isolation and chemical characterization of several peptides in the NT-family from extracts of chicken small intestine.[15, 17] Chromatography of an acid–acetone extract of intestine on Sephadex G-25 revealed the presence of three major peaks of immunoreactivity using the COOH-terminal antiserum (PGL-4), two of which reacted also with antiserum HC-8. There was no recognition at all by the NH_2-terminal antiserum (TG-1). The pattern was similar to that shown for pigeon brain in FIGURE 5; the PGL-4/HC-8 ratios are given under each peak.

Peak I appeared to represent the chicken counterpart to NT, as it displayed several of the biologic properties of NT and exhibited the strongest COOH-terminal homology with NT. The inability of antiserum TG-1 to recognize peak I indicated that the NH_2-terminal part of the molecule differed from that in bovine NT; the relative reactivity of peak I with the COOH-terminal antisera (PGL-4/ HC-8 = 1.3) suggested that an additional difference might exist in the middle of the molecule. When peak I was isolated, it was found to be identical to bovine NT except for three amino acid substitutions (His/Tyr, Val/Glu, Ala/Pro) located in the NH_2-terminal and middle portions of the molecule as shown below.[17] These substitutions involve single changes of the amino acid codon.

The immunochemical properties of the material in peak II (unreactive with TG-1 and giving a measurement 500 times higher with PGL-4 than with HC-8) suggested that it shared only 4–5 COOH-terminal residues with NT. The isolated peptide was shown to be identical to the COOH-terminal half of NT with the amino acid substitutions, Lys/Arg and Asn/Arg as shown below.[29]

Bovine NT　　<Glu-Leu-Tyr-Glu-Asn-Lys-Pro-Arg-Arg-Pro-Tyr-Ile-Leu-OH
Chick I　　　<Glu-Leu-His-Val-Asn-Lys-Ala-Arg-Arg-Pro-Tyr-Ile-Leu-OH
Chick II　　　　　　　　　　　　H-Lys-Asn-Pro-Tyr-Ile-Leu-OH

Immunoreactivity in peak III was found to be quite heterogeneous and to be composed of small peptides of 3–5 residues having much less homology to NT. It is not yet clear whether these substances are members of the NT family or if they only represent cross-reacting substances without biologic significance.

Although the material in peak II did not induce cyanosis or increase the hematocrit of anesthetized rats, it did affect blood pressure. Unlike NT, however, it was hypertensive, instead of hypotensive.[29] It might be suggested then, that peaks I and II are members of the NT family having different biologic capabilities and functions. In support of this notion are results gathered by both RIA and immunocytochemistry that suggest that these two peptides are distributed differently in chicken and quail tissues. Furthermore, immunohistochemical staining of avian intestinal preparations with antisera specific for these two peptides has indicated that both peptides reside within endocrine-like cells of the open type and that they are sometimes found in different cells.[30] Other work in our laboratory has demonstrated that both peptides are highly concentrated in brain synatosomal preparations, and that they can be released from chicken brain-slice preparations by 60 mM potassium-induced membrane depolarization in the presence of calcium.[31]

NT and XP-related Peptides in Amphibia

Xenopsin (XP) is an octapeptide originally isolated from the skin of the African frog, *Xenopus laevis*.[32] Its amino acid sequence, <Glu-Gly-Lys-Arg-Pro-Trp-Ile-Leu-OH, and spectrum of biologic activity suggested the possibility that XP might be the amphibian counterpart of mammalian NT. In order to address this issue,

extracts of skin, brain, and intestine from representative amphibians were subjected to immunochemical, chromatographic, and biologic analyses.[33] The results indicated the dual presence of NT-like and XP-like peptides in *Xenopus laevis, Rana Catesbeiana,* and *Bufo marinus* that could be separated by gel chromatography on Sephadex and by HPLC on μ-Bondapak C18 (e.g., FIG. 6).

Three subfamilies of peptides, one XP-like and two NT-like, could be distinguished on the basis of their distributions in amphibian tissues and their relative cross-reactivities with the different antisera. The NT-related peptides that were recognized by antiserum HC-8 were found primarily in brain and small intestine, while those reactive with antiserum PGL-4 were most concentrated in stomach, liver, and pancreas. Although also present in brain and intestine, immunoreactive XP (iXP) was highest in stomach, pancreas, and *Xenopus* skin.

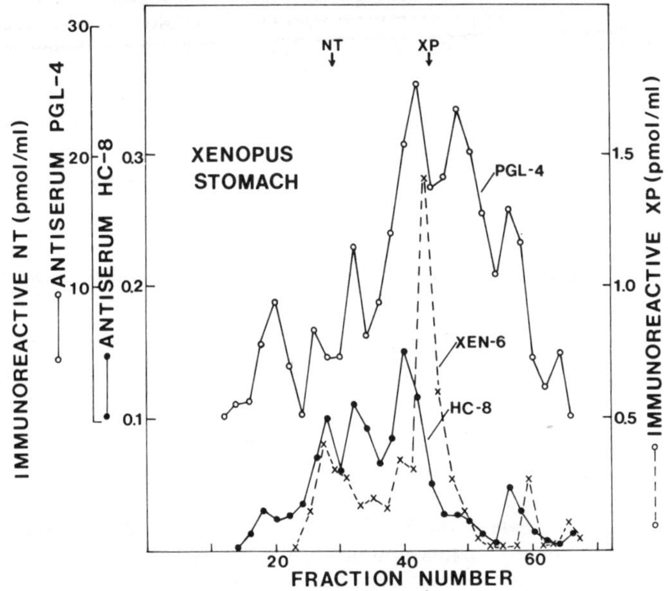

FIGURE 6.   Gel chromatograhy of *Xenopus* gastric extract on Sephadex G-25. The column and conditions were the same as in FIGURE 4. (Taken from reference 33 with permission.)

Partially purified preparations of iNT and iXP obtained from gastrointestinal tissues of *Xenopus* and *Bufo* were found to increase hematocrit and to induce cyanosis in anesthetized rats, effects that are characteristic for NT and XP. Interestingly, the iNT obtained from *Bufo* stomach did not produce cyanosis but instead displayed cholinergic-like effects, promoting salivation and lacrimation. It was similar to the other peptides, however, in that it significantly elevated hematocrit. Thus, although the various peptides in the NT- and XP-families share chemical homologies, they may display a diversity of biologic capabilities.

Although it separated from bovine NT chromatographically, some of the HC-8 reactive material from brain and intestine is likely to represent amphibian counterpart(s) to mammalian NT. The specificity of this antiserum would suggest that these peptides share 5–8 of the COOH-terminal residues of NT. The gastric peptides, which reacted much more strongly with antiserum PGL-4 than with antiserum HC-8, are likely to share only the COOH-terminal 3–5 residues with NT. The XP-related peptides are likely to differ from those similar to NT by the presence of -Pro-Trp-Ile-Leu-OH rather than -Pro-Tyr-Ile-Leu-OH at their COOH-termini. The immunochemical data suggest that all of these peptides share COOH-terminal homologies, which is consistent with their shared ability to increase hematocrit. Although there may be some overlap in biologic capability, it is tempting to suggest that the three subfamilies distinguished so far serve separate biologic roles.

Metabolism of NT in the Rat

Region-specific RIA has permitted the detection and characterization of metabolic breakdown products formed after intravenous injection of synthetic NT into rats.[34] The regional selectivity of the method yielded immediate information concerning the area of the molecule undergoing metabolic cleavage, and the high sensitivity allowed the studies to be performed at near physiologic levels.

In the first experiment, NT was injected via the femoral vein into anesthetized rats (500 pmol/kg) and blood samples were obtained at various times from the femoral artery. After centrifugation, the plasmas were immediately extracted with acid–acetone and after processing, were assayed using an NH_2-terminal (TG-1) and a COOH-terminal (HC-8) directed antiserum. A typical result, shown in Figure 7, indicated that iNT measured with the COOH-terminal antiserum disappeared more quickly (t1/2 ca 0.5 min) than that measured with the NH_2-terminal antiserum (t1/2 ca 5 min). This suggested that the injected NT was being degraded most rapidly by cleavage(s) within its COOH-terminal half, and in order to identify the products, the retrieved plasmas were subjected to high performance liquid chromatography (HPLC). Figure 8 shows the profiles obtained when plasma obtained from 0.5–2.0 min after injection was analyzed by HPLC using a program that was calibrated with various synthetic fragments of NT. The NT injected into the animal was ^{3}H-labeled (equally on Tyr^3 and Tyr^{11} as shown) and the lower half of Figure 8 shows the profile of ^{3}H-radioactivity obtained. The eluates were also assayed for iNT using both the NH_2- and COOH-terminal antisera, and the results are plotted in the upper part of Figure 8. The major metabolites included NT^{1-8} and NT^{1-11}, which were easily detected using the NH_2-terminal antiserum (cross-reactions 20% and 60%), as well as NT^{9-13}, which barely registered in the COOH-terminal assay (cross-reaction, 2%). NT^{9-13} could also be measured using antiserum BDB-5 (cross-reaction, 30%). One peak of ^{3}H-radioactivity was not immunoreactive and was tentatively identified as free tyrosine.

The cleavage pattern deduced from this information is depicted in the upper part of Figure 8. The major clips appear to be tryptic-like (between Arg^8 and Arg^9) and converting-enzyme-like (between Tyr^{11} and Ile^{12}). A postprolinelike cleavage may also occur COOH-terminal to Pro^{10} to yield free tyrosine.

This pattern of metabolism for injected NT appears also to pertain to endogenous NT that gains access to the circulation. Work in humans by Hammer $et.$ $al.$[35] has established that NT^{1-8} and NT^{1-11}, but not NT itself, are elevated in peripheral plasma 45 min after injection of a high-fat meal. Presumably, NT released from the small intestine is metabolized to yield the same fragments identified above. Indeed, Ferris et $al.$,[36] working with anesthetized rats, have observed that lipid perfusion of the small intestine elevates the levels of both NT and NT^{1-8} in

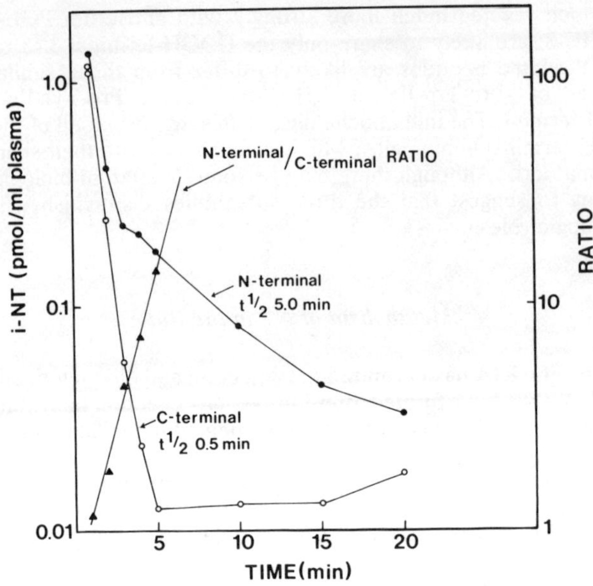

FIGURE 7. Half-life ($t_{1/2}$) of NT immunoreactivity (iNT) after intravenous injection into rat. NT was injected via the femoral vein (500 pmol/kg) and at various times afterward plasma was obtained via the femoral artery. After extraction, iNT was measured using antiserum TG-1 (NH_2-terminal) and antiserum HC-8 (COOH-terminal) (From Aronin, N., *et al.*, manuscript submitted to Peptides, 1982.)

hepatic-portal blood, while not affecting those in the mesenteric artery. It would appear from these results that metabolism of NT can occur very close to its site of release, possibly by enzymes on the endothelium of blood vessels.

When a peripheral human plasma sample obtained 3 hr after breakfast was subjected to HPLC and immunochemical analysis, the results shown in FIGURE 9 were obtained. Using the NH_2-terminal antiserum, TG-1, material eluting in the positions of NT^{1-8}, NT^{1-11} and NT was observed. The Middle- and COOH-terminal-directed HC-8 detected primarily NT but also material eluting near NT^{9-13}. The far COOH-terminal-directed PGL-4 detected NT^{9-13} and multiple other substances that behaved like the variants of NT mentioned earlier. Genuine NT accounted for no more than 1% of the PGL-4 immunoreactivity (COOH-terminal) and approximately 25% of the TG-1 measurements (NH_2-terminal); none of the peaks observed gave equal measurements with all the antisera. From this analysis, we can conclude that plasma iNT can be extremely complex, being composed of NT as well as multiple variants and breakdown products. The need for a two-site assay in order to accurately measure plasma levels of NT is obvious. Region-specific RIA has permitted us to recognize some of the chemical relationships among these different forms of iNT. As these various cross-reacting species are isolated from their source tissues and identified, we will have a fuller understanding of the origins of these peptides and their functions.

CONCLUSIONS

Using region-specific RIA, it is possible to detect a number of peptides that appear to share COOH-terminal homologies with NT. It is proposed that at least some of these peptides belong to a family of NT-related biologically active messengers. This family appears to be rather heterogeneous, being composed of two or three subfamilies of related peptides, one of which appears to resemble XP more than NT. There may be some biologic as well as chemical similarities among these

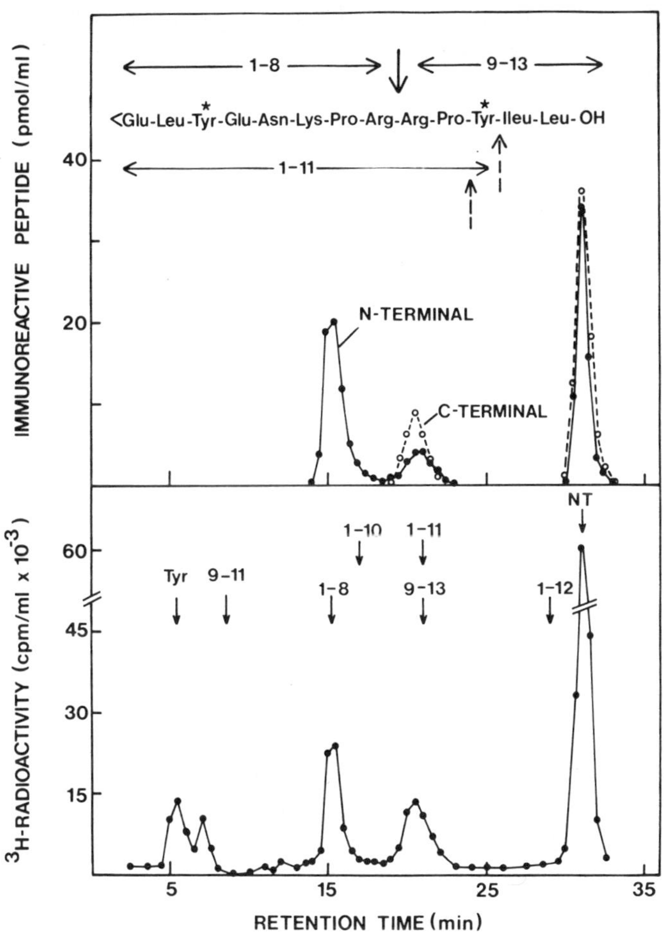

FIGURE 8.   HPLC on circulating metabolites of NT obtained 2 min after intravenous injection of ³H-labeled NT. *Lower:* ³H-radioactivity. *Upper:* NT immunoreactivity expressed as real peptide concentrations applying cross-reactions of 20% (NT$^{1-8}$), 60% (NT$^{1-11}$), 2% (NT$^{9-13}$) and 100% (NT). (From Aronin, *et al.*, manuscript submitted to Peptides, 1982.)

peptides since quite a few have been found to exhibit overlapping pharmacologic activities. The biologic function(s) of the individual peptides might, however, still be quite different judging from the fact that some differences in biologic responses have been observed. The information gathered to date indicates that COOH-terminal-related variants of NT are present in tissues. The existence of peptides

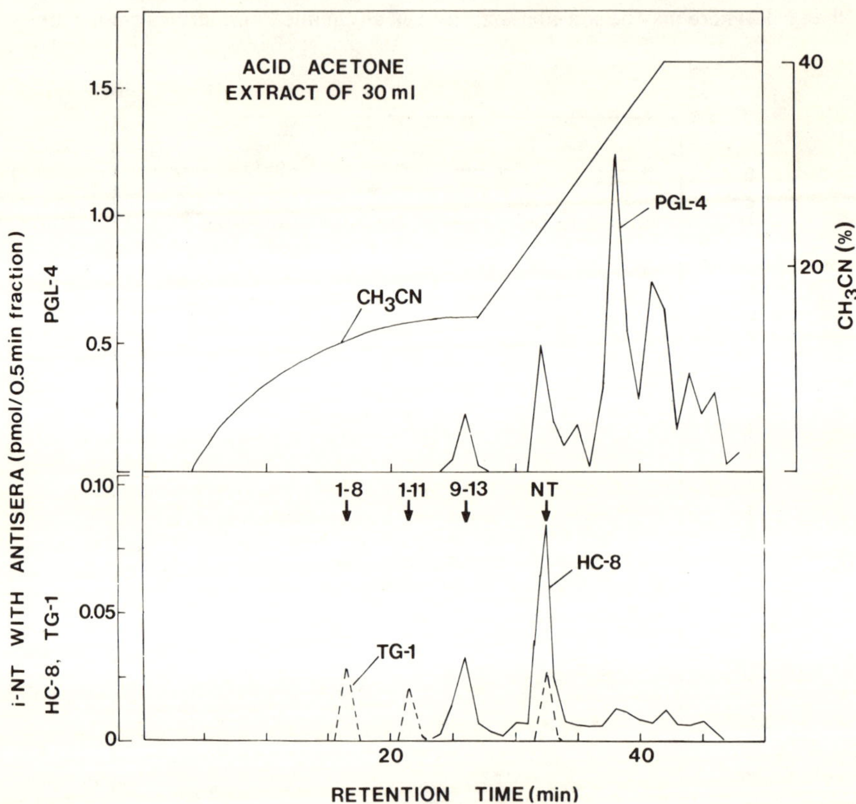

FIGURE 9.   HPLC of extracted human plasma on μ-Bondapak C-18. 30 ml of postprandial plasma (11 AM) was extracted with acid acetone[15] and desalted by adsorption to a Sep-pak cartridge.[33] Column, 2.9 mm × 30 cm. Buffer A, 10 mM potassium phosphate (pH 4.6). Buffer B, 50% CH₃CN. Flow rate, 2.0 mg/min. Gradient, 0–30% B over 25 min using curve 5; then 30–80% B over 25 min using curve 6. Eluates were dried and subjected to RIA for NT using antiserum PGL-4 (*upper*), HC-8, and TG-1 (*lower*). The elution positions for various synthetic peptides are also shown.

resembling the NH₂-terminal or middle portions of NT has not been excluded, however, since the antisera employed may not have permitted their detection. Variants of NT sharing less than five of its NH₂-terminal residues or sharing the middle sequence in NT and differing at both ends might not have been detected by any of our antisera. It is conceivable that portions of NT other than its COOH-terminal region are biologically important in ways we have yet to recognize (e.g., the transport, storage release, or action(s) of NT or related peptides). If so, then it would seem reasonable to suggest that these other vital areas might also be conserved and be shared by some of the NT-like peptides of common ancestry.

It should also be recognized that some of the peptides reacting in the RIA may simply be metabolites of NT and its relatives while others may not be related to NT at all. For this reason, we would suggest that the NT family be defined tentatively as those peptides that fulfill the following criteria:

(a) The peptide should have an obvious structural similarity to NT, having at least four similar or identical residues;

(b) The majority of the similar residues shall relate as involving single changes in the amino acid codon; and

(c) The peptide must be able to display at least one of the pharmacologic properties of NT.

Presence within nerves and endocrine cells is not a requirement for peptides in the NT family. It is also not necessary that the peptide be stored within vesicles and released by calcium-dependent mechanisms as appears to be the case for NT.

Given the large number of peptides already noted to cross-react with antisera towards NT, it has been difficult to decide which substances to isolate and identify. Pharmacologic assays provide a means of selecting some of the biologically active peptides but, depending upon the number of bioassays available, this method might falsely exclude quite a few peptides of interest. Another approach is to select peptides whose levels in biologic fluids are elevated following a physiologic stimulus thought to release NT-related material (e.g., ingestion of fat); however, a means of distinguishing between active peptides and their metabolites might also be necessary for success with this method. One purely chemical approach is to select peptides that react with multiple antisera toward the biologically active region of NT and thus are likely to have biologic significance. Provided that appropriate antisera can be obtained, this method would seem to provide the simplest means of selection. The relative reactivities of each substance with various region-specific antisera could be used as a measure of chemical similarity to NT and peptides displaying the strongest homology to the COOH-terminal region of NT would be given priority.

All of these approaches appear to be worthwhile and could quickly yield much information concerning the chemical and biological relationships among NT-related peptides. As we become more skilled at employing these methods, the task of identifying peptides of the NT family may become routine. In fact, one is tempted to predict that in the future most new peptidic messengers will be discovered and isolated on the basis of cross-reactions in immunoassays for known peptides and not on the basis of their pharmacologic activities.

REFERENCES

1. YALOW, R. S. 1978. Radioimmunoassay: A probe for the fine structure of biologic systems. Science **200:** 1236–1245.
2. YALOW, R. S. & S. A. BERSON. 1959. Assay of plasma insulin in human subjects by immunological methods. Nature **184:** 1648–1657.
3. CATT, K. J., H. D. NIALL & G. W. TREGEAR. 1967. Solid-phase radioimmunoassay. Nature **213:** 825–827.
4. WOODHEAD, J. S., G. M. ADDISON & C. N. HALES. 1974. The immunoradiometric assay and related techniques. Br. Med. Bull. **30**(1): 44–49.

5. MILES, L. E. M., C. P. BIEBER, L. F. ENG & D. A. LIPSCHITZ. 1974. Properties of "two-site" immunoradiometric (labeled antibody) assay systems, *In* Radioimmunoassay and Related Procedures in Medicine. Vol. **1:** 149. International Atomic Energy Agency. Vienna.

6. RUBENSTEIN, K. E., R. S. SCHNEIDER & E. F. ULLMAN. 1972. "Homogeneous" enzyme immunoassay. A new immunochemical technique. Biochem. Biophys. Res. Comm. **47:** 846–451.

7. REHFELD, J. F. 1978. Multiple molecular forms of cholecystokinin. *In* But Hormones. S. R. BLOOM, Ed. Churchill Livingstone. London. pp. 213–218.

8. DIMALINE, R., C. VAILLANT & G. J. DOCKRAY 1980. The use of region-specific antibodies in the characterization and localization of VIP-like substances in the rat gastrointestinal tract. Regulatory Peptides **1:** 1–16.

9. CURD, J. G., N. S. YOUNG & A. N. SCHECHTER. 1976. Antibodies to an NH_2-terminal fragment of B^s globin. J. Biol. Chem. **251:** 1283–1295.

10. BENVENISTE, R. & L. A. FROHMAN. 1978. An immunological approach to the study of protein conformational heterogeneity: Its application to growth hormone. Endocrinology **102:** 198–209.

11. REHFELD, J. F. 1978. Problems in the technology of radioimmunoassays for gut hormones. *In* Gut Hormones. S. R. BLOOM, Ed. Churchill Livingstone. London. pp. 112–119.

12. CARRAWAY, R. E. 1979. Neurotensin and related substances. *In* Methods of Hormone Radioimmunoassay, Second Ed. B. M. JAFFE & H. R. BEHRMAN, Eds. Academic Press. New York. pp. 139–169.

13. HABENER, J. F. & J. T. POTTS JR. 1976. Assay of calcium-regulating hormones. *In* Hormones in human blood. N. H. ANTONIADES, Ed. Harvard University Press. Mass. pp. 551–605.

14. LEEMAN, S. E., E. A. MROZ & R. E. CARRAWAY. 1977. Substance P and neurotensin. *In* Peptides in Neurobiology. H. Gainer, Ed. Plenum. New York. pp. 99–144.

15. CARRAWAY, R. E. & Y. M. BHATNAGAR. 1980. Immunochemical characterization of neurotensin-like peptides in chicken. Peptides **1:** 159–165.

16. CARRAWAY, R. E. & S. E. LEEMAN. 1976. Radioimmunoassay for neurotensin—A hypothelamic peptide. J. Biol. Chem. **251:** 7035–7044.

17. CARRAWAY, R. E. & Y. M. BHATNAGAR. 1980. Isolation, structure, and biological activity of chicken intestinal neurotensin. Peptides **1:** 167–174.

18. BROWN, J. P., K. E. HELLSTROM & I. HELLSTROM. 1981. Clin. Chem. **27:** 1592–1596.

19. CARRAWAY, R., R. A. HAMMER & S. E. LEEMAN. 1980. Endocrinology **107:** 400–406.

20. CATT, K. J. & G. W. TREGEAR. 1967. Solid-phase radioimmunoassay in antibody coated fibers. Science **158:** 1570–1572.

21. CARRAWAY, R. E. & C. F. FERRIS. Unpublished results.

22. CARRAWAY, R., S. E. RUANE & H. R. KIM. 1982. Peptides **3:** 115–123.

23. BHATNAGER, Y. M. & R. CARRAWAY. 1981. Peptides **2:** 51–55.

24. REINECKE, M., R. E. CARRAWAY, S. FALKNER, G. E. FEURLE & W. G. FORSSMANN. 1980. Cell Tissue Res. **212:** 173–183.

25. REINECKE, M., K. ALMASAN, R. CARRAWAY, V. HELMSTAEDER & W. G. FORSSMANN. 1980. Cell Tissue Res. **205:** 383–395.

26. HAMMER, R. A., S. E. LEEMAN, R. CARRAWAY & R. H. WILLIAMS. 1980. Isolation of human intestinal neurotensin. J. Biol. Chem. **225:** 2476.

27. CARRAWAY, R. E. & S. E. LEEMAN. 1976. Characterization of radioimmunoassayable neurotensin in the rat. J. Biol. Chem. **251:** 7045–7052.

28. HIRST, B. H., B. SHAW & L. WILSON. 1982. Neurotensin inhibition of gastric exocrine secretions in the cat. Regulatory Peptides **3:** 289–301.

29. CARRAWAY, R. E. & C. FERRIS. Unpublished results.

30. CARRAWAY, R. E., M. REINEKE & W. G. FORSSMANN. Unpublished results.

31. CARRAWAY, R. E. & M. SCHMIDT. Unpublished results.

32. ARAKI, K., S. TACHIBANA, M. UCHIYAMA, T. NAKAJIMA & T. YASUHARA. 1973. Isolation and structure of a new active peptide, "Xenospin", on the smooth muscle, especially on a strip of fundus from a rat stomach from the skin of *Xenopus laevis*. Chem. Pharm. Bull. (Tokyo) **21:** 2801.

33. CARRAWAY, R., S. E. RUANE, G. E. FEURLE & S. TAYLOR. 1982. Amphibian neurotensin (NT) is not Xenopsin (XP): Dual presence of NT-like and XP-like peptides in various amphibia. Endocrinology **110**: 1094–1101.
34. ARONIN, N. *et al*. Peptides. Submitted for publication.
35. HAMMER, R. A., R. E. CARRAWAY & S. E. LEEMAN. 1982. J. Clin. Invest. July.
36. FERRIS, C. F., R. A. HAMMER & S. E. LEEMAN. 1981. Elevation of plasma neurotensin during lipid perfusion of the rat small intestine. Peptides **2**: 263–266.

DISCUSSION OF THE PAPER

C. B. NEMEROFF (*University of North Carolina, Chapel Hill*): Do you have any data on the purity of synthetic neurotensin? And, secondly, do you have any evidence on the presence of xenopsin in mammalian CNS?

R. E. CARRAWAY (*University of Massachusetts, Worcester*): Most of the peptides that we deal with are made by the solid phase methodology and, of course, there are a number of problems in this regard. One obtains many different varieties of peptides with this procedure. It is often very difficult to separate them from one another. Even though one might do some chromatography, one might not be able to separate peptides that have one amino acid missing, for example, particularly if that change does not make a very striking difference in properties of the substance. HPLC has a fantastic ability to separate peptides and maybe it can uncover some of the heterogeneity that is there, but I have not done a rigorous analysis myself.

Your second question—concerning the xenopsin-related material—in fact, we have been using antibodies towards xenopsin to ask firstly whether xenopsin is the counterpart to neurotensin in amphibian species. We have established that this is not the case, but we have seen that there are xenopsin-like peptides in both the brain and the gut of amphibians as well as neurotensin-related material in these places. Of the amphibians that we have examined only *Xenopus* seems to have high concentrations in the skin, but, all of these amphibians seem to have both xenopsin and neurotensin in brain and gastrointestinal tissues.

We have also utilized our xenopsin radioimmunoassay to ask whether there is xenopsin-related material in the mammal. We have some information suggesting that xenopsin is present in the dog stomach and also in the human stomach.

UNIDENTIFIED SPEAKER: My questions concern the radioimmunoassay in plasma for neurotensin. Would you comment on the use of *trasylol* (*aprotinin*) in the assay? Would you comment on the use of extraction procedures for plasma? Is there any danger that these extraction procedures alter the relative amounts of the different sequences of peptides in plasma, for instance, by cleaving prohormones?

CARRAWAY: In the first place, some of us are not extracting our plasma and are adding it directly to perform radioimmunoassay. There are problems with this, not only specifically in regard to neurotensin, but for any peptide that is being assayed. There can be artifacts, as I have shown you, which even parallel the dose-response relationship for neurotensin. So this can be very tricky. For example, enzymes that degrade the tracer, can artifactually interact in the assay and appear to be neurotensin.

As to inhibitors of these enzymes, we do not add our plasma directly so we have not done much in that regard. I have looked at several inhibitors— phenanthrolene being one as you have seen—and it does appear to stabilize the tracer in certain instances. Phenanthrolene is a chelator of zinc and other metals and it can inhibit the action of carboxypeptidase A, which very readily degrades neurotensin. Therefore it might be a good substance to employ if one is adding plasma directly. One would certainly have to keep the level of phenanthrolene lower than the levels that may interact in the assay being done.

As to chemical changes in the structure of neurotensin during extraction, I do not believe that any of the extraction procedures that we are employing are harsh enough to deamidate or racemize or cause other structural changes. We have validated our extraction procedure by adding neurotensin and retrieving it with the extraction procedure, obtaining about 75 to 80% yield, and chromatography of this material by HPLC indicates that it appears to be intact. I think this is something that is necessary for all of us to do; we have to be wary in this way.

UNIDENTIFIED SPEAKER: Could I just have a further comment? If you perform chromatography of plasma before and after extraction, what are the major changes you see?

CARRAWAY: Neurotensin and other neurotensin-related substances could be associated with carrier substances in plasma and the extraction procedure might dissociate these interactions. If one chromatographs plasma, one may see what appear to be very large molecular forms, which upon extraction become small molecular weight forms. Neurotensin, being basic, is able to associate with negatively charged proteins in plasma and in other tissues as well. If one adds neurotensin to plasma, for example, and then one precipitates proteins, one can precipitate the added neurotensin. If one acidifies plasma to pH 2 and then precipitates the proteins, the neurotensin is not in the precipitate but instead in the supernatant fraction. The extraction procedure which we presently employ allows us to retrieve neurotensin from plasma and to compare it to the synthetic form. If we did not extract, we might find multiple peaks as I said, some larger molecular weight ones which actually are not large molecular weight forms but associations.

EFFECTS OF NEUROTENSIN ON INTESTINAL SMOOTH MUSCLE: APPLICATION TO THE STUDY OF STRUCTURE–ACTIVITY RELATIONSHIPS*

Patrick Kitabgi

Centre de Biochimie du .
Centre National de la Recherche Scientifique
Université de Nice
06034 Nice, Cedex, France

INTRODUCTION

Almost ten years have elapsed since Carraway and Leeman reported the isolation of a hypothalamic peptide they named neurotensin because of its neural origin and its marked hypotensive effect on the rat.[1] Although neurotensin has received less attention than other neuropeptides, it will be apparent throughout this book that a great deal of information has been learned about the tissue distribution, biologic properties, and structure–activity relationships of this interesting peptide. Neurotensin belongs to the continuously growing family of peptides (hypothalamic-releasing factors, somatostatin, the vasoactive intestinal polypeptide, opioid peptides, bombesin-like peptides, gastrin-cholecystokinin-like peptides, substance P), most of them discovered during the past decade, that are present in the central nervous system and peripheral tissues, mainly the gastrointestinal tract. There is now a large body of evidence that these peptides are the messengers of a broad system of intercellular communication. Thus, all of them are synthesized and stored in endocrine and/or nerve cells and appear to act as hormones and/or neurotransmitters. Several aspects of the dual role of neurotensin as a hormone and as a neurotransmitter will be treated in this book. Like almost every hormone and neurotransmitter, these peptides are thought to bind to specific receptors on their target cells. This binding represents the first step of their biologic action and triggers a sequence of cellular events that constitute the biologic response to the peptide messengers. Obviously, studies aimed at characterizing the interaction of these messengers with their receptors, and the subsequent events that follow this interaction should prove essential for understanding their mechanism of action at the cellular and molecular level.

However, although neurotensin exhibits a broad spectrum of biologic effects, only a few *in vitro* systems have been described that are suitable for investigating the mode of action of the peptide. Among such systems are preparations from the cardiovascular system and the gastrointestinal tract. A paper in this book, by Rioux *et al.*, will be devoted to the mechanisms of the cardiovascular effects of neurotensin. The present paper is concerned with the activity of the peptide on the contractility of smooth muscle preparations from various parts of the rat and guinea pig gastrointestinal tracts. Since these preparations are innervated with excitatory and inhibitory nerves of intrinsic and extrinsic origin, both prejunc-

*This work was supported in part by Research Grants ATP 16.75.39 and ATP 71.78.103 from the Institut National de la Santé et de la Recherche Médicale (INSERM) and by a grant from the Fondation pour la Recherche Médicale Française.

37

tional (presynaptic, neurogenic) and postjunctional (postsynaptic, myogenic) effects may be observed with substances that affect smooth muscle contractility. Therefore pharmacologic, biochemical, and electrophysiologic experiments that were designed to determine the sites and mechanisms of action of neurotensin will be reported. Finally, data on structure–activity relationships obtained by evaluating the activity of many neurotensin analogs in bioassay systems derived from gastrointestinal preparations will be presented.

EFFECTS OF NEUROTENSIN ON RAT GASTROINTESTINAL SMOOTH MUSCLE

Rat Stomach

The first report that neurotensin contracted rat-isolated fundus strips was from Rökaeus et al.[2] in 1977. This effect was to be expected from previous reports by Araki et al. in 1973[3] and 1975[4] showing that xenopsin, a peptide isolated from frog skin and closely related to neurotensin in structure, exerted a potent contracting effect on the longitudinal smooth muscle of rat-isolated stomach strips in an organ bath. The findings of Rökaeus et al. were later confirmed by others.[5, 7] TABLE 1

TABLE 1

SOME CHARACTERISTICS OF THE NEUROTENSIN-INDUCED CONTRACTION
IN RAT FUNDUS STRIP

| | | | Antagonists to | | | |
References	Incubation Temperature	EC_{50}	Acetylcholine	Histamine	Serotonin	Desensitization
2	37°C	5 nM	no effect	not tested	no effect	yes
5	30°C	3 nM	no effect	no effect	no effect	yes
7	37°C	11 nM	no effect	no effect	no effect	yes

summarizes some characteristics of neurotensin action on rat stomach strips, as derived from several reports. EC_{50} (concentration of drug that produces half maximal response) values ranged from 3 to 11 nM. The effect of neurotensin was not blocked by antagonists to acetylcholine, histamine, and serotonin, suggesting that the peptide interacted with specific receptors on rat fundus smooth muscle cells. Desensitization of the preparations to neurotensin was reported. The rat fundus preparation was used by Kataoka et al.[5] and Quirion et al.[7, 8] to measure the activity of a variety of neurotensin partial sequences and analogs. The results of these structure–activity studies will be discussed in the last section of this paper.

Rat Duodenum

In their initial report on the discovery of neurotensin, Carraway and Leeman[1] showed that the peptide relaxed rat dudoenum segments in vitro. Antagonists to α- and β-adrenoceptors were without effect on the neurotensin-induced relaxation.[1] Others[2,5,6] also observed the inhibitory effect of neurotensin on the contractility of rat duodenal preparations. However, concentration-response curves were not reported probably because of the slight and variable effect that neurotensin exerts in the rat duodenum.[6]

Rat Ileum

We have previously reported that rat ileal segments relax in response to neurotensin.[9] Since rat ileum preparations usually exhibit low basal tone, the neurotensin-induced relaxation was best observed when smooth muscle was contracted by prior addition of acetylcholine or histamine. TABLE 2 compares some properties of the relaxant effect of neurotensin and adrenaline. Although neurotensin is less effective, it is a hundred times more potent than adrenaline. The lack of effect of the neurotoxin tetrodotoxin, which blocks nerve conduction without affecting smooth muscle cells, strongly suggests that neurotensin interacts at a postjunctional level with smooth muscle cell receptors. These receptors are different from α- and β-adrenoceptors since α- and β-blocking agents do not inhibit neurotensin-induced relaxations in rat ileum (TABLE 2).

Conclusions

Neurotensin affects the contractility of rat gastrointestinal smooth muscle preparations *in vitro*. The peptide contracts the longitudinal muscle of rat fundus and relaxes the longitudinal muscle of rat small intestine, that is, duodenum and ileum. In all these tissues, neurotensin appears to interact at a postjunctional level with specific receptors on smooth muscle cells. However, the mechanism of action of neurotensin, that is, the link between receptor activation and cellular response (contraction or relaxation) has not been investigated in the rat. This may be so partly because preparations such as rat stomach strips, duodenal, or ileal segments are not as suitable as other smooth muscle preparations for electrophysiologic studies, and partly because the myorelaxant effect of neurotensin in rat small intestine is rather weak as compared to that of other agents such as adrenaline. Much more information has been obtained from studies with guinea pig intestinal smooth muscle preparation, as shall be discussed next.

EFFECTS OF NEUROTENSIN ON GUINEA PIG INTESTINAL SMOOTH MUSCLE

Guinea Pig Ileum—Prejunctional Effects

The guinea pig ileum, one of the most widely used smooth muscle preparation in pharmacology, was shown by Carraway and Leeman[1] to contract in response to neurotensin. Several authors have confirmed this observation.[2, 5, 9, 10] FIGURE 1 shows that the addition of neurotensin to a guinea pig ileal preparation induces a concentration-dependent contraction that develops after a lag period of 10–20 seconds. Repeated additions of peptide at short intervals (3 min) result in desensitization of the preparation for neurotensin concentrations higher than 30 nM. FIGURE 2 compares the concentration-response curve of neurotensin with those of several substances known to contract the guinea pig ileum. With an EC_{50} value of 4–5 nM, neurotensin is one of the most potent among these agents; however, it is the least effective. Maximal contraction induced by neurotensin is only 40% of that induced by histamine.

Several authors have reported that antagonists to muscarinic cholinoceptors were without effect on neurotensin-induced contractions in the guinea pig ileum;[1, 5, 10] this is in contrast with our findings.[9] Indeed, we observed that

TABLE 2

COMPARISON OF NEUROTENSIN- AND ADRENALINE-INDUCED
RELAXATIONS IN RAT ILEUM*

Agents	EC_{50}	Maximal† Effect	Tetrodotoxin	Propranolol (6 μM) + Phentolamine (1.5 μM)
Neurotensin	4.4nM	38%	no effect	no effect
Adrenaline	430nM	100%	no effect	inhibition‡

* The effects of neurotensin and adrenaline were measured using rat ileum segments contracted by a maximally effective concentration of acetylcholine (2 μM) as previously reported.[9]

† Maximal effect refers to the maximal drop in tension elicited by agents when added to a rat ileum preparation contracted by 2 μM acetylcholine.

‡ Fifteenfold increase in the EC_{50} value for adrenaline.

atropine blocked about 50% of the contraction elicited by maximally effective concentrations of neurotensin.[9] In addition, it was found that the peptide-induced contraction was totally abolished by tetrodotoxin as shown in FIGURE 3. Since, as already mentioned, tetrodotoxin blocks nerve conduction without affecting smooth muscle cells, we concluded that, in the guinea pig ileum, neurotensin stimulates cholinergic as well as noncholinergic excitatory nerves.[9]

The effect of neurotensin on cholinergic nerves was clearly demonstrated using isolated longitudinal smooth muscle strips from the guinea pig ileum. The longitudinal muscle coat of the guinea pig ileum can easily be dissected out, yielding muscle strips with adherent myenteric plexus.[11] These strips were shown to contract in response to neurotensin and the contraction was inhibited by tetrodotoxin.[12] FIGURE 4 compares the concentration-response curves for neurotensin,

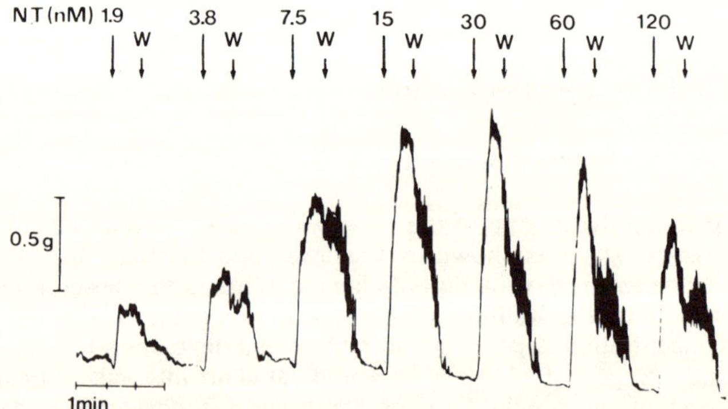

FIGURE 1. Tension recording from a guinea pig ileum preparation exposed to increasing concentrations of neurotensin. A segment of guinea pig ileum was set up for isometric tension recording in a 10-ml organ bath at 37°C in Tyrode solution gassed with 95% O_2–5% CO_2 as previously described.[9] Neurotensin (NT) was added to the organ bath at the concentrations indicated in nM. One minute after addition of peptide, the preparation was washed out (W) and 2 min later a new cycle was started. Note the transient effect of neurotensin and the desensitization that occurs with a high concentration of peptide.

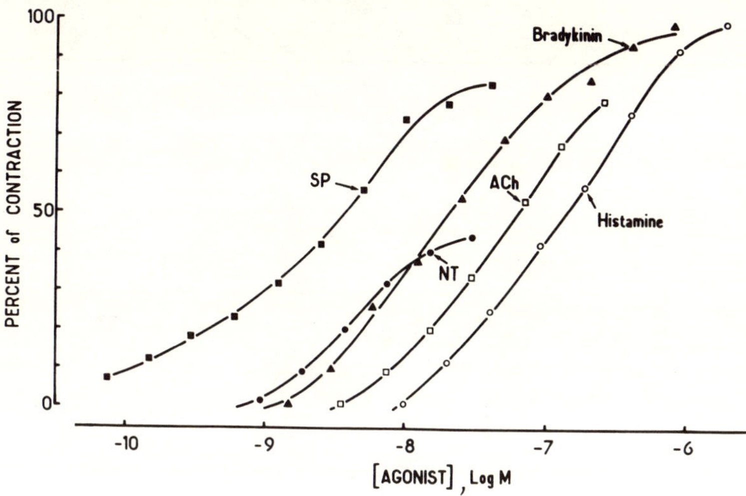

FIGURE 2. Concentration-response curves of contraction by neurotensin (NT), substance P (SP), bradykinin, acetylcholine (ACh), and histamine in guinea pig ileum preparations. Experimental conditions were as previously described.[9] Results are expressed as the percent of maximal contraction induced by histamine.

acetylcholine, and histamine in several incubation conditions.[13] In normal physiologic solution, the maximal contraction induced by neurotensin represents about 20% of the maximal contraction induced by histamine (FIG. 4A). In the presence of the acetylcholine esterase inhibitor neostigmine, neurotensin-induced contractions are potentiated 4-fold with no significant change in the peptide EC_{50} value (FIG. 4B). As expected, neostigmine also potentiates acetylcholine- but not histamine-induced contractions. Addition of the muscarinic antagonist atropine to the physiologic solution results in a complete abolition of neurotensin activity (FIG. 4C). Atropine (0.1 μM) leaves histamine-induced contractions unchanged,

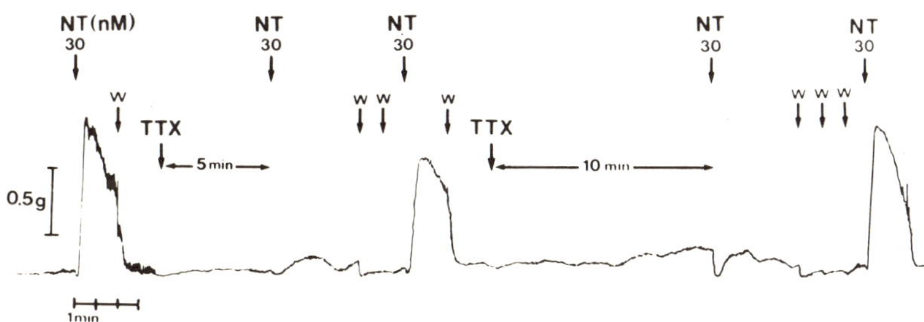

FIGURE 3. Effect of tetrodotoxin on the neurotensin-induced contraction in guinea pig ileum. The ileum was exposed to 300 nM tetrodotoxin (TTX) for either 5 or 10 min before the addition of 30 nM neurotensin (NT). W, washout. Note that during the 10-min exposure to tetrodotoxin, muscle tone increased, and that a subsequent addition of neurotensin induced a transient relaxation. (Reprinted from Kitabgi and Freychet[9] with the permission of the publisher.)

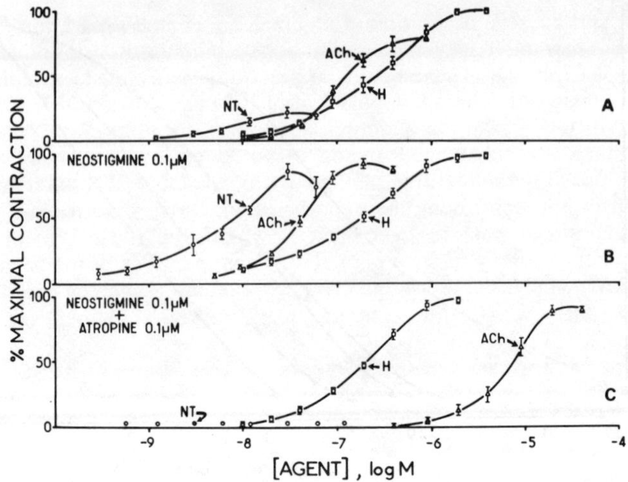

FIGURE 4.   Concentration-response curves for the contraction of isolated longitudinal smooth muscle strips from guinea pig ileum by neurotensin (NT), acetylcholine (ACh), and histamine (H) in: A, normal physiologic (Tyrode) solution; B, Tyrode solution + 0.1 μM neostigmine; and C, Tyrode solution + 0.1 μM neostigmine + 0.1 μM atropine. (Reprinted from Kitabgi and Freychet[13] with the permission of the publisher.)

whereas it increases the EC_{50} of acetylcholine by a factor of 175. In addition, the ganglionic blocking agent hexamethonium had no effect on the contractile response to neurotensin.[13] Taken together, these findings clearly establish that neurotensin stimulates the release of acetylcholine from postganglionic excitatory nerves that possibly originate in the myenteric plexus and innervate the longitudinal muscle coat of the guinea pig ileum. It should be noted that, in longitudinal muscle strips, atropine completely inhibits the response to neurotensin, whereas, in intact segments of ileum, the inhibition by atropine is only partial (about 50%). This could mean that the noncholinergic nerves, which, in addition to cholinergic nerves, are activated by neurotensin in the guinea pig ileum, lose their responsiveness to neurotensin during dissection of the longitudinal muscle coat. These noncholinergic nerves might originate outside the myenteric plexus, for instance in the submucous plexus where the neurotensin receptors would also be located.

The observation that neurotensin stimulates noncholinergic excitatory nerves in guinea pig ileum led us to investigate the nature of the transmitter(s) released from such nerves. Several lines of evidence suggest that substance P, a potent gut contracting peptide, acts as an excitatory neurotransmitter in the intestinal tract.[15, 16] It has been shown recently that substance P is able to induce a specific desensitization of its own receptors on longitudinal smooth muscle cells of the guinea pig ileum, and that such a desensitization results in the inhibition of the contractions produced by electrical stimulation of noncholinergic excitatory nerves.[15] The ability of substance P to specifically inactivate its receptors is a useful property of the peptide in the absence (until very recently) of substance P antagonists. This property was used in order to evaluate the possibility that neurotensin releases substance P from enteric nerves.[17] FIGURE 5 shows that after the cholinergic component of neurotensin action has been abolished by atropine

(Fig. 5, A and B), desensitization of a guinea pig ileum preparation to substance P results in a complete, reversible inhibition of the neurotensin-induced contraction (Fig. 5, C and D). These results suggest that neurotensin stimulates enteric excitatory nerves that upon depolarization release substance P (or substance P-like activity), which acts as a motor excitatory transmitter. Very recently, it was reported that a new substance P analog, D-Pro^2, D-Trp^7, 9-substance P, behaved as a specific, possibly competitive, antagonist to substance P.[18] Such an antagonist might be useful in confirming the hypothesis put forward above.

Several brain gut peptides have been shown to affect the activity of cholinergic excitatory nerves on the guinea pig ileum. Thus, peptides of the gastrin family[19] and thyrotropin-releasing hormone (TRH)[20] stimulate, whereas opioid peptides[21] and somatostatin[22] inhibit the release of acetylcholine from cholinergic nerves of

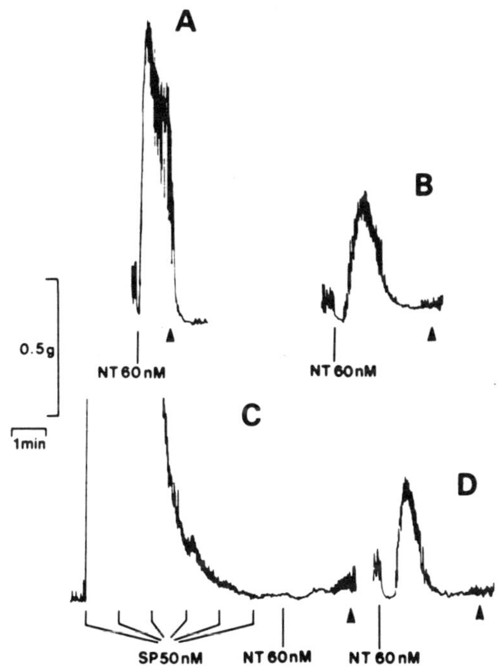

FIGURE 5.   Effect of substance P-induced autodesensitization on the noncholinergic contraction elicited by neurotensin in guinea pig ileum. A: response of a guinea pig ileum segment to 60 nM neurotensin (NT) in normal Tyrode solution. B: same as A in the presence of 100 nM atropine, a concentration which totally blocks the cholinergic component of neurotensin action. C: the muscle preparation in the presence of atropine was desensitized by six additions, one every min, of substance P (SP) at 50 nM (the upper part of the recording is not shown for convenience); muscle tension reached 1.6 g after the first addition of substance P, then gradually returned to basal level and was not affected by the last addition of peptide. Note the absence of contraction after a subsequent addition of neurotensin; in the same conditions, the preparation normally contracts in response to histamine and bradykinin (not shown). D: same as B, 20 min after desensitization and several washouts. In all tracings, arrowheads indicate washouts of the preparation. (Reprinted from Monier and Kitabgi[17] with the permission of the publisher.)

the myenteric plexus. The possibility that Met-enkephalin and somatostatin modulate the neurogenic action of neurotensin was investigated.[23] It was found that the opioid peptide and somatostatin produced a concentration-dependent inhibition of the neurotensin-induced contraction in guinea pig ileum. This inhibitory effect was exerted on both cholinergic and noncholinergic components of neurotensin action, an observation that suggests that Met-enkephalin and somatostatin inhibit the release of substance P (or substance P-like activity) from intestinal nerves. Such an inhibitory effect of Met-enkephalin on the release of substance P from nerve cells has been observed with slices of trigeminal nucleus[24] and with sensory neurons in culture.[25]

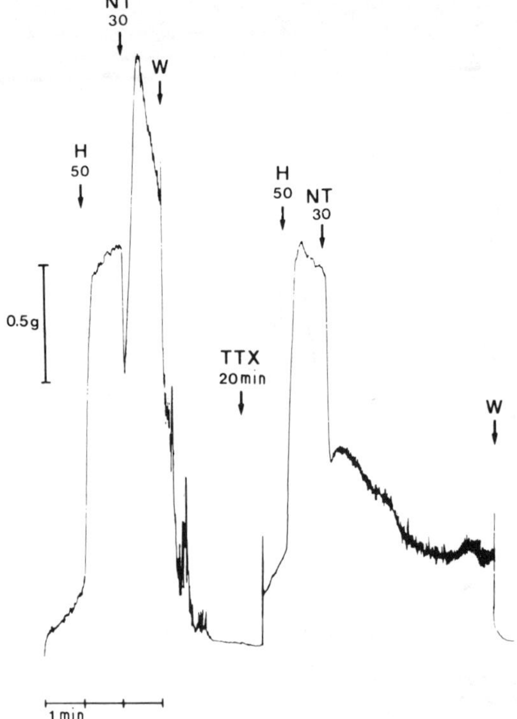

FIGURE 6. Effect of tetrodotoxin (TTX) at 300 nM on the biphasic response induced by 30 nM neurotensin (NT) in a guinea pig ileum contracted by 50 nM histamine (H). The preparation was exposed for 20 min to the toxin during which time recording was stopped. W = washout. (Reprinted from Kitabgi and Freychet[9] with the permission of the publisher.)

Guinea Pig Ileum—Postjunctional Effects

Tracings such as those in FIGURES 1 and 3 show that addition of neurotensin to resting guinea pig ileum preparations induces a contraction. However, we observed that when muscle preparations were contracted either spontaneously or by prior incubation with acetylcholine or histamine, addition of neurotensin produced a biphasic response, that is, a relaxation followed by a contraction.[9] FIGURE 6 shows that tetrodotoxin blocks the contraction phase of the response whereas the toxin does not affect the relaxation phase. Similar results were obtained with isolated longitudinal smooth muscle strips from the guinea pig ileum.[12] Most likely, the neurotensin-induced relaxation results from an interaction of the peptide with receptors on longitudinal smooth muscle cells. Since the longitudinal muscle coat of the guinea pig small intestine is a tissue rich in smooth muscle cells and can be obtained in relatively large amounts, we attempted to investigate the binding of [3H]neurotensin to membranes prepared from smooth muscle cells isolated by collagenase treatment of the muscle tissue.[12] High affinity, specific binding of labeled neurotensin to smooth muscle membranes was observed. Unfortunately, the binding capacity of the membrane preparation was extremely low, which, in addition to the moderate specific activity of tritiated neurotensin, prevented a complete investigation of the peptide-binding characteristics.[12] Nevertheless, it was possible to derive a dissociation constant K_D of 4 nM for neurotensin, a value that compared well with the EC_{50} value of 5 nM obtained from concentration-response curves for the myorelaxant effect of neurotensin in guinea pig ileum.[12]

The mechanism by which some myorelaxant agents, in particular β-adrenoceptor agonists, inhibit intestinal smooth muscle contractility has been shown to be cyclic AMP-dependent. Neurotensin did not increase the cyclic AMP content of ileal longitudinal smooth muscle strips, nor did the peptide increase cyclic-GMP levels, indicating that cyclic nucleotides were not involved in the contractile response of guinea pig ileum to neurotensin.

Guinea Pig Taenia Coli

The taenia coli of the guinea pig cecum is a smooth muscle that has been well characterized with respect to morphologic, biochemical, pharmacologic, and electrophysiologic properties.[27] We have reported that neurotensin exerts a concentration-dependent contraction in strips of taenia coli set up in an organ bath.[9] The EC_{50} value was 4 nM and the maximal contraction induced by the peptide (30–60 nM) amounted to about 80% of that induced by histamine.[9] FIGURE 7D shows the contraction elicited by 7.5 nM neurotensin in a taenia coli preparation. The contraction was rapid in onset and sustained for several minutes. This contractile response was monophasic (FIG. 7, A and B) and was not affected by tetrodotoxin (FIG. 7, B and C). Therefore it appears that the neurotensin-induced contraction in the guinea pig taenia coli is a postjunctional effect that involves a direct interaction of the peptide with smooth muscle cells. This is in contrast to the prejunctional effect observed in the guinea pig ileum.

Strips of taenia coli provide suitable preparations for electrophysiologic studies on smooth muscle. In order to understand the membrane events and ionic mechanisms that underlie the contracting effect of neurotensin, muscle tension and membrane potential and conductance of taenia preparations in physiologic and modified salt solution were recorded using the sucrose gap method.[28] FIGURE 8 shows that neurotensin induces a depolarization in taenia coli. Initially, this depolarization leads to an increase in the frequency of spike discharge in spontaneously firing preparations (FIG. 8, A and B). As the depolarization increases a blockade of action potentials occurs (FIG. 8B). Upon cessation of spikes when depolarization is maximal, tension persists at a somewhat lower level (FIG. 8B). The first component of neurotensin effect (increase in spike frequency) may be referred to as phasic and the second component (depolarization) as tonic. FIGURE 9 shows that the neurotensin-induced depolarization is accompanied with an increase in membrane conductance. Therefore it is reasonable to assume that the interaction of neurotensin with receptors on smooth muscle cells of the taenia coli results in the

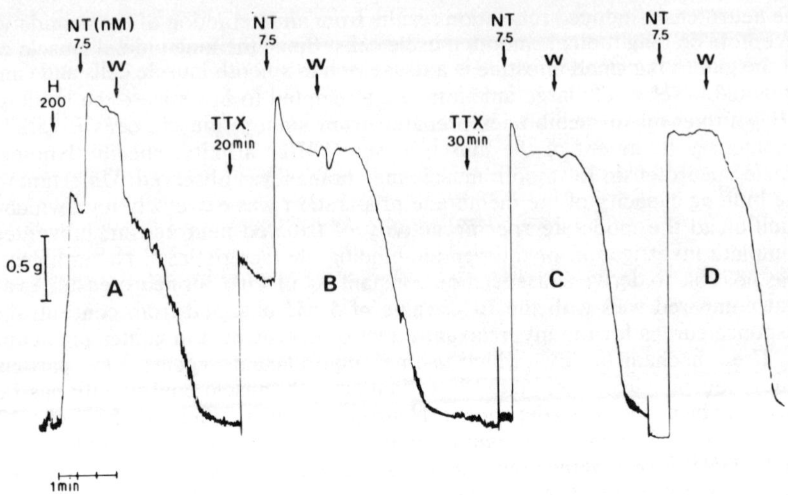

FIGURE 7. Neurotensin-induced contraction in guinea pig taenia coli. A: neurotensin (NT) at 7.5 nM was added after a 1-min exposure to 200 nM histamine. B: the peptide was added after a 20-min expousre to 300 nM tetrodotoxin (TTX). Note the monophasic contraction induced by neurotensin in A and B. C: same as B after a 30-min exposure to tetrodotoxin. D: neurotensin alone. W = washout. Other experimental conditions were as previously described.[9]

opening of ionic channels in the muscle membrane (receptor-operated channels as defined by Bolton[29]). Experiments in modified salt solutions that were designed to investigate the nature of ion species moving through the receptor-operated channels indicated that neurotensin increased both Na and Ca conductances.[28] Given the resting membrane potential in the smooth muscle of the taenia coli and the equilibrium potentials for Na and Ca, it is likely that the neurotensin-induced depolarization is brought about by an inward movement of these ions. The inward calcium current increases intracellular Ca concentration, which leads to contraction. It cannot be excluded that neurotensin also initiates a release of calcium from intracellular stores. All these features of neurotensin action are essentially similar to those reported for other smooth muscle contracting agents such as acetylcholine[29, 30] and angiotensin II.[30]

Guinea Pig Colon

Guinea pig colon preparations in an organ bath often exhibit spontaneous tone and phasic contractions. Recently we showed that neurotensin induced a potent and efficient relaxation in guinea pig proximal colon.[31] Tetrodotoxin or a combination of α- and β-adrenoceptor antagonists had no effect on the peptide-induced relaxation. Neurotensin was effective at concentrations as low as 0.1 nM. The peptide was 100 and 50,000 times more potent than adrenaline and ATP, respectively, in inducing relaxation. It is concluded that neurotensin interacts directly with high-affinity receptors on smooth muscle cells of guinea pig colon.[31] Although there has been no electrophysiologic study of this myogenic effect, there is indirect evidence to suggest that neurotensin hyperpolarizes and hence relaxes colonic smooth muscle cells by increasing a Ca-dependent K permeability in the smooth muscle membrane. This is based on the observation that the bee venom neurotoxin apamin inhibits and converts to a contraction neurotensin-induced relaxation in guinea pig colon (FIG. 10). It is now well established that in a variety of tissues, apamin blocks potassium permeabilities such as those activated in smooth muscle by myorelaxant agents like α-adrenoceptor agonists or ATP.[32] Furthermore, since apamin selectively inhibits calcium-activated potassium permeabilities, it would appear that neurotensin opens a receptor-operated calcium channel, allowing Ca to enter the cell and activate a potassium conductance. This hypothesis is supported by the finding that neurotensin-induced relaxations of the guinea pig colon are reduced or abolished when Ca is lowered or omitted in the physiologic solutions (Kitabgi, unpublished observations).

Conclusions

Neurotensin affects the contractility of guinea pig intestinal smooth muscle *in vitro*. Both prejunctional (presynaptic, neurogenic) and postjunctional (post-

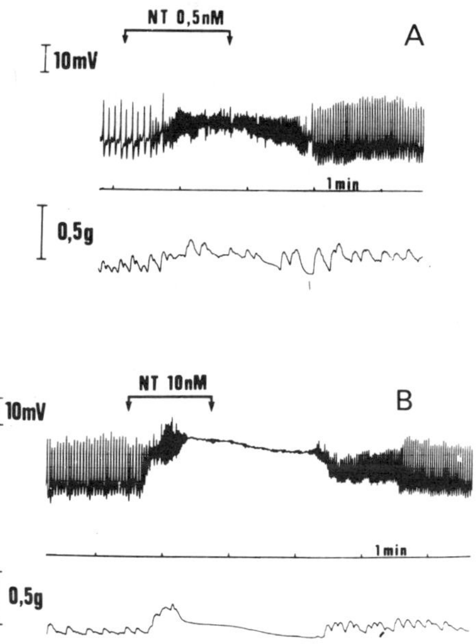

FIGURE 8.   Effect of neurotensin on the electrical and mechanical activity of a spontaneously active guinea pig taenia coli preparation. Single sucrose gap, normal physiologic solution, 37°C. In A and B, the upper tracing represents potential recording and the lower tracing, tension recording. Note the spontaneous spikes and the corresponding phasic contractions. A: effect of 0.5 n*M* neurotensin. Note the increase in spike frequency and the slight depolarization accompanied with an increase in both phasic contractions and underlying muscle tension. B: effect of 10 n*M* neurotensin. Note the strong depolarization, the initial increase in spike frequency, and the subsequent block of spike discharge. These events are associated with an initial increase in phasic contractions followed, upon cessation of spikes, by a tonic contraction. (Reprinted from Kitabgi *et al.*[28] with the permission of the publisher.)

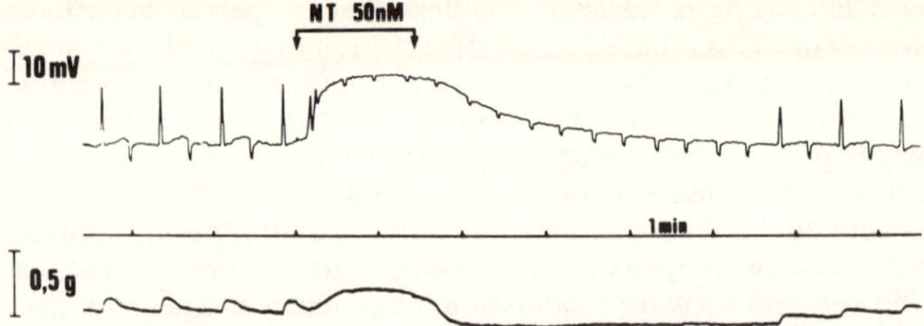

FIGURE 9. Effect of 50 n*M* neurotensin on the electrical and mechanical activity of a guinea pig taenia coli preparation. Double sucrose gap, normal physiologic solution, 20°C. Upper tracing, potential recording; lower tracing, tension recording. The preparation was alternately subjected to hyperpolarizing (downward deflection) and depolarizing (upward deflection) constant current pulses (pulse duration, 2 sec). Note the absence of spontaneous activity (as expected from a preparation incubated at 20°C), the spikes evoked by depolarizing pulses (above threshold), and the associated contractions. During the application of neurotensin, only hyperpolarizing pulses were applied. Note the strong depolarization preceded by a few spikes, the associated tonic contraction, and the decrease in electrotonic potentials evoked by hyperpolarizing pulses that indicates that neurotensin increases smooth muscle membrane conductance. (Reprinted from Kitabgi *et al.*[28] with the permission of the publisher.)

synaptic, myogenic) effects are observed. The prejunctional effects were analyzed in the guinea pig ileum. In this preparation, neurotensin contracts the longitudinal muscle by stimulating cholinergic excitatory nerves and nerves that possibly release substance P. The cholinergic nerves seem to be located within the myenteric plexus whereas the noncholinergic nerves might originate in the submucous plexus where the neurotensin receptors responsible for their activation would also be located. This conclusion is supported by recent autoradiographic studies that showed the localization of neurotensin binding sites in a band in the submucosal plexus layer of the guinea pig ileum (Ninkovic, Jessen, and Hunt, unpublished observations). The excitatory action of neurotensin on both cholinergic and noncholinergic (substance P-releasing) excitatory nerves can be modulated by inhibitory peptides such as Met-enkephalin and somatostatin. Since all these peptides are present in endocrine and/or nerve cells of the gut, we hypothesize that they physiologically modulate the activity of enteric neurons that control intestinal motility in the guinea pig.

The ionic mechanisms by which neurotensin activates these neurons are not known. Electrophysiologic studies with single myenteric neurons from the guinea pig ileum have shown that the peptide stimulated about 50% of these cells.[26] Most neurons that responded to neurotensin were depolarized by the peptide at concentrations in the nanomolar range. The depolarization was associated with an increase in membrane resistance and persisted in Ca-free solution. However, determination of the ionic events underlying the neurotensin-induced depolarization in myenteric nerves was made difficult by the small amplitude of the peptide effect and by desensitization of the preparations following repeated applications of neurotensin.[26] Further studies are needed to understand the mechanism by which neurotensin depolarizes guinea pig myenteric nerves.

The postjunctional effects of neurotensin on guinea pig intestinal smooth muscle were either excitatory as in the taenia coli, or inhibitory as in the ileum and the colon. Other agents that affect smooth muscle contractility such as catecholamines, acetylcholine, and bradykinin also exert excitatory or inhibitory myogenic effects depending on the localization and function of the smooth muscle. The mechanism by which neurotensin generates tension in the guinea pig taenia coli is similar to that described for other excitatory agents. This mechanism involves depolarization of muscle cells by increased permeability to sodium and calcium. It seems likely that neurotensin interacts with smooth muscle membrane receptors that operate calcium and sodium channels. A similar mechanism might also be responsible for the excitatory effect of neurotensin on rat stomach longitudinal smooth muscle. However there is no evidence at present to support this view.

Neurotensin exhibits myogenic inhibitory effects in guinea pig ileum and colon and also in rat small intestine. Myorelaxant agents can be classified in two categories: (1) agents like α-adrenoceptor and P_2-purinoceptor agonists open receptor-operated calcium channels; consequent admission of Ca into smooth muscle cells triggers the opening of potassium channels and this results in a hyperpolarizing outward K current leading to relaxation; (2) agents like β-adrenoceptor and P_1-purinoceptor agonists are thought to induce relaxation by a cyclic-AMP-dependent process that brings about a decrease in cytosolic Ca concentrations. Neurotensin does not seem to belong to the second category, at least with regard to its relaxant action in guinea pig ileum, a tissue in which the peptide had no effects on cyclic AMP levels. More likely, neurotensin belongs to the first category as suggested by the properties of its inhibitory action in guinea pig colon. Study on the mechanisms of the inhibitory effects of neurotensin on intestinal smooth muscle may bear some physiologic relevance with respect to the hormonal role proposed for neurotensin in the digestive tract, in particular as the factor which inhibits gastrointestinal motility *in vivo* following fat ingestion.[14]

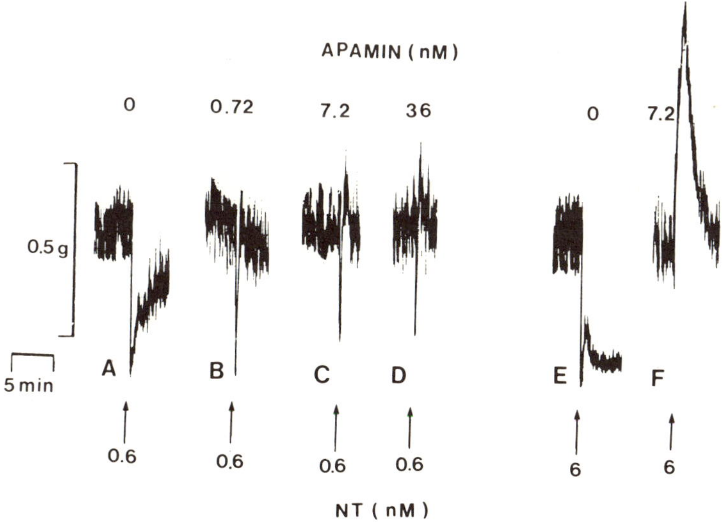

FIGURE 10. Effect of apamin on the relaxation induced by neurotensin (NT) in a guinea pig colon preparation. Apamin was added 4 min before neurotensin. Apamin reduced in a concentration-dependent manner both the magnitude and the duration of the relaxation induced by 0.6 nM neurotensin (A, B, C, D). When neurotensin concentration was increased to 6 nM, apamin at 7.2 nM converted the peptide-induced relaxation into contraction (E, F). (Reprinted from Kitabgi and Vincent[31] with the permission of the publisher.)

It is of interest that both excitatory and inhibitory postjunctional effects of neurotensin on smooth muscle involve the activation of Ca channels, the opening of which is probably linked to the interaction of the peptide with its receptors. These initial steps of neurotensin action (binding to a receptor and activation of Ca channels) might well represent a general mechanism by which neurotensin affects the contractility of gastrointestinal and cardiovascular muscles *in vivo* and *in vitro*, and also other processes such as the secretion of hypothalamic, pituitary, and pancreatic hormones. Further studies are obviously needed to test this hypothesis.

STRUCTURE–ACTIVITY RELATIONSHIPS

The effects of neurotensin on gastrointestinal smooth muscle contractility may find application in the study of structure–activity relationships for the peptide.

TABLE 3

POTENCY OF NEUROTENSIN AND PARTIAL SEQUENCES IN GUINEA PIG ILEUM AND RAT FUNDUS STRIP*

	Guinea Pig Ileum				Rat Fundus	
	References				References	
Peptides	33	10	5	34,35	5	7
Neurotensin (NT)†	100	100	100	100	100	100
NT^{2-13}	50	73	25		95	90
NT^{4-13}	25		25	30	115	85
NT^{6-13}	20	21	15	29	115	78
NT^{7-13}				35		65
NT^{8-13}		21	15	20	85	69
NT^{9-13}				3		11
NT^{10-13}			<1	<0.1	<5	0
NT^{1-12}				<0.1		0.65
Acetyl NT^{8-13}				90		

*Potency values are given relative to neurotensin taken as 100.

† Amino acid sequence of neurotensin: *p*Glu-Leu-Tyr-Glu-Asn-Lys-Pro-Arg-Arg-Pro-Tyr-Ile-Leu-OH.

Indeed, gastrointestinal preparations *in vitro* have been used by several authors for measuring the activity of a number of neurotensin partial sequences and analogs. The preparations that have been mostly used are the rat fundus strip[5, 7, 8] and the segment of guinea pig ileum.[5, 10, 33] In our own studies, we have used the isolated longitudinal smooth muscle strip from guinea pig ileum, incubated in the presence of neostigmine.[34, 35] As shown in FIGURE 4, this preparation exhibits a high responsiveness to neurotensin.

TABLE 3 compares the activity of neurotensin partial sequences on gastrointestinal preparations as reported by several workers. The results confirm the initial findings by Carraway and Leeman[33] that almost all the structural requirements needed for neurotensin to express full activity are contained in the COOH-terminal hexapeptide sequence, that is, Arg-Arg-Pro-Tyr-Ile-Leu (NT^{8-13}, TABLE 3). We will now try to delineate the role played by each of the amino acids present in NT^{8-13}.

About 40 neurotensin analogs (not including partial sequences) have been synthesized so far, essentially by three groups of workers: C. Granier and J. van Rietschoten (Marseille, France),[35] J. Rivier (La Jolla, California),[36] and S. St. Pierre (Sherbrooke, Québec).[37] We will focus on those analogs modified or substituted in the COOH-terminal hexapeptide region of the neurotensin molecule. Their activities in gastrointestinal smooth muscle have been reported in two studies, our own using longitudinal muscle strips of guinea pig ileum,[34, 35] and that of Quirion et al. using rat fundus strips.[8] Data derived from both studies are very similar and may be summarized as follows: (1) There is an almost absolute requirement for L-amino acids in NT^{8-13} since replacement of Arg^9, Pro^{10}, Tyr^{11}, or Leu^{13} by the corresponding D-isomers leads to analogs of very low potency.[8, 34] Only Arg^8 can be replaced by D-Arg with almost no loss of activity;[8] (2) The positive charges on both Arg^8 and Arg^9 are important since substituting these residues with citrulline[34] or replacing Arg^9 by Ala (reference 8) brings about a 6- to 60-fold decrease in potency; (3) An aromatic residue is absolutely required in position 11 since substitution of Tyr^{11} with Phe or Trp yields analogs that display 14 to 100% of neurotensin activity, whereas substitution with nonaromatic D- or L-amino acids produces barely active peptides;[8,34,35,38] and (4) The side-chain methyl groups of Ile^{12} and Leu^{13} appear to be essential for activity since removal of these residues,[8, 34] or their substitution by Ala,[34] results in almost complete loss of activity. These data are compatible with the model originally proposed by Carraway and Leeman[33] for the neurotensin-receptor interactions. This model involves strong ionic interactions between the receptor and the positive charges carried by the side chains of Arg^8 and Arg^9, and hydrophobic interactions of the receptor with Tyr^{11}, Ile^{12}, and Leu^{13}. Additional features of the neurotensin-receptor interactions are stereospecificity with regard to the carboxy terminal pentapeptide and absolute requirement for an aromatic residue in position 11.

It is interesting that essentially similar information was obtained from structure–function relationship studies in which the activities of neurotensin partial sequences and analogs were evaluated in vitro using bioassays with cardiovascular preparations[39, 40] (contractility of isolated guinea pig atrium and rat portal vein), as well as radioreceptorassays with brain tissue[34, 41, 42] (binding to rat brain synaptic membranes), and with cancer cells from human colon[34] (binding to the HT 29 cell line). Therefore, neurotensin receptors display similar structural requirements in a wide variety of tissues from both neural and extraneural origin. An exception appears to be rat mast cells. In this system, neurotensin apparently interacts with a receptor whose structural requirements are quite different from those discussed above.[36] In addition, the affinity of mast cells for neurotensin is two orders of magnitude lower than that reported in other systems.[36] This raises the question as to whether mast cells represent a physiologic target for neurotensin. Alternatively, it may be suggested that rat mast cells have receptors that physiologically interact with a substance different from but structurally related to neurotensin.

It should be pointed out that when neurotensin analogs are tested in vivo, the results often markedly differ from those of in vitro studies. For example, the potency of several neurotensin analogs to induce hypothermia in vivo after intracerebral injection[36, 43, 44] markedly differs (by as much as 3–4 orders of magnitude) from their potency in vitro as measured in various assay systems.[8, 34, 39, 40] Also, neurotensin partial sequences that are active on gastrointestinal smooth muscle preparations in vitro (TABLE 3) were reported to have no effect, in contrast to neurotensin, on the motility of rat intestine in vivo.[45] Rather than differences in neurotensin receptors, we think that these discrepancies may reflect the fact that such factors as distribution in extracellular space, access to receptor sites, inactivation by peptidases and elimination from body fluids are likely to differ *in vivo* and *in vitro*.

We have indicated that most neurotensin analogs that have been tested so far behave similarly in a variety of *in vitro* assay systems. However, there are some noticeable exceptions to this rule. Two analogs, D-Trp[11]-neurotensin and Tyr(Me)[11]-neurotensin have been reported to behave as partial antagonists to neurotensin in rat portal vein and coronary vessels, and as weak agonists in rat fundus and guinea pig atria.[40] Recently we have reported that the analog Trp[11]-neurotensin is as potent as neurotensin in rat tissues, whereas, in contrast, the peptide is about ten times less active than neurotensin in guinea pig tissues.[38] The above observations may be taken as indications that there exist among neurotensin receptors, tissue- and species-related differences or variations that can be detected only with neurotensin analogs bearing appropriate modifications.

Clearly, gastrointestinal preparations represent useful tools to study structure–function relationships for biologically active peptides like neurotensin. It may be expected that detailed knowledge of the structural requirements for the neurotensin–receptor interaction will lead to the design and synthesis of neurotensin analogs with interesting properties such as specific antagonists and tissue- or species-specific analogs. Such analogs might prove extremely useful both as tools for investigating the physiologic role of neurotensin and as potential drugs of pharmacologic, clinical, and therapeutic interest.

ACKNOWLEDGMENTS

I wish to thank G. Visciano for excellent illustration work and M. Valetti for expert secretarial assistance.

REFERENCES

1. CARRAWAY, R. & S. E. LEEMAN. 1973. J. Biol. Chem. **248:** 6854–6861.
2. ROKAEUS, A., E. BURCHER, D. CHANG, K. FOLKERS & S. ROSELL. 1977. Acta Pharmacol. Toxicol. **41:** 141–147.
3. ARAKI, K., S. TACHIBANA, M. UCHIYAMA, T. NAKAJIMA & T. YASUHARA. 1973. Chem. Pharm. Bull. **21:** 2801–2804.
4. ARAKI, K., S. TACHIBANA, M. UCHIYAMA, T. NAKAJIMA & T. YASHUHARA. 1975. Chem. Pharm. Bull. **23:** 3132–3140.
5. KATAOKA, K., A. TANIGUCHI, H. SHIMIZU, K. SODA, S. OKUNO, H. YAJIMA & K. KITAGAWA. 1978. Brain Res. Bull. **3:** 555–557.
6. ARAKI, K., S. TACHIBANA, Y. KATO & T. TAJIMA. 1979. Yakugaku Zasshi **99:** 466–470.
7. QUIRION, R., D. REGOLI, F. RIOUX & S. ST-PIERRE. 1980. Br. J. Pharmacol. **68:** 83–91.
8. QUIRION, R., D. REGOLI, F. RIOUX & S. ST-PIERRE. 1980. Br. J. Pharmacol. **69:** 689–692.
9. KITABGI, P. & P. FREYCHET. 1978. Eur. J. Pharmacol. **50:** 349–357.
10. SEGAWA, T., M. HOSOKAWA, K. KITAGAWA & H. YAJIMA. 1977. J. Pharm. Pharmacol. **29:** 57–58.
11. PATON, W. H. M. & M. A. ZAR. 1968. J. Physiol. **194:** 13–33.
12. KITABGI, P. & P. FREYCHET. 1979. Eur. J. Pharmacol. **55:** 35–42.
13. KITABGI, P. & P. FREYCHET. 1979. Eur. J. Pharmacol. **56:** 403–406.
14. ROSELL, S. & A. ROKAEUS. 1981. Clinical Physiology **1:** 3–20.
15. FRANCO, R., M. COSTA & J. B. FURNESS. 1979. Naunyn-Schmiedeberg's Arch. Pharmacol. **306:** 195–201.

16. FRANCO, R., M. COSTA & J. B. FURNESS. 1979. Naunyn-Schmiedeberg's Arch. Pharmacol. **307:** 57–63.
17. MONIER, S. & P. KITABGI. 1980. Eur. J. Pharmacol. **65:** 461–462.
18. LEANDER, S., R. HAKANSON, S. ROSELL, K. FOLKERS, F. SUNDLER & K. TORNQVIST. 1981. Nature **294:** 467–469.
19. VIZI, E. S., G. BERTACCINI, M. IMPICCIATORE, P. MANTOVANI, J. ZSELI & J. KNOLL. 1974. Naunyn-Schmiedeberg's Arch. Pharmacol. **284:** 233–243.
20. FURUKAWA, K., T. NOMOTO & T. TONOUE. 1980. Eur. J. Pharmacol. **64:** 279–287.
21. WATERFIELD, A. A., R. W. J. SMOCKUM, J. HUGHES, H. W. KOSTERLITZ & G. HENDERSON. 1977. Eur. J. Pharmacol. **43:** 107–116.
22. GUILLEMIN, R. 1976. Endocrinology **99:** 1653–1654.
23. MONIER, S. & P. KITABGI. 1981. Regulatory Peptides **2:** 31–42.
24. JESSEL, T. M. & L. L. IVERSEN. 1977. Nature **268:** 549–551.
25. MUDGE, A. W., S. E. LEEMAN & G. D. FISCHBACH. 1979. Proc. Natl. Acad. Sci. USA **76:** 526–530.
26. WILLIAMS, J. T., Y. KATAYAMA & R. A. NORTH. 1979. Eur. J. Pharmacol. **59:** 181–186.
27. BULBRING, E. & T. B. BOLTON. 1979. Br. Med. Bull. **35:** 209–212.
28. KITABGI, P., G. HAMON & M. WORCEL. 1979. Eur. J. Pharmacol. **56:** 87–93.
29. BOLTON, T. B. 1979. Br. Med. Bull. **35:** 275–283.
30. OHASHI, H., Y. NONOMURA & A. OHGA. 1967. Jpn. J. Pharmacol. **17:** 247–257.
31. KITABGI, P. & J. P. VINCENT. 1981. Eur. J. Pharmacol. **74:** 311–318.
32. JENKINSON, D. H. 1981. Trends Pharmacol. Sci. **2:** 318–320.
33. CARRAWAY, R. & S. E. LEEMAN. 1975. *In* Peptides: Chemistry, Structure, and Biology. R. WALTER & J. MEIENHOFER, Eds. Ann Arbor Science Publishers. Ann Arbor, MI.
34. KITABGI, P., C. POUSTIS, S. GRANIER, J. VAN RIETSCHOTEN, J. RIVIER, J. L. MORGAT & P. FREYCHET. 1980. Mol. Pharmacol. **18:** 11–19.
35. GRANIER, C., J. VAN RIETSCHOTEN, P. KITABGI, C. POUSTIS & P. FREYCHET. 1982. Eur. J. Biochem. **124:** 117–124.
36. RIVIER, J., L. H. LAZARUS, M. H. PERRIN & M. R. BROWN. 1977. J. Med. Chem. **20:** 1409–1412.
37. ST-PIERRE, S., J. M. LALONDE, M. GENDREAU, R. QUIRION, D. REGOLI & F. RIOUX. 1981. J. Med. Chem. **24:** 370–376.
38. CHECLER, F., C. LABBE, C. GRANIER, J. VAN RIETSCHOTEN, P. KITABGI & J. P. VINCENT. 1982. Life Sciences **11:** 1145–1150.
39. QUIRION, R., F. RIOUX, D. REGOLI. & S. ST-PIERRE. 1980. Eur. J. Pharmacol. **66:** 257–266.
40. RIOUX, F., R. QUIRION, D. REGOLI, M. A. LEBLANC & S. ST-PIERRE. 1980. Eur. J. Pharmacol. **66:** 373–379.
41. UHL, G. R., J. P. BENNETT & S. H. SNYDER. 1977. Brain Res. **130:** 299–313.
42. KITABGI, P., R. CARRAWAY, J. VAN RIETSCHOTEN, C. GRANIER, J. L. MORGAT, A. MENEZ, S. E. LEEMAN & P. FREYCHET. 1977. Proc. Natl. Acad. Sci. USA **74:** 1846–1850.
43. LOOSEN, P. T., C. B. NEMEROFF, G. BISSETTE, G. B. BURNETT, A. S. PRANGE JR & M. A. LIPTON. 1978. Neuropharmacology **17:** 109–113.
44. JOLICOEUR, F. B., A. BARBEAU, F. RIOUX, R. QUIRION & S. ST-PIERRE. 1981. Peptides **2:** 171–175.
45. AL-SAFFAR, A. & S. ROSELL. 1981. Acta Physiol. Scand. **112:** 203–208.

DISCUSSION OF THE PAPER

R. QUIRION (*National Institute of Mental Health, Bethesda, MD*): What about histamine as a mediator for the atropine-resistant, neurotensin-induced contraction?

P. KITABGI (*INSERM, Nice, France*): We have tried to see if antihistaminic agents would block or reduce the effect of neurotensin in the guinea pig ileum, and we found no effect of chlorpheniramine, which was the antihistaminic agent that we used in our experiments. So it seems that histamine is not involved in the response to neurotensin in the guinea pig ileum.

J. E. T. Fox (*McMaster University, Hamilton, Ontario*): I realize that you did not show very much of the inhibitory response, but is that neural or is it smooth muscle?

KITABGI: Well, actually tetrodotoxin does not affect the relaxation phase of the biphasic response. It only blocks the contraction phase. So it seems that it is a direct interaction with the smooth muscle.

R. CHIPKIN (*Schering Corporation, Bloomfield, NJ*): If neurotensin acts by releasing acetylcholine and substance P to stimulate the gut, why isn't the maximal response that you get when you add neurotensin equivalent to the effect you see when you give substance P or acetylcholine alone? The maximal effect that you get seems to be much less.

KITABGI: If you remember, I showed a slide using isolated longitudinal strips of smooth muscle. When we incubate the strips with the cholinesterase inhibitor, neostigmine, we get the same maximal response with neurotensin as we obtained with acetylcholine. Probably when neurotensin releases acetylcholine from nerve terminals (this might be also true for substance P) there is degradation at the site of the release, and we do not know how much acetylcholine is released by neurotensin. This is different from the situation when you add exogenous acetylcholine. The preparations are incubated in a 10 ml volume and when we add a concentration of 10^{-8}M acetylcholine it is almost impossible for acetylcholinesterase to degrade all the acetylcholine present. However, when acetylcholine is released synaptically, acetylcholinesterase degrades most of it quickly. When we add the cholinesterase inhibitor, we get the same maximal response.

CHIPKIN: Morphine and opioids are known to inhibit the release of acetylcholine. What effect do opiates have on the effects of neurotensin?

KITABGI: Well, actually we have done a study that I have not shown here, studying the effect of opiate peptides on the contraction induced by neurotensin in the guinea pig ileum. We showed that both Met-enkephalin and β-endorphin could completely block the response to neurotensin. This can be explained because these peptides act on cholinergic nerves in the intestine to inhibit the release of acetylcholine. What is interesting is that we also observed this inhibitory effect on the noncholinergic component, which would indicate that opiate peptides inhibit the release of substance P. This would not be the first example of such an effect.

S. ROSELL (*Karolinska Institute, Stockholm, Sweden*): It is interesting to see that you find that neurotensin acts via nervous pathways in the intestine. We have ample evidence *in vivo* that neurotensin is working via the nervous apparatus in the gastrointestinal tract. Now, you have two effects here: one cholinergic and one substance P effect. The question is whether these are pharmacologic actions or have any physiologic meaning? *In vivo* we cannot block the action of neurotensin on motility which Substance P antagonists, if we give the antagonist intravenously, but we can block some of the effects by giving anticholinergic drugs. So this would perhaps indicate that in the rat at least, the motility effects *in vivo* are mediated by cholinergic motor nerves. Would you care to comment on that?

KITABGI: This is interesting because we have measured the effects of neurotensin on rat ileum *in vitro*, and in this preparation neurotensin does not have any contracting effect. The only response that I see is relaxation. This relaxation is not mediated by nerves; it is a direct effect of neurotensin on the smooth muscle. In the guinea pig neurotensin contracts the taenia coli, but relaxes the colon. Neither of these effects is neurally mediated. They are direct effects on smooth muscle.

Regarding the possibility that what we observed is relevant to physiology, in the guinea pig ileum neurotensin might act as a paracrinehormone. Therefore this might be quite different from the situation that exists when you inject neurotensin intravenously. In that case you are not looking at direct effects on smooth muscle, but instead are injecting neurotensin in the blood. This might give neurotensin access to different smooth muscle layers and nerves that are present in the intestine.

S. St. Pierre (*University of Sherbrooke, Sherbrooke, Quebec*): In your model of the interaction of neurotensin with its receptor you seem to postulate that there exists a hydrogen bond between the carbonyl in the isoleucine moiety. How do you show that this hydrogen bond exists?

Kitabgi: Some evidence exists from a group in Paris who made studies on the conformation of neurotensin in solution by NMR techniques that in the *pentapeptide* there is a hydrogen bond. I am not sure that in the entire neurotensin molecule this hydrogen bond exists.

MECHANISMS OF THE CARDIOVASCULAR EFFECTS
OF NEUROTENSIN*

Francis Rioux, René Kérouac,
Rémi Quirion,† and Serge St-Pierre

Department of Physiology and Pharmacology
School of Medicine
University of Sherbrooke
Sherbrooke, Québec, Canada J1H 5N4

INTRODUCTION

Over the last nine years, several studies have shown that neurotensin (NT), a natural tridecapeptide of nervous and intestinal origin,[1-3] exerts numerous actions on the cardiovascular system of mammals. For instance, in rats,[1,4] rabbits,[5] pigs,[6] goats,[6] and dogs,[7,8] the peptide was found to produce a decrease in blood pressure while in guinea pigs,[9] woodchucks,[10] sheep[6,11] and cats,[12] pressor and biphasic pressor-depressor responses were observed. Changes of heart rate were also described in some species.[6,9,11,13] A few hemodynamic studies with NT have been performed in dogs,[7,8,14] and to a lesser extent, in sheep.[11] Other actions of NT related to the cardiovascular system include the following: increase of capillary permeability and cutaneous vasodilation in rats;[1,15] positive inotropic and chronotropic effects in isolated atria of rats and guinea pigs;[16,17] vasoconstriction in adipose tissue of dogs,[7] in perfused hearts[18] and hindquarters[19] of rats, and in rat isolated portal veins;[20,21] and vasodilatation in dog intestine.[7,14] So far, in humans, NT has been shown to be inactive on the cardiovascular system, although only relatively low, presumably physiologic doses of the peptide have been tested.[22,23]

During the last three years, we have tried to elucidate the mechanism of the cardiovascular effects of NT, particularly in rats and guinea pigs. In this chapter, we will present an update of our major findings in this area. Related data published by other workers will be introduced and discussed.

THE CARDIOVASCULAR EFFECTS OF NEUROTENSIN IN RATS: POSSIBLE ROLE OF HISTAMINE AND SEROTONIN

In pentobarbital-anesthetized rats, intravenous injections of small doses (0.1–1 nmol kg^{-1}) of NT evoked potent hypotensive effects, marked increases of vascular permeability, vasodilation of the small cutaneous vessels, and cyanosis.[1] The hypotensive effect of NT gave rise to acute tachyphylaxis, and its amplitude was apparently dependent upon the initial blood pressure level, being reduced in rats with lower basal levels. The vasodepressor effect of NT was not significantly altered by adrenalectomy, hypophysectomy, or pretreatment of the animals with appropriate doses of atropine sulfate, phenoxybenzamine, or propranolol.

*Research was supported by the Medical Research Council of Canada and by the Canadian Heart Foundation.

† Present address: National Institute of Mental Health, Bethesda, MD.

56

The results obtained with atropine suggest that the hypotensive effect of NT in anesthetized rats was neither mediated by the release of endogenous acetylcholine nor by a direct action of NT on muscarinic receptors. Since phenoxybenzamine and propranolol, α- and β-adrenergic receptor-blocking drugs respectively, as well as acute adrenalectomy did not alter NT-induced hypotension, this peptide is unlikely to depend on endogenous catecholamines or on their receptors to produce this effect in rats.

Based on a report by Chahl[15] showing that changes in vascular permeability induced by NT in rats were blocked by mepyramine, methysergide, and compound

TABLE 1

INFLUENCE OF VARIOUS DRUG TREATMENTS ON THE DECREASE OF MEAN ARTERIAL BLOOD PRESSURE (ΔMABP) EVOKED BY INTRAVENOUS INJECTIONS OF NEUROTENSIN AND OTHER DRUGS IN PENTOBARBITAL-ANESTHETIZED RATS*

Drugs	Doses (μg kg^{-1})	Experimental Conditions	ΔMABP (mm Hg)	
Neurotensin	0.7	Control	-47 ± 4	(11)
	2.0		-59 ± 5	(11)
Neurotensin	0.7	After azatadine + methysergide	-16 ± 2†	(6)
	2.0		-31 ± 4†	(6)
Neurotensin	0.7	After disodium cromoglycate	-23 ± 3†	(8)
	2.0		-28 ± 2†	(8)
Neurotensin	0.7	After compound 48/80 (acute treatment)	0†	(8)
	2.0		$+ 6 \pm 1$†	(8)
	7.2		$+42 \pm 4$†	(8)
Neurotensin	0.7	After compound 48/80 (chronic treatment)	0†	(8)
	7.2		$+40 \pm 4$†	(8)
Bradykinin	3.7	Control	-36 ± 4	(6)
	3.7	After compound 48/80 (acute treatment)	-37 ± 5	(6)
Compound 48/80	42.0	Control	-51 ± 4	(8)
		After neurotensin (acute treatment)	0†	(6)
		After azatadine + methysergide	-25 ± 1†	(6)
		After disodium cromoglycate	-28 ± 2†	(6)

*The results are expressed as means ± SEM. The number of experiments is given in parentheses. Azatadine (4 mg kg^{-1}) methysergide (0.2 mg kg^{-1}) and disodium cromoglycate (62 mg kg^{-1}) were given intravenously (I.V.). The signs + and − indicate an increase or decrease, respectively, of mean arterial blood pressure.

†p < 0.001 versus control values (Taken with permission from Quirion *et al.* 1980. Life Sci., **27:** 1891. Pergamon Press, Ltd.)

48/80, we decided to investigate whether endogenous histamine and serotonin (or 5-hydroxytryptamine) could be responsible for the fall of blood pressure evoked by NT in anesthetized rats. The results of these studies are summarized in TABLE 1. The acute or chronic administration of compound 48/80, a drug known to deplete mast cells of their histamine and serotonin content,[24, 25] completely abolished the hypotensive effect of low to medium doses of NT and revealed the previously unknown hypertensive effects of relatively high doses of this peptide. This observation is illustrated in FIGURE 1. The blockade of NT-induced hypotension in compound 48/80-pretreated animals was specific since the hypotensive effects of bradykinin and of isoproterenol were not affected by this treatment. Prior administration of disodium cromoglycate, a commonly used antiallergic drug that is believed to cause its therapeutic action by stabilizing mast cell membranes,[24, 26] or of a mixture of azatadine, a potent antihistamine drug,[27] and methysergide, a serotonin-receptor-blocking drug, reduced markedly, but did not abolish completely, the hypotensive effects of NT and of compound 48/80 (TABLE 1). Interestingly, rats made unresponsive to NT by repeated intravenous doses of NT

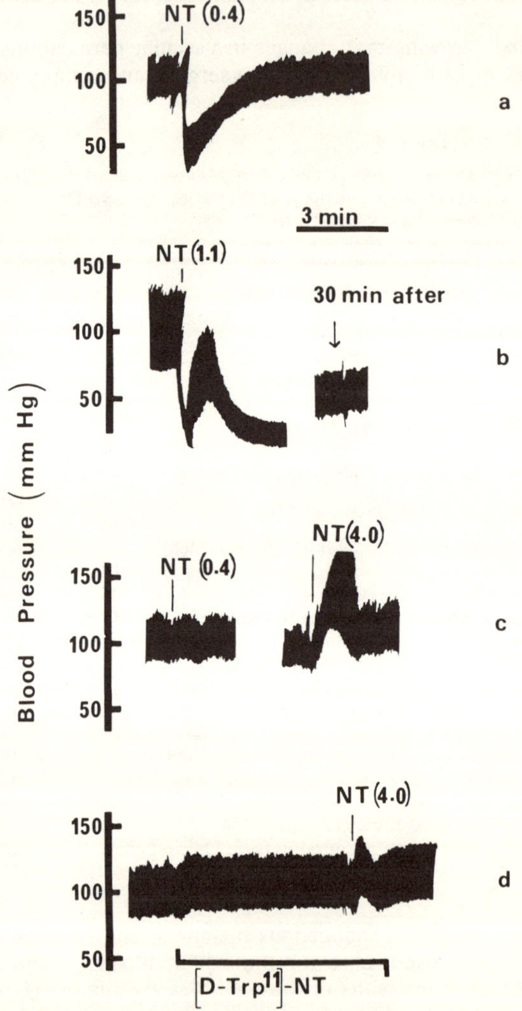

FIGURE 1. Effect of various doses of neurotensin (NT) on the blood pressure of anesthetized rats. (a, b) control rats; (c) rats pretreated with compound 48/80 (chronic treatment); (d) rats pretreated with compound 48/80 (as in c) and infused with [D-Trp11]-NT for 5 min before injecting NT. The doses of NT are in nmol kg⁻¹. [D-Trp11]-NT was infused at a rate of 10 nmol kg⁻¹ min⁻¹. Other details are given in reference 31. (Reproduced with permission from Quirion et al. 1980. Life Sci. 27: 1889. Pergamon Press, Ltd.)

became entirely unresponsive to the hypotensive effect of compound 48/80, thus suggesting a common site of action for NT and compound 48/80. The pressor effect of NT in compound 48/80-pretreated animals was markedly reduced by prior administration of [D-Trp11]-NT, a newly developed NT antagonist[28–30] (FIG. 1). These results provided strong but indirect evidence for the participation of histamine and/or serotonin and possibly of other endogenous vasoactive substances to the hypotensive effect of NT in anesthetized rats.[31]

Recent studies by Chahl and Walker[13] and Oishi and coworkers[32] also supported this interpretation. These researchers also provided evidence for the participation of endogenous catecholamines to the transient pressor effect that sometimes accompanied the hypotensive effect of NT in rats.[13, 31, 32]

The possibility that NT produces hemodynamic as well as other biologic actions by mobilizing endogenous histamine and/or serotonin from mast cells is supported by the following facts: (1) specific binding sites for NT are found in rat mast cells;[33] (2) NT is known to cause mast cell degranulation in human jejunal mucosal biopsy *in vitro*[34] and histamine release from purified rat peritoneal mast cells;[35] (3) rat mast cells are known to contain and to release—upon stimulation—both histamine and serotonin;[25, 36] (4) endogenous histamine appears to be involved in NT-induced hyperglucagonemia and hyperglycemia[37] in rats; and (5) preliminary data from our laboratory showed an increased level of plasma histamine after intravenous injections of NT.[12] Taken together, these results clearly suggest the participation of endogenous histamine and/or serotonin to the hypotensive effect of NT in anesthetized rats. The precise origin of the two mediators is still unknown, but mast cells are likely to be involved.[31, 32, 35] The mechanism subserving the pressor effect of high doses of NT in compound 48/80-pretreated rats remains to be determined.

The next step in our study was to determine whether the hypotensive effect of NT in rats was due to direct vascular effects of histamine and/or of serotonin, or to an action of the autocoids on the central or autonomic nervous system, or to both mechanisms. These hypotheses were tested recently in our laboratory using the pithed rat preparation. In these animals, the spinal cord was completely destroyed thus preventing nervous influences to mask direct vascular effects of drugs.[38] The results of these studies are shown in TABLE 2. In pithed rats, intravenous injections of NT, of serotonin, or of histamine produced an increase of systolic and diastolic arterial blood pressure accompanied by small, variable changes of heart rate while, in anesthetized animals, the three compounds induced marked hypotensive effects. These results suggest that the hypotensive effect of NT in pentobarbital-anesthetized rats is nervously mediated. A likely mechanism would be that histamine, serotonin, and possibly NT itself reduce the activity of the sympathetic nervous system in the anesthetized rats, an effect that could not take place in pithed rats because of the suppression of the sympathetic tone resulting from the destruction of the spinal cord. The participation of cholinergic neurons to the hypotensive effect of NT in anesthetized rats is rather unlikely because neither atropine[1] nor bilateral vagotomy[12] inhibited NT-induced hypotension. The possibility that NT and/or its mediators cross the blood/brain barrier and then cause a decrease of blood pressure by a direct effect on the central nervous system was also considered. However, this hypothesis seems unlikely because we have shown previously that rats made unresponsive to the hypotensive effects of centrally injected NT by repeated intracerebroventricular (lateral brain ventricle) injections of NT still respond normally to intravenous doses of NT.[40] Similarly, rats made unresponsive to the hypotensive effects of intravenous injections of NT by prior chronic treatment with compound 48/80 still exhibited a clear-cut fall of blood pressure when NT was injected into the brain ventricle.

The mechanism by which NT produces its vasopressor effects in pithed rats also attracted our attention because we felt that, in rats devoid of autonomic reflexes, it would be easier to assess the relative contribution of histamine and serotonin to the cardiovascular effects of NT. We also felt that such studies could help us to explain the hypertensive effects previously described with NT in compound 48/80–pretreated animals (TABLE 1) and in conscious rats[10] (TABLE 2).

TABLE 2

CHANGES OF SYSTOLIC BLOOD PRESSURE (ΔSBP), OF DIASTOLIC BLOOD PRESSURE (ΔDBP) AND OF HEART RATE (ΔHR) EVOKED BY INTRAVENOUS INJECTIONS OF NEUROTENSIN (NT), OF HISTAMINE (H), AND OF 5-HYDROXYTRYPTAMINE (5-HT) IN CONSCIOUS, PENTOBARBITAL-ANESTHETIZED, OR PITHED RATS*

Drug	Doses (μg kg^{-1})	Parameters	Anesthetized Rats	Pithed Rats	Conscious Rats
NT	0.5	ΔSBP	-28 ± 7 (8)	$+ 16 \pm 5$ (9)	$+18 \pm 3$
		ΔDBP	-29 ± 8 (8)	$+11 \pm 4$ (9)	$+13 \pm 2$
		ΔHR	$+ 2 \pm 1$ (8)	$+12 \pm 5$ (9)	$+ 3 \pm 1$
NT	1.8	ΔSBP	-44 ± 5 (8)	$+26 \pm 4$ (15)	$+23 \pm 4 - 14 \pm 6$ (8)‡
		ΔDBP	-53 ± 8 (8)	$+23 \pm 5$ (15)	$+19 \pm 3 - 18 \pm 7$ (8)
		ΔHR	$+ 6 \pm 3$ (8)	$+ 7 \pm 2$ (15)	$+ 5 \pm 2 + 8 \pm 2$ (8)
NT	7.2	ΔSBP	ND	$+31 \pm 5$ (6)	$+40 \pm 5 - 18 \pm 5$ (6)
		ΔDBP	ND	$+28 \pm 5$ (6)	$+33 \pm 5 - 23 \pm 6$ (6)
		ΔHR	ND	$+14 \pm 3$ (6)	$-41 \pm 4 + 52 \pm 9$ (6)
H	100	ΔSBP	-37 ± 6 (8)	$+19 \pm 4$ (8)	ND
		ΔDBP	-42 ± 9 (8)	$+11 \pm 2$ (8)	ND
		ΔHR	$+ 1 \pm 1$ (8)	$+ 6 \pm 1$ (8)	ND
5-HT	50	ΔSBP	-51 ± 8 (7)†	$+59 \pm 6$ (9)	ND
		ΔDBP	-62 ± 9 (7)	$+54 \pm 7$ (9)	ND
		ΔHR	$+ 6 \pm 3$ (7)	$- 4 \pm 1$ (9)	ND

*The results are expressed as means $\pm$ the standard error of the mean (SEM). The number of experiments is given in parentheses. The units for ΔSBP, ΔDBP and ΔHR are in mm Hg for the first two and in percent for the latter. ND means "not determined." Each animal received only one dose of agonist.

†This hypotensive effect was generally preceded by transient, small biphasic variations of blood pressure. Conscious animals were prepared as described previously.[39]

‡Biphasic pressor-depressor effects.

Since both histamine and serotonin elicited pressor effects and most likely were involved in the cardiovascular effects of NT in anesthetized rats, we reasoned that these two compounds could also contribute to the hypertensive effect of NT in pithed rats. To test this hypothesis, we looked first at the effect of mepyramine—a compound well known for its ability to block histamine receptors of the H$_1$ type[24]—on the pressor effects of NT, and of other control drugs including histamine, in pithed rats. The results are shown in TABLE 3. Mepyramine (1 mg kg^{-1}) reduced significantly the pressor effect evoked by a medium dose (1.8 μg kg^{-1}) of NT and slightly potentiated that produced by a high dose (7.2 μg kg^{-1}) of the peptide. The tachycardia induced by NT was not altered by the antagonist. The pressor effect of histamine was completely abolished and converted into an hypotensive effect by mepyramine. This antagonist also blocked the slight tachycardia evoked by histamine. Surprisingly, mepyramine was not specific even when used at such a moderate dose because it also reduced significantly the pressor effects evoked by 5-hydroxytryptamine (TABLE 3). The pressor effect of compound 48/80 was slightly but not significantly diminished by mepyramine. Taken together, these results support our previous conclusion concerning the participation of histamine and/or serotonin to the cardiovascular effects of NT in rats. However, due to the lack of specificity of mepyramine, it was not possible to assess the relative contribution of histamine versus serotonin to the pressor effect of NT in pithed rats. Higher doses of mepyramine (5 mg kg^{-1}) significantly inhibited the pressor effects of both NT (7.2 μg kg^{-1}) and compound 48/80 (250 μg kg^{-1}) (two experiments, data not shown). The lack of specificity of several antihistamine drugs is well documented.[24]

Methysergide, a selective antagonist of vascular serotonergic receptors,[24] was tried next. The results are shown in TABLE 4. In rats pretreated with methysergide (2.5 mg kg^{-1}), the pressor effects of NT (intermediate and high doses), of 5-hydroxytryptamine or of compound 48/80 were completely blocked (TABLE 4). Relatively small, residual hypotensive effects were observed with the three drugs after the administration of methysergide. The tachycardia evoked by NT, in contrast to that induced by 5-hydroxytryptamine, was reduced by methysergide. However, this drug did not alter the pressor effect and tachycardia caused by noradrenaline. These results strongly suggest that the pressor effects of NT and of compound 48/80 in pithed rats are largely dependent upon the release of endogenous 5-hydroxytryptamine into the blood and/or in the vicinity of blood vessels, and to the subsequent interaction of this autocoid with its vascular receptors. The possible mast cell origin of the mediator was evaluated by studying the effect of disodium cromoglycate on the pressor effects of NT and of other drugs in pithed rats. As shown in TABLE 5, disodium cromoglycate (DSCG) (10 mg kg^{-1}) blocked almost completely the pressor effects of medium doses of NT and slightly reduced those induced by high doses (7.2 μg kg^{-1}) of this peptide. The pressor effect of 5-hydroxytryptamine was slightly attenuated while that produced by compound 48/80, similarly to that of NT (1.8 μg kg^{-1}), was markedly inhibited by DSCG. The pressor effect of noradrenaline was not altered by the latter drug. The changes of heart rate caused by NT, 5-hydroxytryptamine and noradrenaline were inhibited significantly by DSCG. These results strongly support the idea that mast cells may be an important source for the 5-hydroxytryptamine that contributes to the pressor effects of NT and of compound 48/80 in pithed rats. The results also indicate that

TABLE 3

INFLUENCE OF MEPYRAMINE ON THE CHANGES OF SYSTOLIC BLOOD PRESSURE (ΔSBP), OF DIASTOLIC BLOOD PRESSURE (ΔDBP), AND OF HEART RATE (ΔHR) EVOKED BY INTRAVENOUS INJECTIONS OF NEUROTENSIN (NT), OF HISTAMINE (H), OF 5-HYDROXYTRYPTAMINE (5 HT), AND OF COMPOUND 48/80 (C48/80) IN PITHED RATS*

Dose	Dose (μg kg^{-1})	Experimental Conditions	ΔSBP (mm Hg)	ΔDBP (mm Hg)	ΔHR (%)
NT	1.8	Control	$+26 \pm 4$	$+23 \pm 5$	$+ 7 \pm 2(15)$
NT	1.8	After mepyramine	$+14 \pm 2$†	$+13 \pm 3$	$+13 \pm 5(13)$
NT	7.2	Control	$+30 \pm 4$	$+31 \pm 5$	$+15 \pm 2(15)$
NT	7.2	After mepyramine	$+44 \pm 6$	$+48 \pm 6$†	$+10 \pm 4(6)$
H	100.0	Control	$+19 \pm 4$	$+11 \pm 2$	$+ 6 \pm 1(8)$
H	100.0	After mepyramine	-13 ± 3‡	-14 ± 3‡	$- 9 \pm 3$‡(8)
5-HT	50.0	Control	$+59 \pm 6$	$+54 \pm 7$	$- 4 \pm 1(9)$
5-HT	50.0	After mepyramine	$+18 \pm 3$‡	$+16 \pm 4$‡	$- 4 \pm 1(8)$
C48/80	250.0	Control	$+38 \pm 6$	$+27 \pm 4$	ND (8)
C48/80	250.0	After mepyramine	$+26 \pm 6$	$+23 \pm 6$	ND (8)

*The results are expressed as means $\pm$ SEM. The number of experiments is given in parentheses. Mepyramine (1 mg kg^{-1}) was given intravenously 15 min before the agonist. Only one dose of agonist was given in each animal. The signs + or − indicate, respectively, an increase and a decrease of the parameters under study. ND means "not determined."

†p < 0.05 when compared to the appropriate control.

‡p < 0.05 when compared to the appropriate control.

TABLE 4

INFLUENCE OF METHYSERGIDE ON THE CHANGES OF SYSTOLIC BLOOD PRESSURE (ΔSBP) OF DIASTOLIC BLOOD PRESSURE (ΔDBP) AND OF HEART RATE (ΔHR) PRODUCED BY INTRAVENOUS INJECTIONS OF NEUROTENSIN (NT), OF 5-HYDROXYTRYPTAMINE (5-HT), OF COMPOUND 48/80 (C48/80), AND OF NORADRENALINE (NA) IN PITHED RATS*

Drug	Dose (μg kg^{-1})	Experimental Conditions	ΔSBP† (mm Hg)	ΔDBP (mm Hg)	ΔHR (%)
NT	1.8	Control	$+26 \pm 4$	$+23 \pm 5$	$+ 7 \pm 2(15)$
NT	1.8	After methysergide	-11 ± 2^c	-10 ± 1^c	$- 3 \pm 1^a(11)$
NT	7.2	Control	$+30 \pm 4$	$+31 \pm 5$	$+15 \pm 2(15)$
NT	7.2	After methysergide	$- 18 \pm 3^c$	-21 ± 4^c	$+ 6 \pm 2^b(8)$
5-HT	50.0	Control	$+59 \pm 6$	$+54 \pm 7$	$- 4 \pm 1(9)$
5-HT	50.0	After methysergide	-10 ± 4^c	$- 8 \pm 3^c$	$- 8 \pm 4(9)$
C48/80	250.0	Control	$+38 \pm 6$	$+27 \pm 4$	ND
C48/80	250.0	After methysergide	$- 8 \pm 2^c$	$- 8 \pm 2^c$	ND
NA	0.5	Control	$+31 \pm 3$	$+20 \pm 2$	$+ 5 \pm 2(9)$
NA	0.5	After methysergide	$+30 \pm 4$	$+20 \pm 3$	$+ 5 \pm 2(7)$

*The results are expressed as means $\pm$ SEM. The number of experiments is given in parentheses. Methysergide (2.5 mg kg^{-1}) was injected intravenously 15 min before injecting the agonists. Only one agonist was tested in each animal. The signs + or − mean, respectively, an increase and a decrease of ΔSBP, ΔDBP or ΔHR. ND means "not determined."

†a = p < 0.01; b = p < 0.005; c = p < 0.001 when compared to the appropriate control.

DSCG is a more potent inhibitor of pressor effects evoked by compound 48/80 than of those induced by high doses of NT. The slight blockade of 5-hydroxytryptamine-induced pressor effects by DSCG raises the possibility that intravenous injections of 5-hydroxytryptamine also stimulate the secretion of vasoactive substances from mast cells. This interpretation is consistent with previous reports on the release of histamine by 5-hydroxytryptamine in rat tissues[41] and mast cells.[42]

In conclusion, the pressor effect of NT in rats devoid of autonomic reflexes (pithed rats), appears to involve the participation of endogenous serotonin presumably of mast cell origin. It is interesting to recall that serotonin is particularly abundant in rat mast cells in comparison with those of other species.[43] Since the population of mast cells in some tissues (e.g., skin) vary greatly from one species to another,[43] it is possible that the contribution of mast cell mediators to the cardiovascular effects of NT differ accordingly. For this reason, we have investigated the cardiovascular effects of NT in guinea pigs, a species in which mast cells are less numerous (in some tissues) and more resistant to disruptive compounds than in rats.[43]

THE CARDIOVASCULAR EFFECTS OF NEUROTENSIN IN GUINEA PIGS: POSSIBLE ROLE OF CATECHOLAMINES

In pentobarbital-anesthetized guinea pigs, intravenous injections of NT (1.8 and 5.4 μg kg^{-1}) evoked dose-dependent pressor effects and tachycardia. In most animals, the pressor effect was biphasic, being characterized by an acute, transient (10–15 sec) pressor effect followed by a secondary, more long-lasting (3–6 min), hypertensive effect. This observation is illustrated in FIGURE 2 (top tracing). In some animals, the acute pressor effect of NT was followed by a short-lived decrease and then by a long-lasting (3–6 min.) increase of blood pressure (FIG. 2, middle and bottom tracing). Intravenous infusions of NT also elicited dose-dependent pressor effects and tachycardia in guinea pigs. These results are shown in FIGURES 3 and 4.

The mechanism subserving the cardiovascular effects of NT in guinea pigs was assessed using various drug antagonists and control agonists. The results of studies in which bolus injections of NT were used are summarized in TABLE 6. In animals pretreated with atropine at a dose that abolished by more than 90% the hypotensive effect of carbachol, the pressor effects and tachycardia caused by NT were not affected. On the other hand, pretreatment of the animals with effective doses of pentolinium or of a mixture of phentolamine and propranolol markedly abolished the pressor effects of NT without altering its chronotropic action. A small hypotensive effect was observed with NT in the latter group of animals. Effective doses of indomethacin, of $[\text{Leu}]^8\text{-AT}_{11}$, a well-known angiotensin antagonist, or of a mixture of mepyramine and cimetidine did not modify significantly the cardiovascular effects of bolus injections of NT in guinea pigs. The results obtained with atropine suggest that neither acetylcholine nor its muscarinic receptors are likely to contribute to the pressor and chronotropic effect of NT in guinea pigs, at least in those animals in which purely pressor effects were recorded. However, this conclusion does not apply to those animals in which

TABLE 5

INFLUENCE OF DISODIUM CROMOGLYCATE ON THE CHANGES OF SYSTOLIC BLOOD PRESSURE (ΔSBP), OF DIASTOLIC BLOOD PRESSURE (ΔDBP), AND OF HEART RATE (ΔHR) PRODUCED BY INTRAVENOUS INJECTIONS OF NEUROTENSIN (NT), OF 5-HYDROXYTRYPTAMINE (5-HT), OF COMPOUND 48/80 (C48/80), AND OF NORADRENALINE (NA) IN PITHED RATS*

Drug	Dose (μg kg^{-1})	Experimental Conditions	ΔSBP† (mm Hg)	ΔDBP (mm Hg)	ΔHR (%)
NT	1.8	Control	+26 ± 4	+23 ± 5	+ 7 ± 2(15)
NT	1.8	After disodium cromoglycate	+ 3 ± 2^e	+ 4 ± 2^d	+ 2 ± 1^a(9)
NT	7.2	Control	+30 ± 4	+31 ± 5	+15 ± 2(15)
NT	7.2	After disodium cromoglycate	+22 ± 5	+22 ± 5	+ 6 ± 3^b(9)
5-HT	50.0	Control	+59 ± 6	+54 ± 7	− 4 ± 1(9)
5-HT	50.0	After disodium cromoglycate	+37 ± 4^c	+39 ± 4	+ 6 ± 2^e(9)
C48/80	250.0	Control	+38 ± 6	+27 ± 4	ND
C48/80	250.0	After disodium cromoglycate	+10 ± 4^d	+ 8 ± 3^d	ND
NA	0.5	Control	+31 ± 3	+20 ± 2	+ 5 ± 2(9)
NA	0.5	After disodium cromoglycate	+29 ± 5	+27 ± 5	− 4 ± 1^d(9)

*The results are expressed as means ± SEM. The number of experiments is given in parentheses. Disodium cromoglycate (10 mg kg^{-1}) was given intravenously 15 min before injecting the agonists. Only one dose of agonist was given to each animal. The signs + or − mean respectively, an increase and a decrease of ΔSBP, ΔDBP and ΔHR. ND means "not determined."

†a = p < 0.05; b = p < 0.002; c = p < 0.01; d = < 0.005; e = p < 0.001 when compared to the appropriate control.

biphasic (pressor-depressor) or triphasic responses were recorded because a decreased incidence of these biphasic or triphasic responses was observed after the administration of atropine,[9] thus suggesting that in some animals, NT may activate cholinergic nerves and produce hypotensive effects. Other evidence suggesting the activation of cholinergic nerves by NT are given below. The absence of effect of indomethacin, of [Leu^8]-AT$_{11}$, and of mepyramine and cimetidine on the pressor and chronotropic effects of NT suggest that endogenous prostaglandins or thromboxanes, histamine, and angiotensin are unlikely to be responsible for the cardiovascular effects of NT in guinea pigs. The complete blockade and reversal of the pressor effect of NT observed in animals pretreated with pentolinium or a

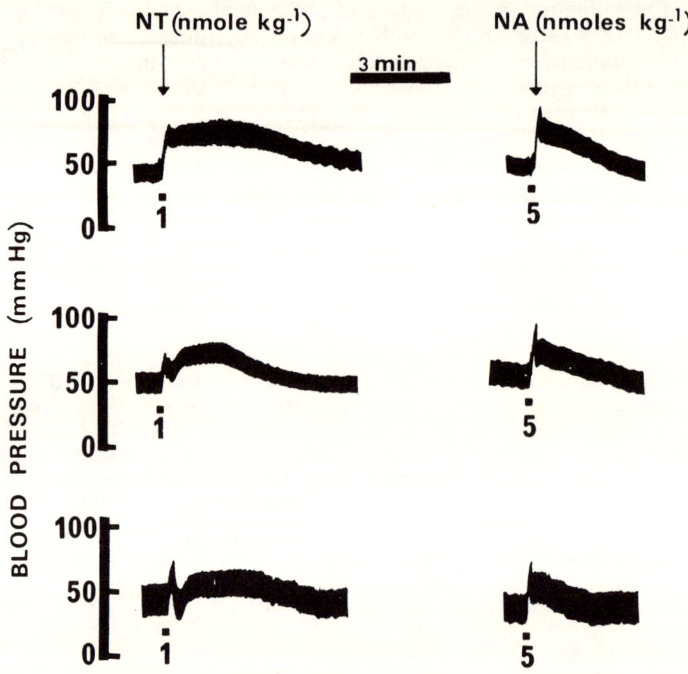

FIGURE 2. Effect of neurotensin (NT) and of noradrenaline (NA) on the blood pressure of pentobartital-anesthetized guinea pigs. NT and NA were injected intravenously.

mixture of phentolamine and propranolol strongly suggests that the pressor effect of NT in guinea pigs results from an activation of the sympathetic nervous system and most likely involves the participation of endogenous catecholamines. On the other hand, the chronotropic effect of NT appears to be due to a direct effect of NT on the heart, most likely at the level of the sinus node. This conclusion is supported by other studies in which we assessed the influence of reserpine, of guanethidine, and of a mixture of phentolamine and propranolol on the pressor effect evoked by intravenous infusions of NT. As shown in FIGURES 5 and 6, the three separate treatments significantly reduced the hypertensive effects of NT without reducing—with the exception of guanethidine—the positive chronotropic effect of NT (FIGS. 7 and 8). The slowly developing inhibitory effect of

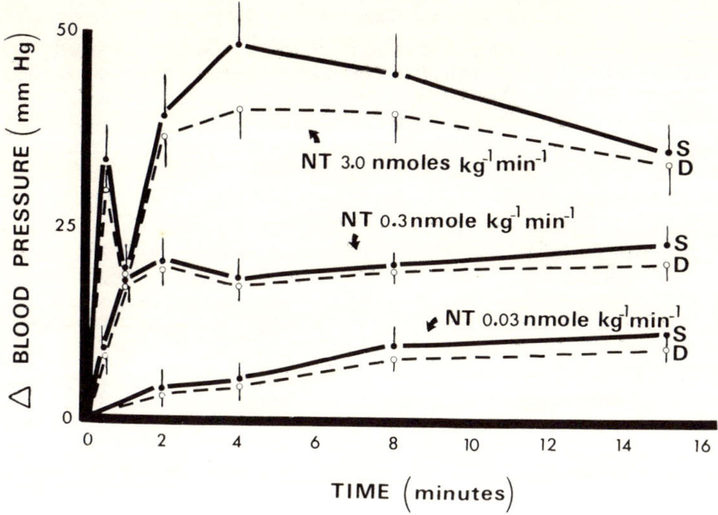

FIGURE 3. Changes of systolic (S) and of diastolic (D) arterial blood pressure produced by intravenous infusions of neurotensin (NT) in pentobarbital-anesthetized guinea pigs. Each point is the mean ± the standard error of the mean (SEM) of 7–12 experiments. (Reproduced with permission from Kérouac *et al.* 1981. Life Sci. **28:** 2477. Pergamon Press, Ltd.)

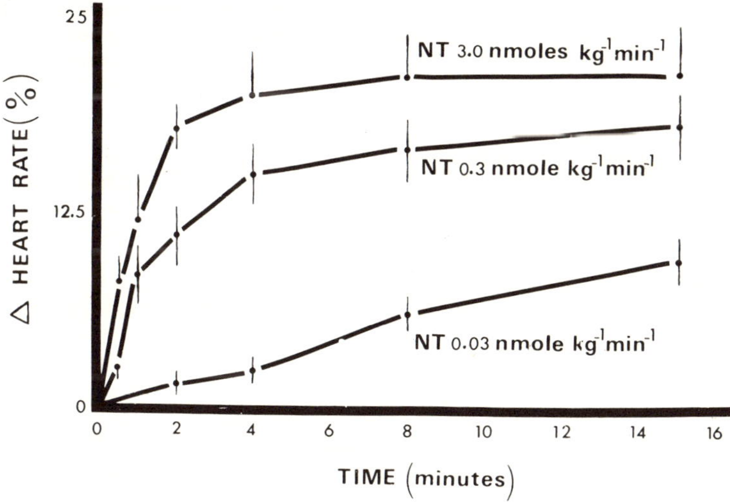

FIGURE 4. Changes of heart rate produced by intravenous infusions of ncurotensin (NT) in pentobarbital-anesthetized guinea pigs. Each point is the mean ± SEM of 7–12 experiments. The changes of heart rate were recorded simultaneously with the changes of blood pressure shown in FIGURE 3. (Reproduced with permission from Kérouac *et al.* 1981. Life Sci. **28:** 2477. Pergamon Press, Ltd.)

Table 6

INFLUENCE OF VARIOUS DRUG ANTAGONISTS ON THE CHANGES OF SYSTOLIC BLOOD
PRESSURE (ΔSBP), OF DIASTOLIC BLOOD PRESSURE (ΔDBP), AND OF HEART RATE
(ΔHR) EVOKED BY INTRAVENOUS INJECTIONS OF NEUROTENSIN (NT) IN
PENTOBARBITAL-ANESTHETIZED GUINEA PIGS*

Dose of NT (μg kg^{-1})	Experimental Conditions	ΔSBP (mm Hg)	ΔDBP (mm Hg)	ΔHR (%)
1.8	Control	+23 ± 2	+23 ± 2	+20 ± 3(12)
1.8	After atropine	+23 ± 4	+24 ± 3	+27 ± 5(12)
5.4	Control	+40 ± 2	+36 ± 2	+30 ± 3(8)
5.4	After atropine	+42 ± 6	+38 ± 4	+36 ± 6(8)
1.8	Control	+27 ± 5	+26 ± 5	+24 ± 4(9)
1.8	After pentolinium	−10 ± 1†	− 8 ± 1†	+20 ± 1(9)
5.4	Control	+42 ± 6	+38 ± 4	+36 ± 6(8)
5.4	After pentolinium	−11 ± 3†	− 7 ± 2†	+30 ± 3(8)
1.8	Control	+23 ± 3	+19 ± 2	+20 ± 6(12)
1.8	After phentolamine + propranolol	− 8 ± 2†	− 9 ± 1†	+23 ± 2(12)
1.8	Control	+30 ± 3	+30 ± 3	+19 ± 2(6)
1.8	After mepyramine + cimetidine	+28 ± 3	+27 ± 3	+26 ± 6(6)
1.8	Control	+25 ± 4	+26 ± 4	+15 ± 4(6)
1.8	After indomethacin	+25 ± 4	+27 ± 4	+16 ± 8(6)
1.8	Control	+24 ± 1	+27 ± 1	+28 ± 5(6)
1.8	After [Leu^8]-AT$_{11}$	+29 ± 3	+32 ± 3	+24 ± 4(6)

*The results are expressed as means ± SEM. The number of experiments is given in parenthesis. Atropine (1 mg kg^{-1}), pentolinium (1 mg kg^{-1}), phentolamine (1 mg kg^{-1}), propranolol (0.5 mg kg^{-1}), mepyramine (5 mg kg^{-1}), cimetidine (5 mg kg^{-1}) and indomethacin (5 mg kg^{-1}) were given intravenously 10 min before repeating the injections of NT. [Leu^8]-AT$_{11}$ (10 ng kg^{-1} min^{-1}) was infused intravenously for 5 min before injecting NT.

† p < 0.001 when compared with the appropriate control. (Taken with permission from Kérouac *et al.* 1981. Life Sci. **28:** 2477. Pergamon press, Ltd.)

guanethidine toward the tachycardia evoked by NT (FIG. 8) is unlikely to be related to an interference with cardiac adrenergic nerves because at the time it took place, the pressor effect of NT, presumably mediated by an increase sympathetic outflow to blood vessels, had already been blocked for several minutes. Further, we see no reason why cardiac sympathetic neurons would be more resistant to the action of guanethidine than those found in blood vessels. Both the pressor effects and tachycardia evoked by an intravenous infusion of NT were increased in the presence of atropine (FIGS. 5 and 7), thus suggesting that cholinergic nerves were indeed activated by NT. Whether cardiac cholinergic nerves were activated by a direct effect of NT on components of the vagal nerves or as a consequence of baroreceptor reflexes initiated by the pressor effects of NT remains to be determined. The potentiation by atropine of the pressor effect evoked by an intravenous infusion of NT suggest that NT activates skeletal muscle cholinergic (dilator) sympathetic fibers that, by releasing acetylcholine, may interfere with the increase of blood pressure produced by the simultaneous activation of sympathetic (vasoconstrictor) adrenergic nerves. Nevertheless, we cannot rule out the possibility that the increased heart rate recorded in the presence of atropine is directly responsible for the increased pressor effect of NT induced by atropine (e.g., as a consequence of an increased cardiac output).

In conclusion, the pressor effect of NT in anesthetized guinea pigs appears to be the consequence of an activation of the sympathetic nervous system while its positive chronotropic effect seems to be due to a direct effect of NT on the heart.

THE CARDIOVASCULAR EFFECTS OF NEUROTENSIN IN DOGS AND SHEEP

In pentobarbital-anesthetized dogs, intravenous injections of NT ($\sim$0.5–16 μg kg^{-1}) induced dose-dependent hypotensive effects with little or no change of heart rate.[7, 8] The maximal hypotensive effect (about 45–50% of the control value) was associated with a decrease of peripheral resistance ($\sim$35%) of stroke volume ($\sim$20%) and only minor alterations of the strength of myocardial contraction ($\sim$−8%) and of heart rate ($\sim$+5%).[8] Slight hypotensive effects were also recorded in dogs receiving intravenous infusions of 18–117 ng kg^{-1} min^{-1} of NT during 10–20 min.[7] In another study, no changes of arterial pressure, of left ventricular pressure or of its first derivative (LV dp/dt), of pulmonary arterial blood pressure, or of cardiac index were recorded following intravenous infusions of 4.5–72 ng kg^{-1} min^{-1} of NT.[14] However, in the same study, the authors reported an increase of blood flow in the "muscularis" of the duodenum, jejunum, ileum, and colon, thus confirming similar observations made previously by Rosell and co-

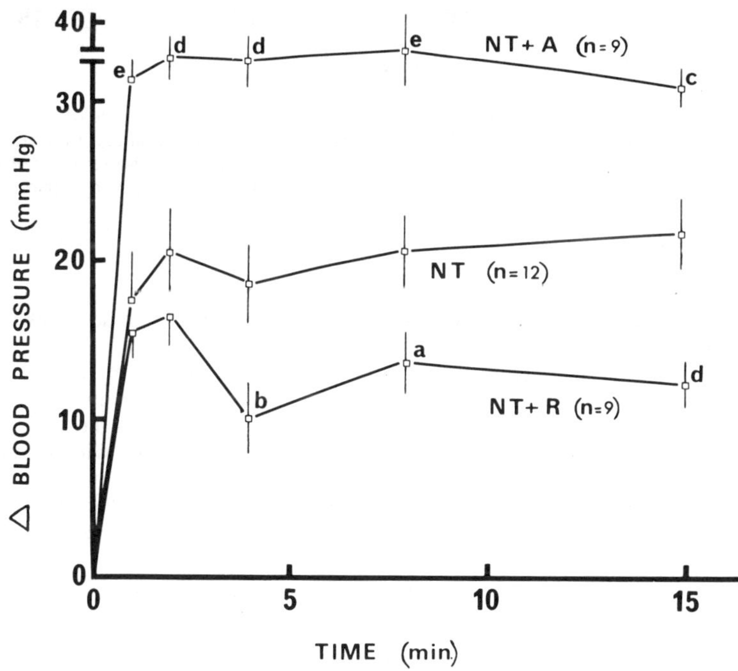

FIGURE 5. Influence of reserpine (R) and of atropine (A) on the changes of mean arterial blood pressure induced by intravenous infusions of neurotensin (NT) (0.3 nmol kg^{-1} min^{-1}) in pentobarbital-anesthetized guinea pigs. Each point is the mean ± SEM. The number of experiments is given in parentheses. a = p < 0.05; b = p < 0.025; c = p < 0.01; d = p < 0.005; e = p < 0.001, when compared to the respective control value.

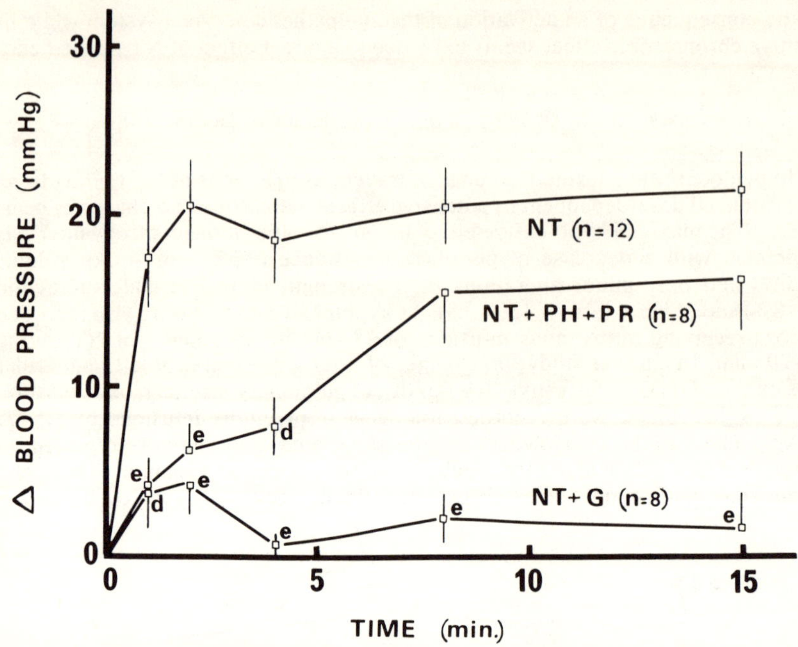

FIGURE 6. Influence of a mixture of phentolamine (PH) and propranolol (PR), and of guanethidine (G) on the changes of mean arterial blood pressure induced by intravenous infusions of neurotensin (NT) (0.3 nmol kg^{-1} min^{-1}) in pentobarbital-anesthetized guinea pigs. Each point is the mean ± SEM. The number of experiments is given in parentheses. d and e: same explanations as in FIGURE 5.

workers.[7] The mechanism of the cardiovascular effects of NT in dogs has not yet been investigated.

In pentobarbital-anesthetized sheep, intravenous injections of NT (1000 ng kg^{-1}) produced biphasic pressor-depressor effects accompanied by tachycardia.[11] Prior administration of reserpine (10 μg kg^{-1} day^{-1} for 9 days) abolished both the hypertensive phase of the response and the tachycardia, without modifying the hypotensive phase of the response. Although these results are only preliminary, there is an indication that in sheep, NT may produce some of its cardiovascular effects by activating the sympathetic nervous system. Further studies will be needed to substantiate these findings.

CONCLUSION

Several studies dealing with the cardiovascular effects of NT in mammals have been summarized in this chapter. NT was shown to produce marked alterations of arterial blood pressure and sometimes also of heart rate in different animal species. These effects varied both qualitatively and quantitatively from one species to another and also within the same species, depending on whether the animals were anesthetized or not,[10] or on the type of anesthetics used.[13] In anesthetized rats, evidence was provided suggesting that endogenous histamine and serotonin, presumably of mast cell origin, contribute extensively to the hypotensive effect of NT, possibly by interfering with the activity of the sympathetic nervous system. In pithed rats, NT-evoked pressor effects that appear to be due entirely to the vascular effects of endogenously released serotonin. In anesthetized guinea pigs, NT produced mostly pressor effects and tachycardia. Pharmacologic evidence was provided indicating that the pressor effect of NT in this species is likely the result of an activation of the sympathetic outflow to the blood vessels, while the positive chronotropic effect of NT seems to be due to a direct action on the heart, presumably at the level of the sinoatrial node. An activation of the sympathetic nervous system may also subserve the pressor effect and tachycardia induced by NT in anesthetized sheep.[11]

The physiologic significance of these studies remains to be assessed. Nevertheless, the following observations strongly support the idea that interactions between NT and the cardiovascular system of mammals may have physiologic and/or pathologic implications: first, NT is active at relatively low doses (50–100 pmol) on the cardiovascular system of several species; second, immunoreactive NT has been found in cardiac nerves;[44, 45] third, evidence derived from

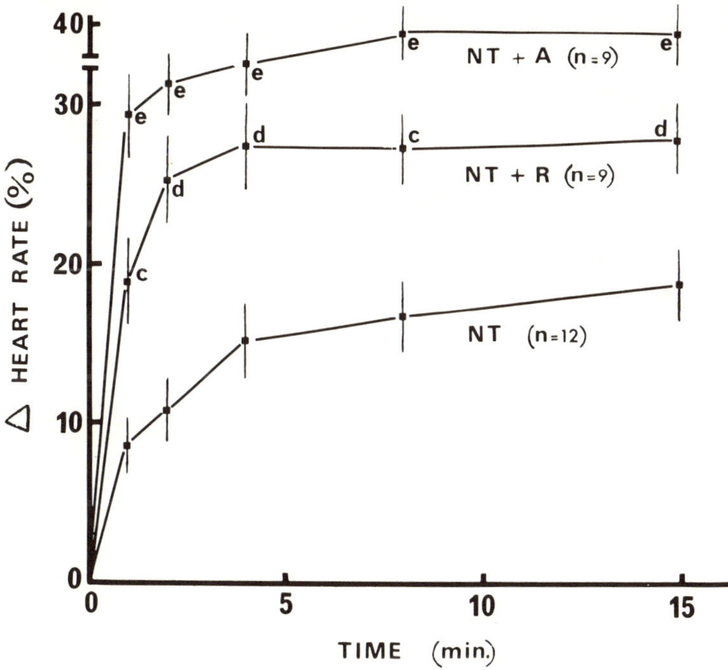

FIGURE 7. Influence of reserpine (R) and of atropine (A) on the changes of heart rate induced by intravenous infusions of neurotensin (NT) (0.3 nmol kg^{-1} min^{-1}) in anesthetized guinea pigs. For other details, see FIGURE 5. The changes of heart rate were recorded from the same animals as those used in FIGURE 5.

structure–activity studies, and particularly from the development of specific NT antagonists,[28–30, 46] strongly suggests the existence of specific receptors for NT in cardiac and vascular smooth muscle cells; fourth, evidence has been obtained that suggests that NT interacts with structures such as autonomic nerves and mast cells that, under some circumstances, modulate cardiovascular functions in mammals; and fifth, NT is present in large quantities in the gastrointestinal system of mammals. It is clear that an eventual release of a large amount of NT from the intestines into the circulating blood may bring about significant variations of blood

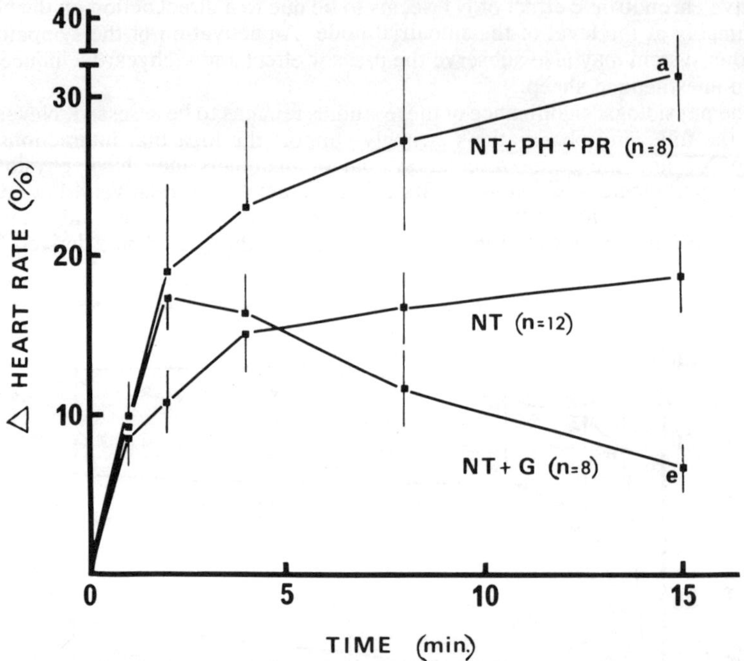

FIGURE 8. Influence of a mixture of phentolamine (PH) and propranolol (PR), and of guanethidine (G) on the changes of heart rate induced by intravenous infusions of neurotensin (NT) (0.3 nmol kg^{-1} min^{-1}) in anesthetized guinea pigs. For other details, see FIGURES 5 and 6. The changes of heart rate were recorded from the same animals as those used in FIGURE 6.

pressure and heart rate in mammals. Further studies are obviously needed before a definite role can be attributed to the various cardiovascular effects of NT.

ACKNOWLEDGMENTS

We thank Dr. Jean Barabé of the Department of Physiology and Pharmacology, University of Sherbrooke, for reviewing the first draft of our manuscript, Mrs. Danielle Laurendeau for typing the manuscript and the tables, and Mr. Marc-André Leblanc for his excellent technical assistance.

REFERENCES

1. CARRAWAY, R. E. & S. E. LEEMAN. 1973. J. Biol. Chem. **248**: 6854–6861.
2. CARRAWAY, R. E. & S. E. LEEMAN. 1976. J. Biol. Chem. **251**: 7045–7052.
3. FERNSTROM, M. H., R. E. CARRAWAY & S. E. LEEMAN. 1980. *In* Frontiers in Neuroendocrinology. E. Martini & W. F. GANONG, Eds. Vol. **6**: 103–127. Raven Press. New York.
4. QUIRION, R., F. RIOUX, S. ST-PIERRE & D. REGOLI. 1979. Life Sci. **25**: 1969–1973.
5. KATAOKA, K., A. TANIGUCHI, H. SHIMIZU, K. SODA, S. OKUNO, H. YAGIMA & K. KITAGAWA. 1978. Brain Res. Bull. **3**: 555–557.
6. TERRIAULT, P., H. ELHILALI, A. CASEY, F. RIOUX, S. ST-PIERRE & A. GERVAIS. 1980. Union Méd. Can. (Abstr.) **109**: 1519.
7. ROSELL, S., F. BURCHER, D. CHANG & K. FOLKERS. 1976. Acta Physiol. Scand. **98**: 484–491.
8. ST-PIERRE, S., R. QUIRION, D. REGOLI, A. GERVAIS, F. LAVIGNE, A. POIRIER, F. JOLICOEUR, A. BARBEAU & F. RIOUX. 1980. Union Méd. Can. **109**: 1447–1455.
9. KEROUAC, R., F. RIOUX & S. ST-PIERRE. 1981. Life Sci. **28**: 2477–2487.
10. MILLER, V. M., A. M. HOFFMAN & F. E. SOUTH. 1981. Fed. Proc. (Abstr.) **40**: 581.
11. GAGNON, L., G. GRIMARD, P. CASEY, S. ST-PIERRE & A. GERVAIS. 1981. Union Méd. Can. **110** (suppl. 1): 59.
12. RIOUX, F., R. KEROUAC & S. ST-PIERRE. Unpublished results.
13. CHAHL, L. A. & S. B. WALKER. 1981. Life Sci. **29**: 2009–2015.
14. BACA, I., U. MITTMANN, G. E. FEURLE, M. HAAS & TH. MÜLLER. 1981. Res. Exp. Med. (Berlin) **179**: 53–58.
15. CHAHL, L. A. 1979. Naunyn-Schmiedeberg's Arch. Pharmacol. **309**: 159–163.
16. QUIRION, R., F. RIOUX & D. REGOLI. 1978. Can. J. Physiol. Pharmacol. **56**: 671–673.
17. STENE-LARSEN, G. & K. B. HELLE. 1979. Comp. Biochem. Physiol. **64C**: 279–283.
18. QUIRION, R., F. RIOUX, D. REGOLI & S. ST-PIERRE. 1979. Eur. J. Pharmacol. **55**: 221–223.
19. RIOUX, F., R. KEROUAC, R. QUIRION & S. ST-PIERRE. 1982. Neuropeptides **2**: 227–231.
20. HELLE, K. B., G. SERCH-HANSSEN, G. JORGENSEN & R. KNUBSEN. 1980. J. Auton. Nerv. Syst. **2**: 143–155.
21. RIOUX, F., R. QUIRION, M. A. LEBLANC, D. REGOLI & S. ST-PIERRE. 1980. Life Sci. **27**: 259–267.
22. ROSELL, S., K. THOR, A. ROKAEUS, O. NYQUIST, A. LEWENHAUPT, L. KAGER & K. FOLKERS. 1980. Acta Physiol. Scand. **109**: 369–375.
23. BLACKBURN, A. M., D. R. FLETCHER, T. E. ADRIAN & S. R. BLOOM. 1980. J. Clin. Endocrinol. Metab. **51**: 1257–1261.
24. GILMAN, A. G., L. S. GOODMAN & A. GILMAN, Eds. 1980. *In* The Pharmacological Basis of Therapeutics, 6th Ed. MacMillan Publishing Company. New York.
25. PARRAT, J. R. & G. B. WEST. 1957. J. Physiol. **137**: 179–192.
26. BROGDEN, R. N., T. M. SPEIGHT & G. S. AVERY. 1974. Drugs **7**: 164–282.
27. TOZZI, S. & F. ROTH. 1974. Agents Actions **4**: 264–270.
28. QUIRION, R., F. RIOUX, D. REGOLI & S. ST-PIERRE. 1980. Eur. J. Pharmacol. **61**: 309–312.
29. QUIRION, R., F. RIOUX, D. REGOLI & S. ST-PIERRE. 1980. Eur. J. Pharmacol. **66**: 257–266.
30. RIOUX, F., R. QUIRION, D. REGOLI, M. A. LEBLANC & S. ST-PIERRE. 1980. Eur. J. Pharmacol. **66**: 273–279.
31. QUIRION, R., F. RIOUX, D. REGOLI & S. ST-PIERRE. 1980. Life Sci. **27**: 1889–1895.
32. OISHI, M., C. INAGAKI, M. FUJIWARA, S. TAKAORI, H. YAJIMA & Y. AKAZAWA. 1981. Jpn. J. Pharmacol. **31**: 1043–1049.
33. LAZARUS, L. H., M. H. PERRIN & M. R. BROWN. 1977. J. Biol. Chem. **252**: 7174–7179.
34. SELBEKK, B. H., O. FLATEN & L. E. HANSSEN. 1980. Scand. J. Gastroenterol. **15**: 457–460.
35. KUROSE, M. & K. SAEKI. 1981. Eur. J. Pharmacol. **76**: 129–136.
36. BHATTACHARYA, B. K. & G. P. LEWIS. 1956. Br. J. Pharmacol. **11**: 202–208.
37. NAGAI, K. & L. A. FROHMAN. 1978. Diabetes **27**: 577–582.
38. GILLESPIE, J. S. & T. C. MUIR. 1967. Br. J. Pharmacol. **39**: 78–87.

39. Quirion, R., F. Rioux & D. Regoli. 1978. Clin. Exp. Hypertension **1:** 267–277.
40. Rioux, F., R. Quirion, S. St-Pierre, D. Regoli, F. B. Jolicoeur, F. Belanger & A. Barbeau. 1981. Eur. J. Pharmacol. **69:** 241–247.
41. Feldberg, W. & A. N. Smith. 1953. Br. J. Pharmacol. **8:** 406–411.
42. Bray, R. E. & P. P. Van Arsdel. 1961. Proc. Soc. Exp. Biol. (NY) **106:** 255–260.
43. West, G. B. 1959. J. Pharm. Pharmacol. **11:** 513–534.
44. Reinecke, M., E. Weihe, R. E. Carraway & W. G. Forssmann. 1981. Abstr. of the European Neurochemistry Meeting. Liège, Belgium. p. 5452.
45. Weihe, E. & M. Reinecke. 1981. Neurosci. Lett. **26:** 283–288.
46. Quirion, R., D. Regoli, F. Rioux & S. St-Pierre. 1980. Br. J. Phrmacol. **68:** 83–91.

Discussion of the Paper

S. Rosell (*Karolinska Institute, Stockholm, Sweden*): I would like to comment on the dose you have used. You are using a dose about 1000 times higher than you need in man, rat, or dog, in order to obtain discrete effects, say, in the gastrointestinal tract or adipose tissue. With these low doses there is no change in blood pressure of heart rate. I wonder if the effects you are showing are pharmacologic or toxic effects of neurotensin rather than physiologic effects?

F. Rioux (*University of Sherbrooke, Sherbrooke, Quebec*): Yes, I agree with you that the doses of NT we used to show an effect on the blood pressure are quite high. I also agree that in some species large increase of neurotensin levels in blood occur without associated changes in heart rate or blood pressure. However, there may be changes of blood flow in selected organs or changes in capillary permeability, which do not lead to significant changes of blood pressure because of the existence of cardiovascular reflexes. In my opinion, it is still too early to decide whether or not the cardiovascular effects of NT have a physiologic relevance.

J. W. Holaday (*Walter Reed Army Hospital,* Washington, D.C.): Does central injection of various neurotensin compounds produce effects that are different from those that you observed after parenteral administration?

Rioux: We have injected neurotensin into the lateral brain ventricle of rats and have observed hypotension, which was not accompanied by changes in heart rate. This hypotensive effect after central injection is apparently mediated by the CNS and not by the periphery secondary to leakage of neurotensin into the periphery. This is certain because in animals treated with compound 48/80 the hypotensive effect of intravenously injected neurotensin is completely blocked, whereas the hypotensive effect after central injection is not. It is reasonable to assume that this central effect could lead to a decrease in the sympathetic nervous system activity.

Holaday: I was wondering specifically about physiologic situations. Have you tested the D-Trp[11]-NT during physiologic alterations of blood pressure?

Rioux: No, we have not. In the guinea pig it does not block the pressor effect of neurotensin. However, we were troubled by the intrinsic activity of this compound.

P. Kitabgi (*INSERM, Nice, France*): The group of Lazarus and Rivier have demonstrated binding of NT to rat mast cells and the structure-activity relationships that were found for these mast cell receptors are extremely unusual, that is, they are very different from the structural requirements that we find for other receptors. Did you do any structure-activity relationship studies in your *in vivo* system where neurotensin is presumed to act through mast cells resulting in histamine release?

RIOUX: Yes, we have done some studies, and our results confirm those of Carraway and Leeman about the importance of the COOH-terminal region of the molecule. Using our system we reached the same conclusion as you did in your studies of the ileal preparation about the importance of the tyrosine residue for receptor activation.

KITABGI: However, you conclude differently than Rivier. Did you try the COOH-terminal pentapeptide of NT which was shown on mast cells to be very active as opposed to other preparations, or did you try the derivative [D-arginine[9]]-neurotensin, which was found to be five times as active as neurotensin in mast cells? In other systems it is about one percent as active.

RIOUX: Yes, that's about what we get also with [D-arginine[9]]-neurotensin in rat stomach strips. However this compound has not been tested for its effects on the blood pressure.

KITABGI: One percent.

RIOUX: About that, yes. As far as the COOH-terminal pentapeptide of NT is concerned, it is about four percent as active as NT as a hypotensive agent.

KITABGI: So our results on the blood pressure are slightly different from those data by Lazarus and coworkers.

RIOUX: Yes.

R. E. CARRAWAY (*University of Massachusetts, Worcester*): I am very interested in the opposite effects that you see in certain species. We also saw similar kinds of responses in anesthetized rats in our earlier work and in particular when we were examining responses to COOH-terminal fragments of neurotensin. We noted that as the peptide is shortened toward its COOH terminus by progression, deleting amino terminal residues, in fact there is a biphasic response for neurotensin. As one shortens the molecule the increase in blood pressure that is seen initially is enhanced. And, I was wondering if you also had the opportunity to look at smaller peptides?

One notion that might explain this kind of biphasic response might be that this represents a cross-reaction of neurotensin-related material which shares COOH-terminal structure with neurotensin; that is, that there may be structural variants of neurotensin that have hypertensive effects and while neurotensin itself has a hypotensive effect. As one shortens the peptide, one may be able to get cross reactions with a different set of receptors. I was wondering if you looked into this possibility and if you saw an enhancement of the hypertensive effect with smaller fragments?

RIOUX: No, we did not. The COOH-terminal pentapeptide fragment and the hexapeptide also are much less potent than neurotensin as far as the hypotensive effect is concerned. We have not seen any increase of the pressor component of the effect. Recently we have also looked at the pressor effect of these fragments in animals pretreated with compound 48/80 and we observed a pattern of activity very similar, to what we observed for the hypotensive effect. Thus the structural requirements for the hypotensive effect appears to be very similar to those for the hypertensive effect in compound 48/80-treated animals.

R. QUIRION (*NIMH, Bethesda, MD*): I just want to add that in their system, Lazarus, Brown and Vale have never been able to induce the release of histamine from mast cells with neurotensin. Also as Dr. Kitabgi noted, their binding characteristics were quite different from what we have observed in other systems. However, Richard Miller has recently shown that neurotensin is the most potent peptide known to induce histamine release from rat mast cells. It would be interesting to test if different structure activity relationships exist in this system.

KITABGI: There has also been a recent report by a Japanese group that neurotensin was able to release histamine from purified mast cells, but it was not very potent. Those dose-response relationships were very unusual with biphasic effects and an EC_{50} range of 10^{-7} M.

QUIRION: Miller and his associates have said that neurotensin is more potent than that and I think he even said that it is more potent than compound 48/80.

D. E. COCHRANE (*Tufts University, Medford, MA*): We have found that neurotensin stimulates the release of histamine from mast cells and does so in a biphasic manner. At high concentrations (10^{-4} M), it is a nonspecific effect of neurotensin.

The effect of neurotensin on the mast cell is first observed at about 10^{-9} M. This effect is markedly influenced by the level of calcium in the bathing solution.

We have found that neurotensin does elevate blood levels of histamine and does so in doses of 0.5–1.0 nmol/kg. This effect is completely abolished by pretreating the animals with 48/80 and significantly reduced by disodium cromoglycate. The effect is quite rapid and easily observed by one minute post injection. We feel that neurotensin has very potent effects on the mast cell although it is certainly not as potent as 48/80. In isolated skin neurotensin is very potent, (in the one picomolar range) in releasing histamine.

RIOUX: We recently observed that in rats the hypotensive effect of neurotensin is completely blocked by infusion of disodium cromoglycate.

S. ST. PIERRE (*University of Sherbrooke, Sherbrooke, Quebec*): In response to Dr. Rosell's comment about the effective concentration of neurotensin in blood, it is my belief that we are probably measuring with RIAs lower levels than are actually circulating in blood; this is because of the nature of the antibodies we are using. I do not think that the effects that Dr. Rioux has described are purely toxic effects of neurotensin on blood vessels and heart, but rather a physiologic effect that really occurs in the organism.

J. E. T. Fox (*McMaster University, Hamilton, Ontario*): We, too, find mast cell involvement in effects of NT. Our end point is gastrointestinal motility. Have you determined whether particular α or β-adrenergic receptors mediate the effects of NT on blood pressure?

RIOUX: All the antagonists we used were tested against their respective agonists. Is that what you mean?

Fox: No. What I want to know is about one catecholamine response. What type of receptor does norepinephrine act (eg. α_1, α_2, β_1 or β_2)? Have you dissected that out with drugs like yohimbine and prazosin?

RIOUX: When we inject noradrenaline and observe a pressor effect, we attribute it to effects on the α receptor, probably both α_1 and α_2. When we observe an increase of heart rate, we attribute it to effects on the β receptors of the heart.

Fox: I just wondered if your blood pressure response to noradrenaline is α_1 or α_2 mediated? Have you tried yohimbine or prazosin as blocking agents?

RIOUX: I cannot say whether it was α_1 or α_2. I am sure that it was the α receptor that is involved with the pressor effect of neurotensin. Neurotensin releases noradrenaline, which acts on α receptors to produce the vasoconstriction leading to hypertensive effects.

THE CENTRAL AND PERIPHERAL DISTRIBUTION OF NEUROTENSIN

J. M. Polak* and S. R. Bloom

Departments of Histochemistry and Medicine
Royal Postgraduate Medical School
Hammersmith Hospital
London W12 OHS England

INTRODUCTION

The distribution and tissue localization of neurotensin has been investigated by our group, using both immunocytochemistry and radiommunoassay with region-specific antibodies (see TABLE 1). For immunocytochemistry at light microscopic level, we employed both the classical immunofluorescence method of Coons, Leduc, and Connolly[1] and a variety of immunoperoxidase methods[2] including the peroxidase-antiperoxidase (PAP) technique[3] and a combination of double immunostaining methods (based on the procedure originally described by Nakane).[4] For the ultrastructural localization of neurotensin, immuncytochemistry was carried out using not only the serial semithin/thin method,[5] but also the newly developed immunogold staining procedure on resin-embedded thin sections.[6] Radioimmunoassay[7] was used either for quantifying immunoreactive neurotensin in a variety of tissues or in combination with chromatography for determining the physicochemical properties of neurotensin employing an antibody that showed greatest avidity for whole neurotensin, some avidity for the larger COOH-terminal peptides, and did not bind with NT^{1-8} or NTH^1.

Using these techniques, we have been able to investigate morphologically and quantitatively the presence of neurotensin-like immunoreactive material (for the sake of brevity referred to as neurotensin) in the brain, spinal cord, gastrointestinal tract, thymus, and adrenal medulla.

BRAIN

In both the human and rat brain, neurotensin is localized in a characteristic complex network of pathways.[8] It is especially concentrated in the hypothalamus, central amygdaloid nucleus, and bed nucleus of the stria terminalis (FIG. 1). Much lower levels of immunoreactivity are seen in the hippocampus and cortex. The longest neurotensin-containing pathway discovered to date is in the stria terminalis.[9] Some of the neurotensin cell bodies found in the central amygdaloid nucleus project via the stria terminalis and terminate in the bed nucleus. Other cell bodies in the amygdala project to the piriform cortex and parts of the ventral striatum. We have recently reported the existence of a very long neurotensin pathway,[10] originating in the hippocampus and projecting through the cingulate cortex to terminate in the frontal cortex (FIG. 2).

*Address for correspondence: Dr. J.M. Polak, Histochemistry Department, Royal Postgraduate Medical School, Hammersmith Hospital, Ducane Road, London, W12 OHS.

TABLE 1

PREPARATION AND CHARACTERISTICS OF THE ANTISERA

| Antiserum | Raised in/against | Region Specificity | Optimal Dilution | | | Sensitivity for Radioimmunoassay (pmol/g wet weight of tissue) | Specificity Absorption Control for Immunocytochemistry (nmol/ml diluted antiserum) |
			LM	EM	RIA		
Neurotensin	rabbit/synthetic prep.	whole molecule	1:4000	1:4000			1
Neurotensin	rabbit/synthetic prep.	whole (COOH-terminal)	—	—	1:80000	0.4	—

Spinal Cord

In the spinal cord, neurotensin is distributed in a characteristic pattern.[11] Both nerve terminals and fibers (of very fine caliber) are found almost exclusively in the upper laminae (I–IV) of the dorsal horn. In this location neurotensin-

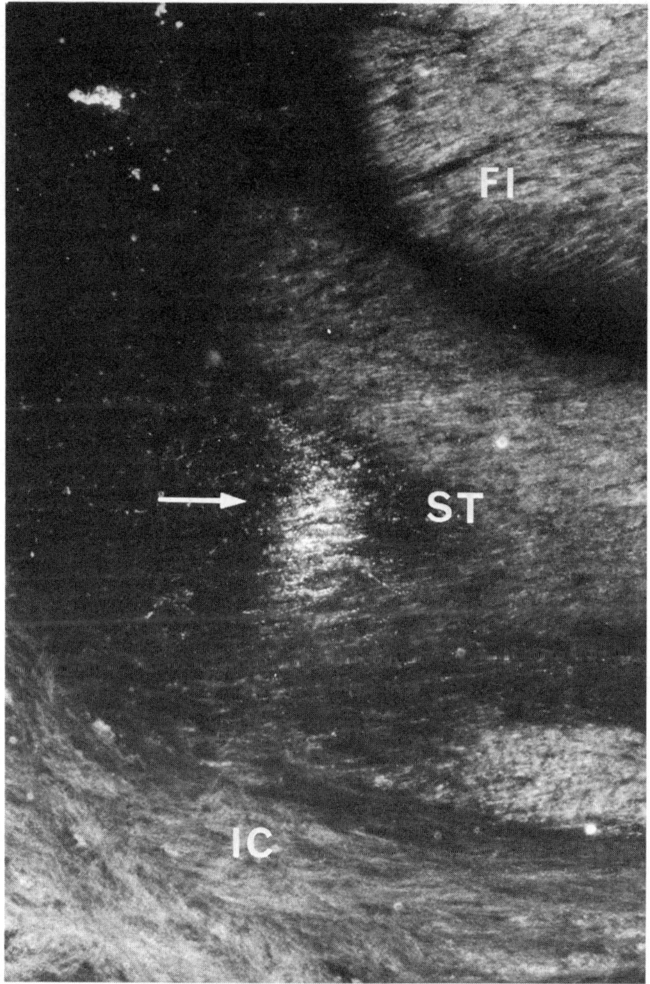

FIGURE 1. Neurotensin fibers in the stria terminalis (St) of the rat brain photographed under dark-field illumination (FI = fimbria; IC = internal capsule).

immunoreactive fibers are arranged in two distinct bands (FIG. 3). The outer or most dorsal band is found in lamina I and the upper portion of lamina II whereas the inner, more densely immunostained, band is localized in lamina II and the border of lamina III. Neurotensin fibers of the dorsal horn run in a rostrocaudal direction, best seen in parasagittal sections of the spinal cord. The origin of the immunoreactive neurotensin found in the spinal cord has not yet been fully validated. However, the lack of change after microsurgical manipulations, (e.g., sciatic nerve sectioning,[12] dorsal rhizotomy, and treatment with capsaicin)[13] designed to determine the possible origin of the peptide from primary sensory neurons of the dorsal root ganglia, points to an intrinsic origin from cell bodies within the spinal cord. Indeed, numerous cell bodies, up to 20 per section, can easily be distinguished in untreated, normal rats (FIG. 4). These cell bodies, bi- or multipolar, are particularly concentrated in lamina II and III and become more evident and numerous following treatment of the animals with the axonal transport blocker, colchicine.[14] A moderate accumulation of neurotensin in immunoreactive fibers is, in addition, noted around the central canal and lamina X as well as in the dorsal lateral funiculi. Very scattered neurotensin fibers can also be seen in the grey matter of the ventral horn.

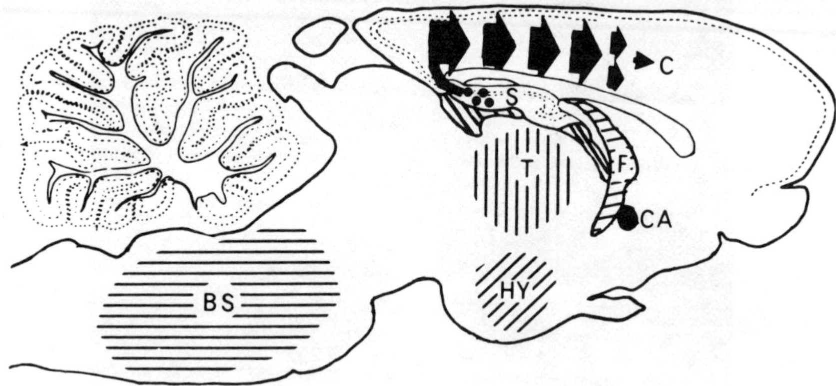

FIGURE 2.   Diagram showing the course of the proposed neurotensin-containing cortical pathway. The fibers originate in the dorsal subiculum (S) in the neurotensin immunoreactive cell bodies (●). The fibers project dorsally and rostrally, running parallel to the midline to terminate in the anterior cingulate cortex (C). BS = brain stem, HY = hypothalamus, T = thalamus, F = fornix, and CA = anterior commissure.

GASTROINTESTINAL TRACT

Neurotensin-like immunoreactive material is found in the gut of representatives of most of the major chordate phyla and even the uro- and cephalochordates.[15] At this stage of evolution, as well as in lower vertebrates, the immunoreactive material is present in scattered endocrine cells. In higher coldblooded vertebrates and birds (e.g., anuran amphibians, lizard, quail, chick), mucosal neurotensin cells are more numerous and found in the lower gut (FIG. 5), although endocrine cells reactive to antisera to neurotensin are also detected in the antrum and upper gut. The identity of the latter with the original 13-amino-acid peptide remains to be elucidated.

In mammals, including man, neurotensin immunoreactive cells are particularly concentrated in the ileum.[16] Neurotensin-containing enteric nerves have been noted in some teleost fishes[17] and birds[18] (upper gut). In the chicken, enteric nerves

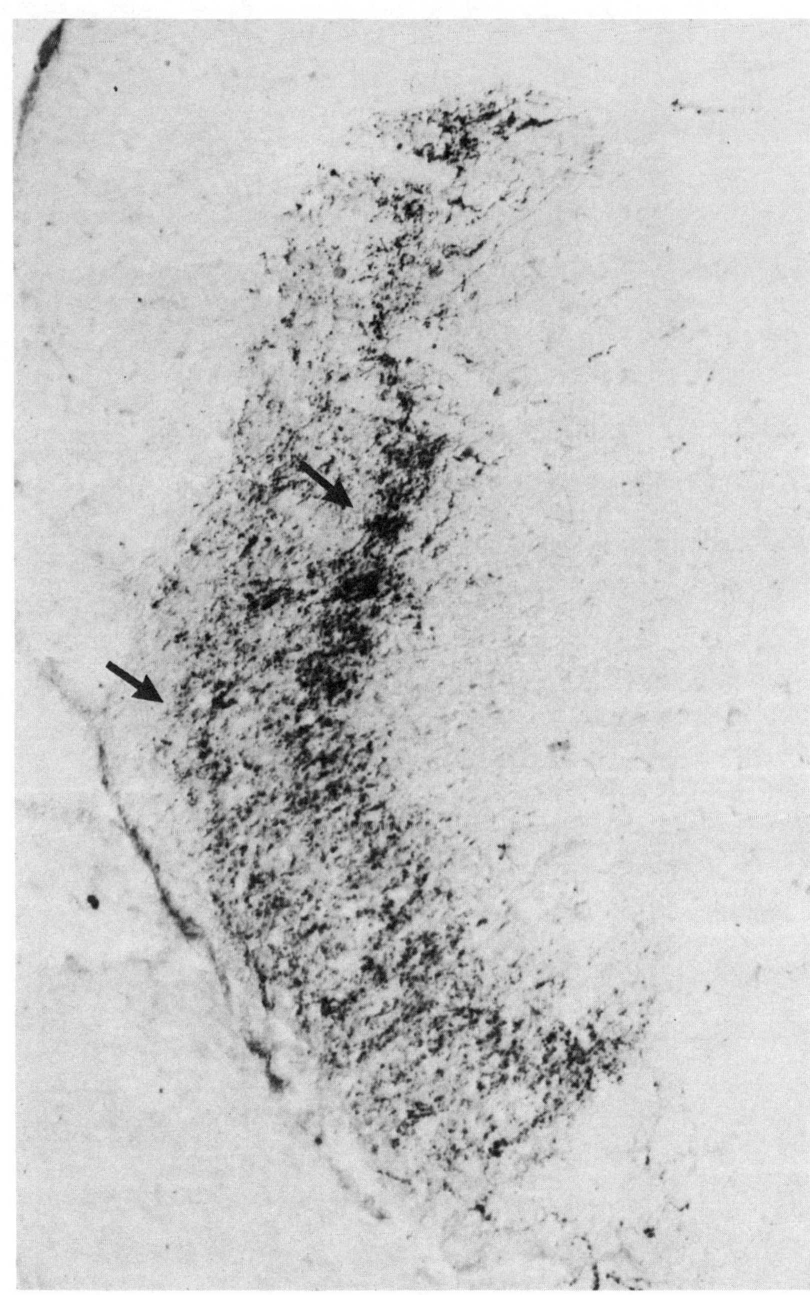

FIGURE 3. Characteristic double band (*see arrows*) of neurotensin immunoreactivity in the dorsal horn of the rat spinal cord. ($\times$ 66.5)

FIGURE 4. Sagittal section of rat spinal cord showing neurotensin-immunoreactive cell bodies. PAP method. ($\times$ 540)

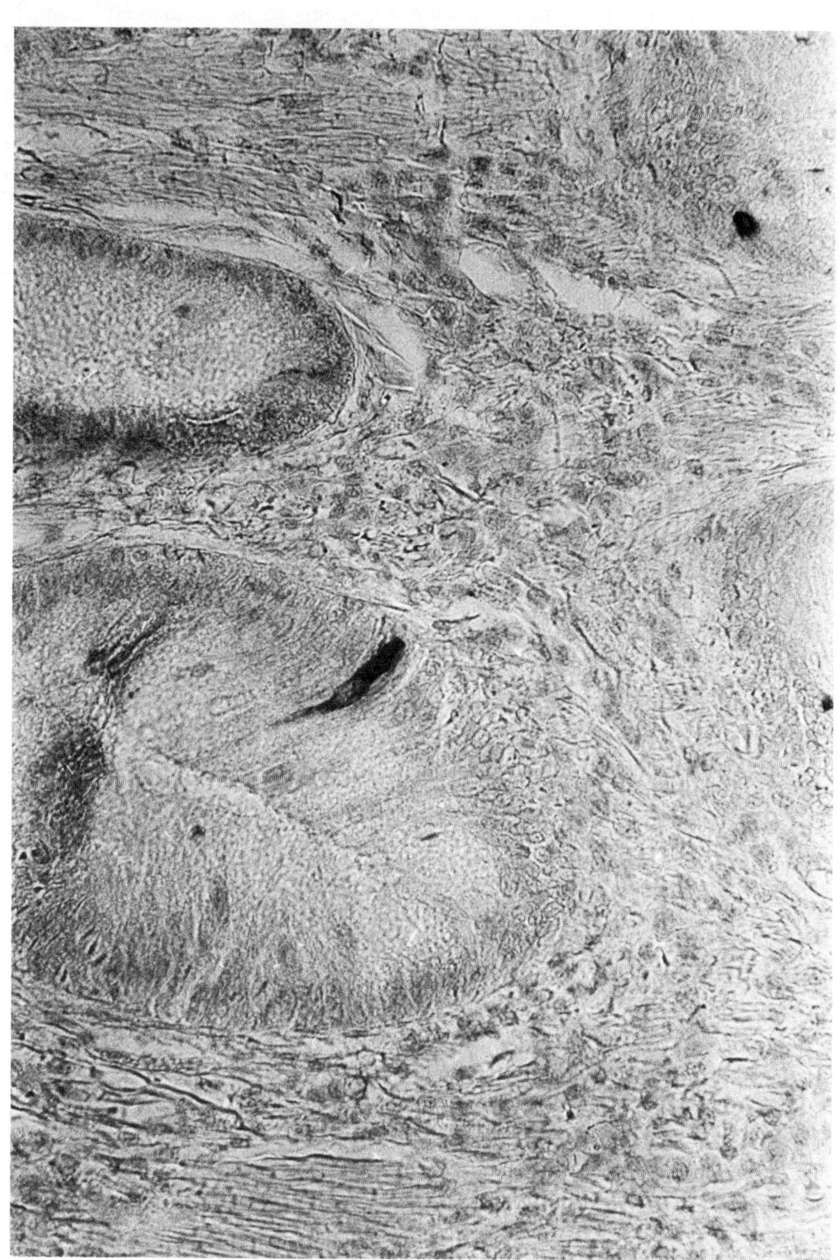

FIGURE 5.   Neurotensin-like immunoreactivity in the *Alligator mississipiens* intestine (courtesy Dr. A. M. J. Buchan). ($\times$ 450)

containing neurotensin are particularly numerous in the embryonic gut, where, in addition, a moderate number of immunoreactive cell bodies are seen.[18] Neurotensin-containing ganglion cells cannot be detected in the adult gut.

In mammals, including man, neurotensin seems to be exclusively found in mucosal endocrine cells[16, 19, 20] (FIG. 6) although there is a single report showing the presence of neurotensin-immunoreactive nerves in the rat intestine.[21] The exclusive epithelial localization of neurotensin is further validated when separated layers of human bowel are individually analyzed for the presence of immunoreactive neurotensin by radioimmunoassay of tissue extracts.[22] The total content of neurotensin-like immunoreactive material is estimated as 86.6 ± 42.5 pmol/g of wet weight whole bowel wall with a $99.3 \pm 1\%$ of this immunoreactive material being recovered from the epithelial layer. The neurotensin-containing cell of the mammalian gut has been extensively studied.[16, 19, 20] It is almost entirely confined to the ileal mucosa and displays the classical features of a mucosal endocrine cell of the "open" type, with a wide base and a narrower apical end, facing the gut lumen (FIG. 6), which ultrastructural studies show to consist of a tuft of characteristic microvilli (FIG. 7a). The secretory granules of the neurotensin (N) cell are large.[23] The application of immunocytochemical methods at the ultrastructural level, in particular the serial semithin/thin method, has permitted the further subclassification of the "large"-granuled cell of the human ileal mucosa into two distinct subtypes,[24] the enteroglucagon (EG) or glicentin cell containing smaller (260 nm $\pm$ 15 mean $\pm$ SEM) and less dense granules than the neurotensin (N) cell (300 nm $\pm$ 14 mean $\pm$ SEM) (FIG. 7a and b).

We have investigated the developmental pattern of neurotensin in the fetal avian[18] and human bowel.[25] In the chicken, neurotensin is first detected at 11 days of incubation, or Hamburger-Hamilton stage 37, in the upper part of the gut. In man, neurotensin is first detected in 12-week-old embryos. At first, neurotensin immunoreactivity is present in the jejunum and ileum, but at later stages the adult (ileal) pattern of distribution becomes evident. Unlike the adult gut, where a predominant single chromatographic peak of neurotensin is detected, the developing human bowel shows two well defined neurotensin immunoreactive peaks. (A Kav 0, B Kav 0.65). The larger peak, A, predominates in younger fetuses whereas peak B, which elutes in the position of synthetic human neurotensin, is the dominant form in the gut of full-term babies and adults (FIG. 8).

Information concerning the number of neurotensin cells in the bowel, in particular in the ileum, has so far been achieved by the use of morphometric analysis of specifically stained tissue sections. This procedure is, however, laborious and inaccurate as it involves theoretical cellular reconstructions. We have recently developed a way of bypassing this problem. It is based on the immunostaining of whole segments of mucosal epithelium (crypts and villi).[26] The neurotensin cells are therefore demonstrated and counted *in toto*, thus avoiding three-dimensional reconstruction of tissue sections (FIG. 9a and b).

THYMUS

Neurotensin-immunoreactive cells are detected in the avian thymus.[27] The cells are predominantly localized in the central part of each thymic lobule (medullary region) (FIG. 10) or at the margins between cortex and medulla. No immunoreactive cells are found in Hassal's bodies. Neurotensin cells, of variable shape (polygonal, elongated) and approximately 10–15 μm at their greatest width, seem to be located at some distance from the blood vessels. On conventional

FIGURE 6. Neurotensin-immunoreactive cell in the human ileum. Note the classical shape of the neurotensin cell with a narrow luminal end (*thin arrow*) and a broader base (*thick arrow*). Lu = lumen. (× 550)

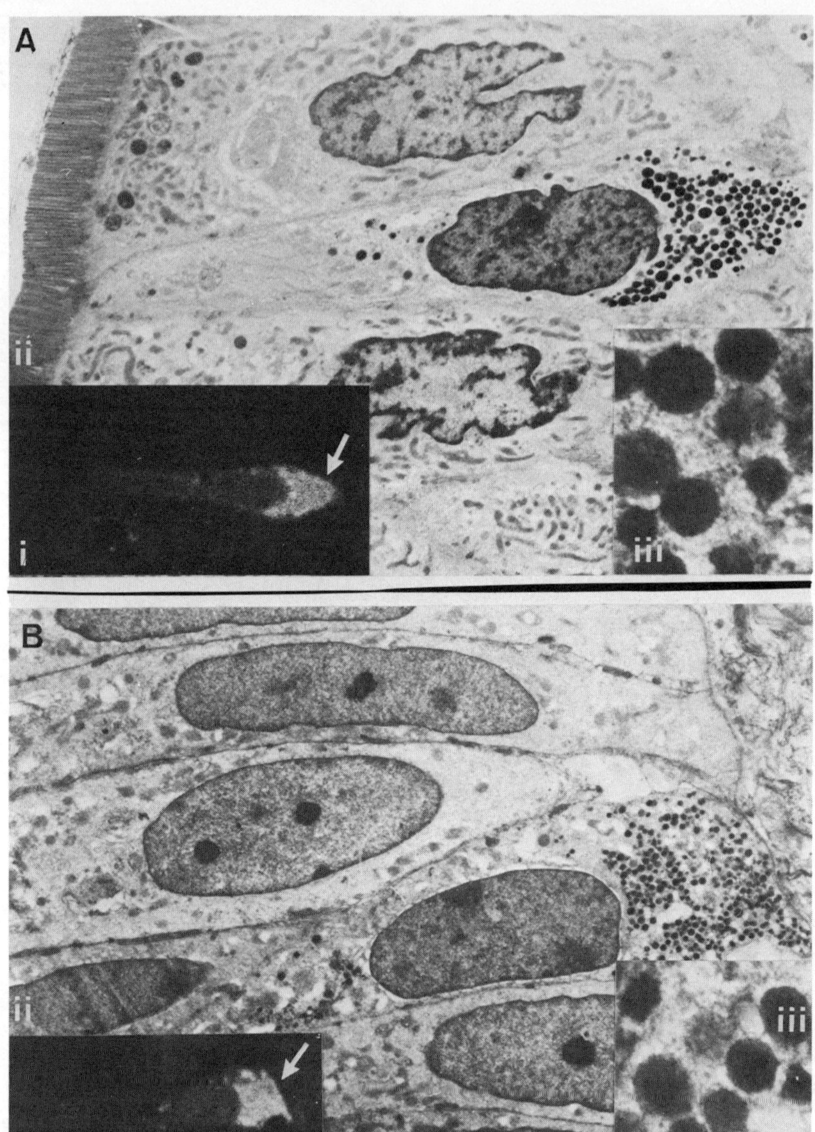

FIGURE 7. Human ileum sequential resin sections (original magnifications are given in parentheses; the figure has been reduced to 65% of original size):

(A) (i) Thick (1 μm) section immunostained for neurotensin (*arrow*). (× 2000) (ii) Serial thin (60nm) section showing the classical large, dense secretory granules (× 6000) details of which can be seen in the inset (iii). (× 45,000)

(B) (i) Thick (1μm) section immunostained for enteroglucagon (*arrow*). (× 2000) (ii) Serial thin (60 nm) section showing the classical large, dense secretory granules (× 6000) details of which can be seen in the inset (iii). (× 45,000) histology, the neurotensin-immunoreactive cells can be easily distinguished from the other much larger, thymic cells displaying lighter cytoplasm. In serially immunostained 3 μm paraffin sections, the cytoplasm of the neurotensin immunoreactive cells can be shown to contain neuron-specific enolase (NSE) (FIG. 11). NSE is an isomer of the glycolytic enzyme enolase[28] and is known to be present in all endocrine cells and nerves of the diffuse neuroendocrine system.[29,30,31]

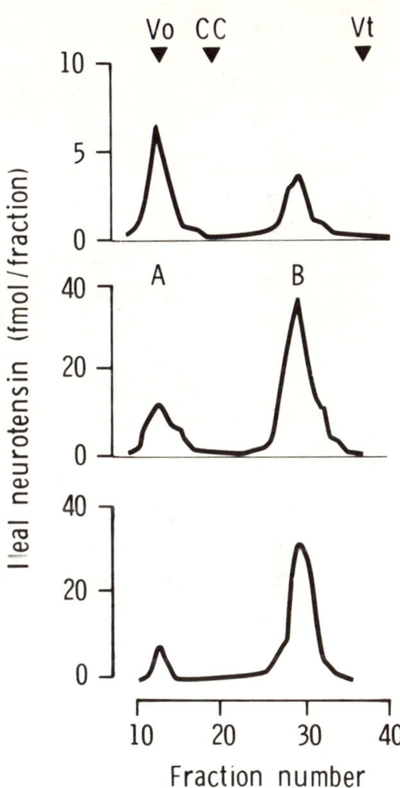

FIGURE 8. Representative gel permeation chromatography profiles of neurotensin immunoreactivity from different regions of fetal and adult gut (water extracts). The top panel is the profile obtained from fetuses aged 12–15 weeks. The central panel is obtained from fetuses aged 31 weeks to term and the bottom panel is obtained from adult extracts.

Neurotensin cells are a completely separate cell type from those thymic endocrine cells reacting to somatostatin antibodies. Neurotensin cells are detected in the neonatal chick (1 day old) and they continue to be numerous in the thymus of chicks up to 2 months old. Neurotensin-like immunoreactive material can also be detected by radioimmunoassay of tissue extracts (46.3 $\pm$ 8.6 pmol/g mean $\pm$ SEM).

FIGURE 9. Thin slice (1 villus thick) of human ileum, microdissected from whole mucosal strips after immunostaining. Note the distribution of neurotensin cells throughout the epithelial layer of the villus (immunofluorescence: A × 180; B × 350).

ADRENAL MEDULLA

Neurotensin immunoreactivity is also demonstrated in the mammalian adrenal medulla.[32-33-34] Particularly high concentrations are found in the cat adrenal gland (15.2 ± 3.6 pmol/g whole gland and 47.9 ± 18.4 pmol/g microdissected medulla mean ± SEM). Chromatographic analysis of this material indicates its identity with the original 13-amino-acid peptide. Immunocytochemistry localizes neurotensin to a subpopulation of noradrenaline-containing cells, separate from those co-storing enkephalin[35] (FIG. 12). This finding is interesting in view of (a) the postulated existence of heterogeneous subpopulations of noradrenaline-containing cells as determined by electron-microscopic[36] and biochemical investigations,[37]

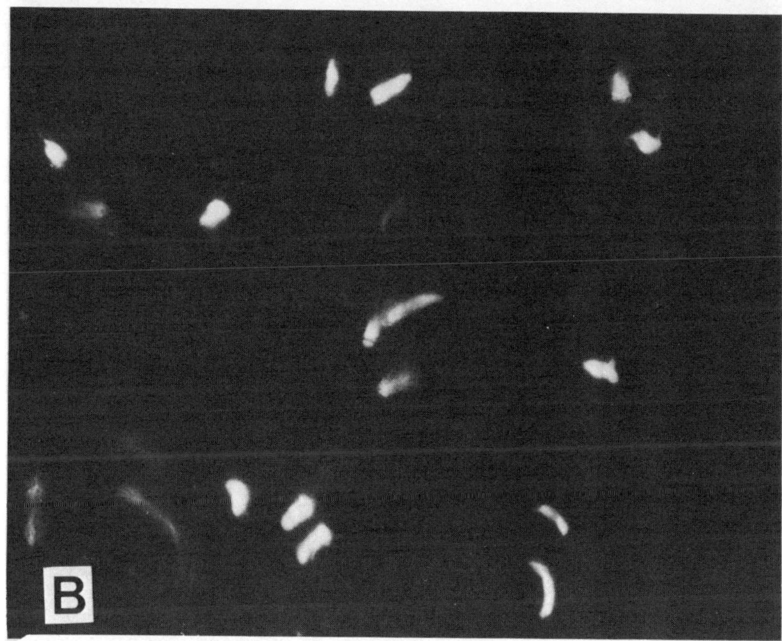

and (b) the recent demonstration of significant concentrations of neurotensin immunoreactive material in a phaeochromocytoma cell line maintained *in vitro*.[38]

CONCLUSION

The finding of neurotensin-like immunoreactive material in tissues outside the brain/gut axis allows the inclusion of this potently acting peptide in the newly recognized group of peripheral regulatory peptides[39] known to exert key roles in a number of bodily functions both in health and disease.

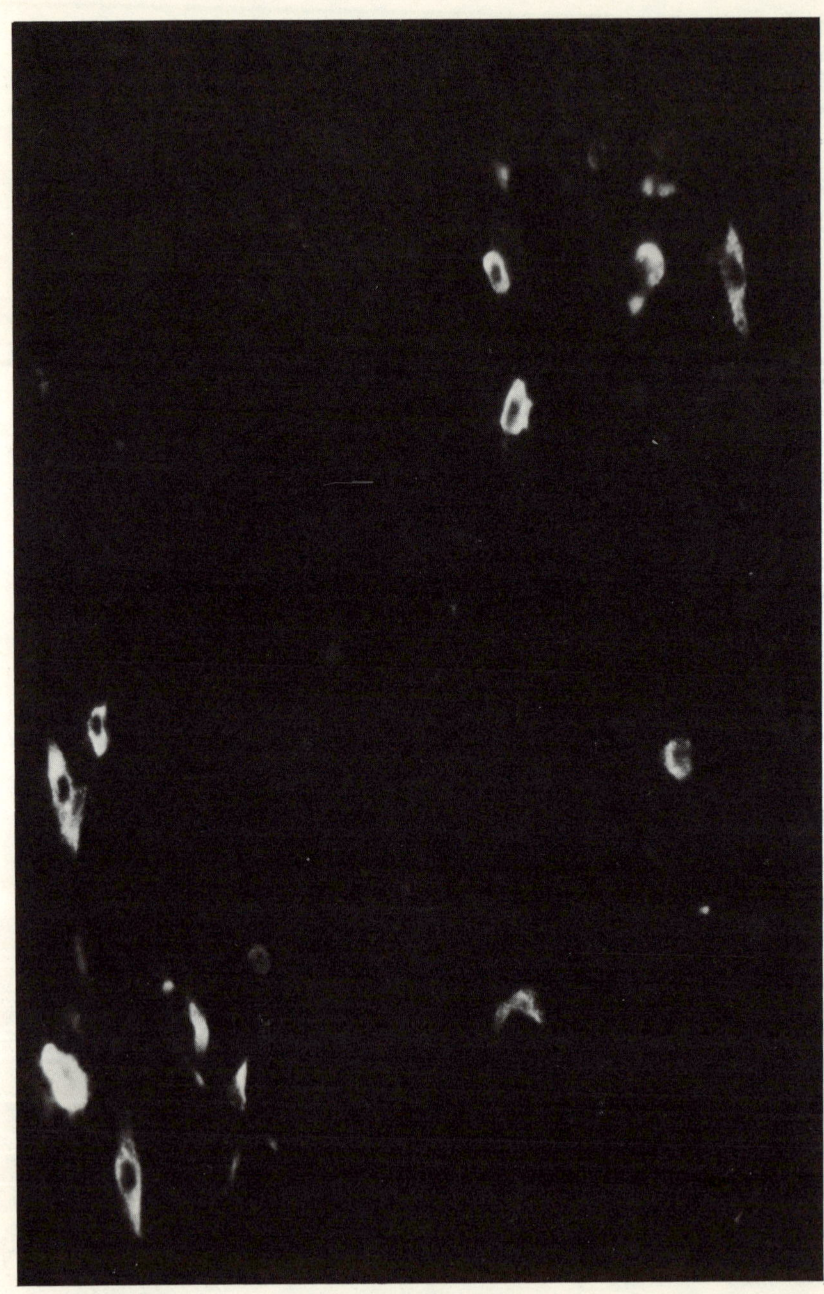

FIGURE 10. Numerous neurotensin-immunoreactive cells in the medulla of the chicken thymus. ($\times$ 400)

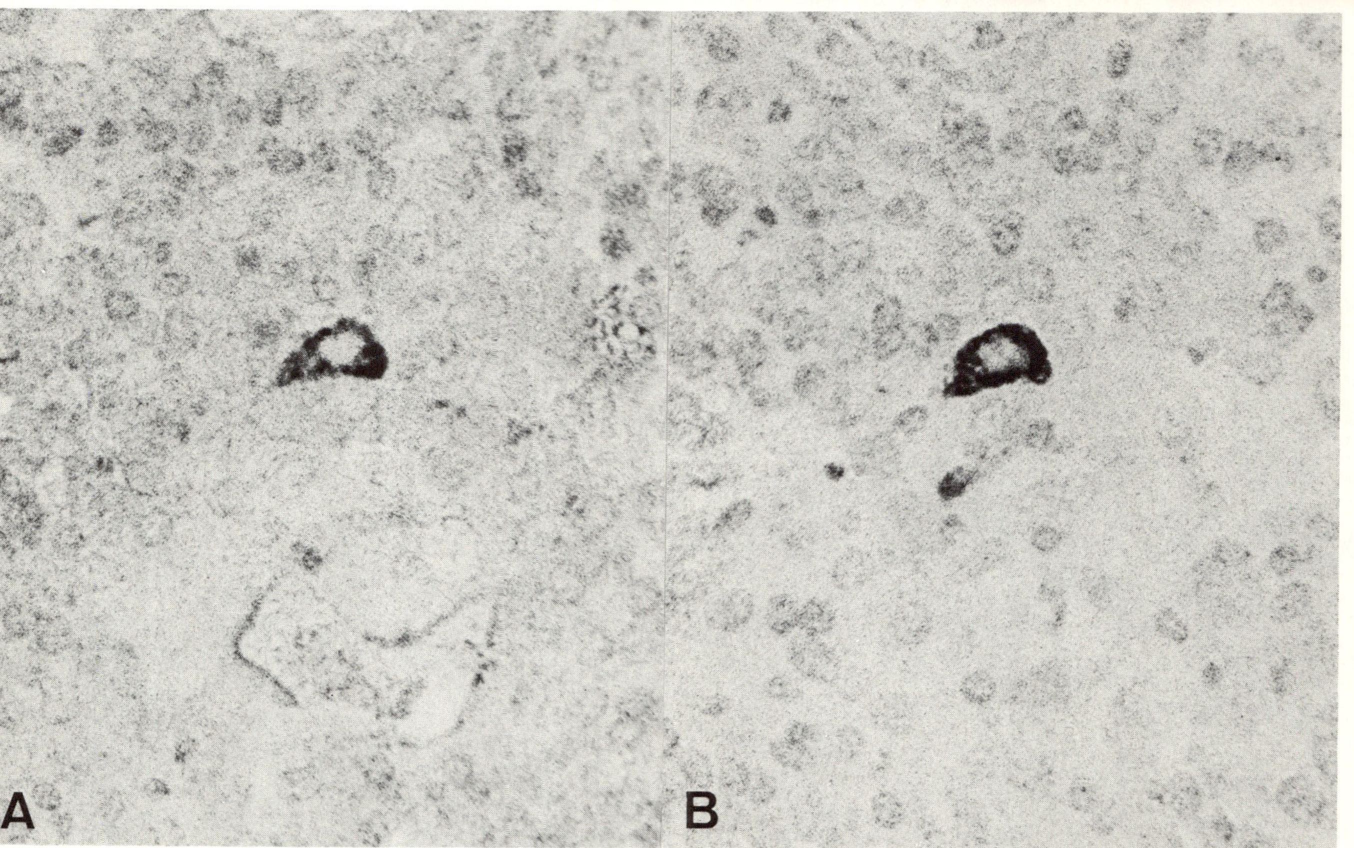

FIGURE 11. Thymus from 5-day-old chicken, freeze dried and vapor fixed in paraformaldehyde. (A) Neurotensin-immunoreactive cell in the chicken thymus, 1μm resin section. (B) Serial section immunostained using antibodies to neuron-specific enolase to demonstrate the neuroendocrine nature of the thymic neurotensin cell. (× 900)

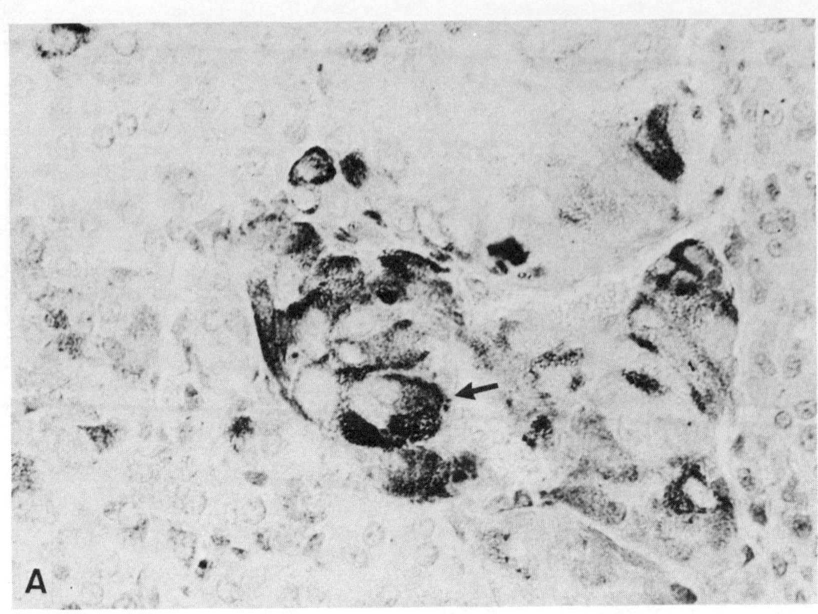

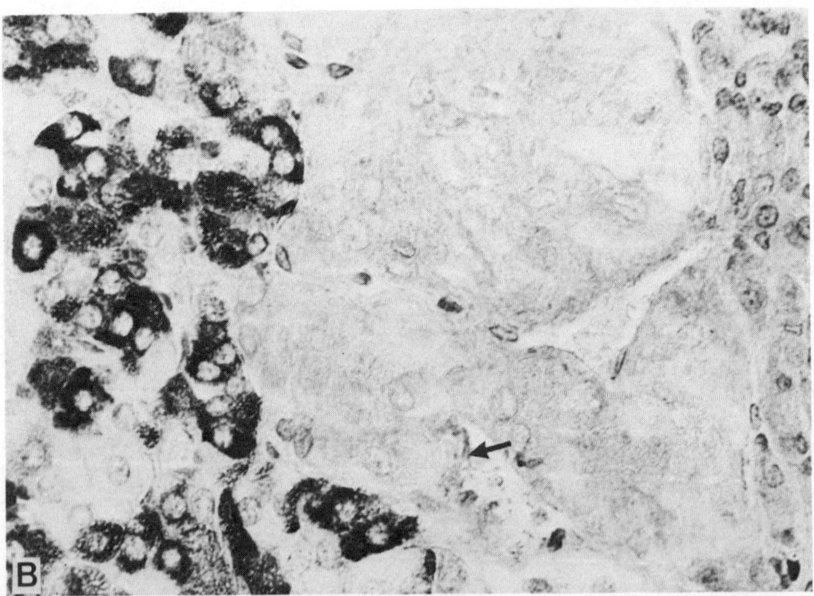

FIGURE 12. Cat adrenal medulla. Two sequential (3 μm) sections to demonstrate the separate cells of origin of neurotensin (A) and of enkephalin (B) immunoreactivity, arrows. (Original magnification: × 580, reduced to 65% of original size)

REFERENCES

1. COONS, A. H., E. H. LEDUC & J. CONNOLLY. 1955. Studies of antibody production. I. A method for the histochemical demonstration of specific antibody and its application to a study of the hyperimmune rabbit. J. Exp. Med. **102:** 49–60.
2. VAN NOORDEN, S. & J. M. POLAK. 1981. Advances in immunocytochemistry. *In* Gut Hormones, 2nd Ed. S. R. BLOOM & J. M. POLAK, Eds.: 80–89. Churchill Livingstone. Edinburgh.
3. STERNBERGER, L. A. 1979. The unlabeled antibody enzyme method. *In* Immunocytochemistry. L. A. STERNBERGER, Ed. Prentice-Hall, Inc. Englewood, NJ.
4. NAKANE, P. K. 1978. Simultaneous localization of multiple tissue antigens using the peroxidase-labeled antibody method: A study on pituitary glands of rats. J. Histochem. Cytochem. **16:** 557–560.
5. POLAK, J. M. A. G. E. PEARSE & C. M. HEATH. 1975. Complete identification of endocrine cells in the gastrointestinal tract using semithin–thin sections to identify motilin cells in human and animal intestine. Gut **16:** 225–229.
6. VARNDELL, I. M., F. J. TAPIA, J. DE MEY, R. A. RUSH, S. R. BLOOM & J. M. POLAK. 1982. Electronimmunocytochemical localization of enkephalin-like main phaeochromocytomas of man and other mammals. J. Histochem. Cytochem. **30:** 682–690.
7. BLACKBURN, A. M., S. R. BLOOM & J. M. POLAK. 1979. A radioimmunoassay for neurotensin in human plasma. J. Endocrinol. **83:** 175–181.
8. ROBERTS, G. W. 1982. Neuropeptides in the limbic system. Distribution, pathways, and possible functions. Ph.D. Thesis. London University.
9. UHL, G. R. & S. H. SNYDER. 1979. Neurotensin: A neuronal pathway projecting from amygdala through stria terminalis. Brain Res. **161:** 522–526.
10. ROBERTS, G. W., T. J. CROW & J. M. POLAK. 1981. Neurotensin: First report of a critical pathway. Peptides **2** (Suppl. I): 37–43.
11. GIBSON, S. J., J. M. POLAK, S. R. BLOOM & P. D. WALL. The distribution of nine peptides in rat spinal cord with especial emphasis on the substantia gelatinosa and on the area around the central canal (lamina X). 1981. J. Comp. Neurol. **201:** 65–79.
12. SEYBOLD, V. & R. ELDE. 1980. Immunohistochemical studies of peptidergic neurons in the dorsal horn of the spinal cord. 1980. J. Histochem. Cytochem. **28:** 367–370.
13. AINSWORTH, A., P. HALL, P. D. WALL, G. ALLT, L. MACKENZIE, S. GIBSON & J. M. POLAK. 1981. Effects of capsaicin applied locally to adult peripheral nerve. II Anatomy and enzyme and peptide chemistry of peripheral nerve and spinal cord. Pain **2:** 379–388.
14. HUNT, S. P., J. S. KELLY, P. C. EMSON, J. R. KIMMEL, R. J. MILLER & J.-Y. WU. 1981. An immunohistochemical study of neuronal populations containing neuropeptides and aminobutyrate within the superficial layers of the rat dorsal horn. Neuroscience **6:** 1883–1898.
15. ROKAEUS, A. 1981. Studies on neurotensin as a hormone. Acta Physiol. Scand. Suppl. **501.**
16. POLAK, J. M., S. M. SULLIVAN, S. R. BLOOM, A. M. J. BUCHAN, P. FACER, M. R. BROWN & A. G. E. PEARSE. 1977. Specific localization of neurotensin to the N cell in human intestine by radioimmunoassay and immunocytochemistry. Nature **270:** 183–184.
17. LANGER, M., S. VAN NOORDEN, J. M. POLAK & A. G. E. PEARSE. 1979. Peptide hormone-like immunoreactivity in the gastrointestinal tract and endocrine pancreas of eleven teleost species. Cell Tissue Res. **199:** 493–508.
18. SAFFREY, M. J., J. M. POLAK & G. BURNSTOCK. 1981. Distribution of VIP, substance P, enkephalin and neurotensin-like immunoreactive nerves in the chicken gut during development. Neuroscience **7** (No.1): 99–131.
19. ORCI, L., O. BAETENS, C. RUFENER, M. BROWN, W. VALE and F. GUILLEMIN. 1976. Evidence for immunoreactive neurotensin in dog intestinal mucosa. Life Sci. **19:** 559–561.
20. SUNDLER, F., R. HAKANSON, R. A. HAMMER, J. ALUMETS, R. CARRAWAY, S. E. LEEMAN & E. ZIMMERMAN. 1977. Immunohistochemical localization of neurotensin to endocrine cells in the gut. Cell Tissue Res. **178:** 313–321.

21. SCHULTZBERG, M., T. HÖKFELT, G. NILSSON, L. TERENIUS, J. F. REHFELD, M. BROWN, R. ELDE, M. GOLDSTEIN & S. SAID. 1980. Distribution of peptide and catecholamine-containing neurons in the gastrointestinal tract of rat and guinea pig: Immunohisto-chemical studies with antisera to substance P, vasoactive intestinal polypeptide, enkephalins, somatostatin, gastrin/cholecystokinin, neurotensin, and dopamine-hydroxylase. Neuroscience 5: 689–744.

22. FERRI, G-L, J. M. POLAK, T. E. ADRIAN, A. M. J. BUCHAN, L. PROBERT, M. A. GHATEI, Y. C. LEE & S. R. BLOOM. 1982. Layer separation of the human bowel for the study of regulatory peptides. Reg. Pep. 3: 70.

23. SOLCIA, E., J. M. POLAK, L. I. LARSSON, A. M. J. BUCHAN and C. CAPELLA. 1981. Update on Lausanne classification of endocrine cells. In Gut Hormones. S. R. BLOOM & J. M. POLAK, Eds.: 96–100. Churchill Livingstone. Edinburgh.

24. POLAK, J. M., A. M. J. BUCHAN, L. PROBERT, F. TAPIA, J. DE MEY & S. R. BLOOM. 1981. Regulatory peptides in endocrine cells and autonomic nerves. Electron immunocyto-chemistry. In Scandinavian Journal of Gastroenterology. J. M. POLAK, S. R. BLOOM, N. A. WRIGHT & M. J. DALY, Eds. Vol. 16 (Suppl. 70): 11–23. Universitetsforlaget. Oslo.

25. BUCHAN, A. M. J., M. G. BRYANT, J. M. POLAK, M. GREGOR, M. A. GHATEI & S. R. BLOOM. 1981. Development of regulatory peptides in the human fetal intestine. In Gut Hormones. S. R. BLOOM & J. M. POLAK, Eds.: 119–124. 2nd Edit. Churchill Livingstone. Edinburgh.

26. FERRI, G-L., A. HARRIS, N. A. WRIGHT, S. R. BLOOM & J. M. POLAK. 1982. Quantification of endocrine cells in whole intestinal crypts and villi. J. Histochem. 14: 692–695.

27. SUNDLER, F., R. E. CARRAWAY, R. HAKANSON, J. ALUMETS & M. P. DUBOIS. 1978. Immunoreactive neurotensin and somatostatin in the chicken thymus. Cell Tissue Res. 194: 367–376.

28. MARANGOS, P. J. & C. ZOMZELY-NEURATH. 1976. Determination and characterization of neuron-specific-protein (NSP) associated enolase activity. Biochem. Biophys. Res. Commun. 68: 1309–1316.

29. BISHOP, A. E., J. M. POLAK, P. FACER, G-L. FERRI, P. J. MARANGOS & A. G. E. PEARSE. 1982. Neuron-specific enolase: A common marker for the endocrine cells and innervation of the gut and pancreas. Gastroenterology. In press.

30. WHARTON, J., J. M. POLAK, G. A. COLE, P. J. MARANGOS & A. G. E. PEARSE. 1981. Neuron-specific enolase as an immunocytochemical marker for the diffuse neuro-endocrine system in human fetal lung. J. Histochem. Cytochem. Vol.29 (No.12): 1359–1364.

31. GU, J., J. M. POLAK, F. J. TAPIA, P. J. MARANGOS & A. G. E. PEARSE. 1981. Neuron-specific enolase in the Merkel cells of mammalian skin: The use of specific antibody as a simple and reliable histologic marker. Am. J. Pathol. 104: 63–68.

32. LUNDBERG, J. M. 1979. Enkephalin, substance P, VIP, somatostatin, gastrin/CCK, and neurotensin in peripheral neurons. Acta Physiol. Scand. 473: 14.

33. TERENGHI, G., I. M. VARNDELL, J. WHARTON, Y. C. LEE, S. R. BLOOM & J. M. POLAK. 1982. Neurotensin-like immunoreactivity in adrenal medulla. J. Pathol. 137: 82.

34. LEE, Y. C., G. TERENGHI, J. M. POLAK & S. R. BLOOM. 1981. Neurotensin in adrenal medulla of the cat. Acta Endocrinol. 97 (Suppl. 243): 84.

35. SCHULTZBERG, M., J. M. LUNDBERG, T. HÖKFELT, L. TERENIUS, J. BRANDT, R. P. ELDE & M. GOLDSTEIN. 1978. Enkephalin-like immunoreactivity in gland cells and nerve terminals of the adrenal medulla. Neuroscience 3: 1168–1186.

36. CHUBB, I.W., W. P. DE POTTER, & A. F. DE SCHAEPDRYVER. 1970. Evidence for two types of noradrenaline storage particles in dog spleen. Nature 228: 1203–1204.

37. BISBY, M. A. & M. FILLENZ. 1971. The storage of endogenous noradrenaline in sym-pathetic nerve terminals. J. Physiol. London 215: 163–179.

38. Tischler, A. S., Y. C. Lee, V. W. Slayton & S. R. Bloom. 1982. Content and release of neurotensin in PC-12 phaeochromocytoma cell cultures: modulation by dexamethasone and nerve growth factor. Regulatory Peptides. In press.
39. Polak, J. M. & S. R. Bloom. 1980. Peripheral localization of regulatory peptides as a clue to their functions. J. Histochem. Cytochem. **28**: 918–924.

Discussion of the Paper

M. Goedert (*Cambridge University, England*): I have a question concerning the adrenal medulla. It might be true for the cat that you find neurotensin in the adrenal medulla, but it is certainly not a general phenomenon. We have looked at guinea pig and rat adrenals with radioimmunoassay, and you certainly do not find any neurotensin in the adrenal of those species.

J. M. Polak (*Royal Postgraduate Medical School, London, England*): Both by RIA and immunocytochemistry we found NT in the cat adrenal.

LIGHT AND ELECTRON MICROSCOPIC LOCALIZATION OF NEUROTENSIN IN THE GASTROINTESTINAL TRACT*

F. Sundler, R. Håkanson, S. Leander, and R. Uddman

Departments of Histology and Pharmacology
University of Lund
Lund, Sweden

INTRODUCTION

Neurotensin is a member of the large and steadily growing family of neuro-hormonal peptides with a localization in both brain and gut.[1-4] Although first discovered in the hypothalamus, neurotensin in the gut represents the bulk of endogenous neurotensin; in the rat, intestinal neurotensin constitutes about 85% of the total body content.[5] Intestinal neurotensin from several mammals has been isolated and sequenced and found to be identical to hypothalamic neurotensin.[6] Within the gut, neurotensin seems to have a dual distribution—in endocrine cells and in neuronal elements. Although neurotensin-storing endocrine cells seem to occur in the gut of all vertebrates, neuronally localized neurotensin appears to be less widely distributed among species. However, as is the case with several other brain-gut peptides, neurotensin also occurs outside these organs. Thus, in the cat, neurotensin immunoreactive nerve fibers have been observed in the prevertebral sympathetic ganglia and in the splanchnic nerves.[7] In the chicken, the thymus has been found to harbor numerous neurotensin-immunoreactive, endocrine-like cells.[8]

DISTRIBUTION OF NEUROTENSIN CELLS IN THE GUT

In mammals the neurotensin-storing endocrine cells occur almost exclusively in the small intestine.[9-15] Generally, the cells are few in the duodenum, moderate in number in the jejunum, and fairly numerous in the ileum (FIG. 1). In certain mammals, including man, neurotensin cells can be seen occasionally also in the large intestine[15] (FIG. 1). In non-mammalian vertebrates, gut neurotensin cells have a more wide-spread distribution.[12, 14, 16, 17] Thus, in birds, neurotensin cells are very numerous in the antrum (the narrow zone joining the gizzard with the duodenum) (FIG. 2), fairly numerous all along the small intestine, and moderate in number in the large intestine.[12, 14, 17]

In the chicken antrum, neurotensin cells vary in the intensity of their immunostaining from very strong to fairly weak[12, 17] (FIG. 2a). Sequential staining for neurotensin and gastrin has revealed that the cells displaying weak-to-moderate neurotensin immunoreactivity exhibit intense gastrin immunoreactivity and vice versa[17] (FIG. 2).

The topographical distribution of neurotensin cells differs somewhat from one

* Grant support was received from the Swedish Medical Research Council (grants no. 4499 and 1007) and A. Påhlsson's Foundation.

94

species to another. In most mammalian species, the cells predominate on the villi. In a few mammals, notably cat and dog, and in lower vertebrates, they are rather uniformly distributed between crypts and villi [12, 14, 16] (FIG. 1).

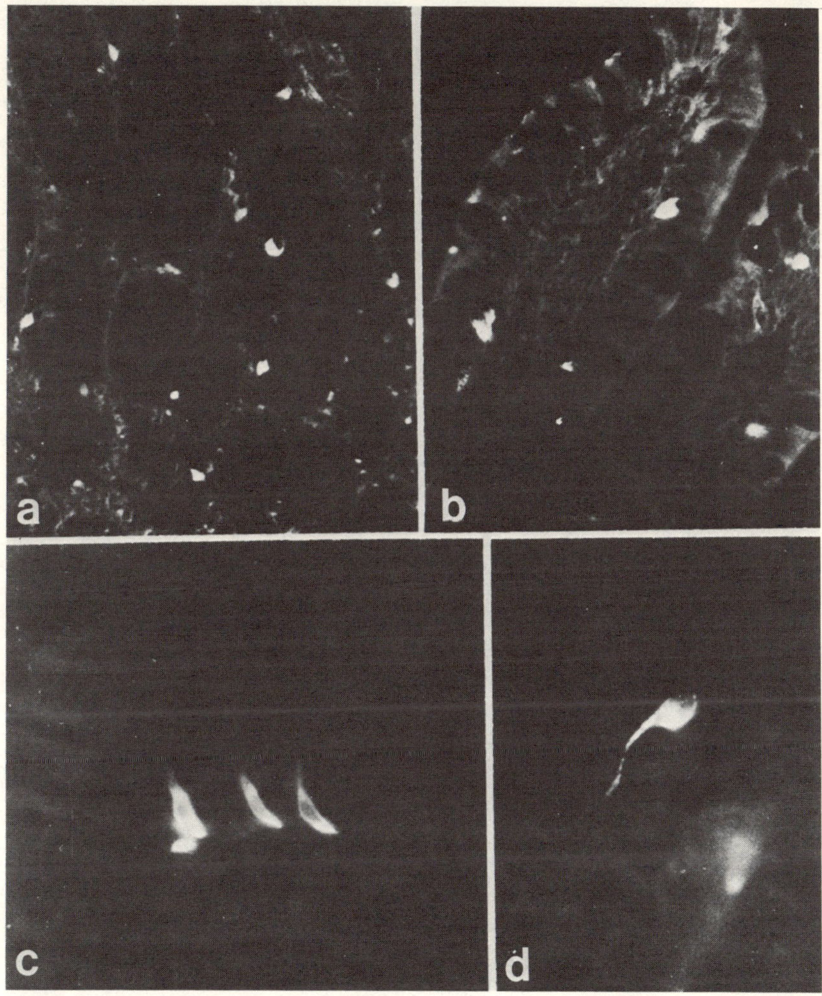

FIGURE 1.   Examples of the varying distribution pattern of intestinal neurotensin cells in different species. (a.) Dog ileum. Neurotensin cells scattered in crypt epithelium (Original magnification: × 200; reduced to 90% of original size); (b.) Human ileum. Neurotensin cells on villi (× 200; reduced as in (a)); (c.) Chicken duodenum. Neurotensin cells at the base of villus (× 250; reduced as in (a)); and (d.) Mouse colon. Single neurotensin cell issuing slender process running close to the base of neighboring cells (× 300; reduced as in (a)).

NEUROTENSIN CELL ULTRASTRUCTURE

The ultrastructure of the neurotensin cell has been studied in a number of species.[10, 18–20] The cells are of the open type, reaching the lumen via an apical

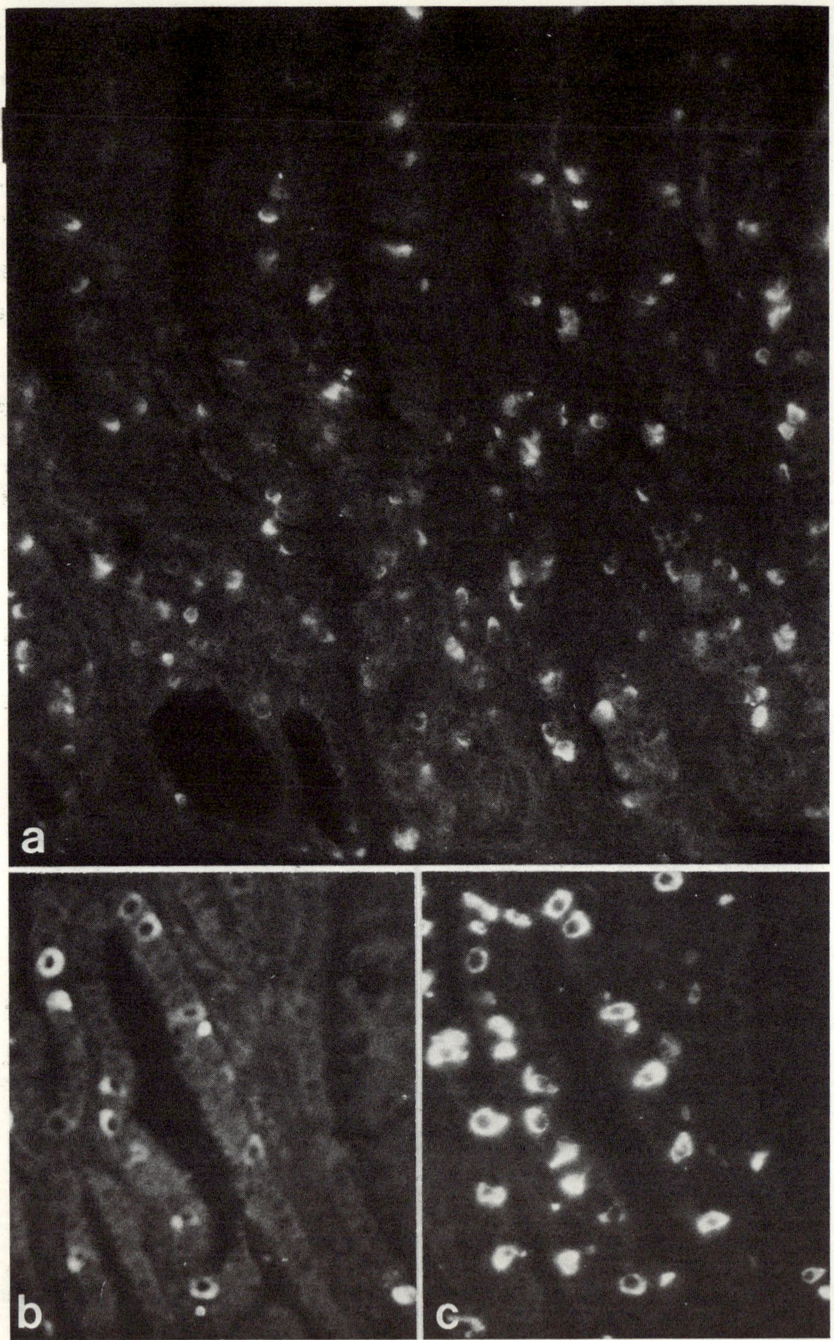

FIGURE 2.   Chicken antrum (gizzard-duodenal junction). (a.) Section covering the whole mucosa showing numerous neurotensin cells in all parts of the glands. Note that the cells display widely varying immunofluorescence intensity (magnification: × 150). (b. and c. Consecutive semithin sections stained for neurotensin (b) and gastrin (c). The neurotensin cells are identical with a subpopulation of gastrin immunoreactive cells (× 300).

process endowed with microvilli. The basal portion of the cell contains numerous secretory granules representing the storage site of neurotensin (Figs. 3 and 4). The granules are fairly large compared to those of most other gut endocrine cells (diameter around 300 nm), round, and of uniformly high electron density. The neurotensin granules are very similar to those of the glicentin cells, a fact that may explain why the neurotensin cells previously could not be distinguished from glicentin cells on pure ultrastructural grounds.[19, 20]

ONTOGENY OF NEUROTENSIN CELLS

The fetal and neonatal development of gut neurotensin cells has been studied in chicken,[12] rat,[12] pig,[21] and man.[22] Generally, neurotensin cells develop later than many other endocrine cell types in the gut. In the chicken, they are first seen in the colon in the 18-day-old embryo and in the antrum and small intestine two days later. The cells reach the adult distribution and frequency of occurrence a few days after hatching. In the rat, neurotensin cells appear in the small intestine immediately before birth reaching the adult distribution and frequency 4–5 days later. In the fetal pig, neurotensin cells are first seen in the distal small intestine at 8 weeks of gestation. They gradually increase in number to reach the adult distribution and frequency of occurrence at birth. In man, neurotensin cells first appear in the distal small intestine at a fetal age of 12 to 13 weeks and display the adult distribution pattern at about 20 weeks of gestation.

NEUROTENSIN IN GASTRO-ENTERO-PANCREATIC (GEP) ENDOCRINE TUMORS

Only few reports are available on the presence of neurotensin in GEP endocrine tumors.[23–25] From these reports it appears that, although neurotensin normally is confined to the gut, only tumors arising in the pancreas are capable of producing neurotensin. We have previously described one pancreatic gastrinoma[23] and one glucagonoma[24] containing neurotensin cells as a minority population. Recently, Blackburn *et al.*[25] reported on the occurrence of immunoreactive neurotensin in 6 out of 21 pancreatic VIPomas.

NEURONAL NEUROTENSIN

Nerve fibers displaying neurotensin immunoreactivity have been observed in low numbers in the intestinal wall of the rat[26] and guinea pig[27] and in the stomach wall of the cat.[7] The nerve fibers seem to be confined to the smooth muscle and the myenteric plexus (Fig. 5a). In the chicken, neurotensin fibers are numerous in the wall of the esophagus, crop, proventriculus, and gizzard[28] (Fig. 5b). In certain species of bony fish, neurotensin fibers are very numerous in the myenteric plexus of the intestine[29] (Fig. 5c).

Neurotensin-containing nerve cell bodies have been observed in the chicken gut,[28] suggesting that gut neurotensin fibers are intrinsic in origin. The localization of neurotensin fibers to gut smooth muscle is interesting in view of the potent actions of neurotensin on gut smooth muscle activity. Thus, neurotensin contracts the isolated guinea pig taenia coli, an effect that seems to be due to a direct action of neurotensin on smooth muscle receptors.[27] Other workers have found neurotensin to stimulate smooth muscle activity in isolated segments of the guinea pig

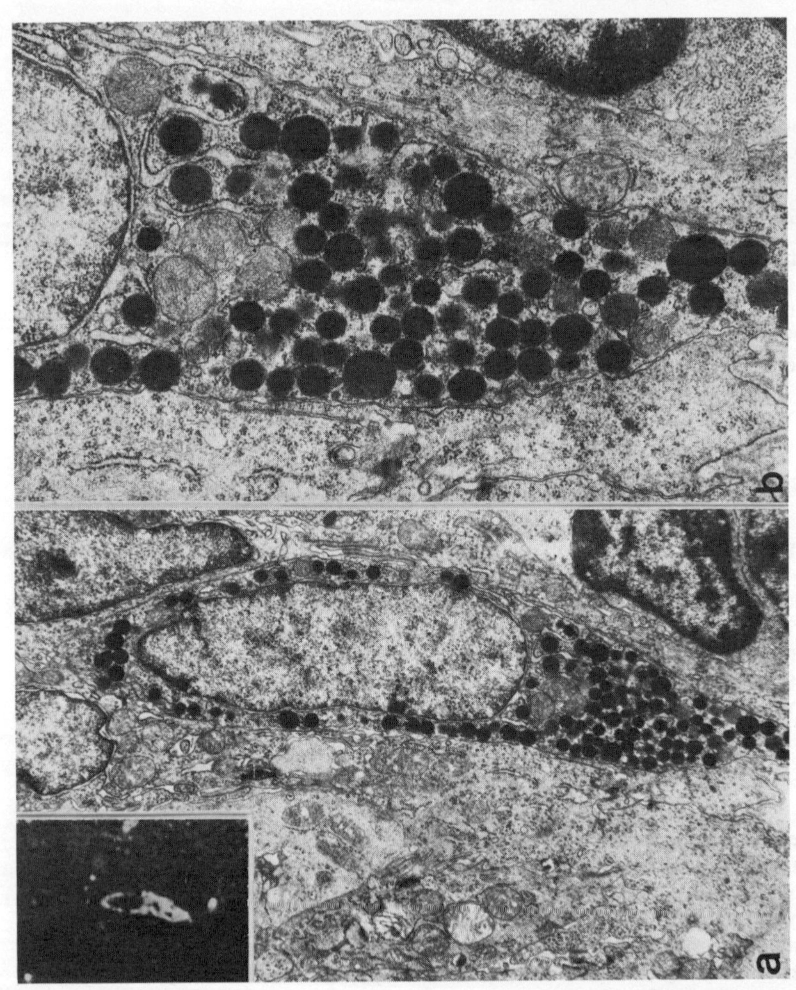

FIGURE 3.   Electron micrographs of neurotensin cell in dog ileum identified by immunocytochemistry performed on the adjacent semithin section (inset). b is a higher magnification of a. The secretory granules are fairly large and highly electron dense. Original magnifications: Inset: × 600; a × 7 500; b × 19 000; reduced to 85% of original size.

ileum through a nerve-mediated, partly cholinergic process.[30] Other preparations such as the rat ileum and guinea pig colon respond to neurotensin by relaxation.[31]

PHYLOGENETIC ASPECTS

Like many other neurohormonal peptides common to the brain and gut, neurotensin (or neurotensin-like peptides) occurs in a wide range of species representing

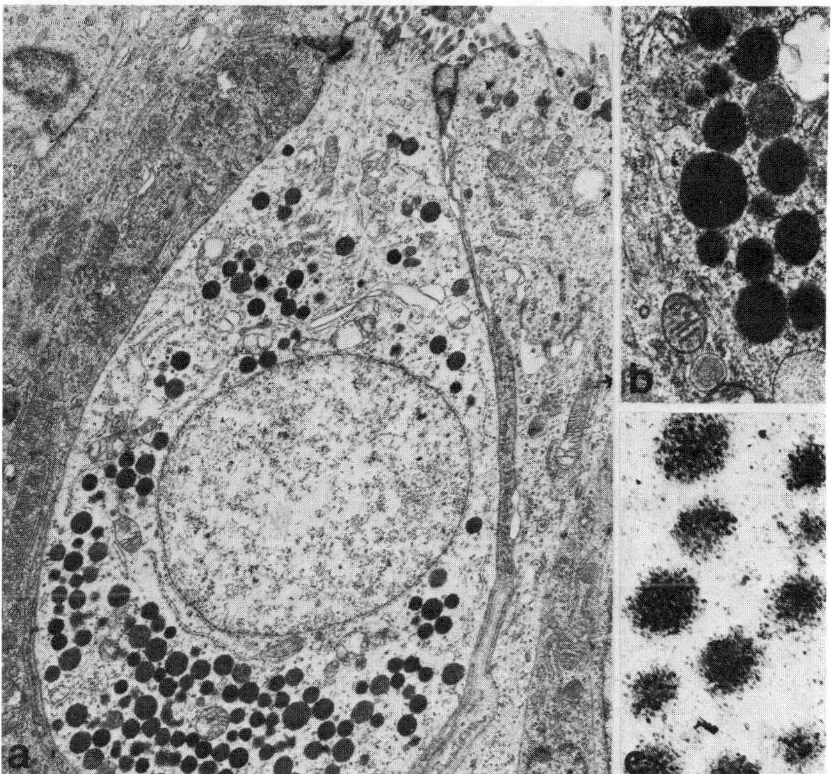

FIGURE 4. Electron micrographs of neurotensin cell in chicken antrum. (a.) The neurotensin cell reaches the lumen via an apical process furnished with microvilli (Original magnification: 8 000; reduced to 85% of original size). (b.) Detail of neurotensin cell cytoplasm showing large highly electron dense granules (× 22 000; reduced as in (a)). (c.) Immunoperoxidase (PAP) staining of ultrathin section. Reaction product is accumulated over the secretory granules (× 27 000; reduced as in (a)).

endocrine cells in the gut have been traced down to the level of deuterostomian invertebrates.[16] Further, there are observations indicating the presence of neurotensin-like peptides in neuronal elements in phylogenetically very old species such as the hydra (*Hydra attenuata*) (FIG. 6), representing one of the most primitive multicellular organisms.[32]

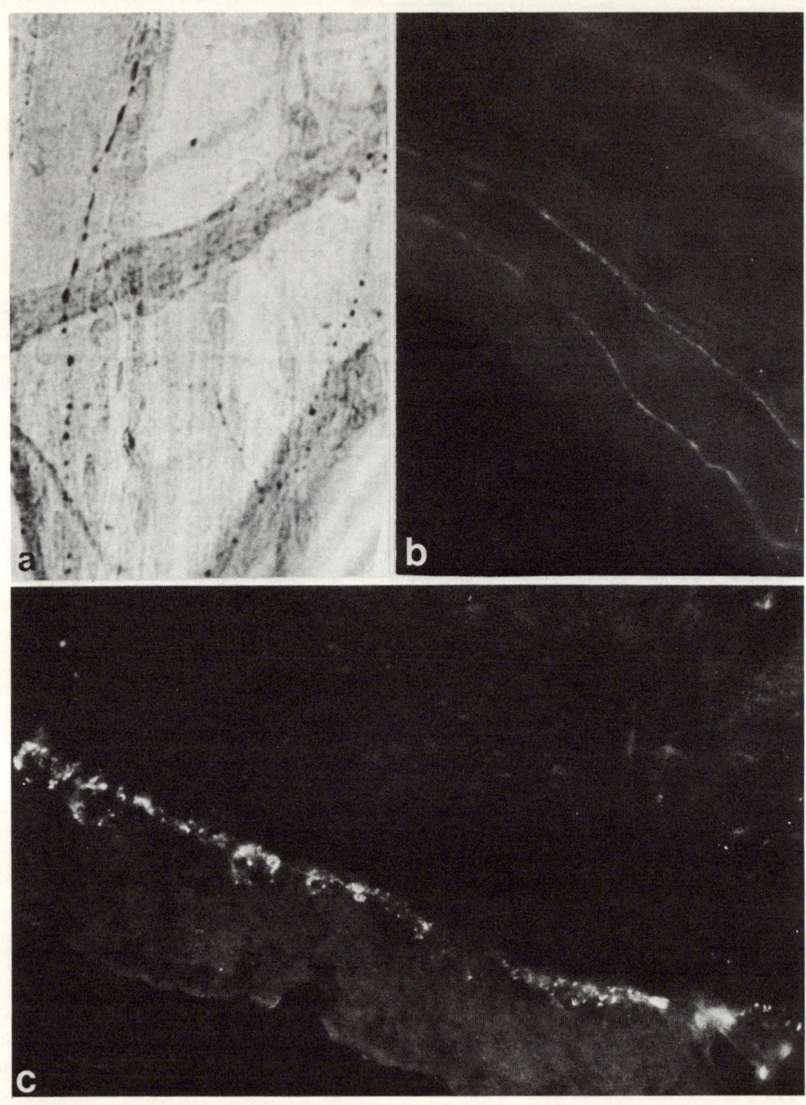

FIGURE 5. Neurotensin immunoreactive nerve fibers in smooth-muscle whole-mounts of guinea pig ileum (a); chicken crop (b) and in section of the smooth-muscle layer of bony fish (*Cottus scorpius*) (c). Neurotensin fibers are particularly numerous in the myenteric plexus of the fish intestine. Original magnifications: a and b × 300, c × 200; reduced to 85% of original size.

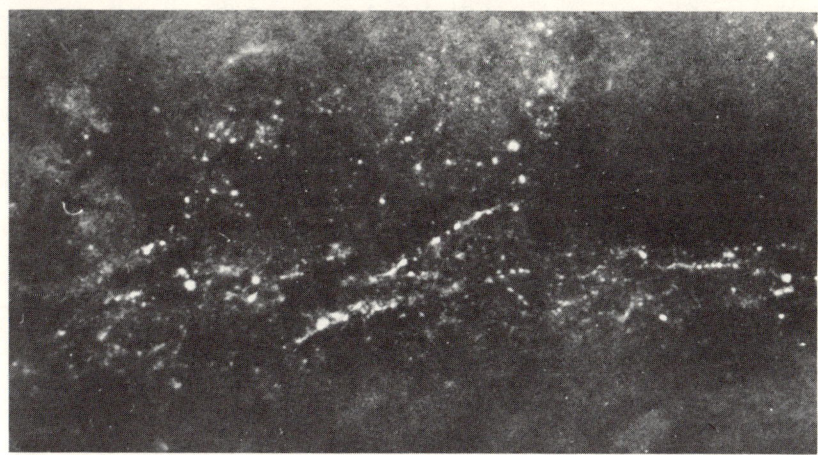

FIGURE 6. Tentacle of hydra. Whole-mount preparation showing network of delicate, beaded neurotensin immunoreactive nerve fibers. Original magnification: × 200; reduced to 90% of original size.

REFERENCES

1. HÅKANSON, R., S. LEANDER, F. SUNDLER & R. UDDMAN. 1981. P-type nerves: Purinergic or peptidergic? *In* Cellular Basis of Chemical Messengers in the Digestive System. M. I. GROSSMAN, M. BRAZIER and J. LECHAGO, Eds. Academic Press. New York, pp. 169–199.
2. HÖKFELT, T. O. JOHANSSON, Å. LJUNGDAHL, J. M. LUNDBERG & M. SCHULTZBERG. 1980. Peptidergic neurones. Nature (London) **284:** 515–521.
3. SAID, S. I. 1980. Peptides common to the nervous system and the gastrointestinal tract. *In* Frontiers in Neuroendocrinology, Vol. 6. L. Martini and W. F. Ganong, Eds. Raven Press. New York. pp. 293–331.
4. SUNDLER, F., R. HÅKANSON & S. LEANDER. 1980. Peptidergic nervous systems in the gut. Clin. Gastroenterol. **9:** 517–543.
5. CARRAWAY, R. & S. E. LEEMAN. 1976. Characterization of radioimmunoassayable neurotensin in the rat. Its differential distribution in the central nervous system, small intestine, and stomach. J. Biol. Chem. **251:** 7045–7052.
6. HAMMER, R. A., S. E. LEEMAN, R. CARRAWAY & R. H. WILLIAMS. 1980. Isolation of human intestinal neurotensin. J. Biol. Chem. **255:** 2476–2480.
7. LUNDBERG. J. M., T. HÖKFELT, A. ÄNGGÅRD, K. UVNÄS-WALLENSTEN, S. BRIMIJOIN. E. BRODIN & J. FAHRENKRUG. 1980. Peripheral peptide neurones: Distribution, axonal transport, and some aspects on possible function. *In* Neural Peptides and Neuronal Communication. E. COSTA & M. TRABUCCHI, Eds. Raven press. New York, pp. 25–36.
8. SUNDLER, F., R. E. CARRAWAY, R. HÅKANSON, J. ALUMETS & M. P. DUBOIS. 1978. Immunoreactive neurotensin and somatostatin in the chicken thymus. A chemical and histochemical study. Cell Tissue Res. **194:** 367–376.
9. ORCI, L., O. BAETENS & C. RUFENER. 1976. Evidence for immunoreactive neurotensin in dog intestinal mucosa. Life Sci. **19:** 559–562.
10. FRIGERIO, B., M. RAVAZZOLA, S. ITO, R. BUFFA, C. CAPELLA, E. SOLICIA & L. ORCI. 1977. Histochemical and ultrastructural identification of neurotensin cells in the dog ileum. Histochemistry **54:** 123–131.
11. HELMSTAEDTER, V., CH. TAUGNER, G. E. FEURLE & W. G. FORSSMANN. 1977. Localization of neurotensin-immunoreactive cells in the small intestine of man and various mammals. Histochemistry **53:** 35–41.

12. SUNDLER, F., R. HÅKANSON, R. A. HAMMER, J. ALUMETS, R. CARRAWAY, S. E. LEEMAN & E. A. ZIMMERMAN. 1977. Immunohistochemical localization of neurotensin in endocrine cells of the gut. Cell Tissue Res. **178:** 313–321.

13. ITO, S., Y. YAMADA, M. HAYASHI, Y. IWASAKI, Y. MATSUBARA & A. SHIBATA. 1979. Neurotensin-positive and somatostatin-positive cells in the canine gut. Tohoku J. Exp. Med. **127:** 123–131.

14. REINECKE, M., K. ALMASAN, R. CARRAWAY, V. HELMSTAEDTER & W. G. FORSS-MANN. 1980. Distribution patterns of neurotensin-like immunoreactive cells in the gastrointestinal tract of higher vertebrates. Cell Tissue Res. **205:** 383–395.

15. SJÖLUND, K., G. SANDÉN, R. HÅKANSON & F. SUNDLER. 1982. Endocrine cells in human intestine. An immunocytochemical study. Gastroenterology. Submitted.

16. REINECKE, M., R. E. CARRAWAY, S. FALKMER, G. E. FEURLE & W. G. FORSS-MANN. 1980. Occurrence of neurotensin-immunoreactive cells in the digestive tract of lower vertebrates and deuterostomian invertebrates. A correlated immunohisto-chemical and radioimmunochemical study. Cell Tissue Res. **212:** 173–183.

17. RAWDON, B. B. & A. ANDREW. 1981. An immunocytochemical survey of endocrine cells in the gastrointestinal tract of chicks at hatching. Cell Tissue Res. **220:** 279–292.

18. HELMSTAEDTER, V., G. E. FEURLE & W. G. FORSSMANN. 1977. Ultrastructural identifi-cation of a new cell type—the N cell as the source of neurotensin in the gut mucosa. Cell Tissue Res. **184:** 445–452.

19. POLAK, J. M., S. N. SULLIVAN, S. R. BLOOM, A. M. J. BUCHAN, P. FACER, M. R. BROWN and A. G. E. PEARSE. 1977. Specific localization of neurotensin to the N cell in human intestine by radioimmunoassay and immunocytochemistry. Nature **270:** 183–184.

20. SUNDLER, F., J. ALUMETS, R. HÅKANSON, R. CARRAWAY & S. E. LEEMAN. 1977. Ultrastructure of the gut neurotensin cell. Histochemistry **53:** 25–34.

21. ALUMETS, J., R. HÅKANSON and F. SUNDLER. 1982. Ontogeny of endocrine cells in porcine gut and pancreas. An immunocytochemical study. Gastroenterology. Submitted.

22. HELMSTAEDTER, V., G. MÜHLMANN, G. E. FEURLE & W. G. FORSSMANN. 1977. Immunohistochemical identification of gastrointestinal neurotensin cells in human embryos. Cell Tissue Res. **184:** 315–320.

23. SUNDLER, F., J. ALUMETS & R. HÅKANSON. 1979. Majority and minority cell popu-lations in GEP and bronchial endocrine tumours. Scand. J. Gastroenterol. **14:** (Suppl. 53): 9–13.

24. ALUMETS, J., F. SUNDLER, S. FALKMER, R. HÅKANSON, O. LJUNGBERG, H. MÅRTENSSON & A. NOBIN. 1982. Neurohormonal peptides in endocrine tumors of the pancreas, stomach, and upper small intestine. An immunohistochemical study of 27 cases. J. Ultrastruct. Pathol. Submitted.

25. BLACKBURN, A. M., M. G. BRYANT, T. E. ADRIAN & S. R. BLOOM. 1981. Pancreatic tumors produce neurotensin. J. Clin. Endocrinol. Metab. **52:** 820–822.

26. SCHULTZBERG, M., T. HÖKFELT, G. NILSSON, L. TERENIUS, J. F. REHFELD, M. BROWN, R. ELDE, M. GOLDSTEIN & S. SAID. 1980. Distribution of peptide- and catechol-amine-containing neurons in the gastrointestinal tract of rat and guinea pig: Immuno-histochemical studies with antisera to substance P, vasoactive intestinal polypeptide, enkephalin, somatostatin, gastrin/cholecystokinin, and dopamine β-hydroxylase. Neuroscience **5:** 689–744.

27. LEANDER, S., R. EKMAN, R. HÅKANSON, F. SUNDLER & R. UDDMAN. 1982. Nerve fibers containing immunoreactive cholecystokinin, gastrin-releasing peptide, neurotensin or β-endorphin in guinea pig intestine. Distribution and possible functions. Neuro-science. Submitted.

28. SAFFREY, M. J., J. M. POLAK & G. BURNSTOCK. 1982. Distribution of vasoactive intestinal polypeptide-, substance P-, enkephalin- and neurotensin-like immuno-reactive nerves in the chicken gut during development. Neuroscience 7: 279–293.

29. LANGER, M., S. VAN NOORDEN, J. M. POLAK & A. G. E. PEARSE. 1979. Peptide hormone-like immunoreactivity in the gastrointestinal tract and endocrine pancreas of eleven teleost species. Cell Tissue Res. **199:** 493–508.

30. KITABGI, P. & P. FREYCHET. 1979. Neurotensin contracts the guinea pig longitudinal ileal smooth muscle by inducing acetylcholine release. Eur. J. Pharmacol. **56:** 403–406.
31. KITABGI, P. & P. FREYCHET. 1978. Effects of neurotensin on isolated intestinal smooth muscles. Eur. J. Pharmacol. **50:** 349–357.
32. GRIMMELIKHUIJZEN, C. P. P., R. E. CARRAWAY, Å. RÖKAEUS & F. SUNDLER. 1981. Neurotensin-like immunoreactivity in the nervous system of hydra. Histochemistry **72:** 199–209.

DISCUSSION OF THE PAPER

R. A. HAMMER (*Northwestern University Medical School, Chicago IL*): In several of the immunofluorescent pictures that both you and Dr. Polak showed there were some fibers that seemed to be extending from the base of the neurotensin-immunoreactive cells in the gastrointestinal tract. I wonder if you would comment on whether you think they are neuronal.

F. SUNDLER (*University of Lund, Sweden*): I think that any fibers you could see on those pictures showing endocrine cells were artifacts. It is very difficult to see any neuronal elements in the mucosa anywhere.

A. S. TISCHLER (*Tufts University Medical School, Boston, MA*): I would like to know whether any consensus exists between Drs. Sundler and Polak with regard to optimal methods for tissue fixation and embedding for demonstration of neurotensin staining. I would also like to add that our experience with human tumors has been that in the course of routine paraffin embedding following formaldehyde fixation most of the neurotensin-like immunoreactivity in human tissues appears to be lost.

SUNDLER: One explanation may be that when one uses paraffin techniques some neurotensin is lost because you have to deparaffinize and use organic solvents.

With the formaldehyde cryostat technique where you either perfuse the animal or soak the tissue in cold buffered formaldehyde, followed by rinsing in buffer and then cryostat sectioning, there is a better preservation of immunoreactive NT. Also the fixation technique devised by Dr. Polak using parabenzoequinone seems satisfactory.

J. M. POLAK (*Royal Postgraduate Medical School, London, England*): I do not think fixation is the only factor. There are other factors which include an antibody suitable for immunocytochemistry, a very good second layer, as well as preservation before you fix, timing of fixation and what kind of fixative. In general, I am in agreement with Dr. Sundler. If you use a good cross linking agent like benzoequinone, you get better results.

M. BERELOWITZ (*University of Cincinnati Medical School, OH*): In collaboration with Jesse Roth's group and Derek Leroith at NIH, we have recently examined extracts of a unicellular ciliate for the presence of somatostatin and neurotensin. In extracts that have been purified on a CIS column, we have demonstrated neurotensin-like immunoreactivity with a COOH-terminal-directed antiserum. On HPLC, we have shown that the neurotensin-like immunoreactivity corresponds to the elution of synthetic neurotensin.

S. ROSELL (*Karolinska Institute, Stockholm, Sweden*): Dr. Sundler, will you comment on the fact that in the dog, the NT-containing cells are in the crypts and in other species on the villi. In our experience in the dog the rise in neurotensin immunoreactivity in blood after fat ingestion is very unpredictable. Could that be explained because of this difference?

SUNDLER: Perhaps it is dependent on the food habits of the animals. It could be because the same distribution is found in the cat and both the cat and dog are carnivores.

P. KITABGI (*INSERM, Nice, France*): Many think that neurotensin might have some paracrine action in the gut. Do you think that it is possible that neurotensin released from endocrine cells (or paracrine cells) diffuses to the muscle layers and acts there?

SUNDLER: The same issues pertaining to the action of gastrin on the lower esophageal sphincter have been discussed. The sphincter contains no gastrin but this peptide may still regulate sphincter pressure. That means that hormonally released gastrin would reach the smooth muscle and have effects there. Thus hormonally released neurotensin may have an effect on smooth muscle, but this is at present a matter of speculation.

R. QUIRION (*NIMH Bethesda, MD*): Has anyone ever looked for neurotensin or other peptides in the kidney?

SUNDLER: Well, we have looked for VIP and substance P, but there is not much of these peptides in the kidney or in the liver. These parenchymatous organs seem to be rather free of neuropeptides.

I. M. MODLIN (*SUNY Downstate Medical Center Brooklyn, NY*): Would both speakers clarify something that I may have misunderstood. Does the same cell contain gastrin, neurotensin, and neurospecific enolase?

SUNDLER: Immunoreactive gastrin and neurotensin occur in the same cells in the chicken antrum. In the chicken intestine neurotensin cells do not seem to contain immunoreactive gastrin.

MODLIN: Do you have any experience with the neurospecific-enolase in the antrum of the chicken?

SUNDLER: I have no experience myself of this enzyme.

POLAK: I can not comment on gastrin, but the coexistence of glycolytic enzymes and regulatory peptides is a generalized phenomenon. Neurospecific enolase is present in all neuroendocrine cells—in the lung, gut and pancreas. We have recently been able to characterize it in every single cell type and there are none that do not have it.

ASPECTS OF NEUROTENSIN PHYSIOLOGY
AND PATHOLOGY

Stephen R. Bloom and Julia M. Polak

Department of Medicine and Histopathology
Royal Postgraduate Medical School
Hammersmith Hospital
London W12 OHS, England

INTRODUCTION

Development of a radioimmunoassay for neurotensin has allowed us to gather information on its pattern of release into the circulation and disturbances produced by gastrointestinal pathology. Secondarily, it enables us to have these concentrations mimicked by exogenous infusion, thereby giving a clue as to their possible biologic consequences. Such studies depend critically on the validity of the radioimmunoassay (see Carraway, this publication). It has been shown that neurotensin can degrade, particularly between the eighth and ninth amino acids, to form NH_2-terminal and COOH-terminal fragments. Further, it is possible that additional neurotensin-like peptides are synthesized and may be part of a quite different biologic system. Thus the stomach has been shown to contain considerable quantities of a COOH-terminal, neurotensin-like fragment (see Leeman and also Carraway, this publication) that may well differ in its NH_2-terminal extension to the sequence of neurotensin itself. We have tried to take as our scientific brief the investigation of the product of those particular endocrine cells found in the mucosa of the ileum (see also chapter by Polak & Bloom, this publication) and the release of the 13-amino-acid peptide originally isolated and named neurotensin by Susan Leeman. This peptide has been synthesized by several different companies and is currently available in large quantities at relatively low prices. Its study is thus quite easy. A considerable number of rabbits have been immunized using the bisdiazotized benzidine technique with bovine serum albumin as a carrier, a coupling ratio of 4 neurotensins to 1 albumin being employed. After about half a year, a number of animals (more than half) produced antisera and the most avid of these was selected for all the subsequent studies. Using a modified lactoperoxidase-iodination technique[1] an assay was developed capable of detecting changes of 4 pmol/l plasma (95% confidence limit) without extraction. The antisera showed no crossreaction with neurotensin fragments 1–8 and 1–9, but did react with fragments 7–13 with approximately 80% of the strength of whole neurotensin. Fragment 8–13, however, only showed a 10% crossreaction. Extracts of stomach showed no neurotensin immunoreactivity with this assay system (< 0.4 pmol/g). When fasting plasma was chromatogrammed, the majority of the immunoreactive neurotensin ($91 \pm 5\%$) eluted in the void volume, while $8.7 \pm 5\%$ eluted from the G-50 column in the position of pure neurotensin.[1] After meal stimulation, however, the entire rise of neurotensin-like immunoreactivity was accounted for by the material eluting in the position of pure neurotensin itself. The nature of the big-molecular-weight neurotensin has not been elucidated, but is is noteworthy that this phenomenon occurs in a number of different radioimmunoassays. With the development of more sensitive antibodies, the material has tended to disappear and has been presumed, in the majority of cases, to be due to nonspecific, large-molecular-

105

weight protein interference. Theoretically it could also be due to the presence of a precursor large-molecular-weight form of neurotensin or, possibly, adherence of immunoreactive fragments of neurotensin to proteins in the circulation. Since the former explanation is the more likely, basal plasma is placed in the standard curve of every assay. Although this shifts the standard curve very slightly, it remains parallel to that seen in the same assays set up in buffer alone.

NEWBORN INFANTS

Neurotensin first appears in the human intestine at the beginning of the second trimester and by the end of the second trimester, it has achieved adult distribution patterns.[2] By birth, the neurotensin concentrations in the ileum are similar to those in the adult.[3] A series of 276 healthy term or preterm human infants was studied, and it was found that at birth in the preterm infant, the cord neurotensin concentration was 21 ± 1 pmol/l, closely similar to the value found in the fasting healthy young adult of 16 ± 4 pmol/l. In a series of infants who, for various reasons, had to be fed intravenously, the neurotensin concentration at 6 days after birth was 21 ± 5 pmol/l, whereas infants that had been fed normally had a value of 66 ± 13 pmol/l at this time.[4] In another series of term infants, the postprandial pattern of neurotensin release was followed through to 24 days after birth. It was found that at the 6th day of life very little postprandial rise was seen, neurotensin remaining high throughout the day. By the 24th day, a distinct postprandial rise and fall had developed with the mean peak postprandial level being approximately two times higher than the mean fasting concentration.[4] The nature of the meal was also studied. In a study of 77 6-day-old, healthy term infants, bottle feeding was compared with breast feeding.[5] The breast-fed infants had a significantly lower neurotensin concentration and also complete absence of postprandial rise. The significance of these differences is at present uncertain but it is of considerable interest that breast and bottle feeding should result in such a distinctly different pattern of digestive response.

ADULT RESPONSES

Plasma neurotensin rises after eating and some response is seen to all meal constituents, though fat is undoubtedly the most potent. Bigger meals produce bigger rises of neurotensin (FIG. 1). The very rapid rise is unexpected for a hormone that is mainly of ileal origin and may indicate either that a small portion of the meal is rapidly moved along the intestine or that a neural pathway is involved. Recent studies with the isolated, perfused rat ileum have demonstrated that carbachol is a potent releasing agent, rivaling the release seen after an intraluminal fatty acid infusion.[6] Neurotensin is also released by an infusion of bombesin in man[7-10] (FIG. 2). While the rise of neurotensin did not reach statistical significance with an infusion of bombesin at 0.5 pmol/kg/min, a shallow dose-response curve was seen above that infusion level. Unfortunately the response is not straightforward, as bombesin also releases a number of other regulatory peptides. In the case of both bombesin and carbachol, concomitant release of somatostatin may explain the failure to observe a classical dose-response curve. Thus the interplay on local regulatory factors within the wall of the ileum may have a very significant role in modulating neurotensin release.

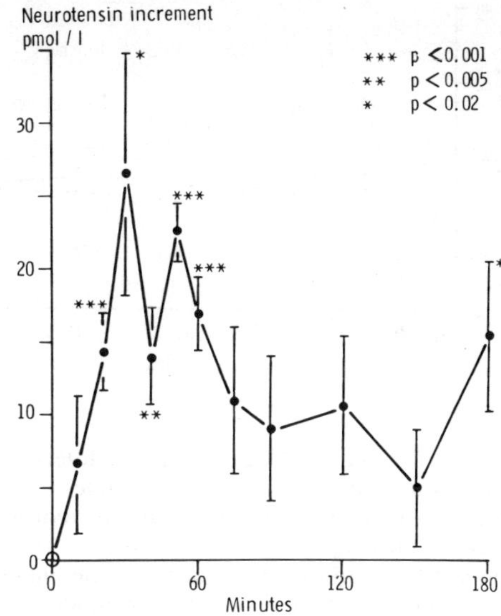

FIGURE 1.   The rise in plasma neurotensin seen following ingestion of a 3000 + calorie evening meal in 9 healthy volunteers. The zero point for each subject was calculated from the mean of 3 basal samples. * = p < 0.02, ** = p < 0.005, *** = p < 0.001 using Student's paired t-test.

PATHOLOGY

Effect of Intestinal Disorders

Neurotensin cells are found in the mucosa of the lower small intestine, and therefore it might be anticipated that any process that led to a greater nutriment load reaching the lower small intestine would increase the postprandial concentration of plasma neurotensin, and this is indeed the case. Thus in a group of 11 patients with active celiac disease (gluten enteropathy), a condition that affects only the mucosa of the upper small intestine, the postprandial neurotensin release was considerably enhanced by comparison with an age- and sex-matched control group of healthy subjects.[11] Interestingly, when a group of patients with celiac disease who had been placed on a gluten-free diet and achieved remission were studied, these subjects had a similar neurotensin increment to the control group. In a series of 59 subjects before or after surgery for duodenal ulcer, the neurotensin

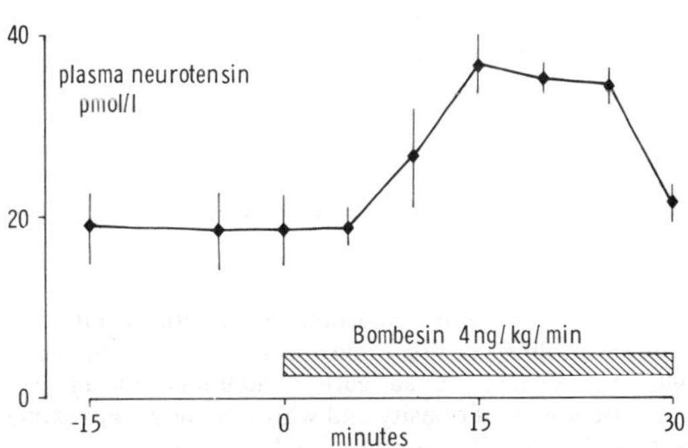

FIGURE 2.   The rise of plasma neurotensin immunoreactivity in 6 healthy fasting volunteers during an infusion of bombesin.

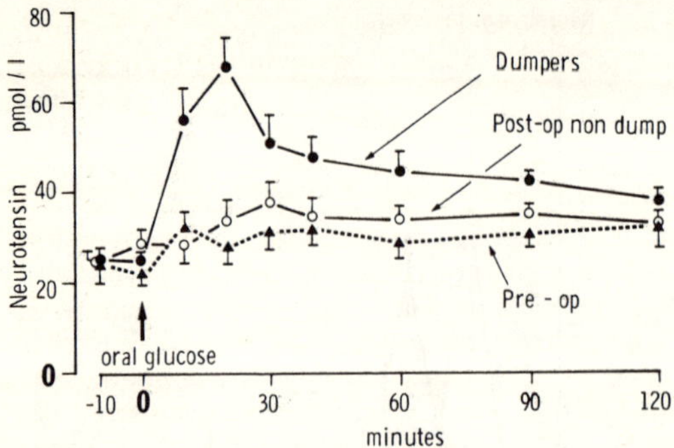

FIGURE 3. Plasma neurotensin immunoreactivity following a 75 g oral glucose load in 20 subjects with a duodenal ulcer awaiting operation, 20 subjects who had received gastrointestinal surgery for duodenal ulcer but who did not develop the dumping syndrome and in 19 subjects who after gastric surgery developed the symptoms of dumping.

release following a 75-g oral glucose tolerance test was studied. This stimulus was chosen not because of the effectiveness of glucose in releasing neurotensin, but because it usually reproduced in the laboratory setting the symptoms of dumping certain individuals suffered in their everyday life. Thus 19 of the subjects were chosen for study because of their complaints of dumping after various forms of upper gastrointestinal surgery for duodenal ulcer. A further postoperative group was carefully matched for type of operation and time postoperatively, but were without symptoms of dumping and these two groups were compared with a preoperative duodenal ulcer group. The post-glucose-peak incremental rise of neurotensin was 43 ± 6 pmol/1 in the dumping group but only 8 ± 5 pmol/1 in the postoperative nondumpers and 4.1 ± 3.5 in the preoperative subjects (FIG. 3).[12] In a further and quite different series of patients studied after upper gastric surgery, a similar, considerably enhanced, postprandial rise of neurotensin was observed.[13]

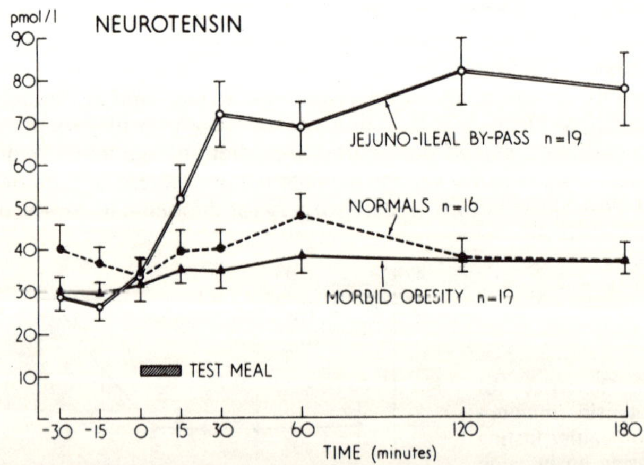

FIGURE 4. Plasma neurotensin immunoreactivity in 19 subjects with morbid obesity (more than 200% of ideal body weight), 16 healthy normal-weight controls, and 19 subjects who approximately a year earlier had undergone a jejunoileal bypass (7″ jejunum to 7″ ileum) in the treatment of obesity and who were now approximately 180% of ideal body weight. The test meal was a 530-calorie mixed breakfast.

A rather similar situation, where surgery leads to food arriving at the ileum in larger quantities than normal, occurs after jejunoileal bypass in the treatment of gross obesity. Following this procedure, the postprandial increment of neurotensin is increased by nearly tenfold (FIG. 4).[14] This situation is of interest as it throws light on the possible role of neurotensin in various postprandial processes. Thus when neurotensin was found to be elevated in the dumping patients, speculation arose on the basis of the pharmacology of neurotensin (producing hypotension, tachycardia, and cyanosis) that it might play a role in the dumping syndrome itself. However, the patients after jejunoileal bypass have higher plasma neurotensin concentrations but have no symptoms of dumping. It is still possible that neurotensin may contribute in a minor way to some aspect of the dumping syndrome, but a major role now seems unlikely. Another suggestion arose from the finding that neurotensin could release insulin in isolated islet preparations, albeit in large concentrations. Thus it was postulated that neurotensin might be the hormonal mediator of the enteroinsular axis, that is, the mechanism by which insulin release is much enhanced when nutriment is taken orally rather than infused intravenously. Oral glucose tolerance tests and intravenous infusions of glucose to precisely match the levels achieved show that after jejunoileal bypass, the enteroinsular axis is almost totally obliterated. Since neurotensin release is greatly enhanced in these circumstances, it seems unlikely that it plays a major role in this axis. More recently, the Italian group of Scopinaro and colleagues have developed an alternate operation of biliopancreatic bypass that has the great advantage of obviating the late weight gain that made the jejunoileal bypass so unpopular in recent years. Here again neurotensin release is greatly increased but showed less tendency to progressive change than after the jejunoileal bypass.[15]

Tumors

In the course of surveying 163 pancreatic endocrine tumors, 6 patients who also had excess production of vasoactive intestinal peptide (VIP) were found to be producing neurotensin. Thus their mean fasting plasma neurotensin was 378 pmol/l.[16] No particular clinical features could be ascribed to the excessive production of neurotensin, but these may have been masked by the fact that the patients were originally referred because of high plasma VIP concentrations. The effect of VIP, producing systemic vasodilation and secretory changes, particularly in the small bowel (thus the diarrhea) may, in some respects, be similar to those of neurotensin itself. Subsequent to the original publication, we have had several further neurotensin-producing tumors (neurotensinomas) but these subjects have still failed to demonstrate any distinctive clinical syndrome. That neurotensin is part of the intrinsic tumor is illustrated by the discovery of high concentrations of neurotensin within both the primary and secondary (hepatic), but not in the surrounding, tissues. Indeed a proportion of the cellular constituents have a distinctive appearance and can be stained for neurotensin immunoreactivity (FIG. 5). In addition to pancreatic endocrine tumors, we have been able to demonstrate that occasional pulmonary tumors can secrete neurotensin.[17] The patient with the highest concentration of tumor neurotensin (8.5 pmol/g) had an adenocarcinoma and an elevated plasma neurotensin concentration of approximately 200 pmol/l. Gel chromatography of both the pancreatic apudomas and the bronchial carcinomas showed the majority of the material to co-elute with pure synthetic neurotensin (FIG. 6).

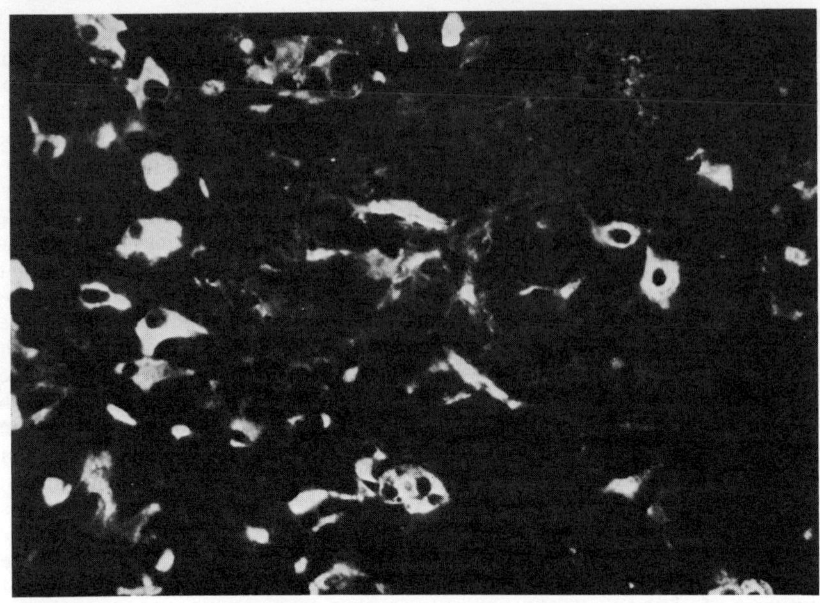

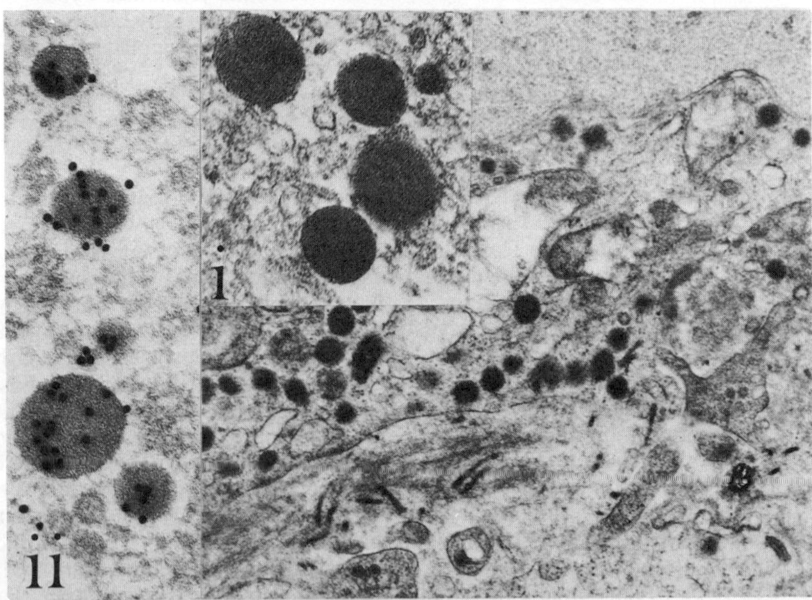

FIGURE 5.   (A) Neurotensin-like immunoreactivity in scattered tumor cells of a pancreatic apudoma (original magnification × 450; reduced to 65% of original size). (B) Same tumor analyzed at the ultrastructural level to show the presence of characteristic secretory granules (× 25,000; reduced as in (A)). Inset i: Higher magnification of granules (× 52,000). Inset ii: Immunogold staining demonstrating the presence of neurotensin immunoreactivity in the secretory granules (× 80,000; reduced as in (A)).

Low-dose Pharmacology of Neurotensin

To ascertain whether a peptide has a hormonal role following release into the circulation, it is necessary to test the peptide in an amount that results in physiologic plasma concentrations. In view of the reported effects of neurotensin on the cardiovascular system, we proceeded with human infusions with considerable care. Surprisingly we detected no effect on blood pressure or pulse rate. Indeed at an infusion level of 2.3 pmol/kg/min (resulting in a rise in plasma neurotensin of 104 ± 10 pmol/l), the subjects were unaware whether they were receiving a

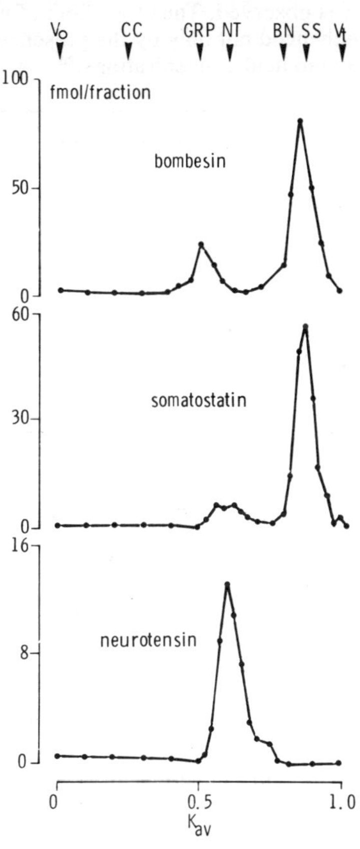

Figure 6. Gel permeation chromatography on Sephadex G-50 column precalibrated with dextran blue (V$_0$), Cytochrome C (CC), gastrin-releasing peptide (GRP), neurotensin (NT), bombesin (BN), somatostatin (SS), and sodium iodide[125] (Vt) as molecular size markers. Bombesin, somatostatin, and neurotensin immunoreactivity is shown from the extract of a single oat cell carcinoma.

neurotensin infusion or the control saline infusion. The metabolic clearance rate was 16 ± 1 ml/kg/min, the half life 3.8 ± 0.2 min, and the apparent space of distribution 88 ± 6 ml/kg.[18] No effect was seen on insulin, glucagon, gastric inhibitory peptide, gastrin, motilin, vasoactive intestinal peptide, TSH, growth hormone, prolactin, LH or FSH. In contrast, plasma pancreatic polypeptide rose by 145 ± 54 pmol/l. Encouraged by reports that neurotensin could affect insulin release, we pursued this aspect in the calf, a particularly friendly experimental animal. Initial experiments showed that a neurotensin infusion of 5 pmol/kg/min (causing a rise of 160 ± 10 pmol/l plasma neurotensin) caused a significant release of insulin only in the presence of hyperglycemia.[19] Since this neurotensin elevation seemed outside the physiologic range, the experiments were repeated at a lower infusion dose of 1 pmol/kg/min, which gave rise to an elevation of neurotensin closely similar to that seen after the calves morning milk drink (18 pmol/l plasma). In order to further mimic the physiologic setting, both glucose and amino acids were infused (in amounts approximately one-tenth of the milk feed), and under these circumstances, neurotensin evoked a rapid increment of insulin of 596 ± 176 pmol/l plasma within 10 min. When this dose of neurotensin was administered in the presence of the glucose infusion alone, an increment of only 69 ± 26 pmol/l was observed. Thus the effect of neurotensin on the beta cell could be seen to be enhanced not only by the presence of hyperglycemia but also by elevated plasma amino acid concentrations.[20]

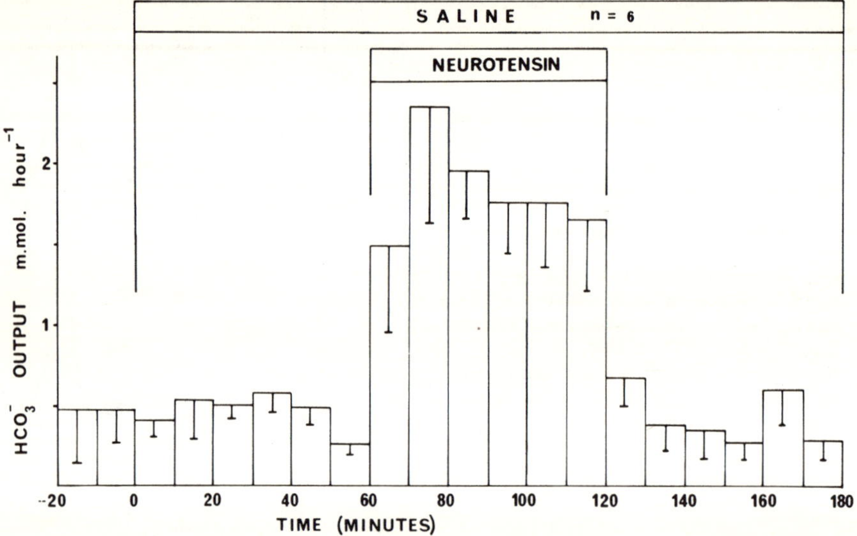

FIGURE 7.   The change in bicarbonate output in duodenal juice in 6 healthy volunteers who received a 1-hour neurotensin infusion of 1.4 pmol/kg/min during a saline infusion. A similar effect was seen during an infusion of low-dose cerulein and secretin (not shown).

In a further series of patients, the effect of neurotensin on gastric function was studied. Thus a neurotensin infusion of 2.4 pmol/kg/min was superimposed on the middle 40 minutes of a submaximal pentagastrin infusion. A small but significant inhibition of gastric acid secretion was observed.[21] A glucose meal with added technetium [99]M was used to assess the effect of neurotensin on gastric emptying as monitored by gastric gamma camera. A rise of plasma neurotensin of 91 ± 17 pmol/l (infusion 2.4 pmol/kg/min) resulted in a considerable delay in gastric emptying so that at 60 min, while 39 ± 4% had left the stomach on the control day, only 24 ± 6% had emptied on the neurotensin infusion day.[21] The effect of neurotensin on pancreatic exocrine secretion was to stimulate bicarbonate output, a feature that was seen both basally and on top of a background of low-dose cerulein and secretin stimulation[22] (FIG. 7). There was a small reduction in trypsin output that might have been secondary to the neurotensin-stimulated release of pancreatic polypeptide, but no significant change in bilirubin output. Experiments in the rat, which have not so far been repeated in man, suggest that neurotensin may be as potent as VIP or GIP in stimulating small intestinal fluid production.[23] This would certainly fit with the high neurotensin found in some patients with the watery diarrhea syndrome.

CONCLUSION

It is clear that the pathophysiology of neurotensin in man is still very incompletely understood. Neurotensin needs to be measured in many more conditions to obtain an overall view of the nature of its responses. The current view of its possible physiologic role (i.e., at the low plasma concentrations of real neurotensin that have been observed) is it may be a weak inhibitor of gastric acid secretion and gastric emptying and a mild stimulant of pancreatic bicarbonate secretion. This seems a very unsatisfactory physiologic end point, particularly as the effects are so weak. One must be suspicious that a more potent action of neurotensin remains to be discovered.

REFERENCES

1. BLACKBURN, A. M. & S. R. BLOOM. 1979. A radioimmunoassay for neurotensin in human plasma. J. Endocrinol. **83:** 175–181.
2. BUCHAN, A. M. J., M. G. BRYANT, J. M. POLAK, M. GREGOR, M. A. GHATEI & S. R. BLOOM. 1981. *In* Gut Hormones. S. R. BLOOM & J. M. POLAK, Eds. Vol 2: 119–124. Churchill Livingstone. Edinburgh.
3. POLAK, J. M., S. N. SULLIVAN, S. R. BLOOM, A. M. J. BUCHAN, P. FACER, M. R. BROWN & A. G. E. PEARSE. 1977. Specific localisation of neurotensin to the N cell in human intestine by radioimmunoassay and immunocytochemistry. Nature **270:** 183–184.
4. LUCAS, A., A. AYNSLEY-GREEN, A. M. BLACKBURN, T. E. ADRIAN & S. R. BLOOM. 1981. Plasma neurotensin in term and preterm neonates. Acta Paediatr. Scand. **70:** 201–206.
5. LUCAS, A., A. M. BLACKBURN, A. AYNSLEY-GREEN, D. L. SARSON, T. E. ADRIAN & S. R. BLOOM. 1980. Breast *vs.* bottle: endocrine responses are different with formula feeding. Lancet **I:** 1267–1269.
6. GILL, S. S., Y. C. LEE, L. O. UTTENTHAL & S. R. BLOOM. 1982. Use of isolated perfused rat ileum to investigate the mechanisms of neurotensin release. Gastroenterology Abstract In press.
7. BLOOM, S. R., M. A. GHATEI, N. D. CHRISTOFIDES, A. M. BLACKBURN, T. E. ADRIAN, P. LEZOCHE, N. BASSO, F. CARLEI & F. SPERANZO. 1979. Release of neurotensin, enteroglucagon, motilin and pancreatic polypeptide by bombesin in man. Gut **10:** A913.
8. ROKAEUS, A. & T. MCDONALD. 1980. Release of neurotensin-like immunoreactivity (NTLI) from rat intestine following administration of bombesin. Acta Physiol. Scand. **110:** 326.
9. MCDONALD, T. J., M. A. GHATEI, S. R. BLOOM, N. S. TRACK, J. RADZIUK, J. DUPRE & V. MUTT. 1981. A qualitative comparison of canine plasma gastroenteropancreatic hormone responses to bombesin and the porcine gastrin-releasing peptide (GRP) Regulatory Peptides 2: 293–304.
10. GHATEI, M. A., R. T. JUNG, J. C. STEVENSON, C. J. HILLYARD, T. E. ADRIAN, Y. C. LEE, N. D. CHRISTOFIDES, D. L. SARSON, K. MASHITER, I. MACINTYRE & S. R. BLOOM. 1982. Bombesin: action on gut hormones and calcium in man. J. Clin. Endocrinol. Metab. In Press.
11. BESTERMAN, H. S., D. L. SARSON, D. I. JOHNSTON, J. S. STEWART, S. GUERIN, S. R. BLOOM, A. M. BLACKBURN, H. R. PATEL, R. MODIGLIANI & C. N. MALLINSON. 1978. Gut-hormone profile in coeliac disease. Lancet **I:** 785–788.

12. BLACKBURN, A. M., N. D. CHRISTOFIDES, M. A. GHATEI, D. L. SARSON, E. H. EBEID, D. N. L. RALPHS & S. R. BLOOM. 1980. Elevation of plasma neurotensin in the dumping syndrome. Clin. Sci. **59:** 237–243.
13. SAGOR, G. R., M. A. GHATEI, G. P. McGREGOR, P. MITCHENERE, R. M. KIRK, & S. R. BLOOM. 1981. The influence of an intact pylorus on postprandial enteroglucagon and neurotensin release after upper gastric surgery. Br. J. Surg. **68:** 190–193.
14. BLOOM, S. R. 1980. Hormonal changes after jejunoileal bypass and their physiological significance. *In* Surgical Management of Obesity. J. D. MAXWELL, J.-C. GAZET, T. R. PILKINGTON, Eds. Academic Press, London. pp. 115–123.
15. SARSON, D. L., N. SCOPINARO & S. R. BLOOM. 1981. Gut hormone changes after jejunoileal (JIB) or biliopancreatic (BPB) bypass surgery for morbid obesity. Int. J. Obes. **5:** 471–480.
16. BLACKBURN, A. M., M. G. BRYANT, T. E. ADRIAN & S. R. BLOOM. 1981. Pancreatic tumors produce neurotensin. J. Clin. Endocrinol. Metab. **52:** 820–822.
17. WOOD, S. M., J. R. WOOD, M. A. GHATEI, Y. C. LEE, D. O'SHAUGHNESSY & S. R. BLOOM. 1981. Bombesin, somatostatin, and neurotensin-like immunoreactivity in bronchial carcinoma. J. Clin. Endocrinol. Metab. **53:** 1310–1312.
18. BLACKBURN, A. M., D. R. FLETCHER, T. E. ADRIAN & S. R. BLOOM. 1980. Neurotensin infusion in man: Pharmacokinetics and effect on gastrointestinal and pituitary hormones. J. Clin. Endocrinol. Metab. **51:** 1257–1261.
19. BLACKBURN, A. M., S. R. BLOOM & A. V. EDWARDS. 1981. Pancreatic endocrine responses to exogenous neurotensin in the conscious calf. J. Physiol. London **314:** 11–21.
20. BLACKBURN, A. M., S. R. BLOOM & A. V. EDWARDS. 1981. Pancreatic endocrine responses to physiological changes in plasma neurotensin concentration in the calf. J. Physiol. London **318:** 407–412.
21. BLACKBURN, A. M., S. R. BLOOM, R. G. LONG, D. R. FLETCHER, N. D. CHRISTOFIDES, M. L. FITZPATRICK & J. H. BARON. 1980. Effect of neurotensin on gastric function in man. Lancet **I:** 987–989.
22. FLETCHER, D. R., A. M. BLACKBURN, T. E. ADRIAN, V. S. CHADWICK & S. R. BLOOM. 1981. Effect of neurotensin on pancreatic function in man. Life Sci. **29:** 2157–2161.
23. MITCHENERE, P., T. E. ADRIAN, R. M. KIRK & S. R. BLOOM. 1981. Effect of gut regulatory peptides of intestinal luminal fluid in the rat. Life Sci. **29:** 1563–1570.

DISCUSSION OF THE PAPER

R. A. HAMMER (*Northwestern University Medical School, Chicago, IL*): I wanted to ask about the possible presence of NT metabolites and about the specificity of your antiserum. I understand that the antiserum you used earlier with Blackburn was an NH_2-terminally directed antiserum. Is that right?

D. T. BLOOM (*Royal Postgraduate Medical, London, England*): That is not in fact, right. It was an antibody that recognized the entire neurotensin molecule. We did not at that stage have fragments of neurotensin available to work out specificity.

V. L. W. GO (*Mayo Clinic, Rochester, MN*): Do you believe, at least in mammals (e.g. dogs or humans), that neurotensin is acting as a classical hormone or rather as a local hormone at least in the stomach and the pancreas? Is this the ileal enterogastrone that we are looking for? I am confused about the antibody that you are using. Dr. Polak showed a slide indicating that the antibody used in the RIA was COOH-terminus directed. In the other work that you and Blackburn published. I thought that was NH_2-terminus directed; presumably that is not correct.

BLOOM: You are right on that supposition that it was not NH_2 terminus directed.

Go: You have infused neurotensin in man and you have calculated the half life. To what degree does fragmentation of the injected neurotensin occur when you start doing chromatography?

BLOOM: We find that half-life in plasma is just under 3.8 minutes, the distribution space about 88 mls per kilogram, and the metabolic clearance rate, if I remember correctly, about 16 ml per kilogram per minute. We do not detect neurotensin[1–8] with this antibody. And that, I imagine, is the major metabolite. We do not know what happens to that as we do not detect it.

From what Dr. Leeman showed earlier, in the rat, the clearance rate of NT 1–8 is much slower than $NT^{1–13}$. I think that Dr. Leeman showed an $NT^{1–13}$ half-life of half a minute and 5 minutes for $NT^{1–8}$, which is 10 times greater. I imagine the same thing is true in man. I imagine that is why some assays currently published are showing much bigger levels of apparent neurotensin levels. I think they're using NH_2-terminal antibodies and picking up $NT^{1–8}$. I think that this analysis is probably fairly well agreed by most groups now.

Go: Do you believe NT is the ileal enterogastrone?

BLOOM: No, I am afraid that I do not. I really do not know what neurotensin is doing physiologically.

A. J. PRANGE (*University of North Carolina at Chapel Hill*): In a way I suppose the two patients with VIPoma are nature's experiment to answer the question whether its possible to put so much neurotensin in the periphery as to drive any across the blood-brain barrier and, if so, what it might do centrally in man? In view of all the many demonstrated effects of centrally administered neurotensin in animals, what do you observe in these patients? Are they lethargic? drowsy? Are they hypothermic or indifferent to pain? Are there brain tumors that produce neurotensin? What would those people be like?

BLOOM: I have not seen any brain tumors that contain neurotensin. Unfortunately these patients were very ill. They had received cytotoxic therapy. They had enlarged livers and massive tumors in the abdomen, and they were not terribly well; on the other hand, they were both going to work and they did not suffer from hypothermia. There did not appear to be any particular effect of the neurotensin. One could put down their clinical syndrome entirely to tumor load and the fact that they had pretty severe diarrhea due to the VIP.

Patients with somatostatinomas do very well and obviously long continued elevations of somatostatin allow various feedback mechanisms to operate and escape to occur. And, unfortunately, it may be the same with neurotensin. So the fact that the patients do not illustrate this or that does not mean that there is no action of neurotensin. We get a very moderate amount of information from such patients.

E. THEODORSSON-NORHEIM (*Karolinska Institute, Stockholm*): We did (with the Endocrinology Department at the University Clinic in Uppsala) a study on endocrine tumors in 32 patients and we have 9 patients who we consider as possible neurotensinomas. We had one rather interesting patient who had a neurotensinoma of the bronchus. We measured the concentration of 12 peptides other than neurotensin, which were all normal. This patient had 500 piomolar concentrations of NT in his blood. He had no symptoms whatsoever indicating any neurotensin effects. We used an NH_2-terminal antibody. He died because of invasion of his tumor into his aorta. Using an NH_2-terminal antibody we have not been able to detect very high levels in patients not only in those having VIPomas, but also

Zollinger-Ellison syndrome or glucagonomas. We had one patient with Verner-Morrison Syndrome who in the beginning had only neurotensin[1-13] in blood, but after repeated streptozotocin treatments he had NT[1-8]. In fact, three of our patients plasma has been chromatographed recently. They are, of course, secreting NT[1-13] but their main immunoreactivity is eluting in the NT[1-8] position.

BLOOM: That might be postsecretional transformation by plasma, mightn't it?

NORHEIM: Yes. And, what supports your opinion is that we have done infusions of both neurotensin[1-13] and [Gln⁴]—neurotensin in humans and they were metabolized with half-lives that you and Dr. Carraway described. Their main product is the NT[1-8] sequence, which has a lot longer half-life.

BLOOM: I would like to support your suggestion that NT is not just found in VIPomas. When we first did our survey of 85 VIPomas and approximately 60 non-VIPomas (glucagonomas), we found NT only in the VIPomas but I think it is an error of small numbers. We have subsequently had, as you had, gastrinomas and glucagonomas producing neurotensin.

NORHEIM: That is what we have found. We have also found it in a bronchial carcinoid, which means that neurotensinomas can occur outside the pancreas. The interesting finding is that we have two patients with midgut carcinoids. We called them possible neurotensinomas because the levels were not as high as in the pancreas.

BLOOM: In the carcinoid syndrome, we found mild elevation of a number of different peptides, for example, pancreatic polypeptide. It does not appear to be very high in concentration in the tumor itself. I am beginning to suspect that this phenomenon is an effect of 5-hydroxytryptamine on the gut mucosa rather than a direct product of the tumor.

I. M. MODLIN *(SUNY Downstate Medical Center, Brooklyn, NY):* Are VIPomas that you have discussed with increased neurotensin adrenal or pancreatic?

BLOOM: Pancreatic. No ganglionomas.

MODLIN: Earlier speakers have discussed release of histamine by neurotensin. This seems inconsistent with our observations and yours of the gastric acid inhibitory effect of neurotensin. I wonder if you could comment on that?

BLOOM: I imagine that the effect on the mast cell is only seen with very much larger concentrations. This may mean that if this effect of NT is physiologic, it must be a local effect and may even be an effect of a neurotensin-like peptide rather than neurotensin itself.

$$\text{LOCALIZATION OF NEUROTENSIN}$$
$$\text{IN THE HYPOTHALAMUS*}$$

David Kahn,† Anna Hou-Yu,‡ and Earl A. Zimmerman‡

†Department of Psychiatry
‡Departmentof Neurology
Columbia University College of Physicians and Surgeons
New York, New York 10032

INTRODUCTION

Within several years of its discovery, neurotensin (NT) was reported to be present in many regions of the central nervous system of the rat. Radioimmunoassays by Carraway and Leeman[1] and Kobayashi *et al.*[2] found relatively large concentrations in extracts of the median eminence, hypothalamus, pituitary, thalamus, and brainstem. In addition, Uhl and Snyder[3] had shown by immunofluorescence that it was found in neurons in many regions of the brain and spinal cord.

As a body of evidence, reviewed elsewhere in this volume, developed that NT was a neural and hormonal modulator, it became apparent that NT was active in several functions related to the hypothalamus, including effects on pituitary hormone secretion. Although the results of these studies differed concerning the patterns of adenohypophyseal secretion following intracerebral or intravenous injection of NT,[4-6] reports that intact hemipituitaries incubated with synthetic NT released prolactin[4] and thyroid-stimulating hormone[7] suggested that NT could have direct effects on anterior pituitary hormone secretion. In this report we review our studies of the distribution of NT in the rat hypothalamus using immunoperoxidase technique and light microscopy. The distribution of NT-like immunoreactivity in several hypothalamic nuclei known to project to the median eminence, and its presence in nerve terminals around the hypophyseal portal capillary system, suggest roles for the peptide in neuroendocrine mechanisms and provide an anatomical basis for further investigation.

METHODS

Tissue Preparation

Brains from 16 adult male and female Sprague-Dawley rats were studied. To enhance visualization of perikarya, 48 hours before sacrifice, rats were injected with 60 μg of colchicine into a lateral cerebral ventricle. Under pentobarbital anesthesia, animals were then perfused by transcardiac puncture with normal saline followed by 10% formalin. After 4 additional hours of immersion fixation in 10% formalin, brain blocks were cut on a sliding microtome cryostat at 80 μm. Brains and pituitaries of two rats were embedded in paraffin, sectioned at 6 μm and mounted on glass slides.

*This work was supported by USPHS grants AM20337 and HD13147

Immunohistochemistry

The peroxidase–antiperoxidase technique of Sternberger[8] was used for both mounted, deparaffinized sections and unembedded tissues, with a primary layer of rabbit antiserum to NT followed sequentially by sheep anti-rabbit serum, peroxidase-rabbit-antiperoxidase complexes, and 3,3′ diaminobenzidine.[9, 10] The unmebedded sections were mounted on glass slides at the end of the procedure.

Antiserum

Antiserum to synthetic NT, HC-8 (11/11/75), provided by Carraway and Leeman, was prepared and characterized as previously reported.[11] By radioimmunoassay, antiserum HC-8 reacts solely with the COOH-terminal regions of NT for full recognition. Furthermore, the HC-8 reacted only with eluates of bovine hypothalamic extracts on Sephadex G-25 chromatography, identical to synthetic NT and not larger or smaller molecular weight substances. In our experiments, this antiserum provided good staining results with minimal background at a dilution of 1:1000 in 0.1 M phosphate-buffered saline, pH 7.2. Control experiments were performed by preabsorbing antiserum with 5 μg synthetic NT (Beckman Instruments, Palo Alto, CA) per ml of 1:1000 dilution, or with equimolar luteinizing hormone–releasing hormone (LHRH) (Peninsula Laboratories, San Carlos, CA).

RESULTS AND DISCUSSION

Immunoreactive nerve cells and fibers were distributed throughout the brain in a manner generally consistent with earlier reports;[3] we limit our report here to the hypothalamus. Reaction products densely filled cytoplasm of small- and intermediate-sized neurons, and could often be traced into axonal and dendritic processes. Colchicine, presumably by blocking axonal transport, greatly enhanced staining of cell bodies. Distal fiber projections were seen best in rats that had not received colchicine. All reactivity was abolished by preabsorption of antiserum with NT, while LHRH had no effect. Although patterns of immunoreactivity were similar in 6 μm paraffin and 80 μm cryostat sections, the intensity and number of reactive neurons, as well as their morphology, were better appreciated in the thicker sections.

Our findings are schematized in the drawings of FIGURE 1. To summarize, many perikarya containing NT-like immunoreactivity were seen in the bed nucleus of the stria terminalis, the medial preoptic area, the periventricular nucleus, the medial-parvocellular portion of the paraventricular nucleus (PVN), the arcuate nucleus, the lateral hypothalamus in the perifornical area, and, outside the hypothalamus proper, the central nucleus of the amygdala. A smaller number of cell bodies was seen in the ventral portion of the anterior hypothalamus, the dorsomedial nucleus, and the posterior hypothalamus. Reactive neurons were not seen in the suprachiasmatic, supraoptic, ventromedial, or mammillary nuclei, nor in the anterior pituitary gland. Reactive fibers were found in the same regions as cell bodies. Additional dense-fiber collections were seen in the lateral aspects of the zona externa of the median eminence, the pituitary stalk, the posterior mammillary nucleus, the stria terminalis, and the most lateral portions of the hypothalamus at the medial edge of the crura cerebri. A few fibers were seen in the premammillary and posterior hypothalamic nuclei and in the posterior pituitary gland.

These results [10] are in broad agreement with regional radioimmunoassays [1, 2] and an immunofluorescence study less complete in regard to the hypothalamus.[3] Unlike Uhl and coworkers,[3] we did not find reactivity in cells of the anterior pituitary. Our findings may represent a false negative due to differences in tissue preparation, particularly loss of antigenicity associated with paraffin procedures. We did not study cryostat sections of pituitary. More recent immunohistochemical studies have confirmed the presence of NT in cells in the anterior pituitary (L. Jennes, personal communication, and M. Goedart, personal communication). Immunoperoxidase technique has been applied to the study of NT in the hypothalamus of one other species, the Japanese quail.[12] Results were similar to the rat, except that fibers were not seen in the zona externa of the median eminence. This may have been a false negative due to the use of 6 μm paraffin sections; in our experience, 80 μm cryostat sections were needed to see fibers well.

In terms of functional significance, our major finding was that several of the regions where NT was concentrated are known to be important in anterior pituitary regulation. These include the cell bodies seen in the medial preoptic area, the periventricular nucleus, the medial parvocellular portion of the PVN and the arcuate nucleus (see FIGS. 2 and 3), as well as the fibers seen in the zona externa of the median eminence (see FIG. 4). The four nuclear groups have been shown by retrograde tracing studies using horseradish peroxidase (HRP) to project to the median eminence.[13, 14] Whether NT-containing cells in one, several or all of these regions are the source of NT fibers in the zona externa is not yet known. The zona externa itself is the region where fibers containing LHRH,[15] thyrotropin-releasing hormone (TRH), somatostatin (SRIF)[16] and vasopressin,[17] to name some known peptide modulators of the anterior pituitary, secrete in the hypophyseal portal capillary plexus. Although it is likely that NT is also secreted into hypophyseal portal blood at this site, it has not been proven by radioimmunoassay of NT in portal plasma.

The median eminence is shown in FIGURE 4. NT fibers are concentrated in the lateral portions. In sagittal sections, fibers were seen best in the rostral median eminence and were oriented parallel to the coronal plane. Like NT, LHRH is distributed laterally in the zona externa,[15] while TRH tends to concentrate medially,[18] and vasopressin and SRIF across the whole extent.[16] The differences in terminal distribution suggest different cells of origin for each peptide.

NT in the pituitary stalk differed from the median eminence in that fibers streamed in the sagittal plane (FIG. 1H), leading to the upper portion of pars nervosa. As described, the pituitary gland itself was studied separately and contained a few NT fibers in the posterior portion, which presumably came from the stalk. In addition to vasopressin and oxytocin, the classic neurosecretory peptides,[19] there are reports of SRIF,[16] TRH,[18] and enkephalin[20] in fibers in the posterior pituitary.

The dense grouping of NT cell bodies in the ventral half of the medial preoptic area (FIG. 1A) is shared by SRIF,[21] TRH,[18] and LHRH,[15] although SRIF and TRH are also seen in the suprachiasmatic subdivision of the preoptic area where NT was absent. Vasoactive intestinal polypeptide (VIP)[9] and enkephalin[22] are other peptides seen in preoptic area; enkephalin is found in the dorsal portion. The suprachiasmatic nucleus itself contains vasopressin[19] and VIP[9] but not NT.

FIGURE 1B shows the layer of NT cell bodies along the entire extent of the wall of the third ventricle in the periventricular nucleus. The pattern continues from this rostral area through the slightly more caudal level within the medial parvocellular portion of the PVN (FIG. 2C). Fibers appear oriented parallel to the wall of the ventricle. This distribution is also characteristic of TRH[18] and SRIF.[21, 23]

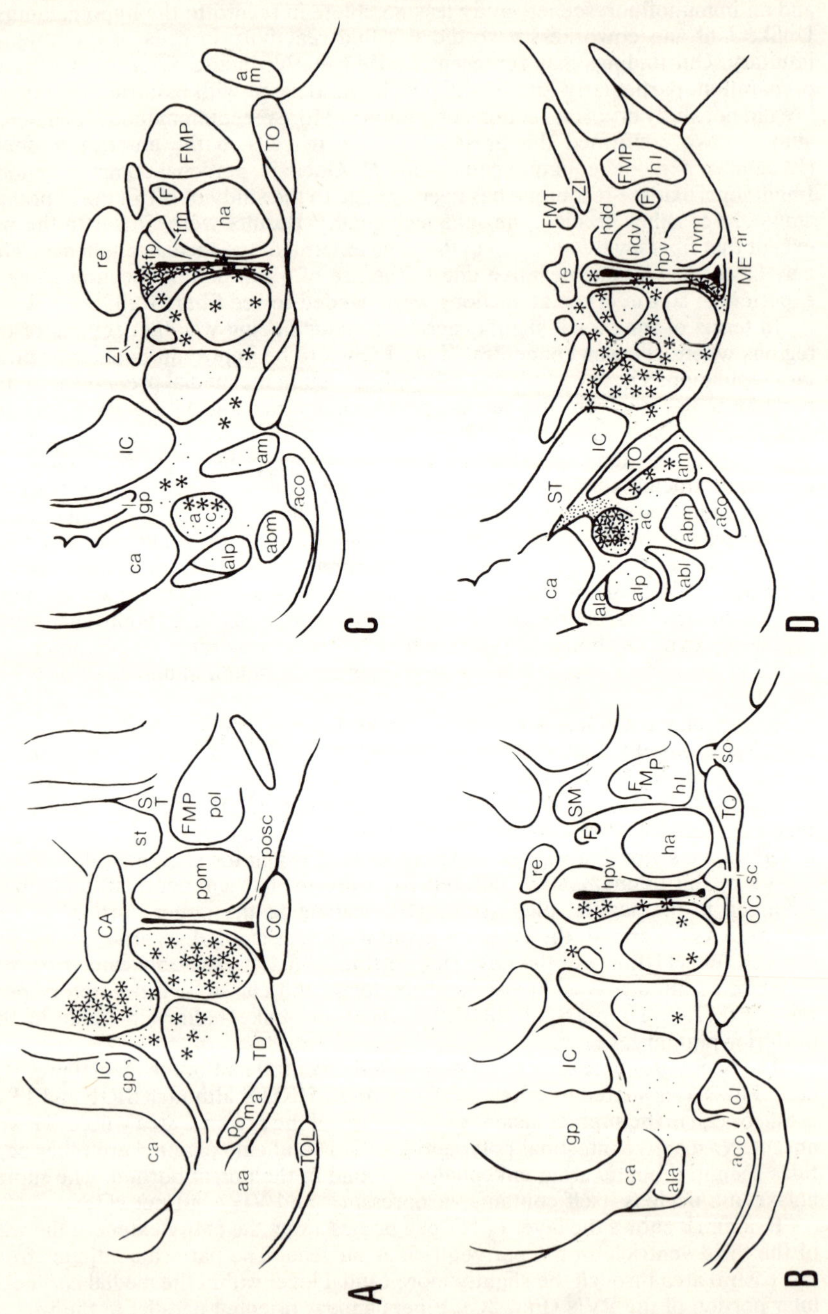
A
st
ST
FMP
pol
CA
pom
posc
CO
IC
gp
TD
pom
a
ca
aa
TOL

B
SM
F
FMP
hl
re
hpv
ha
TO
so
OC sc
IC
gp
ca
ala
aco
ol

C
a
m
FMP
TO
F
fm
ha
re
fp
ZI
IC
gp
ca
alp
abm
aco
am
a c

D
EMT
ZI
FMP
hl
F
hdd
hdv
hpv
hvm
re
ME ar
ST
IC
TO
am
ac
abm
abl
aco
ca
ala
alp

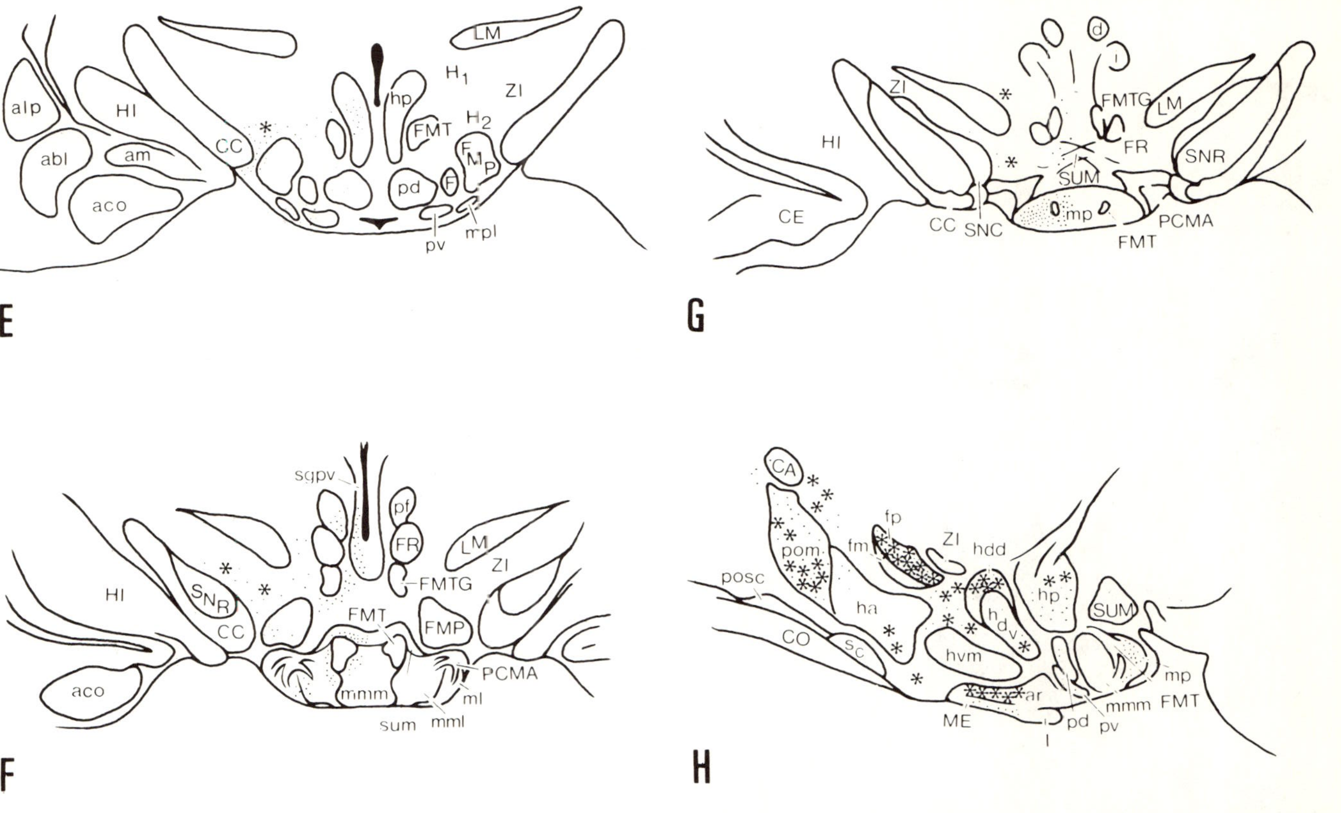

FIGURE 1. Composite plots of the relative local concentrations of neurotensin-containing perikarya (*) and fibers (. . .) in representative coronal (A–G) planes and midsagittal plane (H) of the rat hypothalamus, based on the atlas and nomenclature of Konig and Klippel.[34] (Figure reproduced from Kahn *et al.*[10] with the permission of *Endocrinology*.)

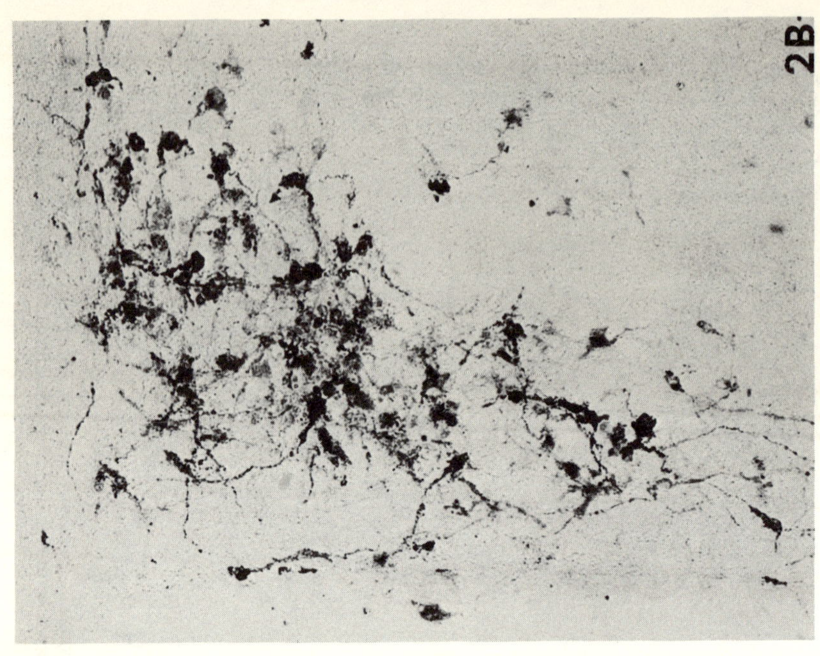
2B

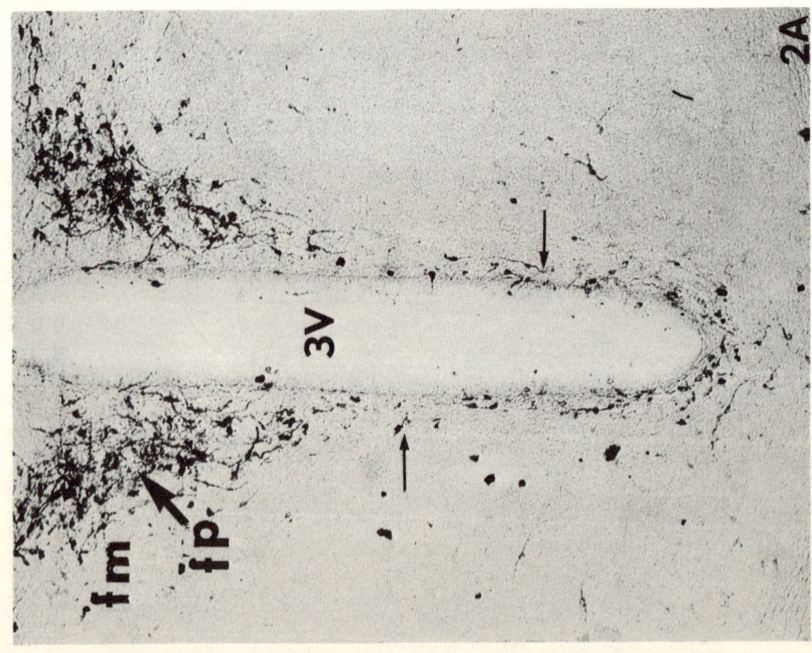
2A
3V
fm
fP

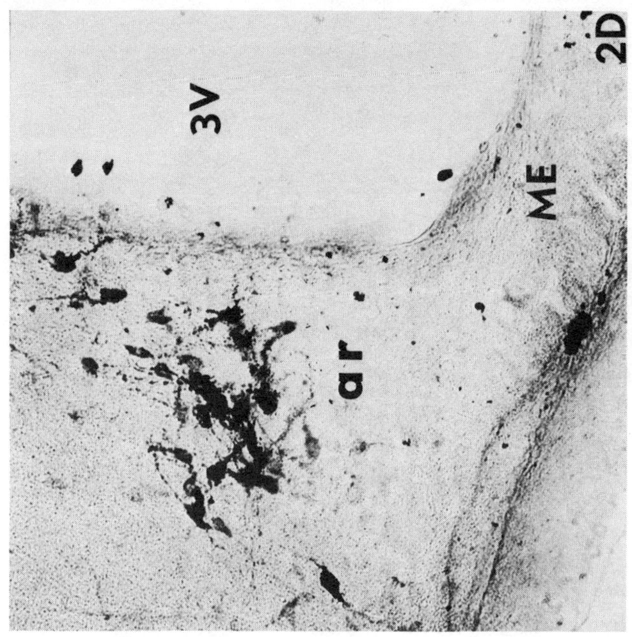

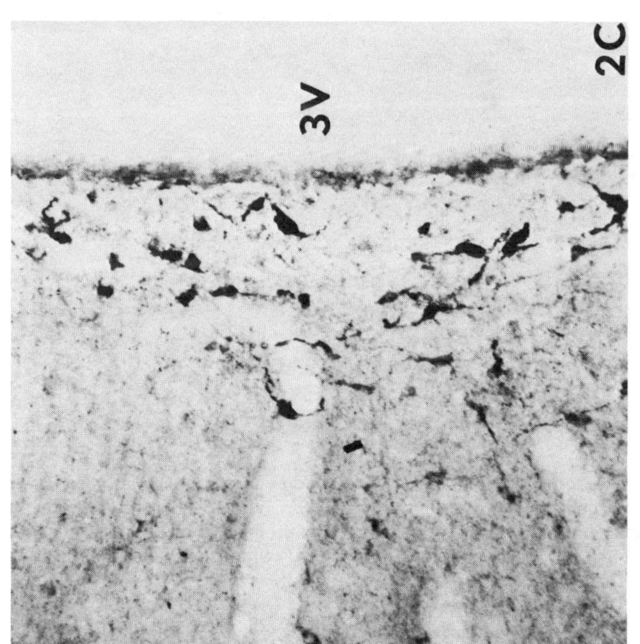

FIGURE 2. Photomicrographs of neurotensin neurons and fibers in coronal sections of the rat hypothalamus. The immunoperoxidase technique was used with 80 μm cryostat sections after intraventricular injection of colchicine. A. Region of the paraventricular nucleus (PVN). Reactive perikarya are clustered in the medial, parvocellular portion [filiformis parvocellularis (fp)], and a few are scattered in the magnocellular part [filiformis magnocellularis (fm)]. Some neurons are seen along the entire extent of the periventricular region (*arrows*). 3V, Third ventricle (×120). B. Higher magnification of the right PVN shown in A, demonstrating cell bodies and fibers. C. Neurons in the PVN in section rostral to A (×250). D. Cells in the arcuate nucleus (ar), caudal and ventral to A at the level of the median eminence (ME). Fiber staining in ME was eliminated by colchicine treatment. Figure reduced to 90% of original size. (Figure reproduced from Kahn *et al.*[10] with the permission of *Endocrinology*.)

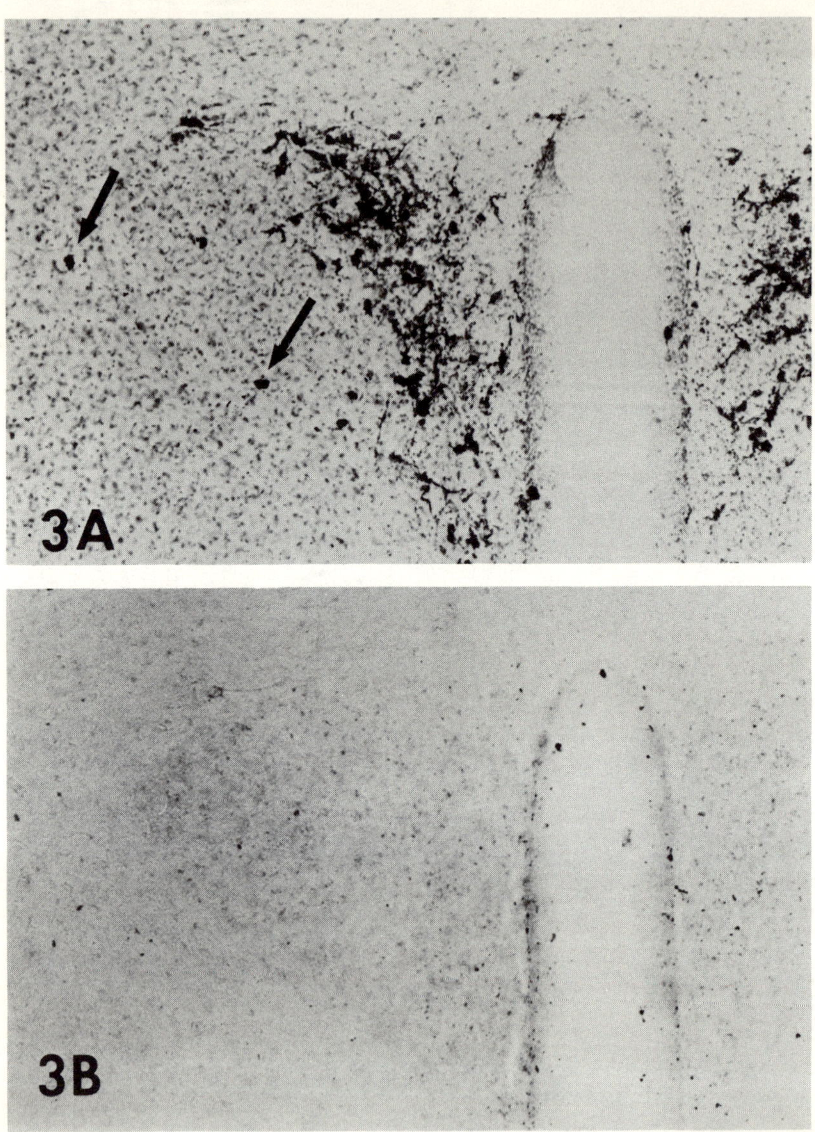

FIGURE 3. A. Immunoreactive neurotensin (NT) cell bodies and fibers in paraventricular nucleus. Note occasional cell body in the lateral region (*arrows*). B. Control experiment performed on an adjacent section. Reactivity is abolished by preincubation of antiserum with synthetic NT (× 120). (Figure reproduced from Zimmerman[19] with the permission of Raven Press.)

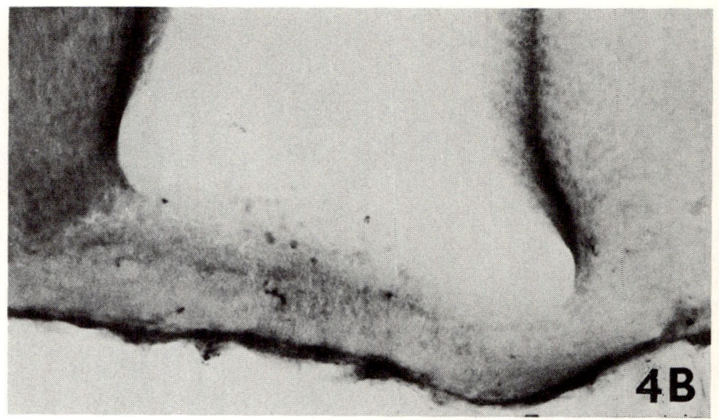

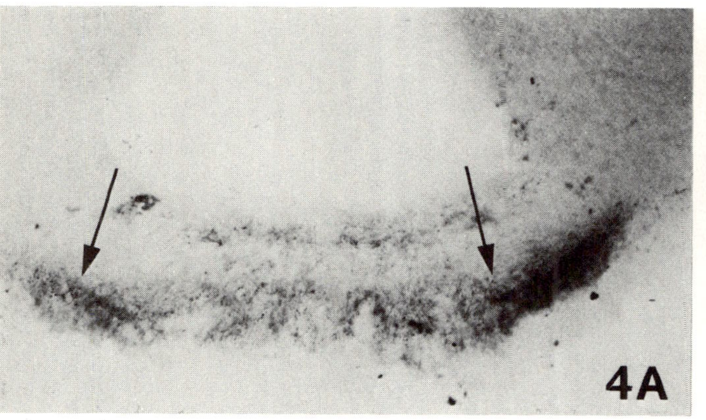

FIGURE 4. A Coronal 80 μm section of the median eminence of a rat that did not receive colchicine. Numerous radially oriented neurotensin fibers are found in the entire sweep of the zona externa, although they are more concentrated in the lateral region (*arrows*). B. Adjacent section, absorption control experiment with NT in which staining is abolished. (Figure reproduced from Kahn *et al.*[10] with the permission of *Endocrinology*.)

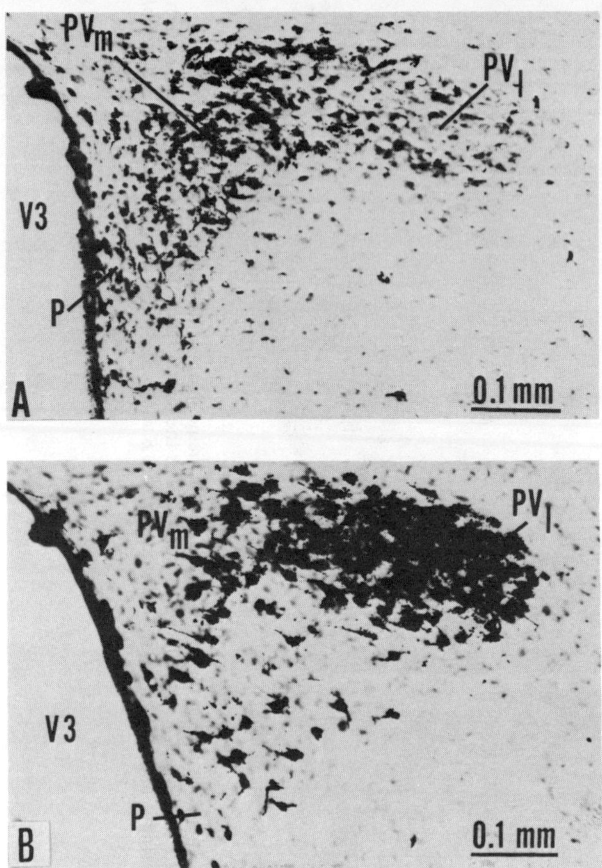

FIGURE 5. Enzymatic labeling of horseradish peroxidase in cells in paraventricular nucleus following injection of the median eminence (A) or the posterior pituitary (B). A. Parvocellular cells heavily labeled in the medial subnucleus (PVm), with a few labeled in the lateral subnucleus (PVl). B. Magnocellular cells demonstrated in PVl, with a few smaller cells labeled in PVm. Compare with FIGURES 4 and 6. PVd, dorsal subnucleus. P, periventricular subnucleus. V3, third ventricle. (Figure modified from Wiegand *et al.*[14] with the permission of *The Journal of Comparative Neurology*.)

FIGURE 6. A. Immunoreactive staining of magnocellular cells in the lateral paraventricular nucleus (PVN) with monoclonal antibody to vasopressin ($\times$ 130). B. Detail of same section, demonstrating a few parvocellular cells (*arrows*) containing vasopressin in the medial PVN ($\times$ 260). Photography by Dr. Hou-Yu, unpublished, using technique described in reference 27.

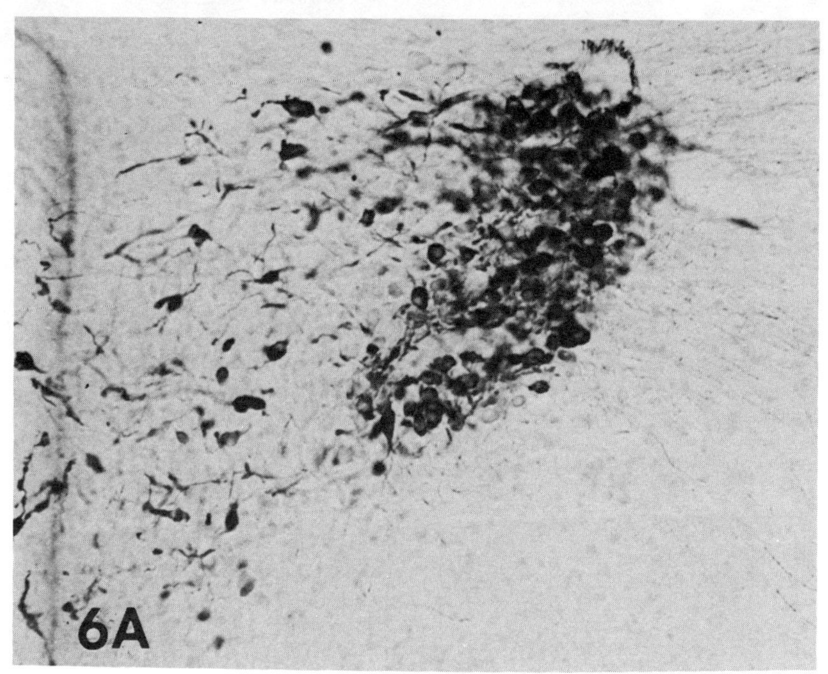

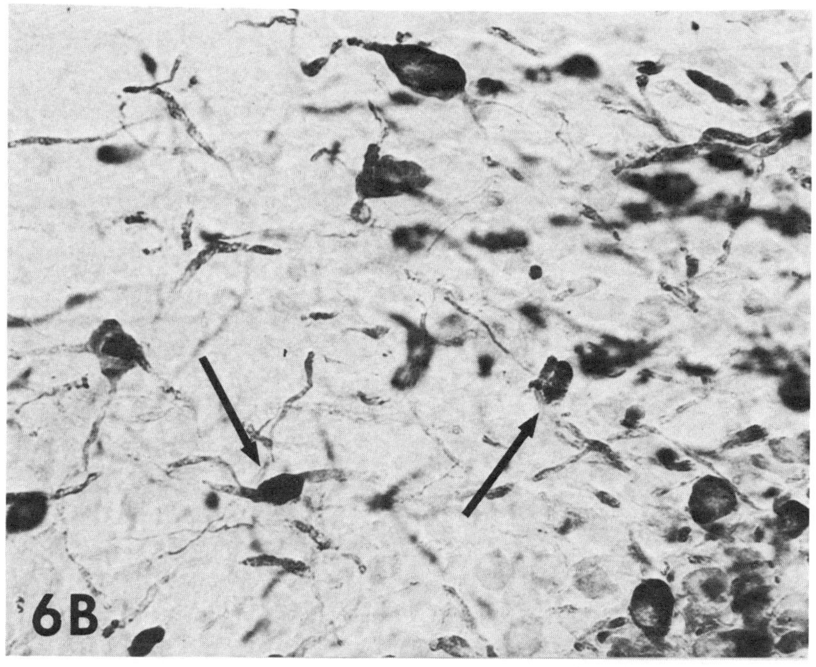

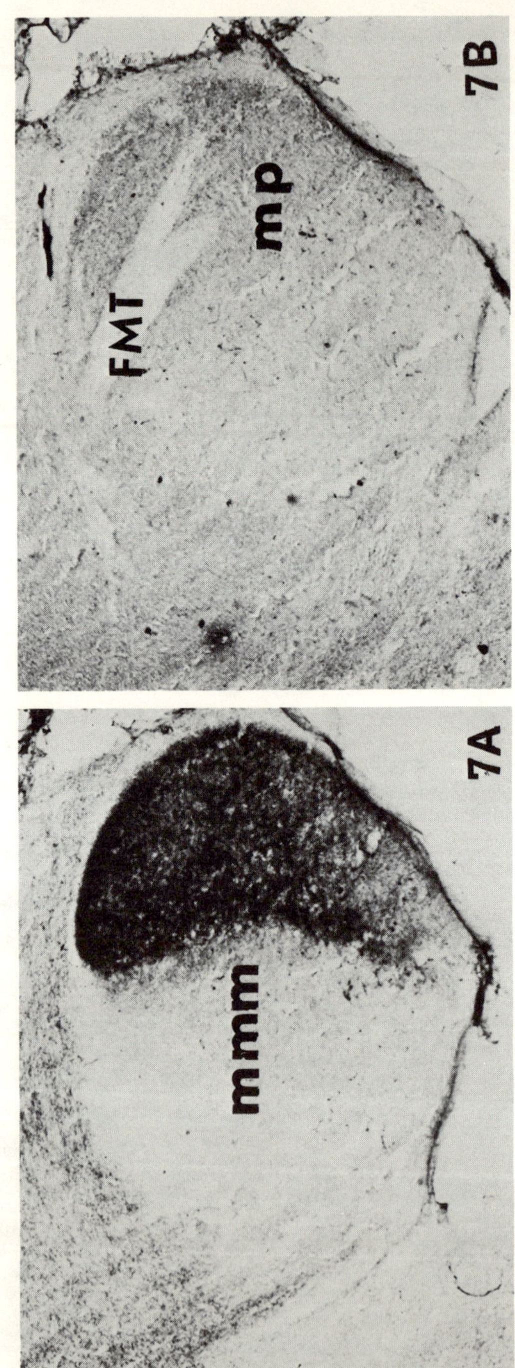

FIGURE 7. Adjacent 80 μm sagittal sections of the mammillary body. A. the entire posterior mammillary nucleus (mp) is outlined by reactive fibers, while the medial mammillary nucleus (mm) is not innervated. B. Absorption of primary antiserum with synthetic neurotensin abolished reactivity. FMT, fasciculus mammillothalamicus ($\times$ 120). (Figure reproduced from Kahn et al.[10] with the permission of Endocrinology.)

Lesion studies have shown that TRH and SRIF cells project from the periventricular nucleus to the median eminence.[16]

The dense distribution of NT-reactive cell bodies in the medial, parvocellular portion of the PVN (FIGS. 2A, 2B, 3) is shared by TRH.[18] SRIF, studied in thinner sections, has a similar, though more medial, distribution.[21, 23] The staining patterns of these three peptides correspond to the medial parvocellular subnucleus identified by HRP tracing as a source of projections to the median eminence[13, 14, 24] (see FIG. 5A). A few positively reacting NT cell bodies were also seen in the lateral PVN (FIG. 3). The medial, parvocellular subnucleus is distinct from the lateral, magnocellular subnucleus, which contains primarily vasopressin and oxytocin,[19] as well as enkephalin,[22] angiotensin II,[25] and cholecystokinin,[26] and projects to the posterior pituitary gland[14, 24] (see FIGS. 5B and 6A). However, some small cells containing vasopressin have been seen in the medial area[27] (see FIG. 6B), and are probably the source of vasopressinergic fibers to the median eminence.[28, 29]

The PVN is also the most likely source of NT fibers seen in the posterior pituitary. HRP tracing from the posterior pituitary, while labeling mainly large cells in the supraoptic nucleus (where NT is absent) and the lateral PVN, does label a few small cells in the medial PVN[14, 24] (see FIG. 5B). NT projections to the posterior pituitary could thus arise either from the mass of positive cells in the medial PVN or, conceivably, from the few positive cells scattered in the lateral area (see FIG. 3). Supporting this suggestion is a recent autoradiographic study that demonstrated a pathway leading from the PVN through the zona externa of the median eminence to the pituitary stalk and the posterior pituitary.[30] In the posterior lobe, NT could either be released into the systemic circulation, modulate release of other peptides in the region, or still interact with the median eminence system through release into short portal capillaries.[31]

The arcuate nucleus (FIG. 2D), which contains NT cell bodies as well as SRIF[21, 23] and TRH,[18] but not LHRH (in the rat) or vasopressin perikarya, has been shown to be an important source of fibers to the median eminence.[13, 14] Adrenocorticotropin and β-lipotropin have also been reported in the arcuate nucleus,[36] but these cell bodies are greater in number and extend laterally in a band under, around, and lateral to the ventromedial nucleus where NT neurons are not found.

NT cell bodies were seen in several areas whose functions are not as well known as the areas discussed above, including the bed nucleus of the stria terminalis, the anterior and dorsomedial nuclei, and the perifornical area. The bed nucleus is interesting in that it not only contains NT cells, but receives NT-afferent fibers from the amygdala via the stria terminalis.[33] Cells of the anterior hypothalamic nucleus as well as the preoptic area may subserve thermoregulatory properties attributed to NT.[38] The function of the dense terminal fiber pattern seen in the posterior mammillary nucleus is unknown (FIG. 7).

CONCLUSIONS

NT has a complex distribution in the rat hypothalamus. The localization of NT in the zona externa of the median eminence and the four hypothalamic nuclei shown to project there suggests an important role for the peptide in the regulation of anterior pituitary hormone secretion. The fibers in the pituitary stalk and the posterior pituitary may function in these and other endocrine mechanisms. Further research is needed to clarify these issues. Lesion studies would help confirm proposed pathways, and portal blood determination of NT as well as ultrastructural immunohistochemistry should be undertaken to define what may take place in the median eminence. It will also be important to explore the relationship of NT to neurons containing other peptides. It may coexist in the same cell with other peptides. For example, angiotensin II[25] and dynorphin[35] immunoreactivity have been found in vasopressin neurons. Whether NT may coexist in some vasopressin or oxytocin neurons is not known. Terminals containing NT may also be found to synapse on other neurons that regulate the pituitary gland, either on their perikarya and associated dendrites, or possibly on their terminals in the median eminence. Such connections, if established, would provide anatomical evidence that NT participates in neuroendocrine mechanisms at the hypothalamic level. Recent preliminary evidence has been provided that small injections of NT into the medial preoptic area release gonadotropins from the pituitary gland (S. E. Leeman, this symposium). These findings suggest that NT in this region may stimulate LHRH neurons known to project from the medial preoptic area to the median eminence[15] to secrete LHRH into hypophyseal portal blood.

REFERENCES

1. CARRAWAY, R. & S. E. LEEMAN. 1976. J. Biol. Chem. **251**: 7045–7052.
2. KOBAYASHI, R. M., M. BROWN & W. VALE. 1977. Brain Res. **126**: 584–588.
3. UHL, G., M. KUHAR & SNYDER. 1977. Proc. Natl. Acad. Sci. USA **74**: 4059–4063.
4. VIJAYAN, E. & S. M. McCANN. 1979. Endocrinology **105**: 64–68.
5. RIVIER, C., S. M. BROWN & W. VALE. 1977. Endocrinology **100**: 751–754.
6. MAEDA, K. & L. A. FROHMAN. 1978. Endocrinology (Abstract) **102**: 389.
7. VIJAYAN, E. & S. M. McCANN. 1980. Life Sci. **26**: 321–327.
8. STERNBERGER, L. A. 1974. Immunocytochemistry. Prentice-Hall. Englewood Cliffs, NJ.
9. SIMS, K. B., D. L. HOFFMAN, S. I. SAID & E. A. ZIMMERMAN. 1980. Brain Res. **186**: 165–183.
10. KAHN, D. A., G. M. ABRAMS, E. A. ZIMMERMAN, R. CARRAWAY & S. E. LEEMAN. 1980. Endocrinology **107**: 47–54.
11. CARRAWAY, R. & S. E. LEEMAN. 1976. J. Biol. Chem. **251**: 7035–7044.
12. YAMADA, S. & S. MIKAMI. 1981. Cell Tissue Res. **218**: 29–39.
13. LECHAN, R. M., J. L. NESTLER, S. JACOBSON & S. RECHLIN. 1980. Brain Res. **95**: 13–27.
14. WIEGAND, S. J. & J. L. PRICE. 1980. J. Comp. Neurol. **192**: 1–19.
15. BARRY, J. 1978. *In* Colloques Internationaux du CNRS, Vol. 280. J. Vincent & C. Kordon, Eds. Editions du CNRS. Paris. pp. 299–414.
16. ELDE, R. & T. HÖKFELT. 1978. Front. Neuroendocrinol. **5**: 1–33.
17. VANDESANDE, F., K. DIERICKX & J. DeMAY. 1977. Cell Tissue Res. **180**: 443–452.
18. LECHAN, R. M. & I. M. D. JACKSON. 1982. Endocrinology. In press.
19. ZIMMERMAN, E. A. 1981. *In* Neurosecretion and Brain Peptides. Martin, J. B., S. Reichlin & K. L. Bick, Eds. Raven Press. New York. pp. 63–75.
20. ROSSIER, J., E. BATTENBERG, Q. PITTMAN, A. BAYON, L. KODA, R. MILLER, R. GUILLEMIN & F. BLOOM. 1979. Nature **277**: 653–655.
21. BENNET-CLARKE, C., M. A. ROMAGNANO & S. A. JOSEPH. 1980. Brain Res. **188**: 473–486.
22. SAR, M., W. E. STUMPF, R. J. MILLER, K.-J. CHANG & P. CUATRECASAS. 1978. J. Comp. Neurol. **182**: 17–38.
23. DIERICKX, K. & F. VANDESANDE. 1979. Cell Tissue Res. **201**: 349–359.
24. SWANSON, L. W., P. E. SAWCHENKO, S. J. WIEGAND & J. L. PRICE. 1980. Brain Res. **198**: 190–195.
25. KILCOYNE, M., D. L. HOFFMAN & E. A. ZIMMERMAN. 1980. Clin. Sci. **95**: 57s–60s.

26. Vanderhaeghen, J. J., F. Lotstra, F. Vandesande & K. Dierickx. 1981. Cell Tissue Res. **221:** 227–231.
27. Hou-Yu, A., P. Ehrlich, G. Valiquette, D. L. Engelhardt, W. H. Sawyer, G. Nilaver & E. A. Zimmerman. 1982. J. Histochem. Cytochem. Submitted.
28. Silverman, A. J. & E. A. Zimmerman. 1982. Annu. Rev. Neurosci. Vol. 6. In press.
29. Swanson, L. W. & P. E. Sawchenko. 1980. Neuroendocrinology **31:** 410–417.
30. Alonso, G. & I. Assenmacher. 1981. Cell Tissue Res. **219:** 525–534.
31. Bergland, R. M. & R. B. Page. 1978. Endocrinology **102:** 1325–1338.
32. Krieger, D. T., A. S. Liotta, M. J. Brownstein & E. A. Zimmerman. 1980. Recent Prog. Horm. Res. **36:** 277–344.
33. Uhl, G., R. R. Goodman, M. Kuhar & S. Snyder. 1978. Adv. Biochem. Psychopharmacol. **18:** 71–87.
34. Nemeroff, C. B., G. Bisette, A. J. Prange, Jr., P. T. Loosen, T. S. Barlow & M. A. Lipton. 1977. Brain Res. **128:** 485–496.
35. Watson, S. J., H. Akil, W. Fischli, A. Goldstein, E. Zimmerman, G. Nilaver & T. B. van Wimersma Greidanus. 1982. Science **216:** 85–87.
36. Konig, J. P. and R. A. Klippel. 1963. The Rat Brain, A Stereotaxic Atlas of Forebrain and Lower Parts of the Brainstem. Williams and Wilkins. Baltimore.

Discussion of the Paper

M.D. Hirsch (*Roche Institute of Molecular Biology, Nutley, NJ*): I am impressed by the tremendous overlap of distributions that you have shown between neurotensin and TRH. There is a very high distribution of TRH in ventromedial nucleus of the hypothalamus and you mentioned that you found no staining for NT there.

D. Kahn (*Columbia University College of Physicians and Surgeons, New York*): By radioimmunoassay there is neurotensin in ventromedial hypothalamus. We did not see it in our immunocytochemistry.

Hirsch: Did you also try to localize NT in other regions, like brain stem nuclei, motor nuclei, or septal nuclei?

Kahn: We looked through other nuclei and our results agree with other reports, particularly those of Uhl.

C.B. Nemeroff (*University of North Carolina, Chapel Hill*): I wanted to report on a finding that was reported by John Hong at NIEHS which I have never understood and I think you have explained it. He reported in "Regulatory Peptides" that if you lesion the arcuate nucleus with monosodium glutamate in neonates, no change in hypothalamic neurotensin concentration was observed, suggesting there is indeed no tubero-infundibular neurotensin pathway. This may have been missed by not using the Palkovitz punch technique and taking out individual nuclei. Considering the concentration of NT in the hypothalamus, perhaps the contribution of the arcuate is very small, relative to the paraventricular nucleus, but if he had taken out individual nuclei he might have seen a difference.

Kahn: I agree with what you said.

DISTRIBUTION OF NEUROTENSIN AND ITS RECEPTOR IN THE CENTRAL NERVOUS SYSTEM

George R. Uhl

Department of Neurology
Johns Hopkins Hospital
Baltimore, Maryland 21205

INTRODUCTION

Neurotensin-containing and neurotensin-receiving cells have an uneven distribution through the central nervous system. The local dispositions of these systems provide clues about which brain neurons and circuits might selectively utilize this peptide, and hints of their possible normal functional significance. Studies of neurotensin circuitry may also suggest pathologic processes involving this peptide and point to possible targets for future neurotensin-related therapeutic agents.

Our understanding of the distribution of neurotensin itself is largely based on radioimmunoassay and immunohistochemical techniques. Such studies must be viewed with some caution; the specificity of antibody-mediated immunorecognition is not absolute. Nevertheless, convergence of evidence obtained using physicochemical techniques to separate differing immunoreactive materials and employing different antisera recognizing different portions of the molecule[1-11] suggests that at least much of the immunoreactivity recognized in radioimmunoassay and immunohistochemical studies is authentic neurotensin.

Neurotensin receptors are demonstrated by binding techniques, both in tissue homogenates and in cryostat-cut sections.[12, 17]. Characterization of binding in both systems reveals similar high affinity, reversible binding with properties expected of physiologic receptor sites. Available neurotensin-related drugs and sequence analogs have potencies in both systems that parallel their potencies in physiologic test systems, but the small number of these neurotensin-related sequences available limits the strength of this argument. Although data from binding studies must be judged carefully, these factors all suggest that the regional localization of neurotensin binding sites presented here reflects a physiologic brain distribution of actual neurotensin receptors.

In this chapter, I summarize and compile the available data concerning the brain distributions of pre- and postsynaptic neurotensin systems that originate from several laboratories.[1, 17] Where possible, I attempt to relate these localizations with brain functions that neurotensin might subserve at each site.

SPINAL CORD

Test-tube-receptor binding and radioimmunoassay studies show low total levels of neurotensin and its receptor in whole cervical spinal cord (see refs. 1, 3, 5, 7, 13; TABLE 1). Both immunohistochemical and receptor autoradiographic studies show that these systems are concentrated in the dorsal horn substantia gelatinosa (see refs. 8, 9, 17, 18; FIG. 1). Much of the neurotensin in this zone may arise locally; neurotensin-rich cell bodies are found in lamina III.[9, 19, 20]

The substantia gelatinosa appears to be an important site for modulation of nociceptive input from the periphery. Substantia gelatinosa interneurons that appear to utilize another peptide, enkephalin, may make synapses on the central processes of primary afferent fibers. Following dorsal rhizotomy, dorsal horn opiate receptors are substantially reduced.[18, 21] In parallel studies, however, no reduction in dorsal horn neurotensin receptors follows rhizotomy.[18] This finding suggests that neurotensin receptors are unlikely to be concentrated on the terminals of primary afferents.

Ultrastructural studies of the gelatinosa interneurons show neurotensin immunoreactivity over small, unmyelinated axons and 0.5–2 μm terminals. These processes make asymmetric synapses onto dendrites and spines, though axoaxonic synapses are not observed.[18, 22] Large, 100 nm granular vesicles within these terminals are labeled by the neurotensin reaction product, while lucent, 40 nm

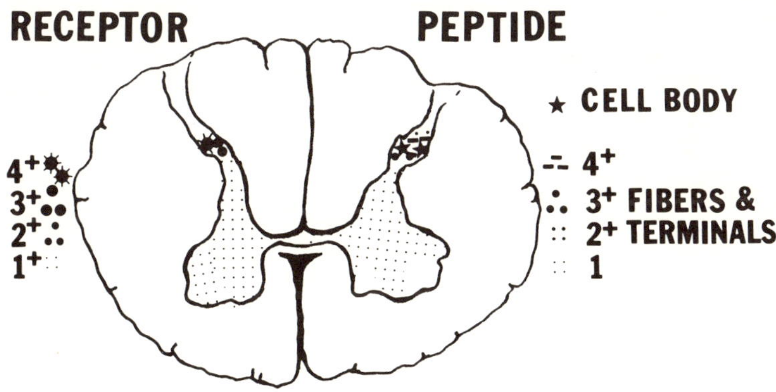

FIGURE 1.   Distribution of neurotensin immunoreactivity (*right side*) and receptor grains (*left side*) in the cervical spinal cord (refs. 8, 18, 20; level from ref. 47).

vesicles contained in the same terminals are not labeled (see ref. 22; Fig. 2). Selective labeling of the larger vesicles in this fashion has also been described in electron microscopic studies of other peptides such as enkephalin and substance P.[23, 24]

Electrophysiologic studies support the presence of neurotensin-receptive elements in the cord in a similar distribution. Units excited by microiontophoretically applied neurotensin are located in lamina I–III, and are more likely to be activated by noxious stimuli than units not affected by neurotensin[25] (but see Henry, this volume).

Neurotensin systems in the spinal cord are thus concentrated in a small zone, the substantia gelatinosa. If neurotensin can produce its analgesic effects at a spinal level (see ref. 18, and Yaksh, this volume), it does not appear to do so by directly inhibiting primary afferent input. Rather, neurotensin interneurons appear to excite other neurons that are preferentially sensitive to noxious stimuli. These non-neurotensin-containing cells in turn modulate central transmission of noxious information. Naloxone antagonism of this effect, further, might suggest that the interposed neuron is enkephalinergic (Yaksh, this volume).

TABLE 1

SUMMARY OF BRAIN DISTRIBUTIONS OF IMMUNOREACTIVE NEUROTENSIN AND NEUROTENSIN RECEPTORS

		IMMUNOREACTIVITY					RECEPTOR
Region	Species Reference Units	Rat Koboyoshi et al.[3] ng/mg protein	Rat Emson et al.[5] pmole/g wet	Calf Uhl & Snyder[1] pmole/g wet	Monkey Kataoko et al.[50] pg/mg wet	Human Cooper et al.[7] pmole/g wet	Calf Uhl et al.[13] (frontal pole=1.0)
Cerebral cortex							
Frontal			4.1	1.8	7.2	0.8	1.0
Cingulate		<0.2	5.3	3.8	15.7	2.1	1.3
Parietal			5.2	2.3	6.3	0.8	0.9
Entorhinal/Parahip- pocampal		0.06	5.8	6.7		2.3	1.8
Uncal						4.9	
Occipital/Striate				2.5	4.0	0.7	1.3
Cortical white matter		0.06		0.7		1.1	0.3
Olfactory bulb		0.06	3.5		12.7	4.4	
Basal forebrain							
Accumbens		1.37	29.2				
Septum		0.68	73.0		5.2		
Caudate & Putamen		0.03	10.2				
Caudate				11.1	8.5	2.9	1.0
Putamen					13.2	2.5	0.9
Preoptic							
Lateral		2.55	124		44.5		
Medial		2.32	143		51.5		
Globus pallidus			12.6	9.3		9.8	0.8
Bed nucleus of the stria terminalis			171.8				
Hippocampus		<0.2	7.1	3.0	13.3	4.4	0.9
Amygdala-whole		0.65		1.9		5.4	1.0
central			106		19.2		
medial			40		18.8		

Hypothalamus						
Anterior		113	17.3			1.4
Medio-basal						
Posterior						
Suprachasmatic	1.28					
Retrochasmatic	0.83					
Arcuate	0.98			70.8		
Median eminence	3.38	128		85.5	42.6	
Lateral		76				
Mamillary		128	11.1	26.0		1.3
Ventral tegmentum/Brainstem		35.5		48.7		
Substantia nigra	0.29	13.3			23.4	
Interpeduncular	0.85			78.9		
Red N.				29.3	10.3	
Pariaqueductal gray	0.75	42.2		74.1	22.4	
Locus coeruleus		54		25.3		
Colliculus-superior			2.7		13.9	0.8
Colliculus-inferior				15.2	16.2	
Pons	0.10		1.3			0.6
Inferior olive						
Cerebellum	<0.02	0.9	<0.32	4.5	0.8	
Thalamus/Epithalamus						
Whole	0.08					
Anterior			4.3	28.0		1.5
Dorsomedial		23.7	2.5	27.4	6.0	2.5
Lateral posterior		5.1				
Lateral geniculate		6.9		7.8	1.9	
Subthalamic				47.7	9.7	
Habenula		47		35.7		
Zona incerta				43.4		
Pineal	<.02					
Spinal Cord–Cervical						
whole	0.14		1.1			0.1
dorsal		23.7				
ventral		2.0				

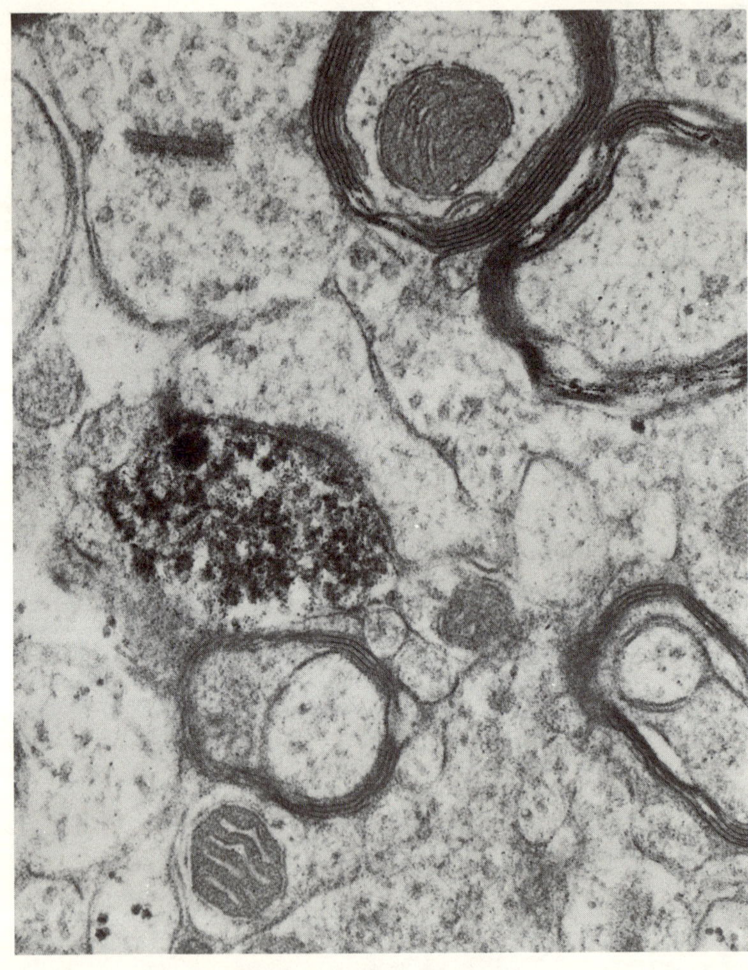

FIGURE 2.   Neurotensin immunoreactivity overlies large granular vesicles in spinal cord dorsal horn terminals (ref. 22; photograph courtesy of Dr. V. Seybolt).

Medulla Oblongata and Pons

Immunoreactivity and receptor binding in the medulla is low but appreciable.[1, 4, 5, 7, 13] The "substantia gelatinosa" zone of the trigeminal nuclear complex contains substantial densities of neurotensin immunofluorescent fibers, terminals, and cell bodies (see ref. 26; Fig. 3). Dense receptor grains are found in this region as well.[17] This concentration of neurotensin systems in another pain-modulating zone again suggests a possible anatomic locus for the peptide's observed antinociceptive activity.[27, 29]

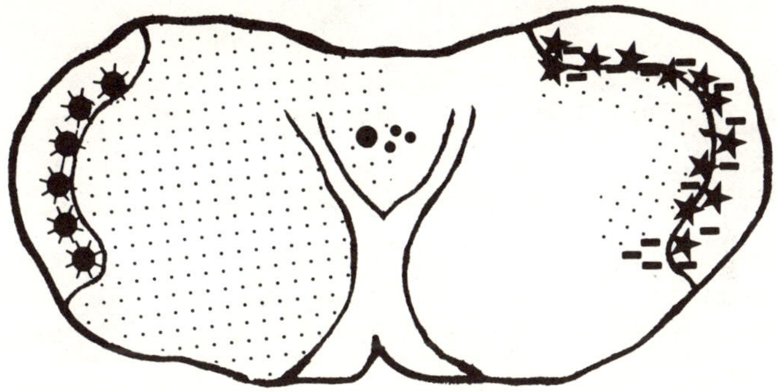

FIGURE 3.   Distribution of neurotensin immunoreactivity (*right side*) and receptor grains (*left side*) at the level of the decussation of the pyramids. Key accompanies FIGURE 1 (refs. 17, 22; level = P 8.0, from ref. 48).

The nucleus of the solitary tract and the floor of the fourth ventricle contain densities of neurotensin receptor grains, cell bodies, and fiber-terminal patterns.[17, 26] These areas play key roles in visceral autonomic regulation. The presence of neurotensin-containing and receiving systems in these regions points to a substantial role for the peptide in autonomic reflexes.

The locus coeruleus shows elevated levels of immunoreactive neurotensin in radioimmunoassay studies.[5, 7] Moderate levels of autoradiographic receptor grains, and fiber and terminal patterns are also observed in histologic studies.[17, 26] Some locus perikarya display neurotensin immunofluorescence as well (FIG. 4; see also refs. 26, 9). This finding could suggest that a small number of locus coeruleus cells may be nonadrenergic, or possibly represent a co-localization of peptide and monoamine transmitter as has recently been reported in the dopamine system.[30] Electrophysiologic study of the locus coeruleus also reveals physiologic neurotensin receptors. Iontophoretically applied neurotensin inhibits almost half of the units tested in the locus, while very few units lying ventral to the locus respond.[75]

The parabrachial nuclei display a dissociation between dense neurotensin cell body and fiber/terminal immufluorescence (FIG. 4), and relatively modest receptor grain densities.[17, 26] Such a dissociation is noted in several brain regions, and may reflect the fact that some cell-body-rich but receptor-poor areas may send projections to zones that are more cell poor and receptor rich.

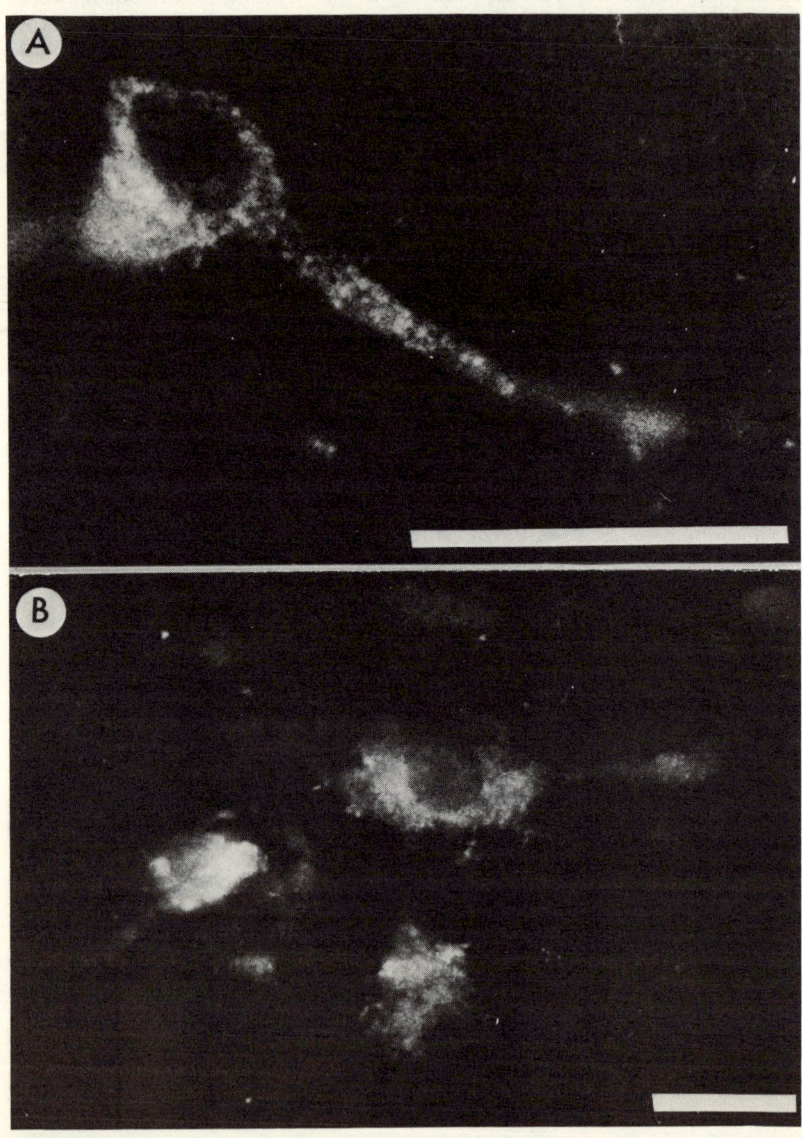

FIGURE 4. Photomicrograph of neurotensin immunoreactive perikarya from the parabrachial nucleus (A), and the locus coeruleus (B). Bars = 25 μm (ref. 26).

MESENCEPHALON

The periaqueductal gray, ventral tegmental area, and dorsal raphe show high levels of immunoreactivity in radioimmunoassay studies.[3, 5, 7] Although all of these regions contain moderate to high densities of fibers and terminals, as well as cell bodies, the corresponding receptor densities vary substantially.

The periaqueductal gray region contains moderate densities of immunoreactive perikarya, fibers, terminals, and receptor grains.[17, 26] This region is broadly implicated in central control of pain and stimulation-produced analgesia and could serve as another site for the antinociceptive action of centrally administered neurotensin.

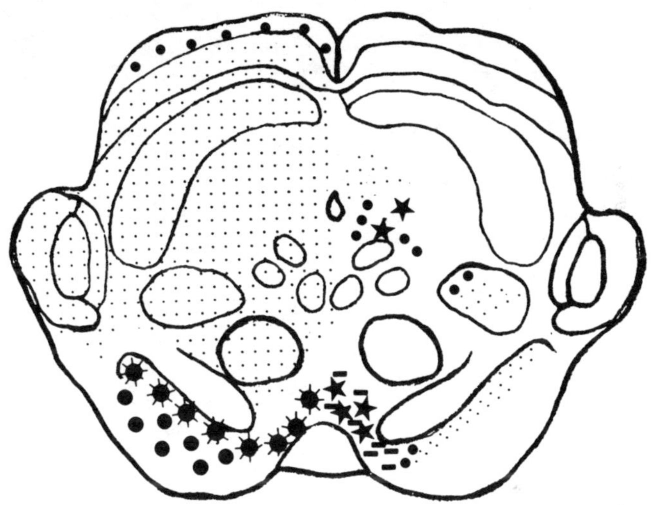

FIGURE 5. Distribution of neurotensin immunoreactivity (right side) and receptor grains (left side) at the level of the interpeduncular nucleus. Key accompanies FIGURE 1. (See refs. 17, 22; level = A 1.8, from ref. 48).

The dorsal raphe possesses concentrations of neurotensin fibers, terminals, and perikarya but has relatively few associated receptor grains.[17, 26] Conversely, the median raphe displays a concentration of receptor grains, accompanied by only moderate densities of fibers and terminals. Such a dissociation may again suggest that the dorsal raphe gives rise to neurotensin-containing projections, while the median raphe may receive such projections.

The ventral tegmental area contains high densities of neurotensin perikarya, fibers, terminals, and receptor grains (see refs. 8, 17, 26; FIG. 5). This density of fiber-terminal immunoreactivity merges with a lower density of fibers and terminals in the substantia nigra.[26] The nigra, however, reveals dense neurotensin-receptor grains, especially over the pars compacta.[17] Nigral lesions with 6-hydroxydopamine dramatically deplete these receptors, suggesting that they are located on the cell processes and perikarya of nigral dopaminergic cells (see ref. 32; FIG. 6). Other evidence for substantial neurotensin influence on these dopamin-

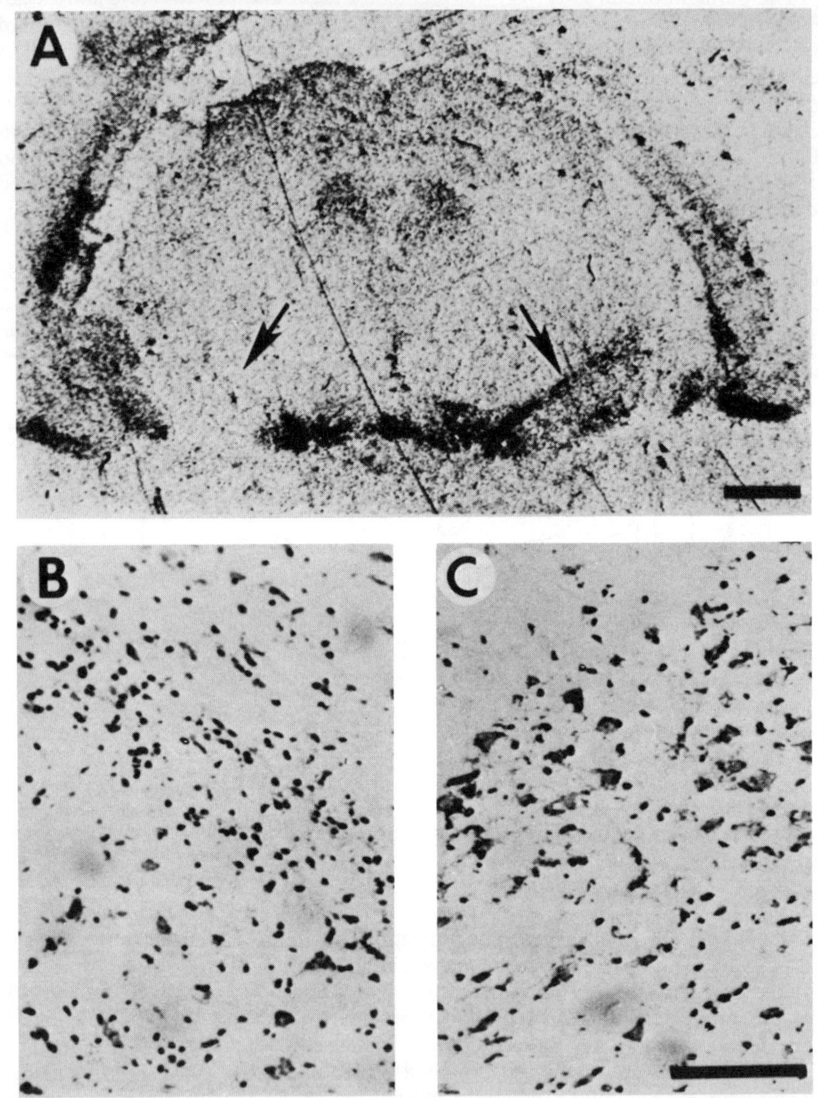

FIGURE 6. Effects of unilateral 6-hydroxydopamine injection on neurotensin receptor grains (A, left side) and pars compacta neurons (B). The uninjected right side of the autoradiogram in A, and the corresponding histologic photomicrograph in C serve as control. Bars = 100 μm (ref. 32; photograph courtesy of Dr. M. Kuhar).

ergic systems comes from electrophysiologic experiments. Microiontophoresis of neurotensin onto compacta cells produces a rapid and powerful excitation.[33]

Several human neurologic disorders involve abnormalities of movement attributable to altered "tone" in the striato-nigral circuits involving these "neurotensinoceptive" dopaminergic cells. Parkinson's disease, clinically characterized by decreased ability to spontaneously initiate movements, is associated with a loss of compacta dopaminergic cells. One may predict that this cellular loss should be accompanied by reductions in nigral neurotensin receptor density. Centrally acting neurotensin agonists could conceivably augment the function of the remaining dopaminergic neurons in Parkinsonian patients, and perhaps supplement currently used dopamine agonists and precursors in treating the disease. Huntington's disease, an inherited degenerative disorder characterized by choreoathetoid adventitious movements, may be attributable to relative overactivity of the nigro-striatal systems as the balancing intrastriatal and striato-nigral inhibitory mechanisms are lost. Intrastriatal neurotensin concentrations are actually increased in Huntington's (ref. 34, Manberg *et al.*, this volume), possibly due to relative preservation of neurotensin systems in the face of striatal shrinkage. Centrally acting neurotensin antagonists could conceivably reduce some of this excess of dopaminergic (and possibly neurotensinergic) tone in Huntingtonian patients, and ameliorate the movement disorder. Tardive dyskinesia is a Huntington-like movement disorder associated with prolonged neuroleptic drug use and presumed dopamine receptor supersensitivity. This disorder might be especially susceptible to a potential neurotensin antagonist that could reduce dopaminergic tone without the increased dopamine receptor blockade caused by currently used agents that may exacerbate the underlying process.

Hypothalamus and Preoptic Area

The hypothalamus contains high concentrations of neurotensin and its receptor. This region seems a likely site for several of neurotensin's neuroendocrine functions, and for its impact on thermoregulatory processes.[19, 34, 37]

Radioimmunoassay studies of microdissected hypothalamic regions suggest that concentrations of neurotensin immunoreactivity are present in many hypothalamic and preoptic nuclei, with especial concentrations in the preoptic and median eminence zones (see refs. 3, 5, 7; Table 1).

The posterior hypothalamus is low in neurotensin immunoreactivity and immunostaining, as are most of the mammillary and premammillary nuclei.[3, 5, 7] In the posterior mammillary nucleus, however, dense fiber and terminal patterns are noted.[6]

At the level of the dorsal and ventral hypothalamic nuclei (Fig. 7), modest levels of fibers, terminals, and receptor grains are noted over the ventral hypothalamic nucleus.[6, 8, 17] The region of the dorsal nucleus displays scattered perikarya, with moderate densities of fibers, terminals, and receptor grains. Moderate densities of neurotensin-positive cell bodies, fibers, and terminals in the lateral hypothalamic nucleus accompany few receptor grains.[6, 8, 17]

The mediobasal hypothalamus and medial eminence contain densities of neurotensin-containing fibers and terminals, while reactive perikarya are found in the adjacent arcuate nucleus (refs. 6, 8; Fig. 8). The basal hypothalamus is not rich in neurotensin receptors, though the median eminence has not been subjected to detailed autoradiographic examination.[17] This density of fibers and paucity of

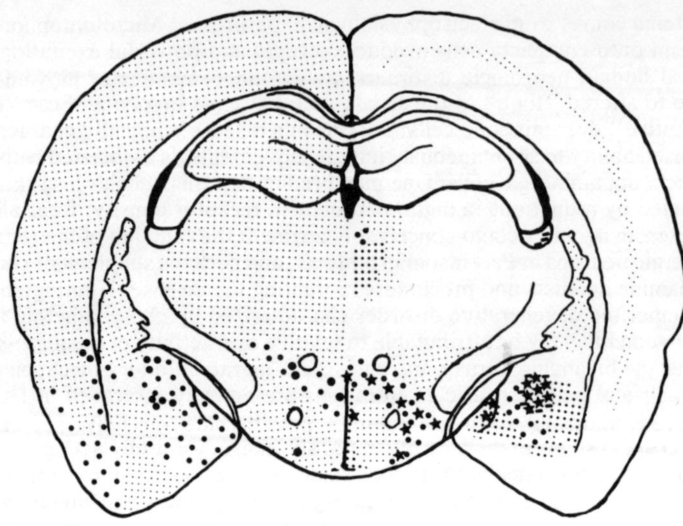

FIGURE 7. Distribution of neurotensin immunoreactivity (*right side*) and receptor grains (*left side*) at the level of the dorsal and ventral hypothalamic nuclei. Key accompanies FIGURE 1. (see refs. 6, 8, 17; level = A 4230, from ref. 49).

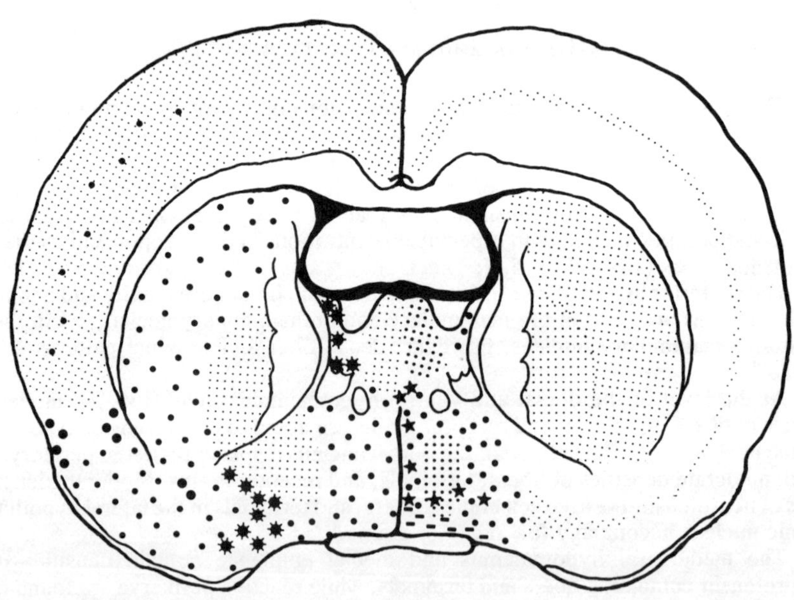

FIGURE 8.   Distribution of neurotensin immunoreactivity (*right side*) and receptor grains (*left side*) at the levels of the anterior hypothalamic nucleus. Key accompanies FIGURE 1. (see refs. 6, 8, 17; level = A 6280, from ref. 49).

receptors would be compatible with the presence of rostro-caudally and ventrally directed neurotensin fibers that funnel into the zona externa of the median eminence.[6] Some of these fibers can be traced immunocytochemically through the pituitary stalk to the level of the posterior pituitary.[6]

Densities of neurotensin-containing cell bodies, fibers, and terminals are found in both medial and lateral preoptic areas (see refs. 6, 8; FIG. 9). At the level of the anterior hypothalamic nucleus, moderate levels of fibers and terminals and receptor densities are found in anterior, periventricular, and portions of the lateral hypothalamic nucleus. Neuronal perikarya are fairly dense in the paraventricular zone, sparsely scattered over the anterior nucleus, and more concentrated in the ventral aspect of the lateral nucleus.

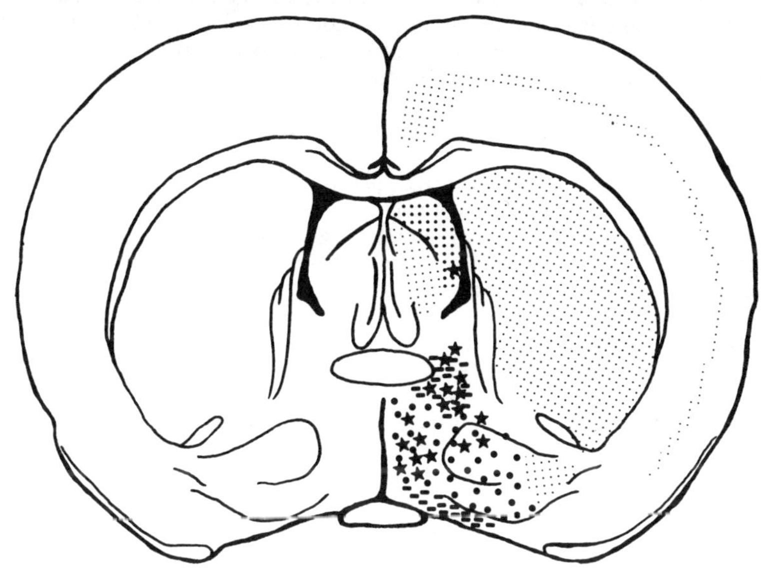

FIGURE 9.   Distribution of neurotensin immunoreactivity (*right side*) (receptor autoradiographic studies not reported at this level) at the level of the preoptic nuclei (refs. 6, 8; level = A 6860, from ref. 49).

THALAMUS/EPITHALAMUS

The thalamus displays modest densities of immunoreactive neurotensin by radioimmunoassay.[1, 3, 5, 7] Low levels of fibers and terminals are more prominent in medial than in lateral zones, and autoradiographic receptor grains are present at low levels.[8, 17] By contrast, homogenate receptor binding studies show relatively higher levels of binding in thalamic preparations.[13, 15] The reasons for this discrepancy are not clear.

The zona incerta displays a modest number of immunoreactive fibers and terminals, but shows a more prominent density of receptor grains.[6, 8, 17]

Basal Forebrain

Several basal forebrain structures show high densities of neurotensin containing and receiving elements. A neurotensin-utilizing pathway connects two of the densest regions, the central nucleus of the amygdala and the interstitial nucleus of the stria terminalis (ref. 33; Figs. 7–10).

In the amgydala, dense cell bodies, fibers, and terminals are concentrated in the central nucleus, although moderate fiber-terminal densities and immunoreactive perikarya are distributed over several other amygdaloid nuclei.[6, 8, 38] Neurotensin receptors are densely clustered just dorsal to, and form a band ventral to, the central nucleus.[17] As discussed above, this dissociation between densities of neurotensin-containing cells and neurotensin receptors may mark the presence of a neurotensin pathway. Observations that fibers from the central amygdala appear to join the stria terminalis, and that the stria contains a dense collection of neurotensin fibers suggest that an amygdaloid efferent pathway travels through the stria terminalis. Lesion studies support this concept; stria immunoreactivity is depleted following electrolytic destruction of the central amgydala, "piles up" caudal to knife cuts interrupting the stria, and disappears rostral to such knife cuts.[38]

One zone of termination of the stria terminalis, the interstitial nucleus of the stria terminalis, also possesses dense neurotensin fibers, terminals, and cell bodies.[8, 38] Autoradiographic receptor grains are concentrated here, and *in vitro* electrical studies in slices of this nucleus suggest that physiologic neurotensin receptivity is present.[39] When applied to the bath, neurotensin excites two-thirds of the sampled neurons in this region.

The nucleus of the diagonal band of Broca and lateral septal nucleus contain densities of receptors, fibers, nerve terminals, and neurotensin-positive perikarya.[8, 9, 17]

In the nucleus accumbens, moderate-to-low densities of neurotensin fibers, terminals, and receptor grains are seen.[9, 17] Local administration of neurotensin into the accumbens markedly decreases locomotor and rearing behaviors elicited by systemic amphetamine.[40] This is another site for interaction between neurotensin- and dopamine-containing systems, since neurotensin presumably exerts this effect by reducing amphetamine-induced dopamine release from the known accumbens dopamine neurons. Interestingly, this neurotensin/dopamine interaction may be reciprocal; neuroleptics increase the neurotensin content of the accumbens.[41] Microiontophoresed peptide in the accumbens elicits inhibition of firing.[42]

Several structures associated with olfaction appear to use neurotensin as a transmitter. The olfactory bulb contains significant densities of receptors, fibers, and terminals as well as scattered cell bodies.[9, 17]

Basal Ganglia

Modest densities of fibers, terminals, and receptors are noted in the caudate, putamen, and globus pallidus, but cell bodies are not observed in this region.[8, 17] Striatal dopamine systems are altered by neurotensin. Striatal dopamine levels decline as brain dopamine metabolites increase following central neurotensin administration.[43, 44]

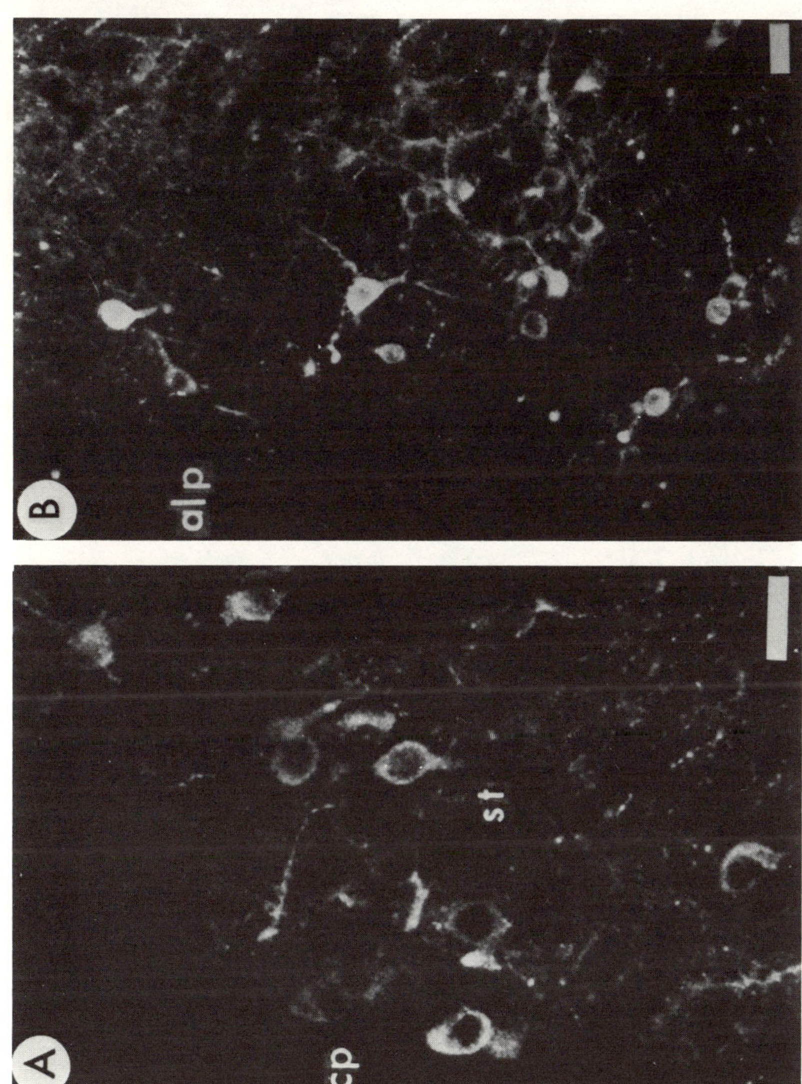

FIGURE 10. Photomicrograph of neurotensin immunoreactivity. (A.) The interstitial nucleus of the stria terminalis (st) and the striatum (cp), and (B). The central amygdaloid nucleus (ac) and the lateral-posterior nucleus (alp) (ref. 38; bars = 25 μm).

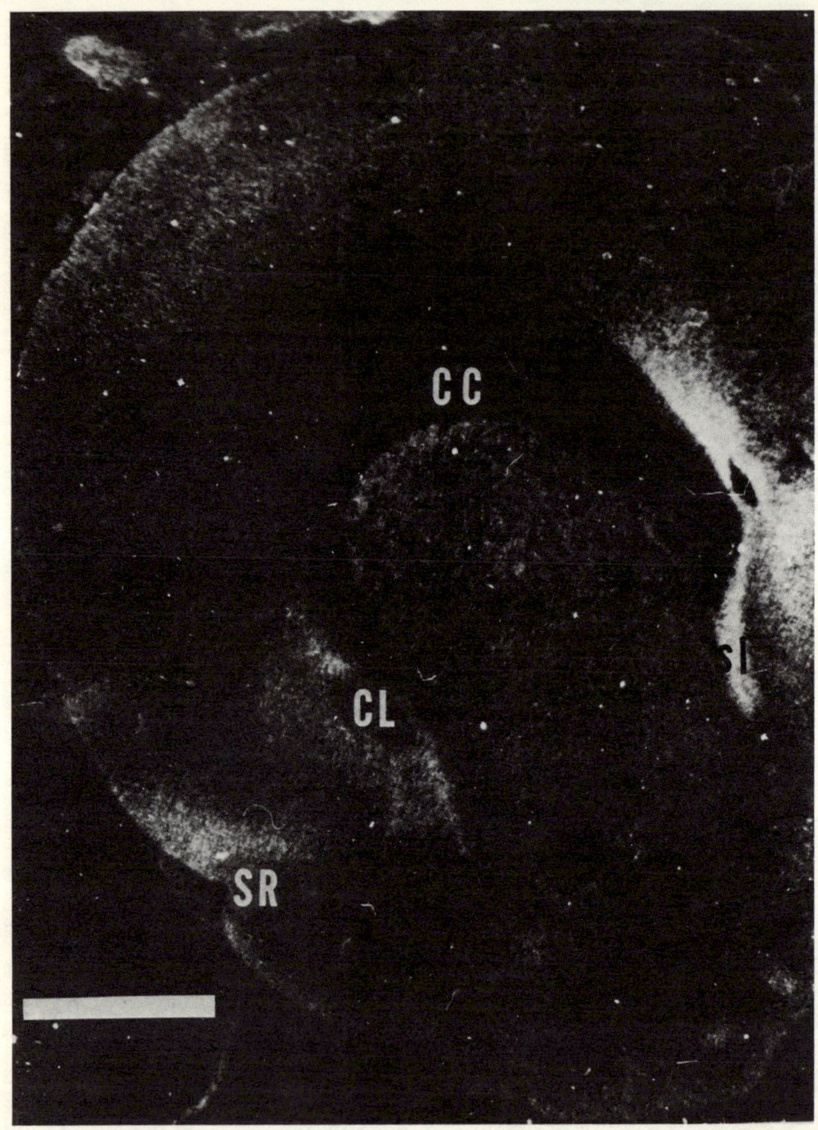

FIGURE 11. Photomicrograph of neurotensin receptor grains in the rat forebrain (see ref. 17; level = 9650, from ref 49, abbreviations in the reference; photograph courtesy of Dr. Michael Kuhar).

CEREBRAL CORTEX

The low cortical levels of neurotensin immunoreactivity and moderate levels of receptor binding are distributed unevenly across cortical regions and through cortical laminae (FIG.s. 7–9). Olfactory, entorhinal, and cingulate cortical areas show relative elevations of both neurotensin-like immunoreactivity and receptor binding.[1, 7, 13] Although neurotensin receptor grains and sparsely-scattered fiber/terminal patterns may be found throughout the layers of the cortex, there is a tendency for both receptors and fiber/terminal patterns to be more concentrated in the middle layers of the cortex.[8, 17] A cortical neurotensin-containing pathway recently described by Roberts, Crow, and Polyak[45] is detailed elsewhere in this volume.

RETINA

Neurotensin is found in the retina by radioimmunoassay and immunohistochemistry, where it is localized to amacrine cells with perikarya in the inner nuclear layer.[46]

CONCLUSION

This central nervous system distribution of neurotensin suggests sites for interplay of the peptide with several brain functions and neurotransmitter systems. Olfactory, visual and nociceptive sensory inputs are apparently intimately interrelated to neurotensin systems. Certain central motor control mechanisms involving the basal ganglia and substantia nigra are rather powerfully influenced by the peptide. Thermoregulatory and neuroendocrine processes are potently modulated by neurotensin systems. Neurotensin must interact with several different neurotransmitter systems in the brain. The substantial evidence reviewed here for interaction with dopaminergic systems and the iontophoretic data suggesting influence on noradrenergic neurons probably represent only a fraction of possible neurotensin effects on other brain neurotransmitter systems.

ACKNOWLEDGMENTS

Personal studies quoted here were performed under the excellent tuteledge of Dr. Solomon Snyder, with the collaboration of Drs. J. P. Bennett, Jr., Michael J. Kuhar, Robert Goodman, and W. S. Young, III. Dr. Kuhar has greatly facilitated preparation of this manuscript and Carol Kenyon generously assisted with manuscript preparation.

REFERENCES

1. UHL, G., & S. H. SNYDER. 1976. Life Sci. **19:** 1827–1832.
2. UHL, G. & S. H. SNYDER. 1980. *In* Role of Peptides in Neuronal Function. J. Barker & T. Smith, Eds. Dekker. New York. pp. 509–543.
3. KOBAYASHI, R., M. BROWN & W. VALE. 1977. Brain Res. **126:** 584–588.
4. CARRAWAY, R. & S. LEEMAN. 1976. J. Biol. Chem. **251:** 7045–7052.

5. EMSON, P., M. GOEDERT, P. HORSFIELD, F. RIOUX & S. ST. PIERRE. J. Neurochem. In press.
6. KAHN, D., G. ABRAMS, E. ZIMMERMAN, R. CARRAWAY & S. LEEMAN. 1980. Endocrinology **107**: 47–54.
7. COOPER, P., M. FERNSTROM, O. RORSTAD, S. LEEMAN & J. MARTIN. 1981. Brain Res. **218**: 219–232.
8. UHL, G., M. KUHAR & S. H. SNYDER. 1977. Proc. Natl. Acad. Sci. USA **74**: 4057–4063.
9. JENNES, L., W. STUMPF, W. BECKMAN, S. BURGESS, D. LUTTINGER, C. NEMEROFF & A. PRANGE. 1981. Neurosco. Abstr. **7**: 223.
10. YAMADA, S. & S. MIKAMI. 1981. Cell Tissue Res. **218**: 29–39.
11. BRECHA, N., H KARTEN & C. SCHENKLER. 1981. Neuroscience **6**: 1329–1340.
12. UHL, G., J. BENNETT & S. H. SNYDER. 1976. Soc. Neurosci. Abstr. **2**: 803.
13. UHL, G., J. BENNETT & S. H. SNYDER. 1977. Brain Res. **130**: 299–313.
14. KITABGI, P., R. CARRAWAY, J. VON RIETSCHAFEN, C. GRAUIER, J. MORGAT, A. MENEZ, S. LEEMAN & P. FREYCHET. 1977. Proc. Natl. Acad. Sci. USA **74**: 1846–1850.
15. LAZARUS, L., M. BROWN & M. PERRIN. 1977. Neuroendocrinology **16**: 625–629.
16. YOUNG, W. & M. KUHAR. 1970. Eur. J. Pharmacol. **59**: 161–163.
17. YOUNG, W. & M. KUHAR. 1981. Brain Res. **206**: 273–285.
18. NIUKOVIC, M., S. HUNT & J. KELLY. 1981. Brain Res. **230**: 111–119.
19. UHL, G. & S. H. SNYDER. 1981. *In* Neurosecretion and Brain Peptides: Implications for Brain Function and Neurological Disease. J. Martin & K. Bick, Eds. Raven Press. New York.
20. SEYBOLD, V. & R. ELDE. 1980. J. Histochem. Cytochem. **28**: 367–370.
21. LAMOTTE, C., C. PERT & S. H. SNYDER. 1976. Brain Res. **112**: 407–412.
22. SEYBOLD, V. & B. MALEY. 1981. Neurosci. Abstr. **7**: 59.
23. HUNT, S., J. KELLY & P. EMSON. 1980. Neuroscience **5**: 1871–1890.
24. PICKEL, V., D. REIS & S. LEEMAN. 1977. Brain Res. **122**: 534–540.
25. MITFEUIC, V. & M. RANDIC. 1979. Brain Res. **169**: 600–604.
26. UHL, G., R. GOODMAN & S. H. SNYDER. 1979. Brain Res. **167**: 77–91.
27. CLINESCHMIDT, B. & J. McGUFFIN. 1977. Eur. J. Pharmacol. **46**: 395–396.
28. NEMEROFF, C., A. OSBAHR, P. MANBERG, G. ERVIN & A. PRANGE. 1979. Proc. Natl. Acad. Sci. USA **76**: 5368–5371.
29. CLINESCHMIDT, B., J. McGUFFIN & P. BUNTING. 1979. Eur. J. Pharmacol. **54**: 129–139.
30. HÖKFELT, T., J. REHFELD, L. KIRBALL, B. IVEMARK, M. GOLDSTEIN & K. MARKEY. 1980. Nature **285**: 476–478.
31. YOUNG, W., G. UHL & M. KUHAR. 1978. Brain Res. **150**: 431–435.
32. PALACIOS, J. & M. KUHAR. 1981. Nature **294**: 587–589.
33. ANDRADE, R. & G. AGHAJANIAN. 1981. Nuerosci. Abstr. **7**: 573.
34. NEMEROFF, C. & P. MANBERG. Personal communication.
35. BISSETTE, G., C. NEMEROFF, P. LOOSEN, A. PRANGE & M. LIPTON. 1976. Nature **262**: 607–609.
36. NEMEROFF, C., A. OSBAHR, P. MANBERG, G. ERVIN & A. PRANGE. 1979. Proc. Natl. Acad. Sci. USA **786**: 5368–5371.
37. MAEDA, K. & L. FROHMAN. 1978. Endocrinology **103**: 1903–1909.
38. UHL, G. 1979. Brain Res. **161**: 522–526.
39. SAWADA, S., S. TAKADA & C. YAMAMOTO. 1980. Brain Res. **188**: 578–581.
40. ERVIN, G., L. BERKEMO, C. NEMEROFF & A. PRANGE. 1981. Nature **291**: 73–76.
41. GOVONI, S., J HONG, H. YANG & E. COSTA. 1980. J. Pharmacol. Exp. Ther. **215**: 413–417.
42. McCARTHY, P., R. WALKER, H. YAJIURA, K. KITAGAWA & G. WOODRUFF. 1979. Gen. Pharmacol. **10**: 331–333.
43. HAUBRICH, D., G. MARTIN, B. PFUEGER & M. WILLIAMS. 1982. Brain Res. **231**: 216–221.
44. WIDERLOV, E., C. KITTS, R. MUELLER, R. MAILMAN, C. NEMEROFF, A. PRANGE & G. BREESE. 1981. Pharmacologist **23**: 1301.

45. ROBERTS, G., T. CROW & J. POLAK. 1981. Peptides 2, suppl. **1:** 37–43.
46. BREDNA, N., H. KARTEN & C. SCHENKER. 1981. Neuroscience **6:** 1320–40.
47. STEINER, T. & L. TURNER. 1972. J. Physiol. (London) **222:** 123–125.
48. PALKOVITZ, M. & L. JACOBOWITZ. 1974. J. Comp. Neurol. **157:** 29–42.
49. KONIG, J. & R. KLIPPEL. 1963. The Rat Brain. Williams & Wilkens. Baltimore.
50. KATAOKO, K., N. MIZUNO & L. FROHMAN. 1970. Brain Res. Bull. **4:** 57–60.

DISCUSSION OF THE PAPER

A.S. TISCHLER, (*Tufts University School of Medicine Boston*): Are any of your data consistent with the possibility that some of the same cells which contain neurotensin might also have neurotensin-receptors, so that neurotensin might serve in those cells as a presynaptic or autoregulatory transmitter?

G.R. UHL (*Johns Hopkins Hospital, Baltimore, MD*): The level of resolution of light microscopic autoradiography with the Young-Kuhar technique is not really sufficient to answer that question. This technique is much better for looking at regions of the brain, detecting localization at that level of resolution.

S.C. FELDMAN (*New Jersey Medical School*): Is there any neurotensin in the hippocampal formation or in the fornix? This would be quite different from many of the other peptides.

UHL: We have seen data earlier in the day to suggest that there are hippocampal cell bodies that provide a projection to the cortex. That is not something that I have found. In fact, I have not personally looked at that area in colchicinized rats.

THE ROLE OF NEUROTENSIN IN THE REGULATION OF CARBOHYDRATE METABOLISM AND IN DIABETES*

Michael Berelowitz and Lawrence A. Frohman

*Division of Endocrinology and Metabolism
Department of Internal Medicine
University of Cincinnati College of Medicine
Cincinnati, Ohio 45267*

INTRODUCTION

Neurotensin (NT) a tridecapeptide originally isolated from bovine hypothalamic extracts[1] has now been identified in extrahypothalamic neural tissues, gastrointestinal tract, and the pancreas.[2] One of the first effects noted after *in vivo* administration of synthetic NT was hyperglycemia associated with depletion of liver glycogen stores and activation of liver glycogen phosphorylase A.[3] NT was without effect on hepatic glycogenolysis *in vitro*[3] so studies were directed toward identification of possible neural and hormonal mediators of these metabolic effects. Rapid elevation of plasma glucagon with inappropriately low insulin levels occurring with NT administration[4] provides a hormonal influence favoring hepatic glucose production. Inhibition of glucagon secretion following NT by somatostatin or restoration of insulin secretion by α-adrenergic blockade or adrenal autotransplantation[5] partially reduced the glucose response suggesting that pancreatic hormones provided only part of the mechanism of NT action. Direct effects of NT on rat islets have been demonstrated *in vitro*.[6] At low glucose levels, NT stimulates release of insulin, glucagon, and somatostatin while release of these peptides stimulated by glucose or arginine is inhibited by NT. It thus appears that NT causes hyperglycemia through a direct effect on pancreatic hormone release combined with α-adrenergic inhibition of insulin secretion. The close similarity between the metabolic effects of NT and those of histamine[7] and the reversal of NT effects by H_1 and H_2 histamine-receptor blockers[7,8] suggest tht the actions of the peptide are mediated through the histamine receptor.

Immunoreactive-NT (IR-NT) has been demonstrated in plasma extracts and plasma IR-NT elevation occurs in response to nutrient stimuli[9] suggesting that NT may influence pancreatic hormone secretion by a hormonal mechanism. IR-NT is present in the pancreas[10] and a direct (paracrine) influence on islet hormone secretion is also possible.

Experimental diabetes mellitus is associated with an increase in pancreatic IR-NT content that is restored to normal following insulin treatment.[10] This suggests an influence of hyperglycemia or circulating insulin on pancreatic NT. The present studies were designed to further evaluate the effects of experimental diabetes on hypothalamic and pancreatic IR-NT content and release.

*This research was supported in part by USPHS grants AM 30686 and AM 30667. Dr. Berelowitz is the recipient of a Career Development Award of the Juvenile Diabetes Foundation.

150

MATERIALS AND METHODS

Male Sprague-Dawley rats weighing 200–250 g were housed in conditions of constant temperature with a 12 hr light/dark cycle and free access to laboratory rat chow and water. Animals were injected via a tail vein with streptozotocin (65 mg/kg) in cold citrate buffer (0.1 M, pH 4.5) or with citrate buffer alone. The occurrence of diabetes was identified by urine glucose concentrations greater than 2 g/100 ml. Diabetic and control rats were studied 2–4 weeks after diagnosis of diabetes as outlined below.

C57 BL/6J ob/ob (obese) and C57 BL/KsJ db/db (diabetes) female mice and normal littermate controls were obtained from the Jackson Laboratory (Bar Harbor, Maine). C57 BL/6J db/db and C57 BL/KsJ ob/ob female mice and normal littermate controls were obtained from mutant research colonies maintained at the Jackson Laboratory. Animals were transported at 5–6 wks of age and then maintained under constant conditions as outlined above. Obese or diabetic mice and littermate controls were sacrificed at various ages from 5 to 24 wks and tissues handled as outlined below.

Hypothalamic Extraction and Incubation

Rats were sacrificed by decapitation with minimal stress and the brain rapidly (<2 min) removed to an ice-cold surface. The hypothalamus, defined by the optic chiasm rostrally, perihypothalamic sulci laterally and mamillary bodies caudally, was removed *en bloc* to a depth of ±1 mm using curved scissors. Tissue blocks removed for extraction were placed in preweighed vials containing ice-cold 2 M acetic acid and snap frozen. After reweighing and boiling, the tissues were homogenized using a Polytron (Brinckmann). Homogenates were centrifuged and the supernatant lyophilized before reconstitution in assay buffer for NT RIA (see below).

Hypothalamic blocks removed for incubation were immediately placed in Kreb's Ringer Bicarbonate buffer containing 14 mM glucose (KRBG) pregassed with 95% O_2: 5% CO_2. Incubations were performed in KRBG at 37°C in a shaking water bath. After a 60-min preincubation to achieve stable basal release, medium was discarded and replaced with fresh medium for a 20-min basal incubation followed by a 20-min incubation in medium containing 60 mM K$^+$.[11] After each period, media were removed, then boiled and stored at −20° until assayed.

Pancreatic Extraction and Perfusion

Obese and diabetic mice and littermate controls were killed in the fasted state by cervical dislocation and the pancreas carefully dissected free of omental fat and removed in its entirety. 2 N acetic acid extraction was performed as described above.

The pancreas from a diabetic and a control rat was vascularly isolated then perfused *in situ* after an overnight fast. The preparation used included the attached segment of duodenum. Perfusions were performed with KRB buffer containing 0.25% bovine serum albumin and 4.6% dextran (mol wt ~ 70,000). The medium was gassed with 95% O_2: 5% CO_2 and maintained at pH 7.4. The flow rate was constant at 2 ml/min.[12] After a period of equilibration, the pancreas was perfused for 10-min periods sequentially with glucose (300 mg/100 ml), then glucose + 19 mM arginine and finally glucose + arginine + lm$M$ theophylline. Effluent from the portal vein was collected into chilled tubes containing bacitracin ($2 \times 10^{-5}\ M$) and Trasylol (1000 KIU/ml) at times indicated in FIGURE 4 and stored at $-20°C$ until assay.

Reverse-phase High Pressure Liquid Chromatography (HPLC)

Mouse pancreas extracts were applied in 0.01 M trifluoroacetic acid (TFA) to disposable octadecasilyl-silica cartridges (Sep-pak C_{18}, Waters Assoc., Inc.) and then eluted in a 10% stepwise gradient of acetonitrile in TFA. Fractions were lyophilized then reconstituted in assay buffer for NT RIA. IR-NT-containing fractions were pooled then applied to a C_{18} μBondapak HPLC column and eluted (2 ml/min) with 27% acetonitrile in TFA followed by a 27–80% linear gradient of acetonitrile in TFA. Fractions (1 min) were collected, lyophilized then reconstituted in assay buffer for NT RIA.

IR-NT was determined by RIA as previously described using rabbit anti-NT serum (1:16,000), ^{125}I-NT prepared by the lactoperoxidase technique and purified by ion-exchange chromatography, synthetic NT standards, and double antibody precipitation of antibody-bound tracer.[13] The assay sensitivity is 20–25 pg/ml with 450–500 pg/ml causing 50% displacement of tracer (from antiserum binding). The antiserum is directed towards the COOH-terminus of the NT molecule as defined by the cross-reactivity of a limited number of available NT analogs and fragments.[13]

Statistical differences were determined using Student's t test and, where appropriate, an analysis of variance.

RESULTS AND DISCUSSION

Hypothalamic IR-NT in Streptozotocin-Diabetic Rats

Hypothalamic IR-NT content in diabetic rats (81.2 ± 7.4 ng/g, mean ± SE, n = 6) was significantly reduced compared to nondiabetic controls (114.2 ± 4.2 ng/g, n = 6, p <0.01) (Fig. 1). Hypothalamic IR-NT content in the nondiabetic rats was comparable to that previously reported in normal rats similarly studied.[10] In a previous study of the effect of diabetes on IR-NT tissue distribution, no change in hypothalamic IR-NT was demonstrated.[10] The reason for this discrepancy is unclear.

We have previously demonstrated in the rat that *in vivo* hormonal or metabolic perturbations predictably influence subsequent *in vitro* release of hypothalamic somatostatin.[11] We thus next studied in the rat the effect of experimental diabetes on *in vitro* hypothalamic IR-NT release and found that basal and 60 mM K^+-stimulated IR-NT release was reduced in diabetic animals compared to nondiabetic controls (FIG. 2).

Administration of synthetic NT within the CNS inhibits pituitary release of growth hormone, thyrotropin, prolactin, and follicle-stimulating hormone.[14] This effect occurs through release of inhibitory hypothalamic factors, possibly dopamine and somatostatin, rather than through a direct effect of NT on the pituitary. The reduction of hypothalamic IR-NT content and release in diabetic animals might thus reflect decreased hypothalamic NT activity and might partly account for abnormal pituitary hormone secretion seen in diabetes.

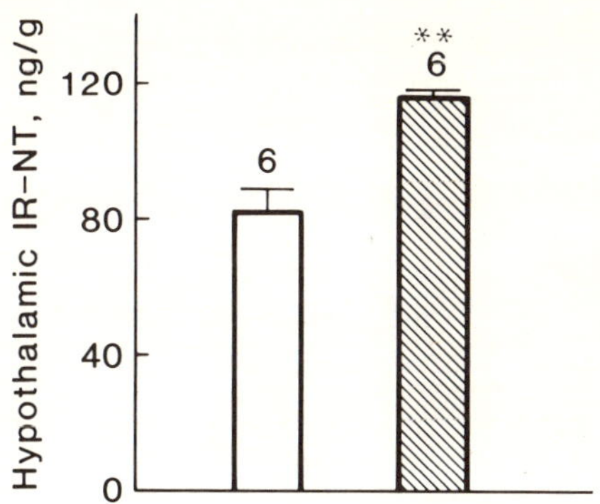

FIGURE 1. Hypothalamic IR-NT concentration expressed as ng/g tissue weight in normal (*hatched bar*) and Streptozotocin-diabetic rats (*open bar*). Results are shown as mean ± SE. The number of animals is given in parentheses. p values: ** <0.01.

Pancreatic IR-NT Content in Spontaneously Diabetic Mice

Mice with the C57 BL/6J background and homozygous ob or db genes demonstrate an "obesity" syndrome—characterized by hyperphagic obesity, moderate hyperglycemia, and marked hyperinsulinemia.[15] In 6J ob/ob and db/db mice, no consistent change in pancreatic IR-NT content (FIG. 3) or concentration (data not shown) was demonstrated compared to lean littermate controls.

The presence of homozygous ob or db genes in mice of the C57 BL/KsJ background results in the development of a "diabetes" syndrome.[15] These animals are metabolically similar to the obese mice until 8–10 weeks of age when pancreatic insulin secretion fails and rapidly increasing hyperglycemia occurs. Pancreatic IR-NT content (FIG. 3) and concentration (data not shown) was consistently elevated from 8 to 24 weeks in such diabetic mice compared to nondiabetic littermate controls with the greatest elevation occurring at 8–10 weeks, the time of pancreatic B-cell failure.

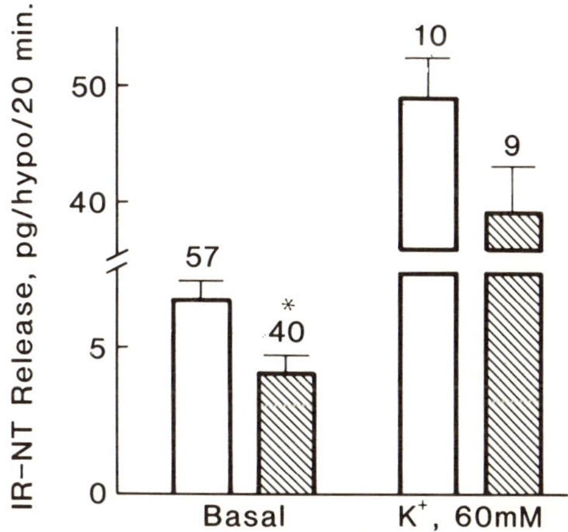

FIGURE 2. IR-NT release from incubated hypothalami of normal (*open bars*) and streptozotocin-diabetic rats (*hatched bars*). Results expressed as pg IR-NT/hypothalamus/20 min are shown as mean ± SE. The number of animals is given in parentheses. p values: * <0.05.

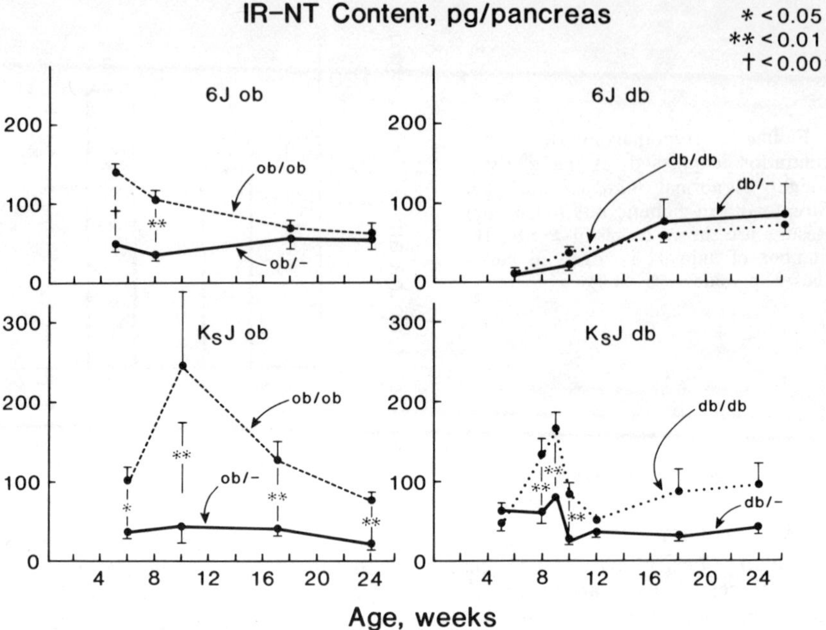

FIGURE 3. Pancreatic IR-NT content in 6J ob and 6J db obese mice and in KsJ db and KsJ ob diabetic mice aged from 5 to 24 wks compared with normal littermate controls. Results expressed as pg IR-NT/pancreas are shown as mean ± SE. Tissue samples from 6–10 animals were studied at each time point. p values: * <0.05, ** <0.01, † <0.001.

The presence of the homozygous ob gene in mice of either the 6J or KsJ backgrounds was associated with elevated pancreatic IR-NT content compared to littermate controls until 8–9 weeks of age, an elevation not seen in mice with the homozygous db gene.

Increased pancreatic IR-NT content in mice with insulinopenic diabetes but not in those with insulin resistance suggests that the presence or absence of insulin itself and not its metabolic sequelae are responsible for the regulation of pancreatic NT. An inhibitory effect of insulin on pancreatic IR-NT has previously been demonstrated in diabetic rats in which elevated pancreatic IR-NT was normalized by insulin therapy.[10] Since NT has direct effects on pancreatic islet hormone secretion, is present in the pancreas, and shows insulin responsiveness, it is possible that the peptide might play a physiological role in islet hormone regulation. By analogy with somatostatin, this role might be local within the islet (paracrine) or via the circulation (as a classical hormone).

Pancreatic IR-NT Release in Streptozotocin-Diabetic Rats

In preliminary experiments, release of IR-NT from the isolated perfused rat pancreas could not be demonstrated under basal conditions or in response to glucose (300 mg/100 ml). The sequential addition of arginine (19 m*M*) and then arginine plus theophylline (1 m*M*) resulted in progressively increasing IR-NT release (FIG. 4). Qualitatively similar IR-NT responses were seen in the diabetic rat, but release was considerably greater in response to both stimuli than in the normal animal (FIG. 4).

Pancreatic IR-NT release could represent an active secretory process that would imply a hormonal role or could merely reflect increased passive release associated with local intrapancreatic NT activity. Elevated IR-NT content and release in diabetic animals suggests an increase in synthesis and secretory activity of the peptide. The physiological significance of this finding remains to be evaluated.

Reverse Phase HPLC of Pancreatic IR-NT

Pancreatic IR-NT eluted on HPLC as two discrete peaks in all groups of mice studied. One peak of IR-NT eluted similarly to synthetic NT (retention time, 15 min) while the second peak, present in all groups of mice, eluted with a retention time of 27 min. Despite the qualitative similarity among all four genetic strains of mice, marked quantitative differences existed. IR-NT coeluting with NT formed the major peak in 6J ob/- and ob/ob mice while the second peak was larger in KsJ db/- and db/db mice. The nature of the second IR-NT peak is unknown at present

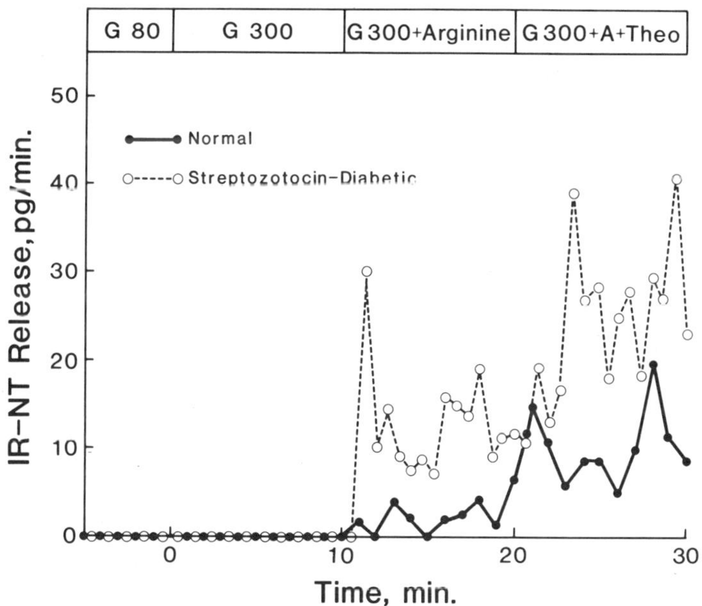

FIGURE 4. Pancreatic IR-NT secretion in response to glucose, glucose and arginine, and glucose, arginine, and theophylline in normal and streptozotocin-diabetic rats. See text for further details.

but may represent a peptide along the biosynthetic pathway of NT that contains a similar COOH-terminal antigenic recognition site.

SUMMARY

These studies of IR-NT in experimental diabetes indicate that changes in NT homeostasis occur within both the hypothalamus and pancreas. Hypothalamic

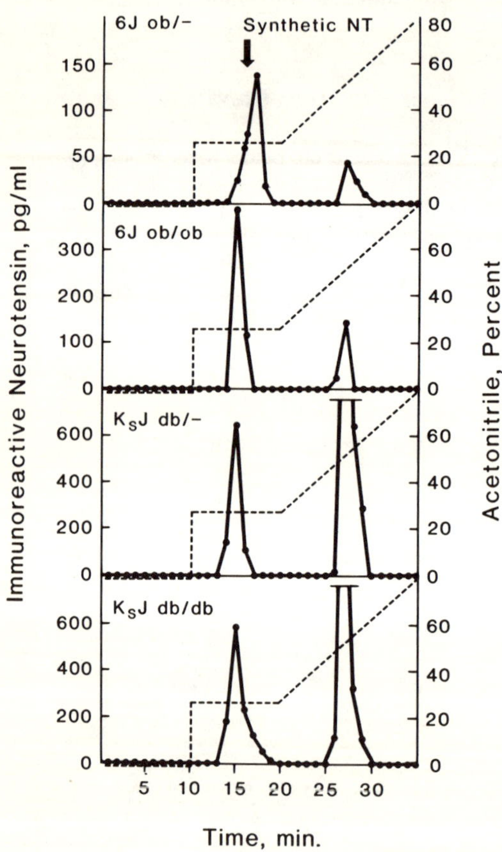

FIGURE 5. Reverse-phase HPLC of Sep-pak-purified extracts from 6J ob/- and ob/ob and KsJ db/- and db/db mice. See text for further details.

IR-NT content and release is decreased in diabetic rats providing a possible explanation for the changes in pituitary hormone secretion seen in such animals. Pancreatic IR-NT content and release is increased in diabetic rats, the major change in content coinciding with the onset of insulin deficiency. Pancreatic IR-NT content is influenced both qualitatively and quantitatively by the genetic background of the animal studied. The physiological role of pancreatic NT in regulation of islet hormone secretion remains to be identified.

ACKNOWLEDGMENT

The authors gratefully acknowledge the excellent technical assistance of
Ms. Nai-Chi Ting.

REFERENCES

1. CARRAWAY, R. E. & S. E. LEEMAN. 1973. The isolation of a new hypotensive peptide, neurotensin, from bovine hypothalami. J. Biol. Chem. **248:** 6854–6861.
2. CARRAWAY, R. E. & S. E. LEEMAN. 1976. Characterization of radioimmunoassayable neurotensin in the rat. Its differential distribution in the central nervous system,
   small intestine and stomach. J. Biol. Chem. **251:** 7035–7044.
3. CARRAWAY, R. E., L. M. DEMERS & S. E. LEEMAN. 1976. Hyperglycemic effect of
   neurotensin, a hypothalamic peptide. Endocrinology **99:** 1452–1462.
4. BROWN, M. & W. VALE. 1976. Effects of neurotensin and substance P on plasma
   insulin, glucagon, and glucose levels. Endocrinology **98:** 819–822.
5. NAGAI, K. & L. A. FROHMAN. 1976. Hyperglycemia and hyperglucagonemia following neurotensin administration. Life Sci. **19:** 273–280.
6. DOLAIS-KITABGI, J., P. KITABGI, P. BRAZEAU & P. FREYCHET. 1979. Effect of neurotensin on insulin, glucagon, and somatostatin release from isolated pancreatic islets.
   Endocrinology **105:** 256–260.
7. BROWN, M., J. VILLARREAL & W. VALE. 1976. Neurotensin and substance P: Effects
   on plasma insulin and glucagon levels. Metabolism **25:** 1459–1461.
8. NAGAI, K. & L. A. FROHMAN. 1978. Neurotensin hyperglycemia: Evidence for histamine mediation and the assessment of a possible physiologic role. Diabetes **27:**
   577–582.
9. ROSELL, S. & Å. ROKAEUS. 1981. Actions and possible hormonal functions of circulating neurotensin. Clin. Physiol. **1:** 3–20.
10. FERNSTROM, M. H., M. A. Z. MIRSKI, R. E. CARRAWAY & S. E. LEEMAN. 1981.
    Immunoreactive neurotensin levels in pancreas: Elevation in diabetic rats and mice.
    Metabolism **30:** 853–855.
11. BERELOWITZ, M., S. L. FIRESTONE & L. A. FROHMAN. 1981. Effects of growth hor-
    mone excess and deficiency on hypothalamic somatostatin content and release and
    on tissue somatostatin distribution. Endocrinology **109:** 714–719.
12. GOTO, Y., R. G. CARPENTER, M. BERELOWITZ & L. A. FROHMAN. 1980. Effect of
    ventromedial hypothalamic lesions on the secretion of somatostatin, insulin, and
    glucagon by the perfused rat pancreas. Metabolism **29:** 986–989.
13. MAEDA, K. & L. A. FROHMAN. 1981. Neurotensin release by rat hypothalamic frag-
    ments *in vitro*. Brain Res. **210:** 261–269.
14. MAEDA, K. & L. A. FROHMAN. 1978. Dissociation of systemic and central effects of
    neurotensin on the secretion of growth hormone, prolactin, and thyrotropin. Endo-
    crinology **103:** 1903–1909.
15. COLEMAN, D. L. 1978. Obese and diabetes: Two mutant genes causing diabetes-
    obesity syndromes in mice. Diabetologia **14:** 141–148.

DISCUSSION OF THE PAPER

E. E. ZIMMERMAN (*Columbia University, New York*): Did you look at neurotensin levels in any other brain areas in the diabetic rats? It would be interesting to
know whether the hypothalamus was different compared to some other regions.

M. BERELOWITZ (*University of Cincinnati College of Medicine, Cincinnati,
OH*): We did not.

D. J. BRAITMAN (*AFRRI, Bethesda, MD*): I was interested in the mechanism of action of neurotensin on the H-1 receptor. Do you think that neurotensin actually acts at the H-1 receptor as an endogenous ligand?

BERELOWITZ: I do not think we can say whether or not neurotensin is acting directly on the H-1 receptor or whether releasing histamine which acts at the H-1 receptor. Certainly the effects of neurotensin on the H-1 receptor in the adrenal medulla have not extensively been studied.

C. B. NEMEROFF (*University of North Carolina, Chapel Hill*): Your data concerning the ICV injections of neurotensin are of great interest to us; i.e. the null effect that you obtained on blood sugar. However, the dose you used—2.5 μg/kg may be a low dose compared to what other groups have used. Recently we have observed reduction in food consumption after central injection of NT and obviously hyperglycemia would be a sufficient explanation. I wonder if you have looked at higher NT doses? I think it is important in view of Brown's data on bombesin, showing that very small quantities centrally produce a very potent hyperglycemia, that may be mediated by adrenal catecholamine release.

BERELOWITZ: I have not looked at that at all. I do not know whether Dr. Frohman has.

L. A. FROHMAN (*University of Cincinnati Medical School, Cincinnati, OH*): We have not gone to doses higher than 5 μg/kg ICV. The main intent of the study was to show that a comparable dose given systemically and centrally produced disparate effects.

S. R. BLOOM (*Royal Postgraduate Medical School, London, England*): I have two questions. First, since the pancreas is 1% islets, when you looked in the isolated islets for neurotensin, was it 100-fold greater in concentration? We have two bits of evidence suggesting that neurotensin may act on exocrine, rather than endocrine, pancreas though it might be found in both. It is endocrine pancreatic tumors that produce neurotensin.

The second question is this: you showed data on the effect in rats of 2.5 and 5 μg/kg of IV neurotensin. Do you know what blood levels were achieved? Were they likely to be physiological?

BERELOWITZ: In answer to the first question, I did not do a quantitative recovery on the islets. We measured immunoactive neurotensin in islets isolated from the rat pancreas by collagenase digestion and kindly provided to us by Dr. Donald Steiner from the University of Chicago. If we multiplied the number of islets by the amount of neurotensin that was present in those islets we did not come up with the sum that equaled the amount of neurotensin in the total pancreas. I do not know whether there were losses in production of the islets, or whether some neurotensin was not present in islets. The information that we have is that there is neurotensin in the islet preparation and that it seemed to be in a subpopulation of cells.

BLOOM: Did you control it against insulin or glucagon, which were presumedly also measured to see whether you observed the same losses of those two hormones?

BERELOWITZ: We did, and there were greater losses of somatostatin and glucagon than of insulin, which is probably understandable from the anatomy of the islets and the way they prepared their preparation. But, those gave much better recoveries than neurotensin did. I think that it is likely that not all the neurotensin is present in the islets.

In answer to your second question: in view of what we now know about plasma neurotensin, I am sure that these were pharmacologic, and not physiologic, levels.

A.S. TISCHLER (*Tufts University School of Medicine, Boston, MA*): What are the cell surface markers that are stained when you use the fluorescence-activated cell sorter to purify islets?

BERELOWITZ: They are not surface markers. Steiner used the technique to separate the cells based on their refractile qualities. The refractile qualities in turn are dependent on the granules within the cells. He has shown that you can separate individual islet cells based on these refractile qualities.

T. L. YAKSH (*Mayo Clinic, Rochester, MN*): Your work on differential NT release after potassium between diabetic and normal rats was very interesting. Were those data expressed in terms of absolute levels of release per unit time, or were those fractional rate constants?

BERELOWITZ: My data on NT release are always expressed as pg per hypothalamic block per 20 min of incubation. It is simply the measurement of the immunoreactive material in the medium at the end of the incubation period after boiling the medium to counteract the enzymes.

YAKSH: Do you account for the difference in NT levels between the diabetic and normal on the basis of the levels that are in the hypothalamus? Have you tried any other releasing agent than 60 mM potassium?

BERELOWITZ: First of all, the amount of peptide released from the hypothalamus in the system that I'm working with—whether I am looking at somatostatin or neurotensin—is a very small fraction of the total hypothalamic content. In normal animals hypothalamic NT content was 120 ng/g. The hypothalamus weighs about 25 milligrams, so you can calculate what the content of neurotensin is in that block. I then showed that release is about 50 to 60 pg over a 20-min period. However, we think that release is a reflection of the amount that is reaching relevant cells.

YAKSH: I agree with that completely. What I'm trying to get at is why there is a difference in release of NT from normal and diabetic hypothalami?

BERELOWITZ: We do not know. The 60 mm potassium is certainly a maximal stimulus. We did not use other stimuli.

THE EFFECTS OF NEUROTENSIN ON ANTERIOR PITUITARY HORMONE SECRETION

S. M. McCann, E. Vijayan, J. Koenig, and L. Krulich

Department of Physiology
University of Texas Health Science Center at Dallas
Dallas, Texas 75235

INTRODUCTION

While screening hypothalamic extracts for the presence of corticotropin-releasing factor, Leeman noticed that the blood pressure was lowered by certain fractions. Carraway and Leeman[1] went on to isolate and determine structure of the new peptide, named neurotensin, which turned out to be a tridecapeptide. Neurotensin has a variety of biological actions and because of its isolation from hypothalamic tissue and its subsequent localization to a variety of hypothalamic areas,[2] it appeared of interest to evaluate its possible effects on the secretion of anterior pituitary hormones.

Considerable work has now been done on this subject. It is our belief that in order to study hypothalamic actions of a putative modulator of pituitary hormone secretion, an important first step is to microinject the peptide either into the third ventricle or into various hypothalamic loci. To determine if indeed these actions might be exerted on the anterior pituitary after possible uptake of the peptide into hypophyseal portal vessels, it is necessary to incubate the peptide with hemipituitaries or dispersed pituitary cells *in vitro*. These results are compared with those obtained by intravenous injection of the peptide. Conscious animals should be used to obviate the effects of anesthesia. A good example of anesthetic modifications in responses is GABA (γ-aminobutyric acid). Results from intraventricular injection of GABA are dramatically altered by anesthesia.[3]

It is also important to employ animals in different hormonal states, because the reaction to a neurotransmitter or peptide may be markedly modified by the hormonal milieu.[4] Much work has now been carried out with neurotensin, but there is considerable disagreement concerning its effects, many of which may be related to the factors mentioned above.

In this communication, we will detail the results that we obtained with neurotensin in conscious animals or following incubation of pituitaries or dispersed pituitary cells *in vitro* with the peptide. The various anterior pituitary hormones have been determined by radioimmunoassay. We have also attempted to determine whether or not the action of neurotensin to suppress prolactin release is mediated directly on the hypothalamic releasing and inhibiting neurons or is mediated indirectly via other transmitters.

MATERIALS AND METHODS

Rats of the Sprague-Dawley strain have been used. We have employed either ovariectomized females or intact males. Before experimentation, third ventricular cannulae were implanted and intrajugular cannulae placed according to previously described procedures to permit microinjection of substances into the ventricle and

160

withdrawal of blood samples from undisturbed rats. Blood samples were removed before and at various times after injection of neurotensin and the volume of blood removed (0.7 ml) was replaced with an equal volume of 0.9% NaCl.

To study the effects of ether stress on the response to neurotensin, animals were anesthetized with ether, and blood samples were removed from the jugular vein through a skin incision. In this case, subsequent blood samples were also obtained through the jugular after etherization of the rat.

A number of compounds were used in the experiments: synthetic neurotensin (Peninsula Laboratories, Inc., San Carlos, CA); the serotonin precursor, L-5-hydroxytryptophan (5HTP, Sigma, St. Louis, MO); a serotonin reuptake blocker, fluoxetine (gift of Lilly Research Laboratory, Indianapolis, IN); an inhibitor of tyrosine hydroxylase, α-methyl-p-tyrosine methyl ester HCl (alpha MT, Regis Chemical Company, Morton Grove, IL); and a dopamine receptor blocker, spiroperidol (a gift of Janssen Pharmaceutica, Berse, Belgium). Compounds were dissolved in 0.9% NaCl except for spiroperidol which was dissolved in 0.1 M tartaric acid.

After centrifugation of blood samples, plasma hormone levels were determined by RIA using kits provided by NIAMDD except in the case of LH (luteinizing hormone), in which ovine antisera to LH supplied by Niswender were used. The results were expressed in terms of the RP-1 and RP-2 reference preparations for all hormones except LH, which was expressed in terms of the NIH ovine LH S-1 standard to be comparable to our previous results. Statistical significance was determined by paired t-test or Student Newman-Keul's multiple comparison procedure for within-group comparisons. Between-group comparisons were performed either by Student's t-test or by analysis of variance as described by Zar.[5]

RESULTS

Ovariectomized Animals

As expected, plasma LH and FSH (follicle-stimulating hormone) titers were elevated in the ovariectomized (OVX) animals when compared to basal levels in normal female rats in this laboratory. Plasma prolactin titers were relatively low in the OVX animals. Levels of plasma GH and TSH were not altered. Plasma hormone levels were not significantly altered by third ventricular injection of 2 μl or intravenous (I.V.) injection of 100 μl of isotonic saline (FIGS. 1–4).

Intraventricular injection of neurotensin produced a lowering of plasma LH concentrations in the OVX females that was first detectable at 5 min and persisted for 30 min following the 0.5 μg dose (FIG. 1).[6] There was a progressive lowering throughout the 1-hr course of the experiment following the 2 μg dose. Intravenous injection of an intermediate dose of 1 μg had no effect on plasma LH.

Intraventricular injection of both doses of neurotensin also markedly suppressed plasma prolactin levels in these OVX females throughout the duration of the experiment (FIG. 2). In this case the 1 μg dose injected intravenously evoked the opposite result with a marked increase in plasma prolactin observable within 5 min that persisted for 30 min after injection (not shown).

In the case of growth hormone (GH), both doses produced a dramatic elevation in plasma growth hormone apparent at 5 min and persisting for 30 min in the case of the lower dose and for 60 min in the case of the higher dose of intraventricular neurotensin (FIG. 3).[7] Plasma GH concentrations were not modified by intravenous injection of the 1 μg dose of the peptide (not shown).

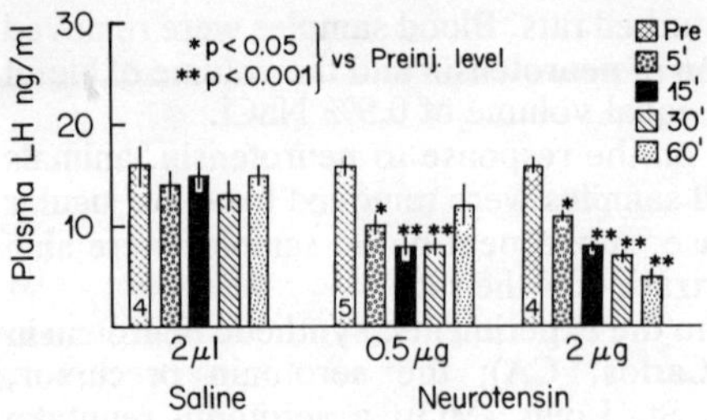

FIGURE 1. Effects of third ventricular injections of 0.9% NaCl or neurotensin on plasma LH levels in OVX, conscious rats. In this and subsequent figures, vertical lines above and below the mean represent the SEM. The number at the base of each column indicates the number of animals per group.

In the case of thyroid-stimulating hormone (TSH), there was no effect of intraventricular injection of either dose of neurotensin; on the other hand, intravenous injection of the 1 μg dose of neurotensin dramatically elevated plasma TSH within 5 min and the effect persisted for 30 min after the injection (FIG. 4).

Male Rats

The effects of neurotensin on prolactin release have been evaluated in the male rat. In this instance, the intraventricular injection of saline produced a slight but not significant lowering of prolactin during the 60 min after injection (FIG. 5).[8] Intraventricular injection of neurotensin caused a dose-dependent decrease in resting plasma prolactin levels. The effect was significant at 15 min after intraventricular injection of the higher 5 μg dose and the levels remained low for 60 min. Although values were lowered by 1 μg of neurotensin, the effect was not significant. Consequently in further experiments in males, the 5 μg dose was used.

Since the resting prolactin concentrations were low, which made it difficult to detect a further lowering, further experiments involved attempts to elevate prolactin levels. In the first experiment, prolactin levels were elevated by activation of CNS serotonin receptors by injection of the serotonin reuptake blocker, fluoxetine (4 mg/kg, I.V.). Thirty minutes later, 5HTP (15 mg/kg, I.V.) was injected to further elevate prolactin. At 15 min after this injection, the initial blood sample was removed and then the animals were injected intraventricularly with either saline or neurotensin. It is apparent that neurotensin markedly inhibited the prolactin-releasing action of 5HTP (FIG. 6A).

In the second experiment of this type, animals were anesthetized with ether and a blood sample was obtained from the exposed jugular vein. Saline or neuro-

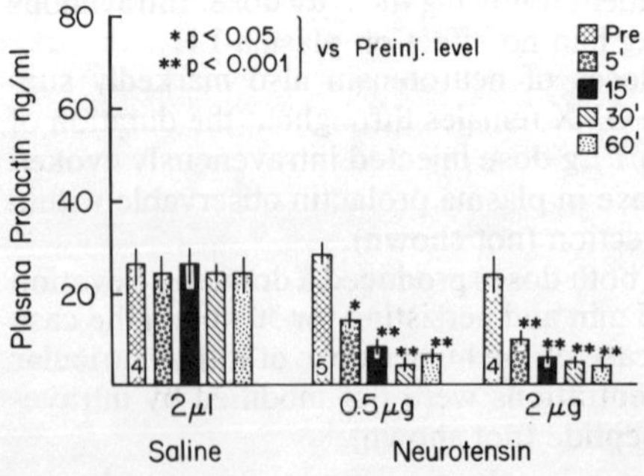

FIGURE 2. Effects of intraventricular injection of saline or neurotensin on plasma prolactin levels of OVX, conscious rats.

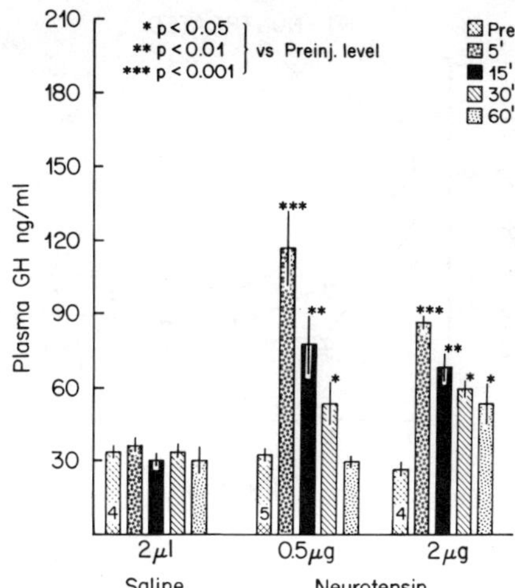

FIGURE 3.   Effects of third ventricular injection of 0.9% NaCl or neurotensin on plasma growth hormone levels in OVX rats.

tensin was injected intraventricularly immediately thereafter. Neurotensin significantly attenuated the prolactin-releasing effect of ether stress (FIG. 6B).

In the next experiment, the animals were injected with α-MT to elevate prolactin via blockade of dopamine synthesis. α-MT (250 mg/kg, i.p.) was injected 60 min before obtaining the initial blood sample and then either neurotensin or saline was injected intraventricularly. In contrast to the ability of neurotensin to counteract stress or 5HTP-activated prolactin secretion, the elevation in prolactin release accomplished by α-MT was not altered by neurotensin (FIG. 7).

In a further experiment, α-MT pretreatment was combined with fluoxetine and

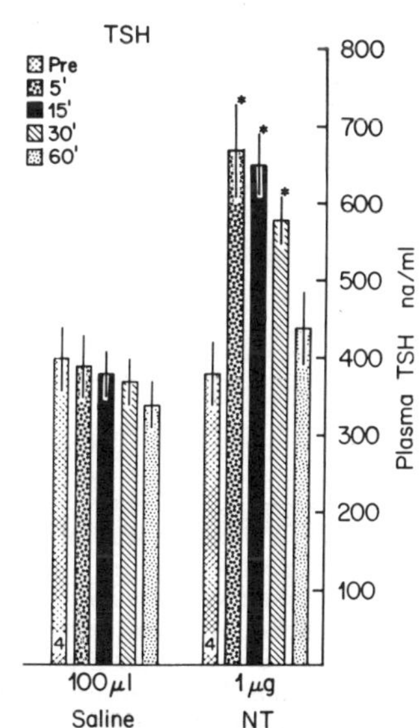

FIGURE 4.   Plasma TSH levels in OVX rats after i.v. pulse injection of saline or neurotensin.

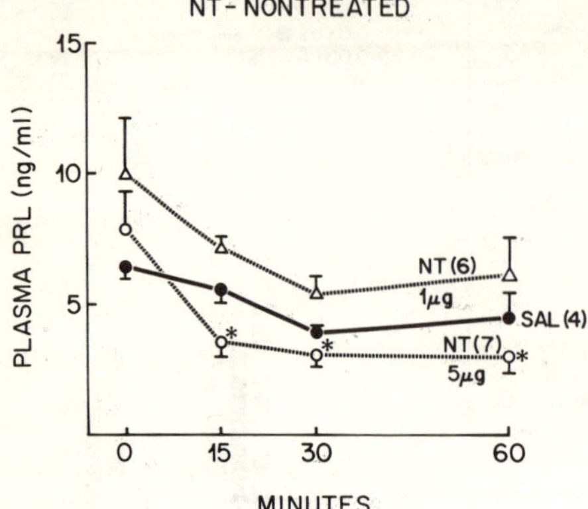

FIGURE 5. The effects of neurotensin on plasma prolactin in conscious, unrestrained male rats. At zero time, saline (SAL) or neurotensin (1 or 5 μg) were injected intraventricularly in a volume of 1 μl. The number in parentheses indicates the number of animals per group. * indicates $p < .05$ and ** indicates $p < .005$.

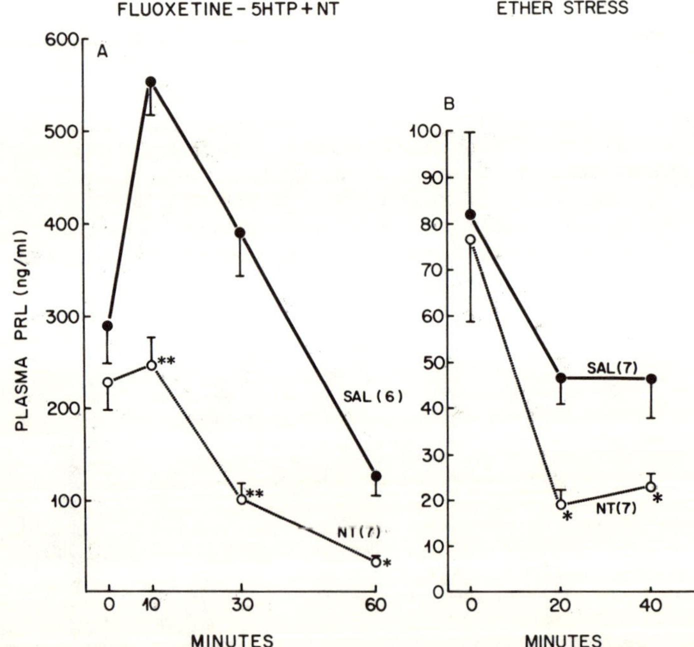

FIGURE 6. (A) The effects of neurotensin on the prolactin-releasing effect of fluoxetine and 5HTP. Fluoxetine was injected 45 min before neurotensin, 5HTP was injected 15 min before the peptide. At zero time, after a blood sample was removed, neurotensin or saline was injected intraventricularly. (B) The effect of neurotensin on the prolactin-releasing effect of ether stress. At zero time, each animal was anesthetized with ether and after a blood sample was removed from the jugular vein, neurotensin or saline was injected intraventricularly. The animal was returned to its cage and allowed to recover. Subsequent samples were also removed at later times under ether anesthesia.

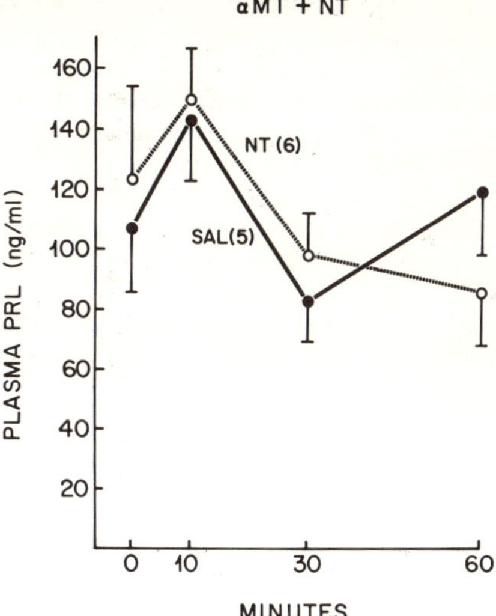

FIGURE 7. The effect of neurotensin on plasma prolactin levels in α-MT-treated rats. α-MT was injected 60 min before intraventricular injection of neurotensin or saline at zero time. Subsequent blood samples were removed at 10, 30, and 60 min later.

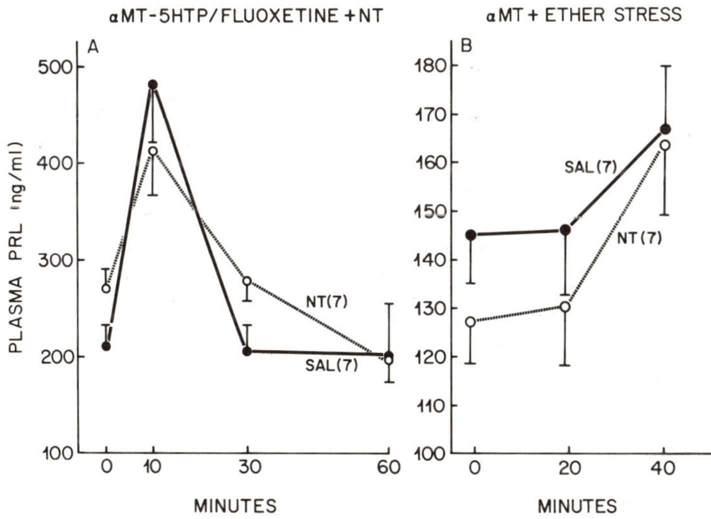

FIGURE 8. (A) The effect of neurotensin on plasma prolactin levels in animals treated with α-MT, fluoxetine and 5HTP. α-MT (250 mg/kg, i.p.) was injected 60 min before neurotensin, fluoxetine (4 mg/kg, i.v.) was given 45 min before NT and 5HTP (15 mg/kg, i.v.) 15 min before neurotensin. At zero time after removal of a blood sample 5 μg of neurotensin was injected intraventricularly. (B) The effect of neurotensin on the prolactin-releasing effect of ether stress in animals pretreated with α-MT. Sixty minutes before exposure to ether, 250 mg/kg of α-MT was injected i.p. At zero time, the animals were anesthetized with ether, a blood sample was removed from the jugular vein and neurotensin was injected intraventricularly. Twenty and 40 min later, the animals were reanesthetized and blood samples removed.

5HTP injection according to the timetable previously described. In this instance, α-MT did not suppress the prolactin release induced by 5HTP. However, it still prevented the prolactin-lowering effects of neurotensin (FIG. 8A). Similarly, α-MT administration failed to decrease the prolactin-releasing effect of ether stress. In this case as well, neurotensin was unable to decrease the elevated prolactin levels (FIG. 8B).

Since α-MT blocks synthesis of both dopamine and norepinephrine, it was important to use a dopamine-receptor blocker to determine that the block of the prolactin-lowering action of neurotensin was via inhibition of dopamine synthesis. The dopamine receptor blocker, spiroperidol (0.3 mg/kg, i.p.), elevated prolactin to levels similar to those obtained with α-MT and it also prevented the prolactin-inhibiting effect of neurotensin administered 60 min later (FIG. 9).

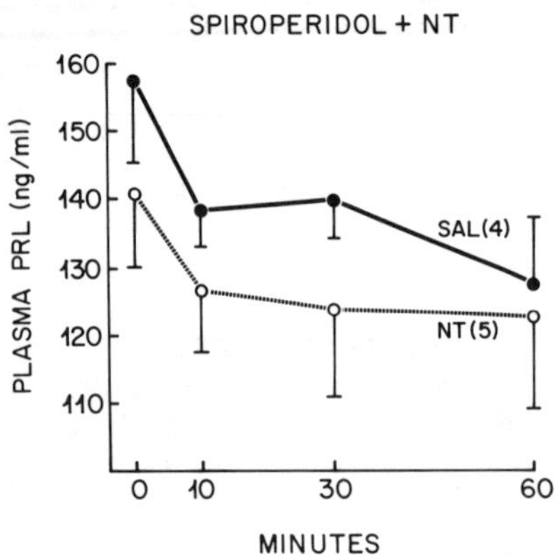

FIGURE 9. The effect of neurotensin on plasma prolactin levels in animals pretreated with spiroperidol. Sixty minutes before the intraventricular injection of neurotensin, spiroperidol was injected s.c. Subsequent blood samples were taken at 10, 30, and 60 min later.

In Vitro *Experiments*

To determine if neurotensin has any direct action on release of pituitary hormones, various doses were incubated *in vitro* with hemipituitaries from OVX females. There was no effect of various doses of neurotensin ranging from 5–500 ng/ml on FSH or LH release into the medium; however, prolactin release was significantly stimulated by doses of neurotensin ranging from 50 to 500 ng/ml (TABLE 1). There was no dose–response relationship, rather the prolactin-releasing effect appeared to be an all-or-none phenomenon. In the case of growth hormone, there was also no significant effect on the release of the hormone *in vitro* (TABLE 2). However, in the case of TSH, significant release occurred with doses of the peptide ranging from 50–500 ng/ml, results similar to those obtained with prolactin. Again, there was no dose response relationship.

DISCUSSION

Our results clearly indicate that neurotensin has powerful effects to modify the release of pituitary hormones. The major action appears to be exerted at the

TABLE 1

EFFECT OF NEUROTENSIN (NT) (5–500 ng) OR LHRH ON GONADOTROPIN AND PRL RELEASE FROM HEMIPITUITARIES INCUBATED *in Vitro* *

Dose (ng/ml)[1]	Hormone Released (ng/ml of medium)		
	LH	FSH	PRL
NT			
5	462.1 ± 54.3	2636.2 ± 268.2	3595.3 ± 548.4[†]
50	413.0 ± 41.4	2416.8 ± 262.6	5117.6 ± 129.0[†]
250	444.6 ± 59.0	2368.4 ± 311.2	4509.7 ± 204.3[‡]
500	396.4 ± 35.3	2446.6 ± 364.6	4888.8 ± 356.4[†]
LHRH 25	2106.4 ± 268.2[†]	10204.6 ± 1804.2[†]	3348.4 ± 268.4
Control	428.7 ± 34.2	2328.6 ± 212.6	2647.2 + 436.3

* Values given are the mean ± SEM.
[†] p <0.001 versus saline control.
[‡] p <0.05 versus saline control.

hypothalamic level. Intraventricular neurotensin clearly suppressed LH but not FSH release in OVX rats. It also suppressed prolactin release, but it augmented the release of growth hormone. There was no effect on the release of TSH or FSH. The actions on LH and growth hormone release are almost certainly exerted on the brain, rather than on the pituitary, in view of the lack of effect of the peptide on the release of these pituitary hormones *in vitro*. The prolactin-lowering action of neurotensin is exerted on the brain, since the direct effect on the pituitary was opposite and a prolactin-releasing action was observed.

A further evaluation was made of the hypothalamic, prolactin-inhibiting action of the peptide and this was found to exist not only in the OVX females but also in intact males. In males, the peptide was also effective in suppressing stress and 5HTP-induced prolactin release.

TABLE 2

EFFECT OF NEUROTENSIN OR LHRH ON GH AND TSH RELEASE FROM HEMIPITUITARIES INCUBATED *in Vitro*

Dose (ng/ml)	Hormone Released (ng/ml medium)	
	GH	TSH
Nt 5	3086.4 ± 412.4 *	1632.3 ± 305.6
50	3346.6 ± 316.4	1996.4 ± 268.0[†]
250	3826.4 ± 264.2	2199.0 ± 390.5[†]
500	4012.8 ± 306.4	2284.6 ± 184.0[‡]
LHRH 25	4438.3 ± 612.4	1468.4 ± 234.6
Control	3724.4 ± 212.6	1275.3 ± 240.6

* Mean ± SEM.
[†] p <0.05 versus saline control.
[‡] p <0.001 versus saline control.

The actions on prolactin release were further analyzed by attempting to determine whether they were exerted directly on prolactin-inhibiting factor (PIF) neurons or, alternatively, by the intermediation of putative synaptic transmitters. The results indicate that blockade of dopaminergic transmission, either by inhibition of synthesis of catecholamines with α-MT or by blockade of its action by receptor blockers, such as spiroperidol, blocked the action of neurotensin. The results indicate that the action of neurotensin was probably exerted via stimulation of the tuberoinfundibular dopaminergic tract, which in turn inhibited prolactin release, either via release of dopamine directly into portal vessels that directly inhibited prolactin release, or, alternatively, by releasing another PIF into portal vessels that inhibited release of the trophic hormone.

We have not examined whether or not the other actions of intraventricular neurotensin on LH and growth-hormone release are mediated directly on the releasing-factor neurons concerned or by the intermediation of another transmitter as in the case of the effects on prolactin.

In addition to actions at the hypothalamic level, it would appear that neurotensin has important actions directly on the adenohypophysis to alter hormone release. It can augment release not only of prolactin but also of TSH by a direct pituitary action. This direct action on the gland to stimulate prolactin release may account for the effect of intravenous neurotensin to stimulate prolactin release in a variety of situations,[9] including those used here.

Neurotensin was ineffective in modifying plasma TSH following its intraventricular injection; however, it was remarkably effective in increasing plasma TSH following its intravenous injection, an action reflected by a high-potency stimulation of TSH release by the glands incubated *in vitro*. Therefore, there is a good possibility that neurotensin may have a direct action on the adenohypophysis to release TSH.

In the meantime, several other studies were reported evaluating the effects of neurotensin on secretion of various pituitary hormones. Rivier *et al.*[9] had earlier shown an increase of plasma prolactin following intravenous administration of the peptide that is probably due to its direct prolactin-releasing action described here. Others have also observed a lowering of prolactin following intraventricular injection of neurotensin.[10,11]

Recently, Maeda, and Frohman[10] obtained significant elevation of plasma TSH following systemic injection of considerably larger doses of neurotensin than employed in our experiments. Gradual lowering of TSH followed intraventricular injection of the peptide in estrogen-primed male rats, anesthetized with urethane. They also found that intraventricular neurotensin blunted the TRH (thyrotropin-releasing hormone)-stimulated TSH release in these circumstances. The discrepancy between their results and those obtained by us may be related to the lateral ventricular rather than third ventricular route of injection, which could have led to actions at other sites in the brain or to the use of anesthetized, estrogen-primed males in their study instead of the unanesthetized OVX animals used in our experiments. We have previously commented on the dangers of using anesthetized animals for these studies. Furthermore, the estrogen-treated male obviously does not conform to any known physiological condition. Maeda and Frohman also reported without presentation of data that neurotensin did not affect TSH release by dispersed adenohypophyseal cells, a result different from the stimulation observed in our experiments utilizing hemipituitaries of OVX females. The explanation for this discrepancy is not apparent but could be related to their use of dispersed cells in which receptors for neurotensin could have been removed by the dispersion procedure.

It is possible that some of the actions of neurotensin described above may be mediated by its effect on the hypothalamic temperature-regulating centers since intracerebroventricular administration of the tridecapeptide lowered body temperature.[12] This action was not blocked by antimuscarinic, antinoradrenergic, or anti-opiate agents. Similarly, depletion of brain serotonin by *p*-chlorophenylalanine did not alter the neurotensin-induced hypothermia. Depletion of brain catecholamines by 6-hydroxydopamine instead resulted in significant potentiation of the hypothermia as did pretreatment with haloperidol to block dopamine receptors. Furthermore, in rats with selective depletion of brain dopamine but not norepinephrine, neurotensin-induced hypothermia was augmented. This suggests an interaction of dopamine and neurotensin in the hypothalamic temperature-regulating centers as well as in control of prolactin release. Interestingly, neurotensin-induced hypothermia was antagonized by TRH. Furthermore, intra-hypothalamic administration of neurotensin reduced TSH levels in OVX animals. Thus, there appears to be a relationship between neurotensin and TRH.

From the results of all these studies, it is apparent that neurotensin has powerful effects to alter prolactin, LH, and GH release by hypothalamic actions. The action to alter prolactin may be mediated by the dopaminergic system. In addition, there appear to be important actions at the pituitary level if indeed neurotensin reaches the gland in high concentrations via the hypophyseal portal vessels. In that event, it should stimulate not only TSH but also prolactin release by a direct action on the pituitary cells.

There is considerable controversy concerning the precise effects of the peptide. This may be related to the employment of anesthetized animals in some studies leading to results only applicable to the anesthetized state. Another problem is the use of animals in different hormonal conditions. In our own work we found no difference in the prolactin-lowering action of the peptide in OVX as compared to intact male rats.

There is no information that indicates whether or not these actions of neurotensin have physiological significance. The fact that it is found in the hypothalamus, in areas concerned with the control of adenohypophyseal function, suggests a role. Studies with inhibitory analogs of neurotensin or, alternatively, with antisera directed against the peptide should determine the physiological significance of neurotensin in control of anterior pituitary function.

SUMMARY

The present experiments were conducted to determine the effects of neurotensin on secretion of a variety of anterior pituitary hormones. Conscious rats with indwelling cannulae in the third ventricle and external jugular vein were used and the effects on plasma hormone levels measured by radioimmunoassay. Neurotensin was found to decrease plasma prolactin levels in overiectomized females, normal males, and males in which prolactin levels had been elevated by ether or by a combination of fluoxetine and 5-hydroxytryptaphane. The prolacitn-lowering effect was blocked by α-methyl-tyrosine to inhibit catecholamine synthesis and by the specific dopamine receptor blocker, spiroperidol.

In ovariectomized females, neurotensin was also capable of suppressing LH and elevating growth hormone following its intraventricular injection. Intravenous injection of the peptide elevated prolactin but had no effect on the release of the other pituitary hormones.

When hemipituitaries of ovariectomized rats were incubated *in vitro*, neurotensin elevated prolactin and TSH release into the medium. The minimal effective dose to elevate prolactin and TSH release was 50 ng/ml. Release of gonadotropins and growth hormone was unaffected.

It is concluded that neurotensin inhibits prolactin release by a CNS, presumably hypothalamic action, to stimulate the tuberoinfundibular dopaminergic neurons. The dopamine released then inhibits prolactin release either by a direct action on the pituitary or by release of another prolactin-inhibiting factor. In addition, the peptide has a direct prolactin-releasing action on the pituitary. Neurotensin can inhibit LH and stimulate growth hormone presumably by a hypothalamic action since there was no effect on the release of these pituitary hormones by glands incubated *in vitro*. Although the peptide had no effect on TSH release following its intraventricular injection, it stimulated prolactin release by pituitaries incubated *in vitro*. The physiological significance of these results is not yet established; however, the presence of the peptide in regions concerned with pituitary control suggests that it may play a physiological role.

REFERENCES

1. CARRAWAY, R. & S. E. LEEMAN. 1973. The isolation of a new hypotensive peptide, neurotensin, from bovine hypothalami. J. Biol. Chem. **248:** 6854.
2. KAHN, D., A. HOU-YU and ZIMMERMAN, E. 1982. Localization of neurotensin in the hypothalamus. Ann. N. Y. Acad. Sci. **400:** 117–131. This volume.
3. PASS, K. A. & J. G. ONDO. 1977. Effects of γ-aminobutyric acid on prolactin and gonadotropin secretion in unanesthetized rats. Endocrinology **100:** 1437–1442.
4. VIJAYAN, E. & S. M. McCANN. 1978. The effect of systemic administration of dopamine and apomorphine on plasma LH and prolactin concentrations in conscious rats. Neuroendocrinology **25:** 221–235.
5. ZAR, J. H. 1974. *In* Biostatistical Analysis. Prentice Hall. Englewood Cliffs, New Jersey.
6. VIJAYAN, E. & S. M. McCANN. 1979. *In vivo* and *in vitro* effects of substance P and neurotensin on gonadotropin and prolactin release. Endocrinology **105:** 64–68.
7. VIJAYAN, E. & S. M. McCANN. 1980. The effects of substance P and neurotensin on growth hormone and thyrotropin release *in vivo* and *in vitro*. Life Sci. **26:** 321–327.
8. KOENIG, J. I., M. A. MAYFIELD, S. M. McCANN & L. KRULICH. 1982. On the prolactin-inhibiting effect of neurotensin: The role of dopamine. Neuroendocrinology. In press.
9. RIVIER, C., M. BROWN & W. VALE. 1977. Effect of neurotensin, substance P and morphine sulfate on the secretrion of prolactin and growth hormone in the rat. Endocrinology **100:** 751.
10. MAEDA, K. & L. A. FROHMAN. 1978. Dissociation of systemic and central effects of neurotensin on the secretion of growth hormone, prolactin, and thyrotropin. Endocrinology **103:** 1903.
11. TACHE, Y., M. BROWN & R. COLLU. 1979. Effects of neuropeptides on adenohypophyseal hormone response to acute stress in male rats. Endocrinology **105:** 220–224.
12. NEMEROFF, C. B., G. PISETT, P. G. NAMBERG, A. J. OSBAR, G. R. BRISY & A. G. PRANG. 1980. Neurotensin-induced hypothermia: Evidence for interaction with dopaminergic systems in the hypothalamic pituitary thyroid axis. Brain Res. **195:** 69–84.

DISCUSSION OF THE PAPER

C. B. NEMEROFF (*University of North Carolina, Chapel Hill*): I want to make a suggestion about the effects of neurotensin that you have observed. It is clear that in the rat that is exposed to stress growth hormone secretion is inhibited and prolactin secretion is elevated. What you told us today is that when you inject neurotensin into CSF, it essentially counteracts that hormonal profile of stress. Growth hormone is elevated, prolactin is reduced. I think this resonates with certain of our findings on the anti-gastric ulcer effects of neurotensin injected centrally.

S. M. McCann (*University of Texas at Dallas*): That is a very good possibility. Of course, it lowers body temperature and that may have something to do with these effects on the endocrine system, particularly the effects on TSH.

M. Goedert (*Cambridge University, England*): Has anybody tried to measure neurotensin levels in hypothalamo-hypophyseal portal blood?

McCann: I do not know of any such measurements, but I would predict that NT is there. It seems to be in the external layer of the median eminence. It can be found in pituitary cells. Snyder's group found neurotensin in pituitary cells by immunofluorescence. However, they do not indicate which pituitary cells they are.

M. D. Hirsch, (*Roche Institute, Nutley, NJ*): Do you know whether there is an interaction between neurotensin and TRH in terms of prolactin release?

McCann: Actually we have not done anything on the interaction between neurotensin and TRH.

E. A. Zimmerman (*Columbia University, New York*): Would you like to comment on the differences in the lack of response between intraventricular neurotensin on plasma LH levels in your lab and the data that Dr. Leeman presented with local injections in the preoptic area, which released LH?

McCann: What were those conditions, Dr. Leeman?

S. E. Leeman (*University of Massachusetts, Worcester*): The data were from ovariectomized animals. Stereotaxic injections of neurotensin into the medial preoptic area caused an elevation of peripheral levels of LH. However, we also confirmed your finding that injecting it into the ventricle lowers plasma LH levels.

McCann: I think that what is happening in the field of peptide research is a recapitulation of what happened in catecholamine research. Results depend on the conditions, where you put the agent, the hormonal state of the animal, etc. Were your animals anesthetized?

Leeman: They were anesthetized with thiourethane.

S. R. Bloom (*Royal Postgraduate Medical School, London*): You said that you had tested a vast number of different peptides. How do you rate neurotensin for potency relative to all the others?

McCann: We have not done a complete exploration of potency. VIP is extremely potent. Four nanograms of VIP raises plasma LH. CCK is also extremely potent. A few nanograms injected intraventricularly will reduce plasma LH and elevate growth hormone. Neurotensin is not one of the most potent peptides. However, the point is not how much we administer, but what arrives at the synaptic cleft and that probably is a very high concentration. Thus, 0.5 micrograms of neurotensin is probably not too high a dose to have physiological relevance.

EFFECTS OF NEUROTENSIN ON HYPOTHALAMIC AND PITUITARY HORMONE SECRETION *

Lawrence A. Frohman† Kiyoshi Maeda, Michael Berelowitz,
Marta Szabo, and Jennifer Thominet

*Division of Endocrinology and Metabolism
University of Cincinnati College of Medicine
Cincinnati, Ohio 45267
Division of Endocrinology and Metabolism
Michael Reese Hospital, and
University of Chicago
Chicago, Illinois 60637*

INTRODUCTION

Neurotensin is a tridecapeptide originally isolated from bovine hypothalamus [1] and subsequently found to have a widespread distribution within the central nervous system (CNS) and in selected areas of the gastrointestinal tract. Among the earliest reports of the effects of neurotensin on endocrine function were its actions on the anterior pituitary [2,3] and the pancreatic islets.[4,5] In our initial studies on the endocrine pancreas, it was apparent that the route of administration was of critical importance since cerebroventricular, in contrast to systemic, injection of the peptide did not affect levels of plasma glucagon, insulin, or glucose.[5]

We therefore initiated a series of studies to determine whether the pituitary hormone-releasing effects of neurotensin involved central or peripheral mechanisms and to explore the possible mediating role of other centrally active compounds. This report summarizes the experience from our laboratory related to the effects of neurotensin on growth hormone (GH), thyroid-stimulating hormone (TSH), and prolactin (PRL) release and on the mechanism of its action.

MATERIALS AND METHODS

In Vivo *Experiments*

Studies were performed in male Sprague-Dawley rats weighing 250–300 g, maintained in a temperature-controlled, light-cycled room and given standard laboratory food. Animals were pretreated with estradiol valerate (200 μg, s.c.) for three or four days before study. In some experiments a PE-10 catheter was implanted into the lateral cerebral ventricle 5–7 days before study. Test materials for intraventricular injection were dissolved in 0.24% fast-green solution and administered in a total volume of 10 microliters.[6-9]

Animals were anesthetized with urethane (1.5 g/kg i.p.) and permitted to stabi-

* These studies were supported in part by USPHS Grants AM 18722, AM 30667, and AM 30686 and a Research Career Development Award (M.B.) by the Juvenile Diabetes Foundation.

† Address correspondence and reprint requests to: Lawrence A. Frohman, M.D., Division of Endocrinology and Metabolism, University of Cincinnati College of Medicine, 231 Bethesda Avenue, Cincinnati, Ohio 45267.

lize for a 60–120-minute period. After an initial blood sample was obtained from a jugular vein, the test material was injected into the opposite jugular vein or through the indwelling intraventricular catheter and samples collected at varying intervals over the subsequent 30 minutes. Blood samples were heparinized, chilled, centrifuged at 4°C, and plasma stored at $-20°C$. GH, PRL, and TSH were measured by radioimmunoassay (RIA) as previously described[8,10] using materials provided by the NIAMDD Pituitary Hormone Distribution Program.

In Vitro *Experiments*

Pituitary Cell Culture and Incubation

Primary monolayer cultures of enzymatically dispersed adenohypophyseal cells were prepared from pituitaries of adult male Sprague-Dawley rats as previously described.[11,12] After four days in culture, cell monolayers were incubated in Krebs Ringer bicarbonate containing 14 mM glucose, 1% bovine serum albumin, Fr V, and 10 microliters of the test agent at 100 times the desired final concentration. Incubations were performed at 37°C in a humidified atmosphere of 94% air–6% CO_2 for four hours. Hormone released into the incubation medium was measured by radioimmunoassay.

Hypothalamic Incubation

Hypothalamic tissue was obtained from male Sprague-Dawley rats as previously described.[13,14] Quartered hypothalamic fragments or, in subsequent experiments, intact hypothalami were incubated in Krebs-Ringer bicarbonate buffer containing glucose and bacitracin in a 37°C shaking water bath under a 95% O_2–5% CO_2 atmosphere. Tissue was preincubated for 20–40 minutes to permit the establishment of a stable release rate. The medium was replaced and fragments incubated for another 20 minutes in the presence of test substances. After removal of tissue, medium was boiled, centrifuged, and the supernatants stored at $-20°C$ until assayed. Somatostatin RIA was performed at previously desired.[13,15]

Characterization of Neurotensin

Neurotensin was kindly provided by Dr. Jean Rivier, Salk Institute, La Jolla, California (Lot no. 14-286-50) and purchased from Bachem, Inc. and Peninsula Laboratories. High-performance liquid chromatography (HPLC) was performed on a 0.4 × 30 cm Waters C_{18} analytical column using a Varian LC 5000 system. Samples were injected through a 100 microliter manual injector loop and eluted isocratically with 27% or 30% acetonitrile in 0.01 M trifluroacetic acid at a flow rate of 2 ml per minute and an operating pressure of 120–130 atmospheres. The retention time of neurotensin was monitored by UV absorbance at 210 nm. 0.5- or 1.0-minute fractions were collected and measured for somatostatin and, as previously described, neurotensin.[16]

Analysis of Data

Student's t test, analysis of variance, and the new Duncan's multiple range test were used for statistical analyses.

RESULTS AND DISCUSSION

The systemic injection of neurotensin at a dose of 30 μg/kg resulted in marked increases in plasma levels of GH, PRL, and TSH (FIG. 1). Peak values of GH and PRL occurred at 5 minutes while those of TSH occurred at 15 minutes. The elevations of all three hormone levels persisted for the 30-minute experimental period. Because of the previous suggestion that some of the effects of neurotensin on pancreatic islet hormones are mediated by histamine,[17] we examined the possible mediation by this neurotransmitter on the pituitary hormone-releasing effects of neurotensin. Intravenous administration of histamine, (5 mg/kg) resulted in marked increases in plasma GH, PRL, and TSH. Diphenhydramine, a histamine H_1 receptor antagonist, did not significantly alter plasma GH, PRL, or TSH levels when injected alone at a dose of 5 mg/kg i.v. However, the injection of diphenhy-

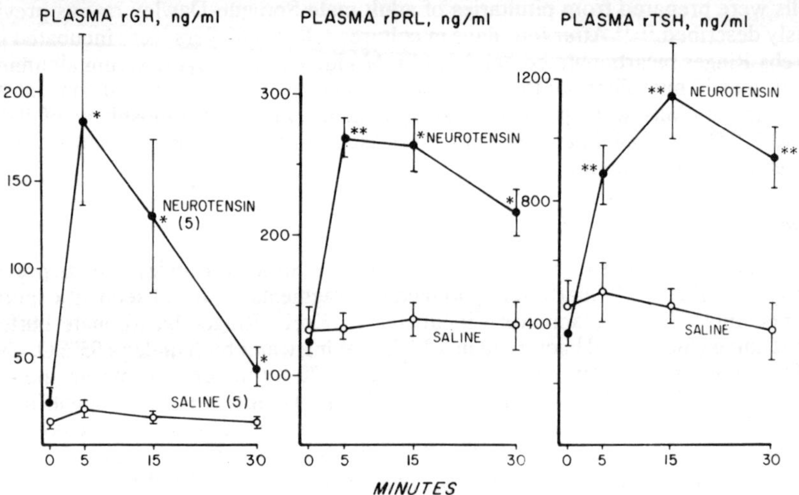

FIGURE 1. Effect of intravenous administration of neurotensin, (30 μg/kg) on plasma GH, PRL, and TSH levels in estrogen-primed, urethane-anesthetized male rats. Number of animals per group is indicated in parentheses. Shown are the mean ± SEM. * = p <0.05; ** = p <0.01. (Reprinted from Maeda & Frohman[6] with permission from *Endocrinology*.)

dramine 10 minutes before the systemic administration of histamine completely suppressed the GH and TSH responses and partially suppressed the PRL response. In contrast, the same dose of diphenhydramine was ineffective in blocking the hormonal responses to neurotensin (FIG. 2). In other experiments, a larger dose of diphenhydramine (10 mg/kg) also was ineffective in suppressing the effects of neurotensin.

From these studies we concluded that the effects of systemically administered neurotensin on the anterior pituitary were not histamine mediated. These data are at variance with a previous report[3] in which diphenhydramine did inhibit GH and PRL responses to neurotensin. The basis for this discrepancy is not entirely clear but may be related in part to varying degrees of stress induced by the neurotensin injection since diphenhydramine has been reported to inhibit stress-induced prolactin release.[18]

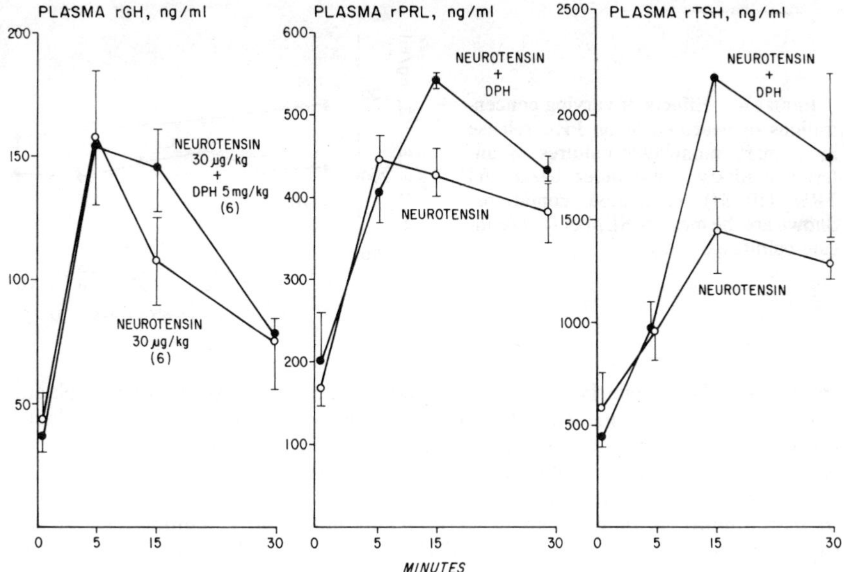

FIGURE 2. Effects of neurotensin alone and in the presence of diphenhydramine (DPH) on plasma GH, PRL, and TSH levels in estrogen-primed, urethane-anesthestized male rats. Symbols as in FIGURE 1.

We therefore examined the possibility that the effects of neurotensin might be due to direct stimulatory actions on the pituitary. Using the dispersed adenohypophyseal primary monolayer culture system, we examined the role of neurotensin on the release of the same three hormones. Neurotensin at a concentration of 10^{-6} M did not alter GH release (FIG. 3). This cell preparation was responsive to stimulation by an ectopic growth-hormone-releasing factor (2.5 μg protein of a partially purified extract of a carcinoid tumor[19] and was suppressed by somato-

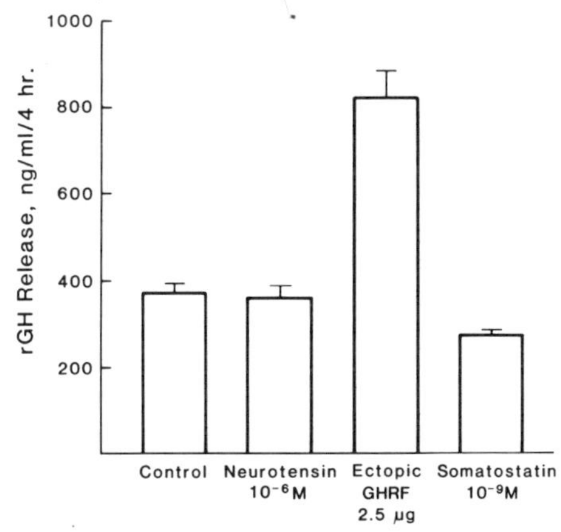

FIGURE 3.   Effects of neurotensin, ectopic GH-releasing factor (GHRF), and somatostatin on GH release by primary monolayer cultures of anterior pituitary cells. Shown are the mean ± SEM of four replicate cultures. The effects of both ectopic GHRF and SRIF differed from that of control by p <0.01.

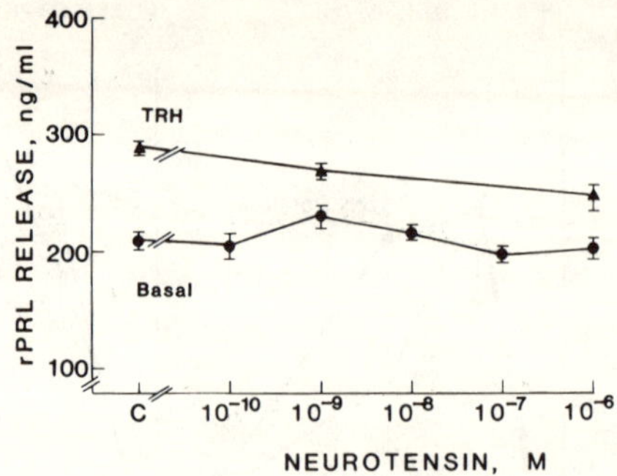

FIGURE 4.   Effects of varying concentrations of neurotensin on PRL release by primary monolayer cultures of anterior pituitary cells under basal and TRH $(10^{-8}M)$ stimulated conditions. Shown are the mean ± SEM of four replicate cultures.

statin, 10^{-9} M. Similarly, neurotensin was ineffective in stimulating prolactin release in concentrations from 10^{-10} to 10^{-6} M under conditions in which prolactin release could be stimulated by TRH (FIG. 4). In several experiments, no consistent effect of neurotensin was observed on TSH secretion.

In contrast to the effects of systemically injected neurotensin, intracerebroventricular administration of comparable or smaller doses of the peptide resulted in a significant decrease in plasma GH, PRL, and TSH levels at 15 minutes when compared with saline-injected controls (FIG. 5). The suppressive effects on GH and TSH were observed with a dose of 5 μg at both 15 and 30 minutes, whereas effects on PRL were produced by both 2 and 5 μg doses, but only at 15 minutes.

The effect of neurotensin on stimulated secretion of the three hormones was assessed by intracerebroventricular administration of the peptide one minute be-

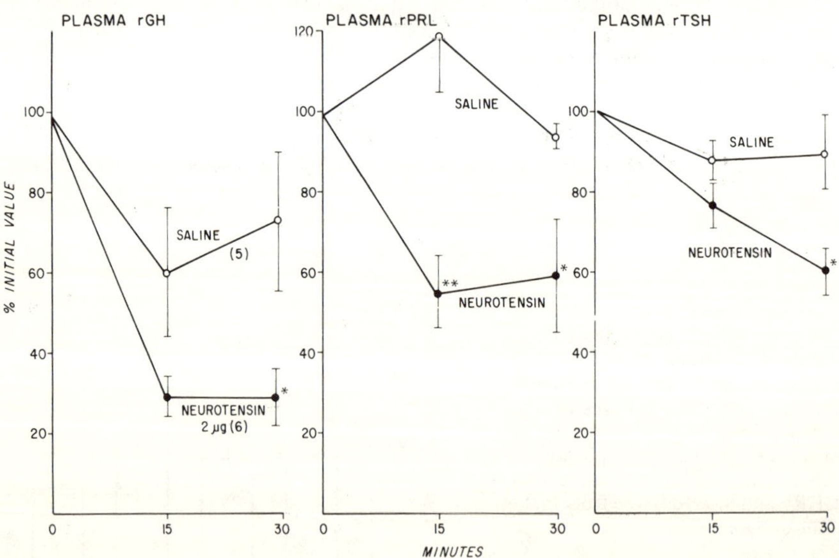

FIGURE 5.   Inhibitory effect of intracerebroventricular neurotensin administration on plasma GH, PRL, and TSH in estrogen-primed urethane-anesthestized rats. Legends as in FIGURE 1.

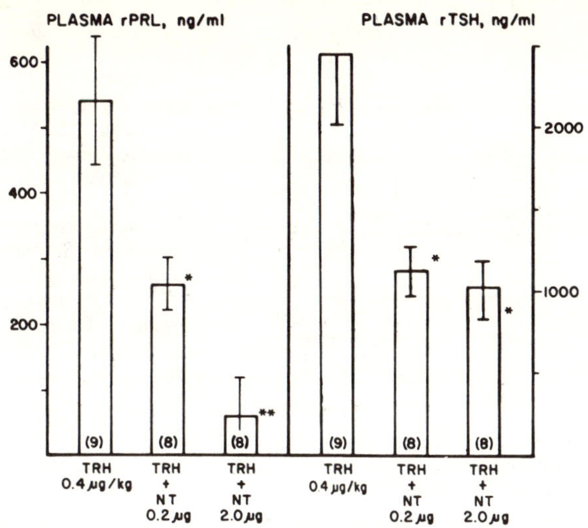

FIGURE 6. Suppression of TRH-stimulated plasma PRL and TSH secretion by intracerebroventricular administration of neurotensin. Shown are plasma PRL levels 5 minutes and TSH levels 15 minutes after the injection of TRH. Symbols as in FIGURE 1. (Reprinted from Maeda & Frohman[6] with permission from *Endocrinology*.)

fore the i.v. injection of TRH (thyrotropin-releasing hormone) or chlorpromazine. The increases in plasma PRL and TSH after TRH administration (0.4 μg/kg) were significantly blunted by the injection of neurotensin and the suppression of PRL secretion was dose related (FIG. 6). In animals injected centrally with neurotensin (0.5 or 2.0 μg) immediately before chlorpromazine (0.2 mg/kg i.v.), the stimulatory effect on GH release was completely inhibited (FIG. 7). Both doses of neurotensin

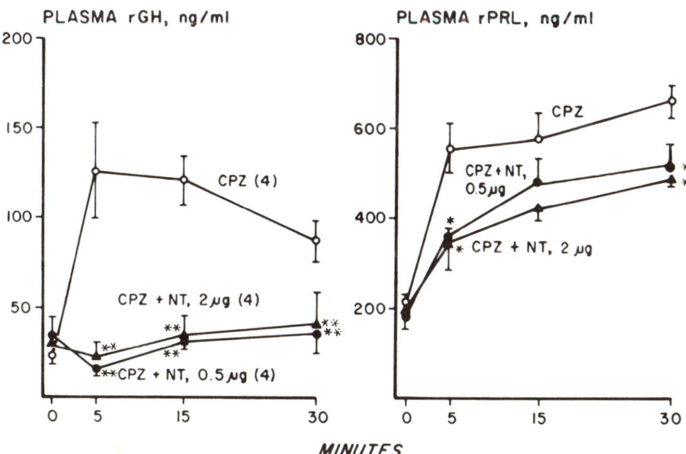

FIGURE 7. Effect of centrally administered neurotensin (NT) on chlorpromazine (CPZ) induced GH and PRL secretion. The dose of CPZ was 0.2 mg/kg. Symbols are as in FIGURE 1. (Reprinted from Maeda & Frohman[6] with permission from *Endocrinology*.)

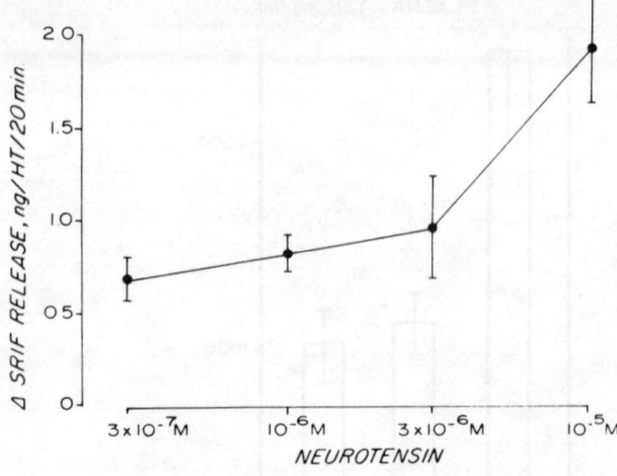

FIGURE 8. Effect of neurotensin (Rivier) on the release of somatostatin (SRIF) from quartered hypothalamic fragments. The mean ± SEM of four incubations is shown. The ordinate depicts somatostatin release in the presence of neurotensin minus that during the preceding control incubation. Somatostatin release during the control incubation was 0.53 ± 0.03 ng/hypothalamus/20 minutes. (Reprinted from Maeda & Frohman[13] with permission from *Endocrinology*.)

exhibited a partial suppression of chlorpromazine-stimulated PRL release. Administration of diphenhydramine at a dose of 5 or 10 mg/kg was ineffective in blocking the hormonal responses to neurotensin.

To investigate the possibility that the inhibition of GH and TSH secretion was mediated by somatostatin, quartered hypothalamic fragments were incubated in the presence of varying concentrations of neurotensin and we observed a dose-related increase in somatostatin release (FIG. 8).[13] More recent extensions of these studies involved the use of both intact mediobasal hypothalamic fragments and cerebral cortex and the substitution of a more sensitive somatostatin RIA.[15,20] Increases in medium somatostatin were observed during incubation of both hypothalamic and cerebral cortical tissue in the presence of neurotensin, 10^{-7} to 10^{-5} M (FIG. 9).

In the course of these studies, our initial supply of neurotensin (Rivier, Lot 14-286-50) became depleted. In repeat experiments using neurotensin obtained from two other sources (Peninsula Laboratories and Bachem), we were unable to demonstrate any stimulation of somatostatin release from cerebral cortex. Cross-reactivity studies with the three neurotensin preparations in the somatostatin RIA revealed the presence of cross-reacting material (equivalent to 1.7×10^{-4}) in Rivier neurotensin but not the other two preparations (FIG. 10). This degree of cross-reactivity appeared sufficient to account for the stimulatory effect observed in both hypothalamic and cerebral cortex incubations.

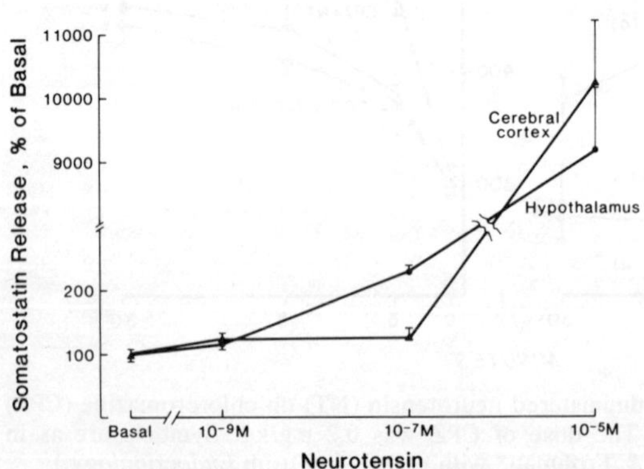

FIGURE 9. Effect of neurotensin (Rivier) on somatostatin release from intact hypothalamus and from cerebral cortex fragments. Shown are the mean ± SEM of six replicate incubations.

Chromatography of Rivier neurotensin on HPLC revealed an elution pattern identical to that of the other neurotensin preparations and the single peak of neurotensin immunoreactivity coeluted with the 210 nm absorbance peak (FIG. 11). There was no somatostatin radioactivity in this peak indicating that the cross-reactivity was not intrinsic to neurotensin. The repurified neurotensin, when tested in the hypothalamic incubation system at a concentration identical to that of the original preparation, no longer stimulated the release of somatostatin into the incubation medium (FIG. 12). With the small amounts of remaining material, we have attempted to determine the identity of the somatostatin-like immunoreactivity and have concluded that part, though probably not all, can be attributed to somatostatin itself.

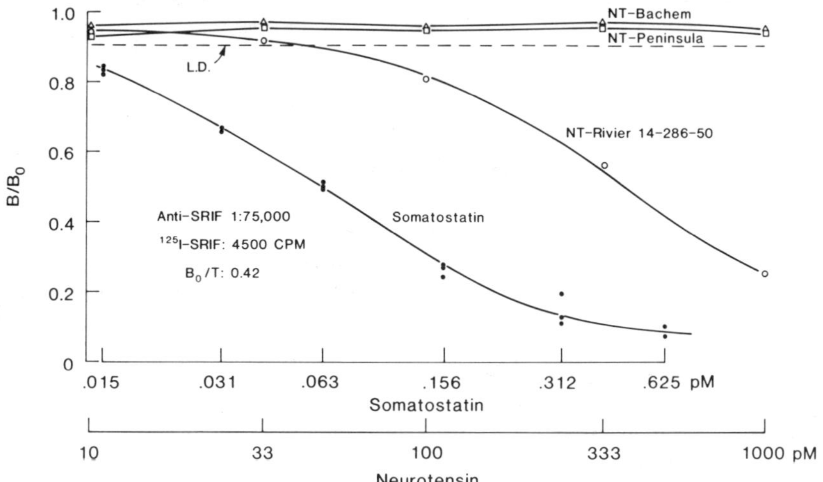

FIGURE 10.   Cross-reactivity of various neurotensin preparations in the somatostatin RIA. L.D. = least detectable dose as determined by computer analysis of weighed log–logit curve.

There was insufficient purified neurotensin available to repeat the *in vivo* injection studies. However, the quantity of somatostatin (340 pg) contained in the original neurotensin preparation that was injected into the lateral cerebral ventricle (2 μg) would be inadequate to suppress GH and TSH secretion *in vivo*.

SUMMARY AND CONCLUSIONS

Our studies have demonstrated that injection of neurotensin into the systemic circulation results in an increase in GH, PRL, and TSH secretion by an effect that is not histamine-mediated and that does not appear to be a direct action on the pituitary. Whether these effects are secondary to alterations in pituitary blood flow or to other actions of the peptide remains to be determined. Within the central nervous system, however, neurotensin appears to inhibit the release of these three

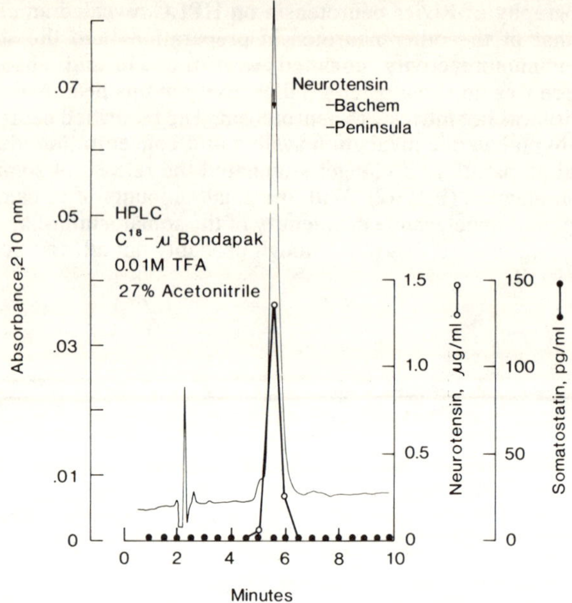

FIGURE 11. Chromatography of neurotensin (Rivier) on HPLC. Flow rate was 2 ml/minute. Individual fractions were assayed for somatostatin and neurotensin immunoreactivity as described.

pituitary hormones under both basal and stimulated conditions. Our previously described effect of neurotensin on somatostatin release appears to be the result of the coexistence in the preparation utilized of a compound(s) with somatostatin-like immunoreactivity and not to an effect on endogenous somatostatin release.

Whereas an explanation for the CNS effects of neurotensin on GH, PRL, and TSH secretion cannot be conclusively provided, the results are all consistent with a stimulatory effect of neurotensin on dopamine release. Stimulation of dopamine release in either anterior hypothalamus, which is a locus involved in the regulation

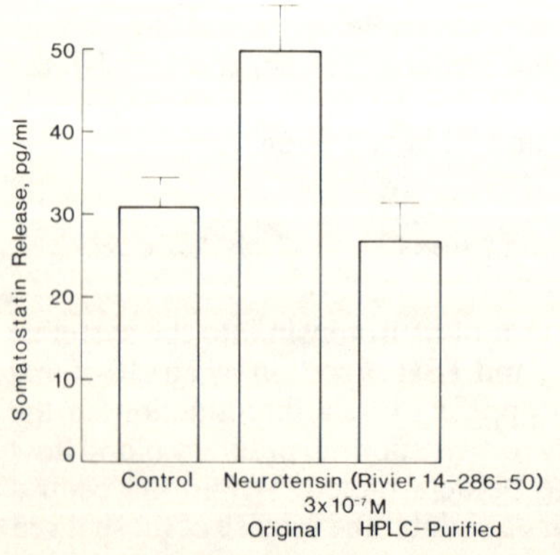

FIGURE 12. Effect of neurotensin (Rivier) before and after repurification on somatostatin release by hypothalamic fragments *in vitro*. Shown are the mean ± SEM of six replicate incubations. The effect of the unchromatographed neurotensin preparation was significantly greater than that of control, $p < 0.05$.

of somatostatin secretion, or in extrahypothalamic brain could in turn stimulate mediobasal hypothalamic somatostatin release,[13,21] thereby inhibiting GH and TSH secretion. A direct effect of neurotensin on the tuberoinfundibular system could result in release of dopamine from the median eminence leading to suppression of PRL secretion. Further studies to evaluate this possibility are currently in progress.

REFERENCES

1. CARRAWAY, R. & S. E. LEEMAN. 1973. The isolation of a new hypotensive peptide, neurotensin, from bovine hypothalami. J. Biol. Chem. **248:** 6854–6861.
2. MAKINO, T., R. CARRAWAY, S. E. LEEMAN & R. O. GREEP. 1973. *In vitro* effecters of a newly purified hypothalamic tridecapeptide on rat LH and FSH release. Program of the Society for the Study of Reproduction. (Abstract 26).
3. RIVIER, C., M. BROWN & W. VALE. 1977. Effect of neurotensin, substance P, and morphine sulphate on the secretion of prolactin and growth hormone in the rat. Endocrinology **100:** 751–754.
4. BROWN, M. & W. VALE. 1976. Effects of neurotensin and substance P on plasma insulin, glucagon, and glucose levels. Endocrinology **98:** 819–822.
5. NAGAI, K. & L. A. FROHMAN. 1976. Hyperglycemia and hyperglucagonemia following neurotensin administration. Life Sci. **19:** 273–280.
6. MAEDA, K. & L. A. FROHMAN. 1978. Dissociation of systemic and central effects of neurotensin on the secretion of growth hormone, prolactin, and thyrotropin. Endocrinology **103:** 1903–1909.
7. SZABO, M. & L. A. FROHMAN. 1975. Effects of porcine stalk median eminence and prostaglandin E_2 on rat growth-hormone secretion *in vivo* and their inhibition by somatostatin. Endocrinology **96:** 955–961.
8. SZABO, M. & L. A. FROHMAN. 1976. Dissociation of prolactin-releasing activity from thyrotropin-releasing hormone in porcine stalk median eminence. Endocrinology **98:** 1451–1459.
9. ALTAFFER, F. B., F. DeBALBIAN VERSTER, S. HALL, C. J. LONG & P. D'ENCARNACAO. 1970. A simple and inexpensive cannula technique for chemical stimulation of the brain. Physiol. Behav. **5:** 119–121.
10. FROHMAN, L. A. & L. L. BERNARDIS. 1968. Growth hormone and insulin levels in weanling rats with ventromedial hypothalamic lesions. Endocrinology **82:** 1125–1132.
11. RICK, J., M. SZABO, P. PAYNE, N. KOVATHANA, J. G. CANNON & L. A. FROHMAN. 1979. Prolactin-suppressive effects of two aminotetralin analogs of dopamine: Their use in the characterization of the pituitary dopamine receptor. Endocrinology **104:** 1234–1242.
12. SZABO, M., L. CHU & L. A. FROHMAN. 1982. Biologic effects of an ectopic growth-hormone-releasing peptide in cultured adenohypophyseal cells: Comparison with growth-hormone-releasing activity of porcine hypothalamus. Endocrinology. In press.
13. MAEDA, K. & L. A. FROHMAN. 1980. Release of somatostatin and thyrotropin-releasing hormone from hypothalamic fragments *in vitro*. Endocrinology **106:** 1837–1842.
14. BERELOWITZ, M., S. L. FIRESTONE & L. A. FROHMAN. 1981. Effects of growth hormone excess and deficiency on hypothalamic somatostatin content and release and on tissue somatostatin distribution. Endocrinology **109:** 714–719.
15. BERELOWITZ, M., K. MAEDA, S. HARRIS & L. A. FROHMAN. 1980. The effect of alterations in the pituitary-thyroid axis on hypothalamic content and *in vitro* release of somatostatin-like immunoreactivity. Endocrinology **107:** 24–29.
16. MAEDA, K. & L. A. FROHMAN. 1981. Neurotensin release by rat hypothalamic fragments *in vitro*. Brain Res. **210:** 261–269.

17. NAGAI, K. & L. A. FROHMAN. 1978. Neurotensin hyperglycemia: Evidence for his-
     tamine mediation and assessment of a possible physiologic role. Diabetes **27:**
     577–582.
18. LIBERTUN, C. & S. M. McCANN. 1976. The possible role of histamine in the control
     of prolactin and gonadotropin release. Neuroendocrinology **20:** 110–120.
19. FROHMAN, L. A., M. SZABO, M. BERELOWITZ & M. E. STACHURA. 1980. Partial
     purification and characterization of a peptide with growth-hormone-releasing activity
     from extrapituitary tumors in patients with acromegaly, J. Clin. Invest. **65:** 43–54.
20. KRONHEIM, S., M. BERELOWITZ & B. L. PIMSTONE. 1976. A radioimmunoassay for
     growth hormone release-inhibiting hormone: Method and quantitative tissue distribu-
     tion. Clin. Endocrinol. **5:** 619–630.
21. NEGRO-VILAR, A., S. R. OJEDA, A. ARIMURA & S. M. McCANN. 1978. Dopamine and
     norepinephrine stimulate somatostatin release by median eminence fragments *in
     vitro*, Life Sci. **23:** 1493–1487.

DISCUSSION OF THE PAPER

E. A. ZIMMERMAN (*Columbia University, New York*): I still keep thinking of
those terminals in the zona externa of the median eminence and wonder if it might
not well be that NT exerts some action on the pituitary itself. One might look at
the effect of TRH stimulation in the dispersed cells and see if neurotensin inhibits
that effect.

L. A. FROHMAN (*University of Cincinnati College of Medicine, OH*): At 10^{-6} M
NT, but not at lower concentration, there was a suggestion of some NT-induced
inhibition of the TRH effect on PRL release. The question, of course, that is
unanswered is: What might the concentration of neurotensin be in portal stalk
blood? 10^{-6} M equals 1.5 μg/ml and that seems high compared to the levels of
LHRH or somatostatin for example.

E. SINGER (*University of Massachusetts, Worcester*): You mentioned that Dr.
Uhl did not show binding sites in the middle preoptic area. We did binding studies
in that area and we pick up a reasonable number of binding sites. K_0's are in the
nanomolar range and density of binding sites for a high affinity site is about 100
femtomoles per mg of protein.

FROHMAN: I think that is a very important piece of information.

D. KAHN (*Columbia University, New York*): Do you have thoughts on what
neurotensin might do secondarily to pituitary hormone release by its primary
effects on body temperature or blood pressure when you administer it to an
animal?

FROHMAN: Certainly the hypotensive effects of a peptide could play a role,
though these kinds of changes have not been seen after administration of hypotensive agents. I do not think that we can really say.

M. D. HIRSCH (*Roche Institute of Molecular Biology, Nutley, NJ*): You showed
a lack of hormonal release by neurotensin in pituitary tissue culture. Did you also
try incubating the culture with chlorpromazine to see what the neurotensin effect
would be?

FROHMAN: These cell cultures are grown for 4 days in standard nutrient media.
By this point there has been enough of an absence of dopamine effect that the
administration of dopamine receptor blockers does not cause any stimulation of
prolactin release. If one adds dopamine to the system and then adds the blocker,
one can of course block the effects of dopamine. But we did not use chlorpromazine because this agent by itself is inactive in a 4-day cell culture preparation.

THE ROLE OF NEUROTENSIN IN THE UPTAKE AND DISTRIBUTION OF FAT*

Sune Rosell

Department of Pharmacology
Karolinska Institute
S-104 01 Stockholm, Sweden

INTRODUCTION

It has been shown repeatedly that gastric acid secretion and gastrointestinal motility are inhibited when ingested fat reaches the small intestine.[1] Experimental evidence of long standing indicates that this is caused by a hormone released from the intestinal mucosa, which Kosaka and Lim named enterogastrone.[2] Several gastrointestinal peptides have been proposed as enterogastrone candidates and neurotensin is one of them because of its actions on the gastrointestinal canal and the fact that ingestion of fat causes a pronounced increase in the plasma concentration of immunoreactive neurotensin.[3, 4]

The "enterogastrone hypothesis" is based on the results of experiments in which the effect of extracts of the small intestine on gastric acid secretion and intestinal motility was studied. These extracts probably contained a mixture of peptides and therefore no single substance is likely to be responsible for all the actions ascribed to enterogastrone. It's true that neurotensin may function in part as an enterogastrone, but rather than discussing which hormone may be *the* enterogastrone, it may be more fruitful to discuss the actions of the new gut hormones, including neurotensin, on their own merits and separately from the enterogastrone hypothesis.

INGESTION OF FAT AND NEUROTENSIN

The ingestion of food has marked effects on the concentration of immunoreactive neurotensin in plasma (p-NTLI).[5, 6] Thus, in healthy volunteers who had a meal consisting of an aperitif of 40 ml of whisky, a white bread roll and 4 g of butter, fried potatoes, steak and béarnaise sauce, wine, and coffee, the mean concentration of immunoreactive neurotensin rose from 65 ± 12 pM (n = 6) to 233 $\pm$ 86 pM at 45 min. The concentration fell thereafter but remained elevated in all subjects for the 2 hr 45 min during which sampling continued.[6] Indeed, increase in immunoreactive neurotensin in blood can be observed after quite a small breakfast consisting of a sandwich and 100 ml of yoghurt (FIG. 1).

Evidently the increase in p-NTLI is due to release of neurotensin from its storage site in the distal part of the small intestine. Such a release has been demonstrated in dogs. The terminal 20 cm of the ileum of anesthetized dogs was perfused *in situ* with a buffered solution and blood was collected from the carotid artery, femoral vein, and a mesenteric vein draining the terminal ileum.[7] There was a positive V-A difference in p-NTLI across the terminal ileum, indicating that neurotensin is released into the blood from the ileum.

*This study has been supported by the Swedish Medical Research Council, Grant no. 3518.

The consistent pattern of change in the concentration of p-NTLI associated with food ingestion suggests that circulating neurotensin or a neurotensin metabolite is serving some endocrine function in relation to eating. To elucidate that possibility, it is of importance to know if one or more of the principal nutrients in a mixed meal have a major influence on the concentration of p-NTLI. Bloom *et al.*[8] found that p-NTLI rose from 22 to 32 pM after 75 g of oral glucose. However, in comparison with fat, glucose seems to be of minor importance as far as release of neurotensin is concerned.[3] This is evident by comparing the effects of isocaloric amounts of glucose, fat, and amino acids. Following ingestion of fat (Intralipid) there is a significant increase in the plasma concentration of immunoreactive neurotensin from 42 ± 5.2 pM to a peak value of 109 ± 27 pM at 90 min with return of the plasma level to the starting value by 180 min, whereas isocaloric test meals comprising amino acids (Vamin) or glucose did not change the p-NTLI significantly (Fig. 2). Thus, fat seems to be the most important nutrient in a meal in terms of effect on the concentration of p-NTLI. Evidently, fat does not have to be absorbed, since intravenous administration of Intralipid in the same dose as is ingested orally does not influence the concentration of immunoreactive neuro-

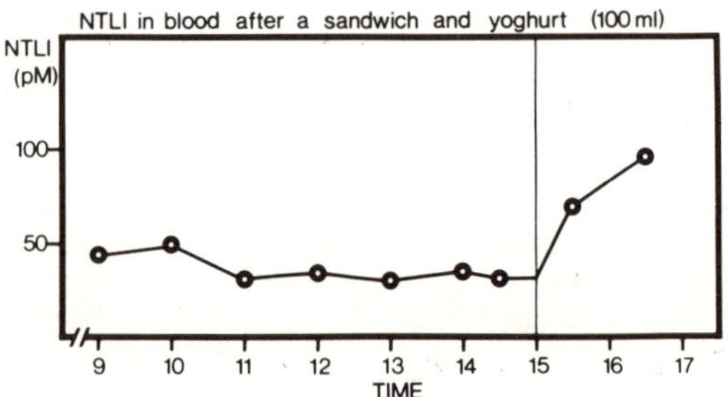

FIGURE 1. Concentration of p-NTLI in a healthy human subject. A sandwich with cheese and butter and 100 ml yoghurt was ingested at 15 hr.

tensin in blood (Fig. 3). The response of p-NTLI to orally administered fat seems to be rather specific. The volume of the ingested food does not affect the plasma concentration appreciably.[3] Similarly, antral distension in seven healthy subjects did not change the concentration of p-NTLI in venous blood or in portal blood.[9]

Presumably, neurotensin is released when the neurotensin-containing N-cells in the small intestine are exposed to the ingested fat. In some subjects, an increased plasma concentration of NTLI is already noticed 10 min after ingestion of fat in the form of fluid. This might seem to be a short period for the fluid to pass into the small intestine and cause a release of neurotensin that is detected as a rise in p-NTLI. However, in fasting healthy subjects, chyme appears in the small intestine 100 cm distal to the ligament of Treitz from 10 to 30 min after ingestion of a standard meal.[10] Therefore, experimental data are compatible with the view that the ingested fat has to reach the N-cell-containing part of the small intestine in order to cause a release of neurotensin. This hypothesis is further supported by the finding that intraduodenal administration of small amounts of oleic acid results in an increased concentration of p-NTLI.[11] Thus, 5 ml of oleic acid administered intraduodenally to healthy subjects induced a significant increase in p-NTLI with a peak concentration at 30 min. The maximal response of p-NTLI occurred after instillation of 20 ml oleic acid.

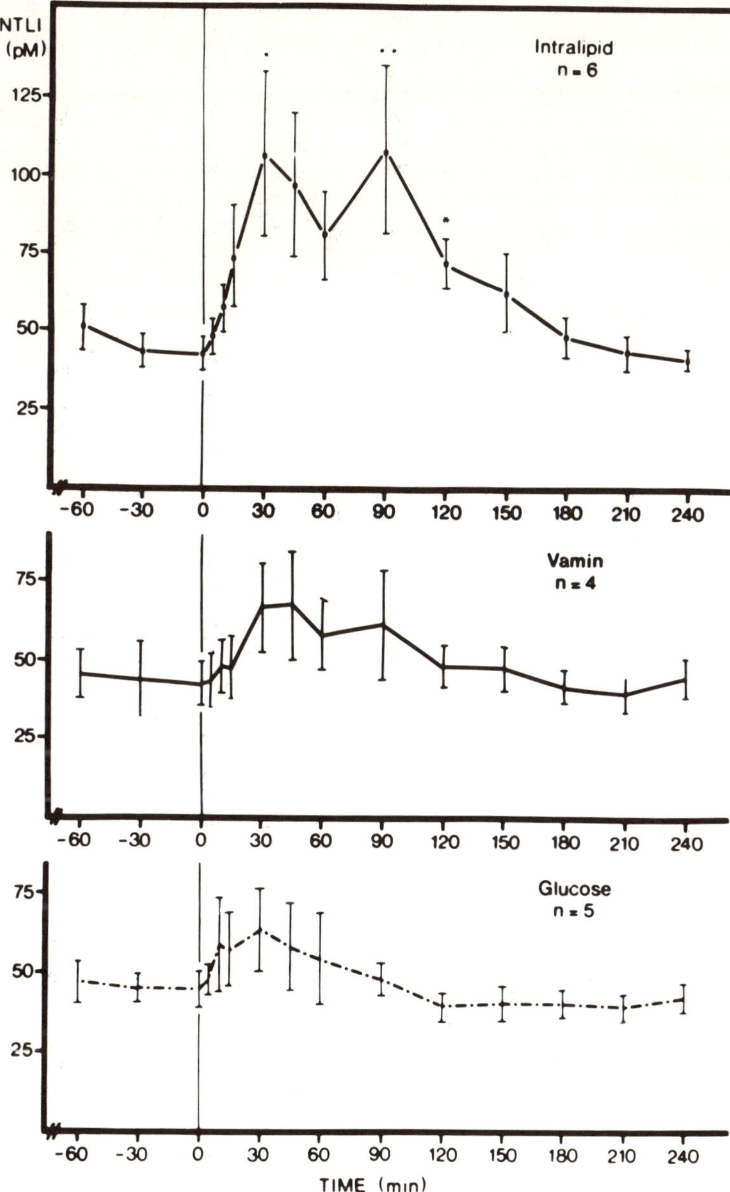

FIGURE 2. Concentrations of p-NTLI after oral administration of Intralipid, Vamin, or glucose. Number of subjects = n. The vertical bars indicate ± SEM. * = p <0.05; ** = p <0.01 (from ref.3).

Alternatively, a hormone or a nervous reflex from the proximal part of the gastrointestinal tract may be implicated in the fat-induced elevation of p-NTLI. These possibilities have been investigated in rats. Instillation of fat into the stomach did not increase the p-NTLI if a ligature was placed around the intestine proximal to the ligament of Treitz thus exposing only the stomach and duodenum to fat. However, instillation of fat distal to the ligature resulted in an increase of p-NTLI within 5 min (Rökaeus & Al-Saffar. To be published.) Thus, the release of neurotensin does not seem to be triggered from the proximal part of the gastrointestinal tract in the rat but rather to a direct exposure to fat of the N-cells in the distal part of the intestine. However, it is still possible that a local hormone or a local nervous pathway acting of the N-cells are involved in the release.

GASTROINTESTINAL EFFECTS OF EXOGENOUS NEUROTENSIN AND INGESTION OF FAT

Gastric Acid Secretion

In Pavlov pouch dogs, intravenous administration of neurotensin causes a considerable inhibition of gastric acid secretory responses to pentagastrin, a test meal, or insulin hypoglycemia[12] and the finding that exogenous neurotensin may depress gastric acid secretion has been confirmed in man.[13]

In this context it is interesting that fat in the small intestine also inhibits gastric acid secretion in man (the enterogastrone hypothesis), and it may be suggested that the fat-induced inhibition is mediated by neurotensin. Thus, oleic acid (5–40 ml) given intraduodenally causes a dose-related inhibition with a maximal effect on pentagastrin-stimulated acid secretion after 20 ml.[11] The maximal inhibition appears 30–60 min after intestinal application of fat, which is not easily reconciled with an inhibitory nervous reflex, but rather to a hormonal mechanism. In addition there is a firm temporal relation between the inhibition of gastric acid secretion by oleic acid and the increase of the plasma NTLI concentration (FIG. 4). Furthermore, a significant correlation is found between the inhibitory effect of oleic acid and the integrated NTLI response (FIG. 5). These results are compatible with the hypothesis that the release of neurotensin provides the temporal information from the intestine to the stomach and thereby ensures the inhibition of gastric acid secretion postprandially. Neurotensin may have a physiological role as a hormone with enterogastrone functions.[3, 4]

As a matter of fact, several experimental findings show a parallelism between the inhibition of gastric acid secretion caused by exogenously administered neurotensin and by fat in the small intestine. Thus, infusion of neurotensin in dogs inhibits pentagastrin-induced gastric acid secretion from vagally innervated, but not from vagally denervated, fundic pouches[12, 14] and proximal gastric vagotomy in duodenal ulcer patients abolishes the inhibition of gastric acid secretion induced by fat in the small intestine (FIG. 6). Moreover, histamine-induced gastric acid secretion is not blocked by neurotensin (dog)[12] and intestinally administered oleic acid does not block betazole (histamine analog) induced gastric acid secretion (man).[15] On the other hand, exogenous neurotensin inhibits insulin-induced gastric acid secretion and intestinally administered oleic acid inhibits secretion induced by sham feeding.[12, 15] Both insulin and sham feeding cause gastric acid secretion by a vagal mechanism.

By requiring an intact innervation (FIG. 6), neurotensin acts differently from

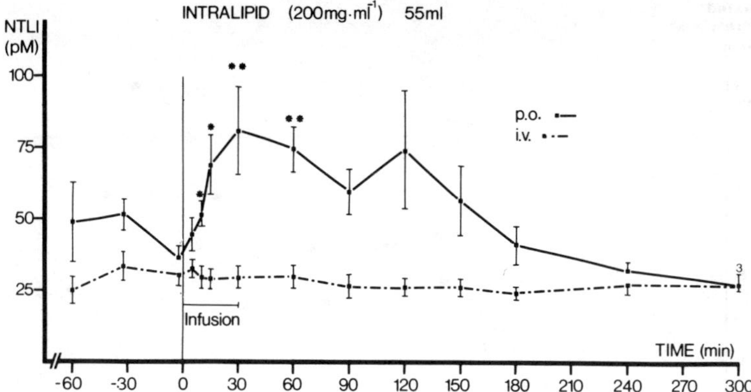

FIGURE 3.   Concentrations of p-NTLI after intravenous infusion and oral administration of fat (from B. Wiklund *et al.*, to be published.)

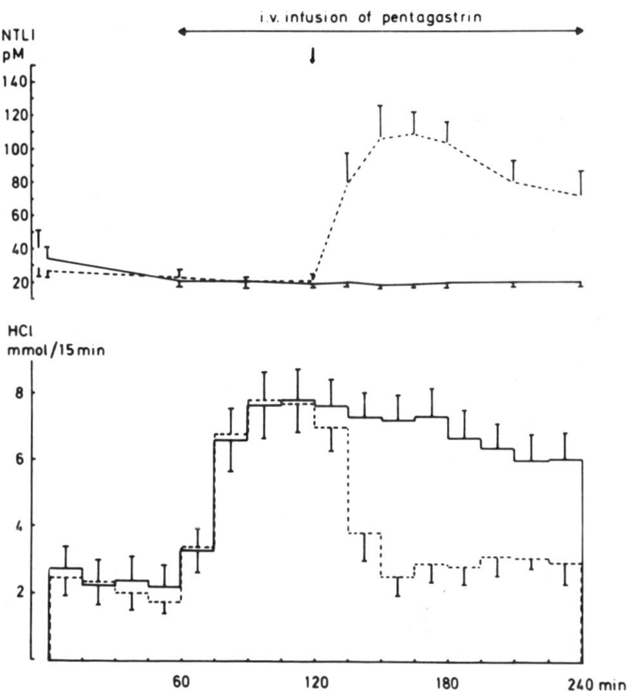

FIGURE 4.   Lower panel: Mean gastric acid output in mmol/15 min ± SEM after intraduodenal administration of 20 ml saline (*solid line*) or 20 ml oleic acid (*broken line*) on submaximal pentagastrin-stimulated gastric acid secretion in 10 healthy individuals. ↓ = Intraduodenal administration. Upper panel: Concentration of p-NTLI, mean ± SEM (from ref. 11).

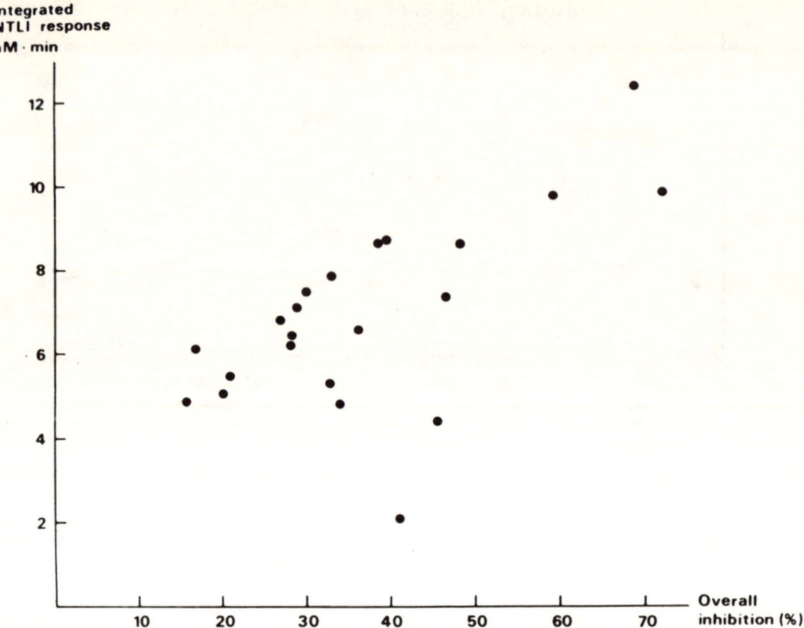

FIGURE 5.   Overall inhibition of gastric acid secretion (%) plotted against p-NTLI response (nM × min) in 22 healthy subjects tested on 5, 10, (n = 6) or 20 (n = 10) ml oleic acid; r = 0.61 (from ref. 11).

several other peptides that inhibit gastric acid secretion. Secretin effectively inhibits the gastric acid secretion in denervated, but also in innervated, fundic pouches.[16] Cholecystokinin in low doses inhibits the gastrin-induced acid secretion from vagally denervated pouches, but potentiates the acid secretion caused by vagal stimulation.[17] Gastric inhibitory polypeptide, GIP, is also a potent inhibitor of acid secretion from vagally denervated mucosa, but produces only a moderate reduction of vagally induced secretion from the innervated stomach.[18] Neurotensin seems to be the only known peptide that is dependent on an intact vagal innervation of the acid-secreting gland in order to exert its inhibitory effect.

GASTROINTESTINAL MOTILITY

Neurotensin changes gastric motility in dogs at low infusion rates, 3–6 pmol × kg^{-1} × min^{-1}.[19] In human subjects, the pressure of the lower esophageal sphincter (LES) is significantly reduced from 13.7 ± 1.3 mm Hg upon infusion of 12 pmol × kg^{-1} × min^{-1} (Gln^4)-neurotensin intravenously (FIG. 7).[20] Within 4 min after the cessation of the infusion, the LES pressure is back to the preinfusion value. The decrease in the LES occurs at a concentration of p-NTLI of approximately 50 pM, which is considerably lower than that reported to occur postprandially.[6] Therefore, the decrease in LES pressure seems to occur at physiological blood concentration of p-NTLI.

Nebel and Castell[21] have shown that ingestion of a fatty meal causes a pronounced decrease in the LES pressure. In view of the fact that p-NTLI is increased under such circumstances, neurotensin or a neurotensin metabolite might be responsible for the decrease in the LES pressure postprandially. Several hormones released after food intake, such as gastrin, secretin, cholecystokinin, and glucagon have been shown to influence the LES pressure,[21–26] but it has not been established whether these hormones are of any physiological significance in the regulation of the LES pressure.[27–30]

The motility effects also involve other parts of the gastrointestinal tract. In rats, intravenous infusion of neurotensin or (Gln^4)-neurotensin, 1.8 pmol $\times$ kg^{-1} $\times$ min^{-1}, abolishes the interdigestive migrating myoelectric complexes (MMC), which are replaced by increased spiking activity along the whole length of the small intestine.[31] The changes in myoelectric activity are observed within 2–4min after commencement of the infusion and the activity returns to control 5–15 min after the end of the infusion period. The neurotensin sequences NT$^{9–13}$, NT$^{8–13}$,

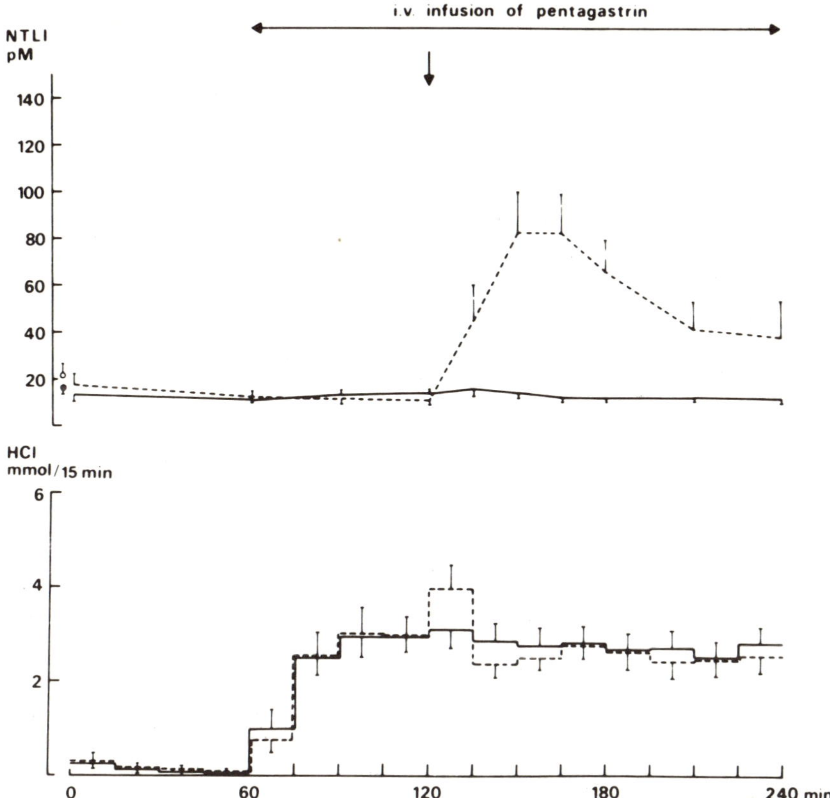

FIGURE 6. *Lower panel:* Mean gastric acid output in mmol/15 min $\pm$ SEM during submaximal pentagastrin administration after intraduodenal administration of 20 ml saline (*solid line*) or 20 ml oleic acid (*broken line*) in seven duodenal ulcer patients after proximal gastric vagotomy. $\downarrow$ = Intraduodenal administration. *Upper panel:* Concentration of p-NTLI, mean $\pm$ SEM (from ref. 11).

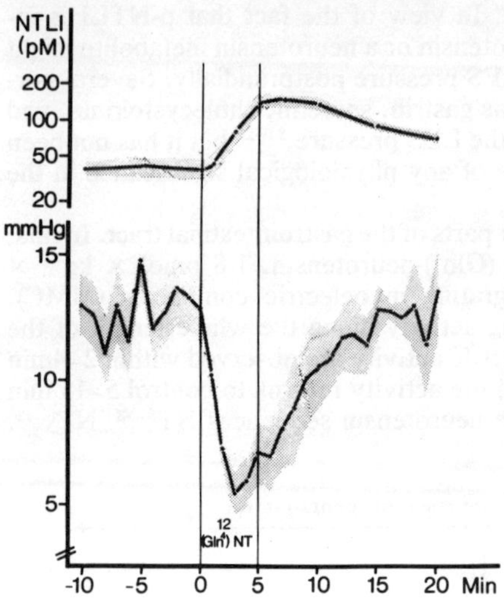

FIGURE 7. p-NTLI and LES-pressure in response to i.v. infusion of 12 pmol × kg⁻¹ × min⁻¹ of (Gln⁴)-neurotensin for 5 min (from ref. 20).

NT^{4-13}, NT^{1-9}, and (Gln^4)-NT^{1-11} do not induce any changes in the small intestinal electric activity whereas NT and (Gln⁴)-neurotensin have indistinguishable actions. Thus, intact neurotensin or (Gln⁴)-neurotensin sequences rather than smaller NT fragments are necessary to induce changes in small intestinal myoelectric activity.

Similar results are obtained in healthy human subjects. Thor *et al.*[32] measured the pressures in antrum, proximal, and distal duodenum, respectively. During infusion of (Gln⁴)-neurotensin (6 pmol × kg⁻¹ × min⁻¹) the peristaltic activity is completely abolished from a resting value of 3.4 ± 0.6 contractions/5 min.

Blackburn *et al.*,[13] reported that neurotensin decreases the gastric-emptying rate in humans, and in rats the peptide has been shown to prolong the transit time of the chyme in the stomach as well as in the small intestine.[33] It appears that the motility changes induced by neurotensin as measured as changes in myoelectric activity or pressure waves result in a retardation of the transport of the chyme (FIG. 8). Consequently, neurotensin may facilitate thorough digestion by intraluminal enzymes and nutrients may be more readily exposed to the intestinal mucosa, which might in turn enhance absorption.

Ingestion of fat induces changes in the motility of the gastrointestinal canal that are similar to those observed after intravenous infusion of neurotensin.

Nebel and Castell[21] have shown that ingestion of a fatty meal causes a pronounced decrease in the LES pressure in man. It is also a well-known clinical observation that patients suffering from reflux esophagitis should avoid fat. Moreover, fat is a potent inhibitor of MMC.[34] Since ingestion of fat causes a pronounced increase in p-NTLI that seems to be sufficient to induce a change in the motility pattern in the small intestine from a fasted- to a fed-type,[3] it is tempting to suggest that neurotensin may be a hormone that carries the information to the upper part of the gastrointestinal canal to alter the motility pattern after a fatty meal in order to establish optimal conditions for digestion of fat. It is well-known that digestion of fat requires longer time than digestion of carbohydrates and proteins. In this context, it is interesting to note that the neurotensin-containing cells are located in the distal part of the small intestine.[35, 36, 37] Thus, as long as fat reaches the distal part of the small intestine, neurotensin will be released and in turn retard the transport of the chyme. Consequently, neurotensin may keep the chyme from being transported to the large intestine until it is thoroughly digested.

Vascular Actions of Neurotensin in Adipose Tissue

Apart from its gastrointestinal actions, neurotensin constricts the vascular bed of canine subcutaneous adipose tissue in doses that do not affect blood pressure, heart rate, or skeletal muscle blood flow.[38, 39] This selective vasoconstrictor effect

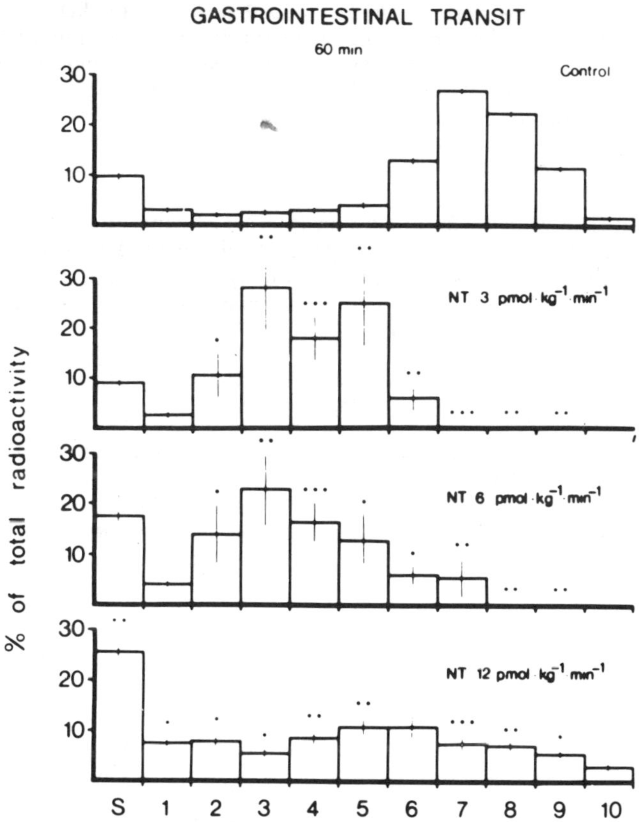

FIGURE 8.   Gastrointestinal transit as measured by a radioactive marker ($Na_2Co^{51}O_4$) deposited intragastrically. I.v. infusion of neurotensin at 3–12 pmol $\times$ kg^{-1} $\times$ min^{-1} during intervals of 60 min. Three pmol $\times$ kg^{-1} $\times$ min^{-1} of neurotensin efficiently retarded transit in the intestinal segments (p <0.001). At 6 pmol $\times$ kg^{-1} $\times$ min^{-1} of neurotensin a slightly farther transport of the radioactive marker along the bowel was registered. Also, a tendency for inhibition of the gastric emptying rate was registered. At 12 pmol $\times$ kg^{-1} $\times$ min^{-1} neurotensin efficiently inhibited the gastric emptying rate (p <0.01) concomitant with gross spreading, proximal pooling and farther migration of the radioactive marker (from ref. 33).

in subcutaneous adipose tissue suggests that neurotensin regulates not only postprandial gastrointestinal function but also the postprandial uptake and storage of substrates in adipose tissue, since these processes are dependent upon substrate supply, that is, blood concentration of substrates and blood flow.[40] This vasoconstriction in adipose tissue has been evaluated in human subjects.[41] Subcutaneous adipose tissue blood flow was measured after intravenous infusion of (Gln[4])-neurotensin in amounts that resulted in plasma concentrations of NTLI resembling those obtained after a fatty meal. Adipose tissue blood flow was monitored by the local clearance technique using ^{99m}Tc-pertechnetate. The disappearance rate from the abdominal adipose tissue began to decrease five to ten minutes after the start of the (Gln[4])-neurotensin infusion. The reduction in clearance from abdominal fat persisted during the observation period of thirty minutes after the end of the (Gln[4])-neurotensin infusion and in two subjects it was even accentuated during this period. The calf blood flow was unchanged following (Gln[4])-neurotensin infusion, indicating unaltered muscle blood flow, which is in agreement with the results of animal studies.[38] Heart rate and blood did not change. Interestingly, the vasoconstriction following neurotensin was most pronounced in lean subjects, while subjects with a higher body fat content responded poorly, if at all. It remains to be shown whether the degree of vasoconstrictor response in adipose tissue to neurotensin may be a factor of importance in the development of adiposity.

This study shows that (Gln[4])-neurotensin at NTLI concentrations in the range obtained after a fatty meal[6] causes a significant selective reduction of blood flow in human abdominal adipose tissue. Since the uptake of substrates depends on their delivery to adipose tissue,[40] neurotensin may regulate the postprandial uptake of substrates in adipose tissue.

CONCLUDING REMARKS

Available experimental data show that circulating neurotensin or a neurotensin metabolite, satisfies the rather rigid physiological criteria established for a true hormone of the endocrine type. The release of neurotensin into the blood circulation is produced by a physiologically meaningful event (ingestion of food). Moreover, the gastrointestinal effects and the vasoconstriction in adipose tissue following infusion of neurotensin or (Gln[4])-neurotensin are produced at p-NTLI concentrations obtained after the ingestion of food. Furthermore, the sites of action of neurotensin are remote from the storage site, so it has to reach them via the blood stream.

The present review of published and unpublished data indicates that infusion of neurotensin mimics the adjustments of the gastrointestinal canal seen after the ingestion of fat. The physiological role of neurotensin or a neurotensin metabolite may be to *coordinate* blood flow, motility, and acid secretion in the gastrointestinal canal in order to, for example, optimize the digestion of fat. In addition, it may also regulate, at least in part, the postprandial distribution of absorbed fat by adjusting the blood flow in certain regions of adipose tissue, which in turn may be of importance for the development of adiposity. Apart from catecholamines, neurotensin is the only known endogenous substance that has been shown to selectively regulate blood flow in adipose tissue on a short-term (minute-to-minute) basis.

The data indicate that neurotensin does not act directly on the cells in the way that a classical hormone (e.g. insulin) does but rather on the nervous system. This is in agreement with the hypothesis that the function of neurotensin is to coordinate the activity of various organs rather than to regulate cellular activity directly. This mechanism of action may explain why surgical extirpation of the small intestine containing the neurotensin storage cells is compatible with the maintenance of life. On the other hand, the consequences of such an extirpation may be a loss of coordination of digestive processes, resulting in, for example, gastritis, diarrhoea, and adiposity.

REFERENCES

1. BABKIN, B. P. 1950. Secretory mechanism of the digestive glands. 2nd edit. P.B. Hoeber Inc. New York.
2. KOSAKA, T. & R. K. S. LIM. 1930. Demonstration of the humoral agent in fat inhibition of gastric secretion. Proc. Soc. Exp. Biol. **27:** 890–891.
3. ROSELL, S. & A. RÖKAEUS. 1979. The effect of ingestion of amino acids, glucose, and fat on circulating neurotensin-like immunoreactivity (NTLI)—in man. Acta Physiol. Scand. **107:** 263–267.
4. ROSELL, S. 1980. Substance P and neurotensin in the control of gastrointestinal function. *In* The Peptides: Integrators of cell and tissue function. F. E. Bloom, Ed. Raven Press. New York. pp. 147–162.
5. BESTERMAN, H. S., D. L. SARSON, A. M. BLACKBURN, J. CLEARY, T. R. E. PILKINGTON & S. R. BLOOM. 1978. The gut hormone profile in morbid obesity and following jejuno-ileal bypass. Scand. J. Gastroenterol. **13** (Suppl. 49): 15.
6. MASHFORD, M. L., G. NILSSON, Å. RÖKAEUS & S. ROSELL. 1978. The effect of food ingestion on circulating neurotensin-like immunoreactivity (NTLI) in the human. Acta Physiol. Scand. **104:** 244–246.
7. MASHFORD, M. L., G. NILSSON, Å. RÖKAEUS & S. ROSELL. 1978. Release of neurotensin-like immunoreactivity (NTLI) from the gut in anaesthetized dogs. Acta Physiol. Scand. **104:** 375–376.
8. BLOOM, S. R., A. M. BLACKBURN, H. S. BESTERMAN, D. L. SARSON & J. M. POLAK. 1979. Measurement of neurotensin in man: A new circulating peptide hormone affecting insulin release and carbohydrate metabolism. Diabetologia **15:** 220.
9. SCHÖÖN, J.-M., S. R. BLOOM & L. OLBE. 1980. The effect of antral distension in healthy subjects on betazole-stimulated gastric acid secretion and plasma concentration of immunoreactive neurotensin. Scand. J. Gastroenterol. **15:** 277–282.
10. CLAIN, J. E., V. LIANG & J. R. MALAGELADE. 1978. Inhibitory role of the distal small intestine on the gastric secretory response to meals in man. Gastroenterology **74:** 704–707.
11. KIHL, B., Å. RÖKAEUS, S. ROSELL & L. OLBE. 1981. Inhibition of pentagastrin-stimulated gastric acid secretion and rise in the plasma concentration of neurotensin-like immunoreactivity (NTLI) by intraduodenal oleic acid in man. Scand. J. Gastrorenterol. **16:** 513–517.
12. ANDERSSON, S., D. CHANG, K. FOLKERS & S. ROSELL. 1976. Inhibition of gastric acid secretion in dogs by neurotensin. Life Sci. **19:** 367–370.
13. BLACKBURN, A. M., S. R. BLOOM & R. G. LONG. 1980. Effect of neurotensin on gastric function in man. Lancet **1:** 987–989.
14. ANDERSSON, S., S. ROSELL, L. SJÖDIN, D. CHANG & K. FOLKERS. 1980. Inhibition of acid secretion from vagally innervated and denervated gastric pouches by (Gln^4)-neurotensin. Scand. J. Gastroenterol. **15:** 253–256.
15. KIHL, B., Å. RÖKAEUS, S. ROSELL & L. OLBE. 1982. The effect of intraduodenal instillation of oleic acid on plasma neurotensin-like immunoreactivity and on gastric acid secretion stimulated by betazole and sham feeding in man. Scand. J. Gastroenterol. **17:** 633–639.
16. SJÖDIN, L. & S. MIURA. 1974. Secretin-induced inhibition of acid secretion before

and after vagal denervation of canine gastric pouches. Scand. J. Gastroenterol. **9:** 185–190.

17. SJÖDIN, L. 1972. Effects of secretin, cholecystokinin, and caerulein on gastric secretion in response to sham feeding in dogs. Acta Physiol. Scand. **85:** 99–109.
18. PEDERSEN, R. & J. C. BROWN. 1972. Inhibition of histamin-, pentagastrin-, and insulin-stimulated canine gastric secretion by pure gastric inhibitory polypeptide. Gastroenterology **62:** 393–416.
19. ANDERSSON, S., S. ROSELL, U. HJELMQUIST, D. CHANG & K. FOLKERS. 1977. Inhibition of gastric and intestinal motor activity in dogs by (Gln^4)-neurotensin. Acta Physiol. Scand. **100:** 231–235.
20. ROSELL, S., K. THOR, Å. RÖKAEUS, O. NYQVIST, A. LEWENHAUPT, L. KAGER & K. FOLKERS. 1980. Plasma concentration of neurotensin-like immunoreactivity (NTLI) and lower esophageal sphincter (LES) pressure in man following infusion of (Gln^4)-neurotensin. Acta Physiol. Scand. **109:** 369–375.
21. NEBEL, O. T. & D. O. CASTELL. 1972. Lower oesophageal sphincter pressure changes after food ingestion. Gastroenterology **63:** 778–783.
22. GILES, G. R., M. C. MASON, C. HUMPHERIES & C. G. CLARK. 1969. Action of gastrin on the lower esophageal sphincter in man. Gut **10:** 730–734.
23. CASTELL, D. O. & L. D. HARRIS. 1970. Hormonal control of gastroesophageal sphincter strength. N. Engl. J. Med. **282:** 886–889.
24. COHEN, S. & W. LIPSHUTZ. 1971. Role of the gastrin supersensitivity in the pathogenesis of lower esophageal sphincter, hypertension in achalasia. J. Clin. Invest. **50:** 1241–1247.
25. RESIN, H., D. H. STERN, R. A. L. STURDEVANT & J. J. ISENBERG. 1973. Effect on the COOH-terminal octapeptide of cholecystokinin on lower esophageal sphincter pressure in man. Gastroenterology **64:** 946–949.
26. JAFFER, S. S., G. M. MAKHLOUF, B. A. SCHORR & A. M. ZFASS. 1974. Nature and kinetics of inhibition of lower esophageal sphincter pressure by glucagon. Gastroenterology **67:** 42–47.
27. GROSSMAN, M. I. 1973. What is physiological? Gastroenterology **65:** 994.
28. COHEN, S. 1974. What is physiological? An answer! Gastroenterology **66:** 479.
29. STURDEVANT, R. A. L. 1974. Is gastrin the major regulator of lower esophageal sphincter pressure? Gastroenterology **67:** 551–553.
30. HENDERSON, J. M., G. LIDGARD, D. H. OSBORNE, D. C. CARTER & R. C. HEADING. 1978. Lower esophageal sphincter response to gastrin—pharmacological or physiological? Gut **19**(2): 99–102.
31. AL-SAFFAR, A. & S. ROSELL. 1981. Effects of neurotensin and neurotensin analogues on the migrating myoelectrical complexes in the small intestine of rats. Acta Physiol. Scand. **112:** 203–208.
32. THOR, K., S. ROSELL, Å. RÖKAEUS & L. KAGER. 1982. (Gln^4)-neurotensin changes the motility pattern of the duodenum and proximal jejunum from a fasting-type to a fed-type. Gastroenterology. **83:** 569–574.
33. HELLSTRÖM, P. M., G. NYLANDER & S. ROSELL. 1982. Effects of neurotensin on the transit of gastrointestinal contents in the rat. Acta Physiol. Scand.
34. SCHANG, J. C. & K. A. KELLY. 1980. Inhibition of canine interdigestive proximal gastric motility by cholecystokinin octapeptide (CCK-OP). Gastroenterology **78:** (5 part 2): 1253.
35. ORCI, L., O. BAETENS, C. RUFENER, M. BROWN, W. VALE & R. GUILLEMIN. 1976. Evidence for immunoreactive neurotensin in dog intestinal mucosa. Life Sci. **19:** 559–562.
36. SUNDLER, F., J. ALUMETS, R. HÅKANSON, R. CARRAWAY & S. E. LEEMAN. 1977b. Ultrastructure of the gut neurotensin cell. Histochemistry **53:** 25–34.
37. HELMSTAEDTER, V., G. E. FEURLE & W. G. FORSSMANN. 1977. Ultrastructural identification of a new cell type—the N-cell as the source of neurotensin in the gut mucosa. Cell. Tissue Res. **184:** 445–452.
38. ROSELL, S., E. BURCHER, D. CHANG & K. FOLKERS. 1976. Cardiovascular and metabolic actions of neurotensin and (Gln^4)-neurotensin. Acta Physiol. Scand. **98:** 484–491.

39. ROSELL, S., Å. RÖKAEUS, D. CHANG & K. FOLKERS. 1978. Indirect vascular actions
       of (Gln[4])-neurotensin in canine adipose tissue. Acta Physiol. Scand. **102:** 143–147.
40. HAGENFELDT, L. & J. WAHREN. 1968. Human forearm muscle metabolism during
       exercise II. Uptake, release and oxidation of individual FFA and glycerol. Scand. J.
       Clin. Lab. Invest. **21:** 263–276.
41. LINDE, B., S. ROSELL & Å. RÖKAEUS. 1982. Blood flow in human adipose tissue after
       infusion of (Gln[4])-neurotensin. Acta Physiol. Scand. **115:** 311–315.

DISCUSSION OF THE PAPER

J. W. HOLADAY (*Walter Reed Army Medical Center, Washington DC*): Is there
any evidence that in clinical situations such as achlorhydria, there might be an
excess of neurotensin? Conversely where there is too much acid secretion could
there be NT deficiencies?

S. ROSELL (*Karolinska Institute, Stockholm, Sweden*): Neurotensin may have
some bearing on the defective inhibition of gastric acid secretion in duodenal ulcer
patients.[11]

HOLADAY: ACTH and β-lipotrophic hormone have some lipolytic activity.
Does NT have any direct lipolytic activity?

ROSELL: Neurotensin has no direct lipolytic activity. All we have found in
adipose tissue is the vascular effects of NT. This is not a direct effect of neurotensin or neurotensin metabolites, since if we give neurotensin intra-arterially to
adipose tissue, we do not observe these effects. Neurotensin has to circulate to
produce this effect. So, it seems as if neurotensin (or a neurotensin metabolite) is
releasing something which in turn is causing vasoconstriction. It is not an adrenergic effect.

E. D. NIKOLAIDES (*Warner-Lambert Company, Ann Arbor, MI*): Have you
looked at the levels of CCK during infusion of neurotensin or vice versa?

ROSELL: We have not done that, and I do not know of any relevant data.

R. A. HAMMER (*Northwestern University Medical School, Chicago, IL*): I want
to raise several points about your experiments with lipid instillation in the duodenum and the subsequent inhibition of acid secretion. One question that I have is
why you look at the integrated response over a 3-hour period rather than the point
to point response? If this is a hormonal control by neurotensin (or a measured
metabolite) on acid secretion, you should be able to see a response at each point
over time rather than to have to do an integrated response over 4 hours. You may
minimize the effect by doing that.

ROSELL: Yes. An increase is also apparent if point to point analysis is used.

I would like to point out that we do not know the receptor for neurotensin on
the parietal cell. It may respond to neurotensin, a fragment of neurotensin, or it
could be that neurotensin releases something which in turn causes inhibition of
gastric acid secretion. It is very difficult to relate the inhibition of gastric acid
secretion to plasma concentration of neurotensin immunoreactivity as long as we
do not know the receptor ligand.

HAMMER: The additional complicating factor is that, because Ronan has shown
that NH$_2$-terminal metabolites have a much longer half-life in circulation, they may
have an accumulative effect on NTLI (NT-like immunoreactivity), without being
reflected in biologic activity.

S. R. BLOOM (*Royal Postgraduate Medical School, London, England*): We
have infused NT[1–8] in man and looked at the pentagastrin-stimulated gastric acid
and do not find any effect. I do not think that makes any difference to your results,

because I think the natural form is released from the ileum just as if you had infused it into the left arm. Since your assay is the same in both situations, though it may be reading somewhat higher, it nonetheless reflects the same reality of either NT infusion into the arm or NT infusion naturally by the ileum.

ROSELL: It is interesting that you did not find anything with NT[1-8].

BLOOM: I think, we are too readily concluding that fat is the major mediator of neurotensin release. After a meal, glucose is very rapidly absorbed. It never gets down to the ileum nor indeed do amino acids. The fact that they do not in the natural situation release neurotensin may be merely because they do not get to the ileum. When we have looked at patients with dumping syndrome after glucose, glucose is a potent releaser of neurotensin, rather like fat. Is there any isolated tissue work that might support the idea that many nutrients would release NT if locally placed in the ileum?

ROSELL: No, I do not know of any, but I think these are the type of studies that should be done.

J. E. T. Fox: (*McMaster University, Hamilton, Ont., Canada*): I am interested in the fact that all your motility effects were dependent on an intact innervation. That is very similar to the work that we have been doing in dogs. Are the receptors adrenergic or cholinergic?

ROSELL: The receptors appear to be cholinergic rather than adrenergic.

Fox: Is that only in rats?

ROSELL: Correct. We do not know anything about man or dog in that respect.

C. F. FERRIS (*University of Massachusetts Medical Center, Worcester*): For the past year, we have been looking at the release of neurotensin from rat small intestine during continuous perfusion of the small intestine with a micellular solution of *oleic* acid. We have taken blood from the superior mesenteric vein as well as the femoral artery in measuring arteriovenous differences. We have been characterizing the release of NTLI with high pressure liquid chromatography as well as region-specific antiserum. While there is certainly an increase in neurotensin in the portal circulation of the rat during lipid perfusion, we are unable to find any neurotensin elevation in the general circulation. However, there may be an elevation in the NH_2-terminal NT metabolites, which raises the question as to whether NT is acting as a classical hormone, or as a paracrine agent?

ROSELL: We have obtained similar results. However, in some experiments the neurotensin concentration in peripheral blood is increased after lipid administration. This was measured with HC8 antiserum, which recognizes the COOH-terminal.

FERRIS: We find that there is an elevation of HC8 immunoreactivity though it is not neurotensin on HPLC.

ROSELL: Work in my laboratory indicates an increase of neurotensin in peripheral blood in some experiments: I do not think that these are paracrine effects.

FERRIS: Is it possible that the neurotensin could be working through the enteric plexus?

ROSELL: Yes.

H. S. KOOPMANS (*Columbia University, New York*): We agree with you that lower gut hormones are important in general regulatory effects. We have done a chronic study in which we move a small segment of the ileum up to the duodenum so that it gets overstimulated with nutrients. We find several hypertrophic effects. We find an increase in the size of the pancreas, stomach, and small intestines even below the site of the transposed tissue. We believe that there is an inhibitory effect from lower gut factors. However, in our preparations we are not sure whether the effect is caused by neurotensin, enteroglucagon or some other lower gut peptide. Some of your experiments have the same type of phenomena; if you infuse fat you are also going to release enteroglucagon. And, I wonder if you might comment on whether other peptides are involved.

ROSELL: Certainly other peptides are involved and the matter is more complicated than I have described today.

NEUROTENSIN: REGIONAL DISTRIBUTION, CHARACTERIZATION, AND INACTIVATION

P. C. Emson, M. Goedert, B. Williams, M. Ninkovic, and S. P. Hunt

MRC Neurochemical Pharmacology Unit
MRC Centre
Cambridge, England

Introduction

Neurotensin is a basic tridecapeptide that was originally isolated from bovine hypothalamus by Leeman and her colleagues.[1] Subsequently this group has also isolated and sequenced neurotensin from both human and bovine small intestine.[2,3] Radioimmunological and immunocytochemical studies have demonstrated that neurotensin-like immunoreactivity (NTLI) is localized to specific cells in brain,[4-6] anterior pituitary,[7,8] and gut.[9-11] In the mammalian brain, neurotensin is concentrated in vesicles within synaptic terminals[5,7,12] and it can be released from regions rich in NTLI by depolarizing stimuli.[13,14] Furthermore, neurotensin binding sites have been demonstrated by autoradiographic[15] and biochemical techniques[16-18] and shown to be localized in regions where NTLI is concentrated. Together with the observations that neurotensin has direct effects when applied to single neurons in regions where binding sites are localized[19,20] these data are consistent with the view that neurotensin may be a neurotransmitter in the central nervous system.[21] In the gastrointestinal tract, neurotensin is localized in endocrine-like cells of the gut mucosa with highest concentrations in the ileum. It is one candidate for the so far elusive hormone enterogasterone.[9-11]

In order to investigate possible functions of neurotensin and to characterize the NTLI in the brain, pituitary gland, and peripheral tissues of the rat, we have developed antisera specific for the carboxy-terminal and amino-terminal sequences of NT_{1-13}.[6,24] These antisera have also been used to follow the degradation of neurotensin by rat brain synaptic membranes. It would be an additional support for a neurotransmitter role of neurotensin in the central nervous system if neurotensin could be shown to be rapidly degraded by synaptic membranes as this would provide a means for rapid inactivation.[21]

Regional Distribution and Characterization
of Neurotensin-like Immunoreactivity

Sequence-specific neurotensin antisera were raised by conventional methods using either carbodiimide or glutaraldehyde to conjugate neurotensin to a protein carrier. The specificities of two rabbit neurotensin antisera characterized in our laboratory are indicated in Table 1. One (R 174) is amino-terminal specific, a second (R 99) is carboxy-terminal directed as is a commercially available antibody (IMN) obtained from the Immunonuclear Corporation, Stillwater, Minnesota. Details of the radioimmunoassays are given in Emson *et al.*[6] Both radioimmunoassays detected equivalent amounts of NTLI in the rat brain, anterior pituitary,

---

* Please direct all correspondence to P. C. Emson.

198

TABLE 1

COMPARATIVE IMMUNOREACTIVITY OF NEUROTENSIN AND NEUROTENSIN FRAGMENTS

Peptide	COOH-terminus directed Antiserum "Immunonuclear" antiserum *	COOH-terminus directed Antiserum Rabbit R-99 *	NH_2-terminus directed Antiserum Rabbit R-174 *
Neurotensin (NT_{1-13})	1.00	1.00	1.00
NT_{1-12}	<0.001	<0.001	0.74
NT_{1-10}	<0.001	<0.001	0.77
NT_{1-7}	<0.001	<0.001	0.24
NT_{1-6}	<0.001	<0.001	0.035
$C_{11} NT_{3-13}$	1.00	1.00	<0.001
$C_5 NT_{9-13}$	0.33	ND+	<0.001
$C_4 NT_{10-13}$	<0.001	<0.001	<0.001

*The amount of peptide (mol. ratios) needed to inhibit by 50% the binding of ^{125}I-neurotensin to the respective antiserum. The relative potency of neurotensin was taken arbitrarily as 1.00.

+ ND = not determined

and gastrointestinal tract. Serial dilution of brain, gut, or pituitary extracts also showed that in the RIA, the NTLI in these extracts diluted in parallel with synthetic neurotensin standards, suggesting that the majority of rat NTLI is antigenically similar to bovine and human neurotensin. High-performance liquid chromatography (HPLC) on C_{18} reverse-phase columns also showed that the majority of NTLI in the rat brain, gut, or pituitary elutes at the position expected for the synthetic bovine neurotensin standard (FIGS. 1 and 2). These observations

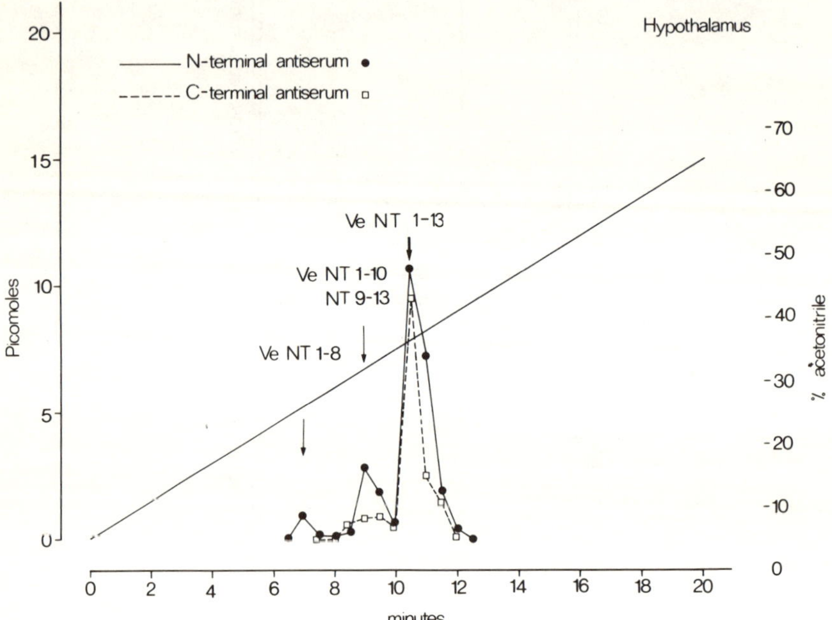

FIGURE 1. Separation of neurotensin-like immunoreactivity from rat hypothalamus using reverse-phase HPLC on μBondapak C_{18} columns using a 5–65% gradient of acetonitrile in 10 mM ammonium acetate buffer, pH 4.0. Note the major peak of both NH_2- and COOH-terminal NTLI eluting at the position expected for synthetic neurotensin (Ve NT_{1-13}). As well as the major peak with both NH_2- and COOH-terminal immunoreactivity, there are smaller amounts of NH_2- and COOH-terminal immunoreactivity that may represent fragments of neurotensin. The elution position of the neurotensin fragments NT_{1-8}, NT_{1-10}, and NT_{9-13} are indicated.

suggest that rat neurotensin is identical to the other characterized mammalian neurotensin. However, although the majority of NTLI in the rat brain and pituitary is indistinguishable from synthetic neurotensin, HPLC showed the presence in extracts of brain of some NTLI material having only amino- or carboxy-terminal immunoreactivity (FIG. 1). The elution positions of this neurotensin immunoreactivity suggests these may be fragments of neurotensin that elute earlier from the HPLC column (FIG. 1).

The availability of sequence-specific antisera encouraged us to look at the distribution of NTLI in the rat brain using either radioimmunoassay or immunohistochemical techniques. Our radioimmunoassay studies revealed high concentrations of NTLI in the bed nucleus of the stria terminalis, central nucleus of the amygdala, hypothalamus, anterior pituitary, and spinal cord (TABLE 2). In the

TABLE 2

REGIONAL DISTRIBUTION OF NEUROTENSIN-LIKE
IMMUNOREACTIVITY IN THE RAT*

	NTLI pmol neurotensin/g wet weight[†]
Cerebral Cortex and Hippocampus	
Frontal cortex	4.1 ± 1.2 (11)
Parietal cortex	5.2 ± 2.8 (10)
Hippocampus	7.1 ± 2.8 (7)
Subiculum	9.2 ± 2.8 (4)
Basal Forebrain	
Nucleus accumbens	29.2 ± 3.8 (5)
Nucleus caudate putamen	10.2 ± 3.8 (7)
Globus pallidus	12.6 ± 5.8 (8)
Bed nucleus of the stria terminalis	171.8 ± 18.6 (6)
Central amygdaloid nucleus	106.0 ± 31.2 (6)
Hypothalamus	
Lateral preoptic area	124.9 ± 12.0 (6)
Medial preoptic area	143.5 ± 13.2 (7)
Lateral hypothalamic nucleus	76.0 ± 12.9 (7)
Anterior hypothalamic nucleus	113.4 ± 5.1 (7)
Arcuate nucleus/median eminence	128.1 ± 11.6 (7)
Dorsomedial hypothalamic nucleus	123.9 ± 9.5 (7)
Ventromedial hypothalamic nucleus	87.7 ± 22.3 (7)
Mammillary body	128.1 ± 11.6 (7)
Brain Stem and Spinal Cord	
Locus ceruleus	54.0 ± 11.4 (5)
Periaqueductal gray	42.2 ± 7.2 (5)
Nucleus of the solitary tract	49.8 ± 6.4 (7)
Trigeminal nucleus	35.16 ± 2.2 (12)
Cervical dorsal horn	23.7 ± 5.6 (4)
Cervical ventral horn	2.0 ± 0.8 (5)
Pituitary	
Anterior lobe	118.2 ± 5.2 (20)
Neurointermediate lobe	28.7 ± 1.6 (20)
Gastrointestinal tract	
Esophagus	2.9 ± 0.2 (20)
Stomach	0.7 ± 0.04 (20)
Duodenum	0.4 ± 0.05 (20)
Jejunum	14.4 ± 1.0 (20)
Ileum	81.1 ± 5.3 (20)
Colon	5.5 ± 0.3 (20)
Pancreas	>0.1 (5)

*Determinations were carried out using NH_2-terminal specific antiserum (R-174).
[†] Value are means ± SE of the number of determinators in parenthesis.

gastrointestinal tract, high levels of NTLI were detected in the ileum with only low levels in the stomach, duodenum, jejunum, and colon. Low levels were also found in several peripheral tissues, such as the liver and the heart. It remains to be seen whether neurotensin is present in cells or in nerve fibers within these tissues. A recent immunohistochemical study suggests that in the heart, NTLI is present in postganglionic parasympathetic nerve fibers.[25] In immunohistochemical studies, our antisera revealed a similar distribution of NTLI when compared with the radioimmunoassay data. Our histochemical observations accorded well with previous histochemical studies.[12, 21, 26] Of particular interest was the demonstration of NTLI in the anterior lobe of the pituitary gland and in the dorsal horn of the spinal cord.

ANTERIOR PITUITARY NEUROTENSIN

The majority of NTLI in the rat anterior pituitary gland separates on HPLC and on gel chromatography at the position expected for intact NT_{1-13} and this molecule has both amino- and carboxy-terminal immunoreactivity. The nature of the small amount of NTLI eluting before NT_{1-13} on the HPLC and after NT_{1-13} on Sephadex G25 columns is not known (FIG. 2). Immunohistochemical studies (FIG. 3) show that the NTLI is present in specific cells in the anterior pituitary gland. The establishment of rat anterior pituitary neurotensin as authentic neurotensin led us to consider other species, and NTLI has been detected in guinea pig, bovine, and postmortem human anterior pituitaries.

The influence of various endocrine manipulations on the anterior pituitary neurotensin content is discussed elsewhere in this volume (Goedert, Lightman, and Emson). These manipulations suggest an involvement of neurotensin with the hypothalamus/anterior pituitary/thyroid axis as chemical thyroidectomy (using propylthiouracil) or administration of thyrotrophin-releasing hormone (TRH) drastically reduced rat anterior pituitary neurotensin content. If neurotensin is an anterior pituitary hormone, we should expect NTLI to be released from the anterior pituitary and in recent experiments using rat anterior pituitary slices *in vitro* we have been able to demonstrate a potassium-evoked release of NTLI[8] (FIG. 4). It is not known under what conditions this anterior pituitary neurotensin can be released to have a possible hormonal function. It is not known either which peripheral tissues could be possible targets for circulating neurotensin.

SPINAL CORD NEUROTENSIN AND ITS BINDING SITES

Intraventricular or intracisternal injection of neurotensin produces a naloxone-insensitive antinociceptive response to both chemical and thermal stimuli.[27, 28] Moreover, studies by Yaksh (personal communication) indicate that spinal administration of neurotensin produces analgesia although a further study does not support this contention.[29] Given the possible involvement of neurotensin in spinal analgesia, it was interesting to find NTLI-and neurotensin-receptor binding sites to be localized in the substantia gelatinosa of the dorsal horn of the spinal cord.[30, 31] The NTLI-positive neurons have a characteristic distribution of processes in layer I and in part of layer II of the dorsal horn of the spinal cord.[32] Electron-microscopic immunohistochemistry revealed that the NTLI was concentrated in nerve endings and localized in vesicles within the nerve endings (Fig. 5). The neurotensin-containing processes make synaptic connections predominantly with the dendrites and dendritic spines of other gelatinosa neurons and not with the small-diameter axons of sensory fibers. This localization suggests that NT

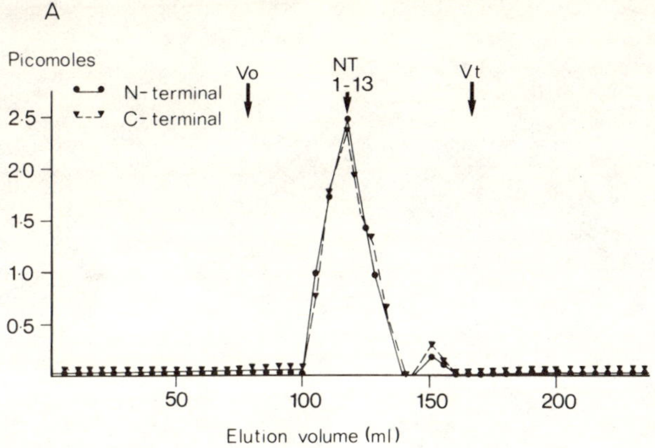

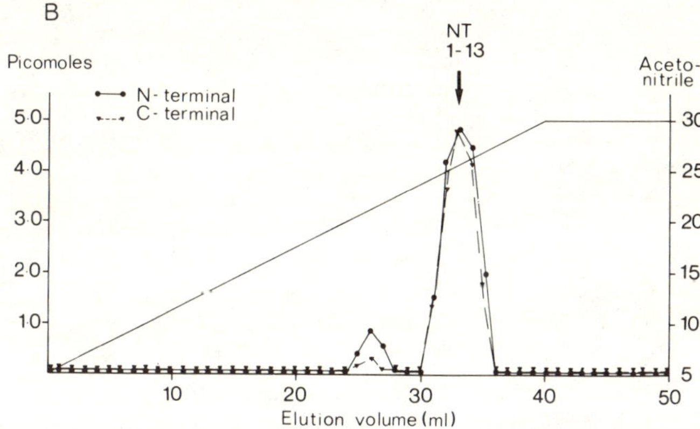

FIGURE 2. Fractionation of the neurotensin-like immunoreactivity in acetic acid extracts of rat anterior pituitary gland on a Sephadex G-25 column (A) and on a high-performance liquid chromatography (HPLC) reverse-phase column (C_{18} μBondapak, Waters Associates) (B).

A. Lyophilized tissue extracts were reconstituted in 4–6 mls and applied to a Sephadex G-25 (1.6 × 90 cm), then equilibrated and eluted at room temperature with 0.1 M acetic acid. Fractions of 3 ml were collected at a flow rate of 15 ml/hr. The fractions were lyophilized, resuspended in buffer and assayed using the neurotensin amino-terminal- and carboxy-terminal-directed antisera. The column was calibrated by blue dextran (void volume, V_o), synthetic neurotensin 1–13, 1–8, and 8–13 and ^{22}NaCl (total volume V_T).

neurons are more likely to act on other gelatinosa neurons than to directly influence primary afferent axons. Consistent with this observation is the finding that the substantia gelatinosa sites binding ^{3}H-neurotensin are not decreased following

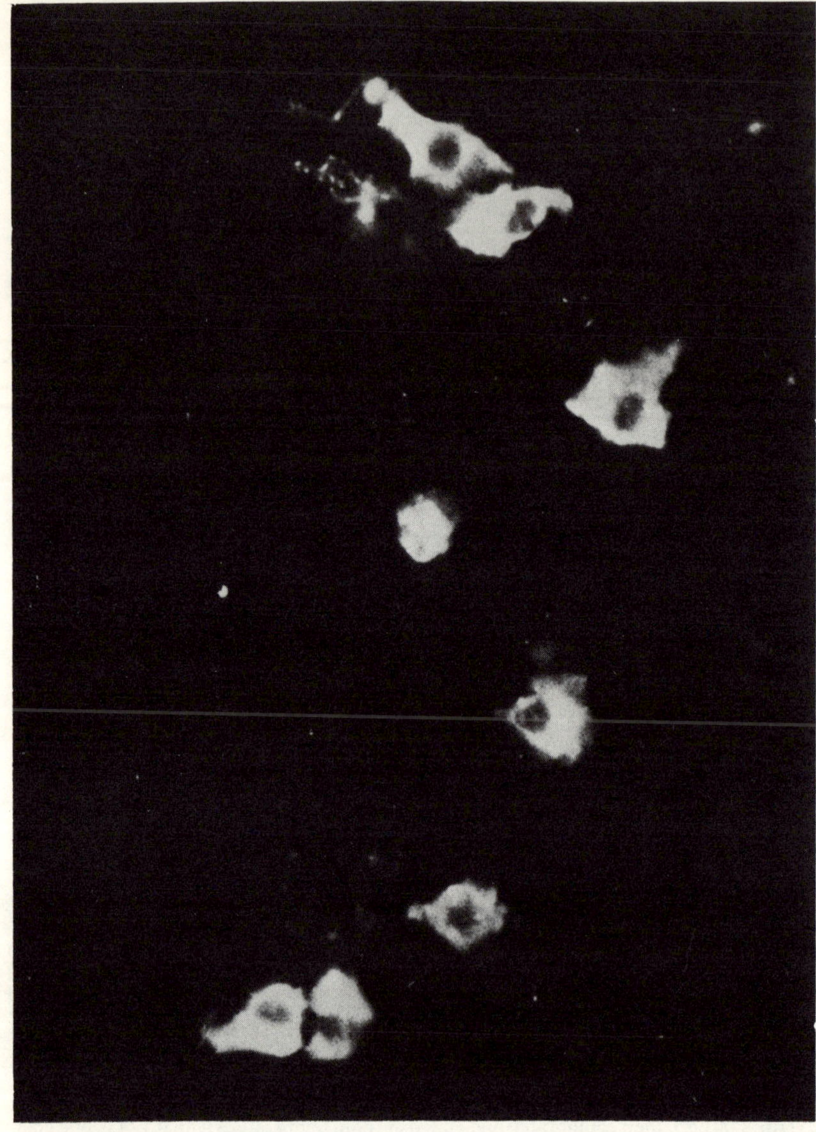

FIGURE 3.   Neurotensin-containing cells in the anterior pituitary of the rat demonstrated by immunofluorescent technique essentially as described by Uhl *et al.*[7]

dorsal rhizotomy (FIG. 6). In contrast to this, a proportion of opiate-receptor binding sites are localized on primary afferent fibers (FIG. 6). The physiological role of these spinal cord, neurotensin-containing interneurons remains to be established. Electrophysiological studies show that neurotensin applied to layer I–III neurons has excitatory effects,[20] and consistent with this observation are the asymmetric synapses made by the neurotensin neurons (See FIG. 5).

INACTIVATION OF NEUROTENSIN

The mechanisms involved in the inactivation of neurotensin after its synaptic release are unknown, although the rapid disappearance of NTLI after incubation with brain and hypothalamic soluble and particulate fractions has been reported.[33] In our initial studies, we also investigated the breakdown of synthetic neurotensin (0.2 μM) using soluble (S_1) and particulate (P_2) fractions of rat brain. Breakdown of neurotensin was monitored by using the amino- and carboxy-terminal specific antisera. These studies showed that the majority (at least 80%) of neurotensin-degrading activity was present in the soluble (S_1) fraction (TABLE 3). Furthermore, they also showed that the carboxy-terminal region of neurotensin was much more rapidly degraded as the half-time of disappearance of the carboxy-terminal immunoreactivity was approximately half of that of the amino-terminal immunoreactivity (data not shown). Careful washing of the P_2 pellet indicated that the remaining NTDA was membrane bound and could not be displaced by exposure to either hypo- or hypertonic media.

In order to characterize the neurotensin-degrading enzymes present in the rat brain membrane, membranes were extracted using 1% Triton X-100 in order to solubilize membrane-bound proteins. This procedure completely removed all the

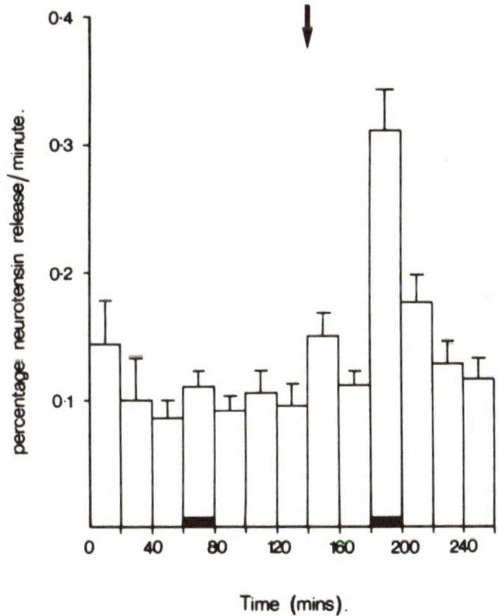

FIGURE 4. Calcium-dependent release of neurotensin-like immunoreactivity (NTLI) from rat anterior pituitary evoked by depolarizing concentrations of potassium. Anterior pituitaries (corresponding to about 60 mg of tissue) were transferred to plastic chambers containing 2 mls of Krebs solution and incubated at 37°C under constant oxygenation for 20-minute intervals. At the end of each interval, the incubation medium was removed onto ice, and subsequently NTLI was determined with the amino-terminal directed assay using a neurotensin standard curve prepared in the same Krebs solutions. At the end of the experiments, the tissue was collected, extracted in 0.1 M HCl, lyophilized, and NTLI determined in a separate assay. The results are expressed as fractional, release-rate constants (defined as the percentage of total tissue NTLI content released per minute) and represent ± SEM (n = 6). Solutions to the left of the arrow contained 0.5 mM EGTA and no $CaCl_2$, those to the right contained no EGTA and 1.8mM $CaCl_2$. The black bars indicate the replacement of 40 mM NaCl with 40 mM KCl.

NTDA from the membrane preparation (TABLE 3). The 1% Triton X-100 extract was applied directly to a DEAE-Sephacel column (5 × 40 cm) and eluted with 50 mM Tris-HCl buffer pH 7.4 containing 0.1% Triton X-100 followed by a linear gradient of sodium chloride (25 mM–500 mM) in 50 mM Tris-HCl buffer pH 7.4 containing 0.1% Triton X-100. The broad band of NTDA from this column separation was pooled, concentrated and then applied to a hydroxyapatite column. Stepwise elution of the hydroxyapatite column using increasing concentrations of

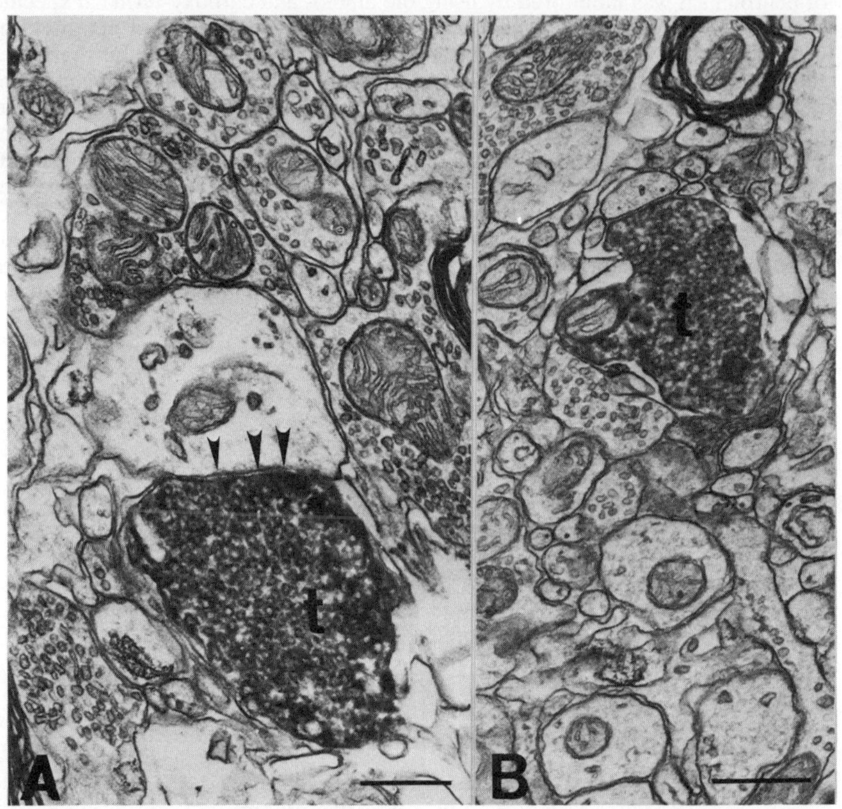

FIGURE 5. Electron microscopic immunohistochemical localization of neurotensin-containing axon terminals in layer II[1] of the dorsal horn. Arrows, synapse. Scale bar = A, 0.45 μm; B, 0.50 μm.

potassium phosphate buffer pH 7.8 (see FIG. 7) eluted two peaks of NTDA. (One peak (Fraction I) was eluted with 25 mM phosphate buffer and the second peak (Fraction II) was eluted by 50 mM phosphate buffer). Incubation of fraction II with synthetic neurotensin followed by HPLC and amino-acid analysis of the products revealed that peak II contained an enzyme-cleaving neurotensin between amino acids 10 (proline) and 11 (tyrosine) of the tridecapeptide (FIG. 8). In contrast, peak I contained a proline endopeptidase-cleaving neurotensin on the carboxyl side of both the proline at position 10 and the proline at position 7. Fraction I also

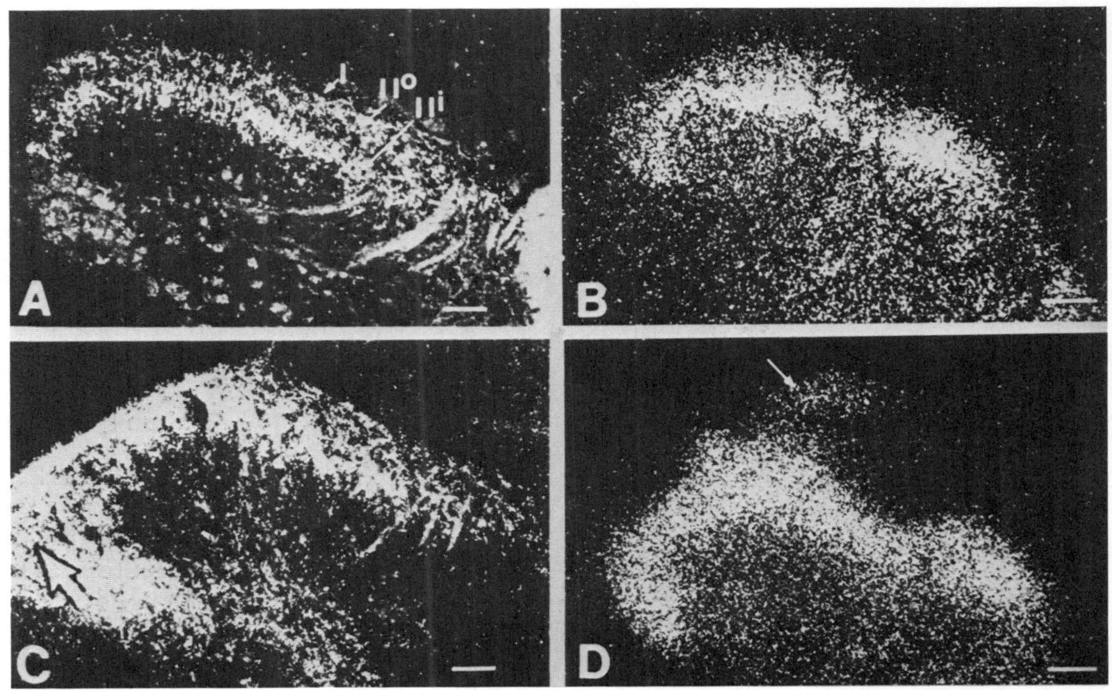

FIGURE 6.   Correspondence between the distribution of neuropeptides and their receptors in the superficial layers of the rat dorsal horn. Neurotensin (A) and Met-enkephalin (C) like immunoreactivity were localized by immunoperoxidase staining. The autoradiographic distributions of [^{3}H]-neurotensin and opiate ([^{3}H]-etorphine) binding sites are shown in B and D, respectively. C: arrow indicates presence of Met-enkephalin-like immunoreactivity within the dorsal white matter—an area that does not show a corresponding density of opiate receptors. D: arrow shows the presence of opiate receptors on incoming dorsal root fibers.

Scale bar = 100 μm.

TABLE 3

PURIFICATION OF NEUROTENSIN AND SUBSTANCE-P-DEGRADING ACTIVITIES FROM RAT BRAIN

Fraction	Neurotensin-degrading Activity (nmol/mg protein/hour*)	Substance-P-degrading Activity (nmol/mg protein/hour[†])	Proline Endopeptidase Activity (nmol/mg 7-AMC protein/hour[‡])
Homogenate	45.1	30.07	104.4
Pellet 40,000 $\times$ g	5.05	4.35	51.6
1% Triton X-100 solubilized	5.40 (1.0)[§]	3.29 (1.0)	45.6 (1.0)
Pellet $\times$ 100,000 g	<0.1	<0.1	<0.6
DEAE-Sephacel	30.92 (5.72)	24.7 (7.5)	254.4 (5.6)
Hydroxyapatite fraction I	54.5 (10.92)	43.6 (13.25)	1330.3 (29.5)
Hydroxyapatite fraction II	109.0 (20.18)	150.0 (45.5)	0

* Neurotensin-degrading activity as determined by the COOH-terminal specific radioimmunoassay.
[†] Substance P degrading activity as determined by the NH_2-terminal specific radioimmunoassay.
[‡] Proline endopeptidase activity determined using fluorescent substrate Z-Glycyl-Prolyl-7-AMC.
[§] Numbers in parentheses refer to the purification factor relative to the Triton X-100 solubilized activity.

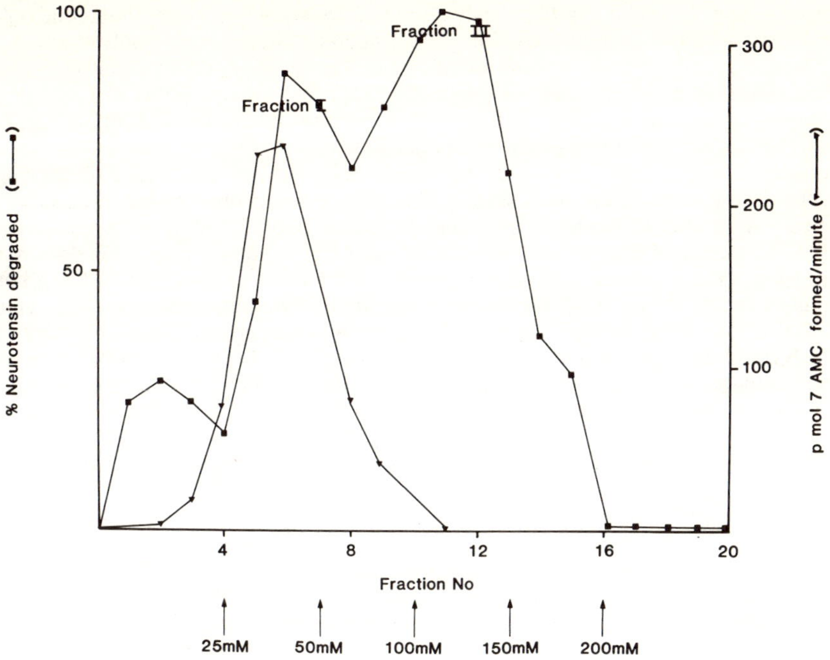

FIGURE 7. Separation of neurotensin-degrading activity on hydroxyapatite gel. Neurotensin-degrading activity was eluted with increasing concentrations of potassium phosphate buffer pH 7.8 containing 0.1% Triton X-100. Neurotensin-degrading activity was monitored with a COOH-terminal-directed RIA (see TABLE 1) and proline endopeptidase activity using the synthetic substrate Z-Gly-Pro-7-AMC (see text for details).

contained a dipeptidyl carboxypeptidase activity producing the fragment isoleucyl-leucine.

These initial observations suggested that at least two of the neurotensin-degrading enzymes might represent proline endopeptidases.[34, 35] It was therefore of some interest to test the synthetic proline endopeptidase substrates benzyloxycarbonyl-glycyl-L-proline-4-methylcoumarinyl-7-amide (Z-Gly-Pro-7-

FIGURE 8. High-performance liquid chromatography of the products of digestion of neurotensin with enzyme fraction II. The peaks indicated NT_{1-10}, NT_{11-13}, and NT_{1-13} were identified by amino acid analysis.

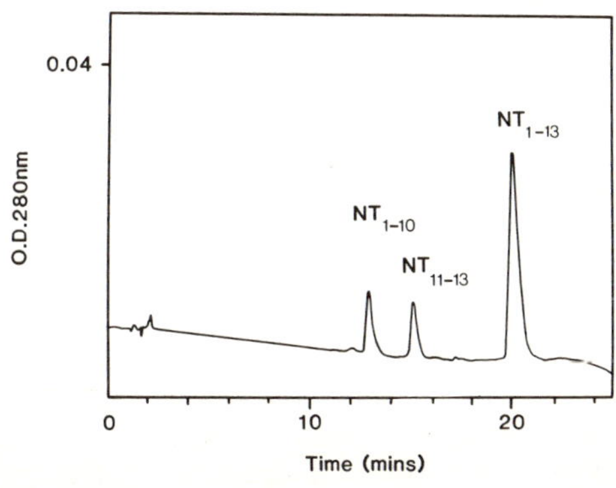

AMC) and benzyloxycarbonyl-L-lysyl-L-proline-4-methoxy-2-napthylamide (Z-Lys-Pro-4-MNA) as substrates for these fractions. Interestingly, only fraction I contained activity towards both synthetic proline endopeptidase substrates and fraction II did not degrade either substrate (TABLE 3). The availability of a sensitive fluorescent assay for proline endopeptidase using the Z-Gly-Pro-7-AMC substrate allowed us to confirm that proline endopeptidase activity is present in the brain membranes and that it was isolated in parallel with the neurotensin-degrading activity in fraction I. Previous studies have shown that proline endopeptidase will cleave a number of peptides including substance P, bradykinin, and angiotensin I.[34] The nature of the enzyme present in fraction II is not yet known. It does not recognize the synthetic proline endopeptidase substrate (Z-Gly-Pro-7-AMC), but it does cleave at one of the proline sites (position 10) in neurotensin. This enzyme fraction also degrades substance P (Table 3) and comparison of the activities of both fractions I and II towards substance P (0.2 μM) indicates that both fractions will degrade substance P (as detected by an amino-terminal-directed substance P assay, Lee et al. 1980[36]).

In this respect it is interesting to note that although proline endopeptidases will not biologically inactivate substance P (as they do not degrade the biologically active carboxy-terminal sequence[37]), they will inactivate neurotensin.[16–18] Given a possible involvement of proline endopeptidase in neurotensin inactivation, it was of interest to investigate the affinity of neurotensin for the enzyme (fraction I) and to compare neurotensin with other peptides as inhibitors of the proline endopeptidase activity as measured using the fluorescent substrate Z-Gly-Pro-7-AMC.[34]

Kinetic studies of fraction I with the synthetic substrate showed that the fraction I proline endopeptidase had a K_m of 11.4 μM for the synthetic substrate (FIG. 9). Furthermore, study of the properties of fraction I proline endopeptidase with respect to temperature, pH, and inhibitors (TABLE 4) indicated that this enzyme was very similar to the previously characterized soluble proline endopeptidase.[35, 36] Neurotensin was a potent competitive inhibitor of proline endopeptidase (FIG. 10) and has the highest affinity of the proline-containing neuropeptides so far tested (TABLE 5). Interestingly, partial sequences of neurotensin and substance P do act as inhibitors of proline endopeptidase, and as would be expected, the sequences containing the proline groups (SP_{1-8}, NT_{8-13}) are much better inhibitors than the nonproline-containing sequences (substance P_{5-11}, NT_{1-6}) (TABLE 5). Previous workers have reported very high rates for proline endopeptidase activity using substance P as a substrate (comparable to those found for acetylcholinesterase[38] and given the very much higher affinity for neurotensin for this enzyme it was of some interest to see if the proline endopeptidase was associated with synaptic membranes where it might contribute to the inactivation of neurotensin. By using a sucrose gradient separation of a lysed synaptosomal fraction (prepared as described by Cotman et al. 1981[39]), we obtained light and heavy synaptic membrane fractions. These fractions were enriched in both proline endopeptidase and neurotensin-degrading activity (TABLE 6) whereas the myelin and mitochondrial fractions contained only traces of activity. This observation suggests that proline endopeptidase is in the correct anatomical localization to contribute to the inactivation of neurotensin. In order to ascertain if proline endopeptidase does significantly contribute to the breakdown of proline-containing peptides by rat brain membranes, rat brain membranes were incubated with neurotensin (0.2 μM) in the presence of a specific antibody raised against soluble rat brain proline endopeptidase.[40] This antibody substantially stabilizes neurotensin against brain membrane degradation suggesting that proline endopeptidases are likely to be involved in neurotensin inactivation. It is, of course, not yet clear if

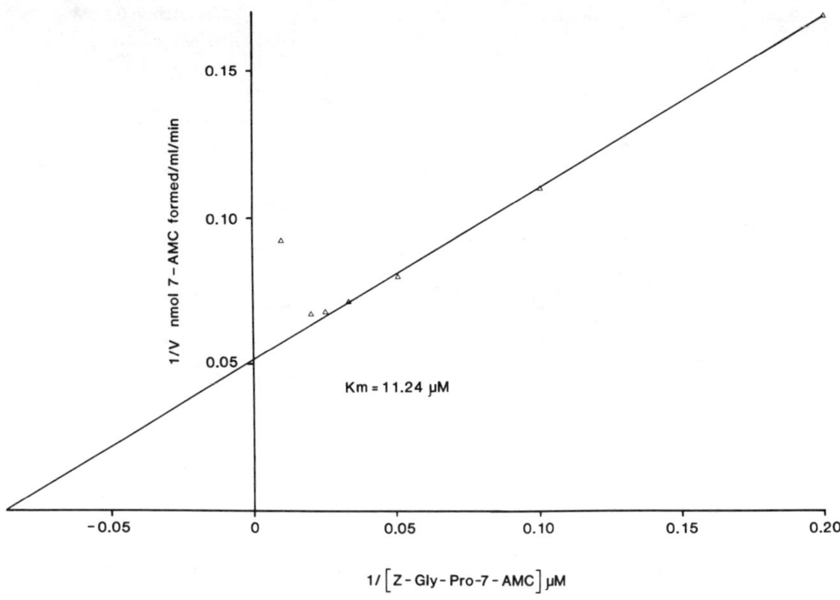

FIGURE 9.   Lineweaver-Burk plot illustrating the effects of increasing concentrations of Z-Gly-Pro-7-AMC on the Triton-X-100 solubilized proline endopeptidase activity (Fraction I). Above 40 μM Z-Gly-Pro-7-AMC concentration, substrate inhibition occurs reducing the rate of formation of the fluorescent product.

TABLE 4

EFFECT OF INHIBITORS ON RAT BRAIN PROLINE ENDOPEPTIDASE (SOLUBLE ENZYME) AND ON THE PROLINE ENDOPEPTIDASE ACTIVITY EXTRACTED BY TRITON X100 TREATMENT FROM RAT BRAIN MEMBRANES

	Soluble Proline Endopeptidase*		Triton X-100 "Solubilized" Proline Endopeptidase Activity	
	Concentration	% Inactivation	Concentration	% Inactivation
Iodoacetate	0.9	30	1.0	6
p-chloromercuri- benzoate	0.9	100	1.0	97
p-chloromercuri- benzoate	0.09	90	0.1	96
N-ethylmalemide	0.9	100	1.0	85
Phenylmethane- sulfonylfluoride	0.45	34	1.0	0
Dipsopropyl- fluorophosphate	0.0067	93	0.01	83
Dipsopropyl- fluorophosphate	0.267	100	0.1	97

* Data from Andrews *et al.* 1980[33] and Kato *et al.* 1980.[34]

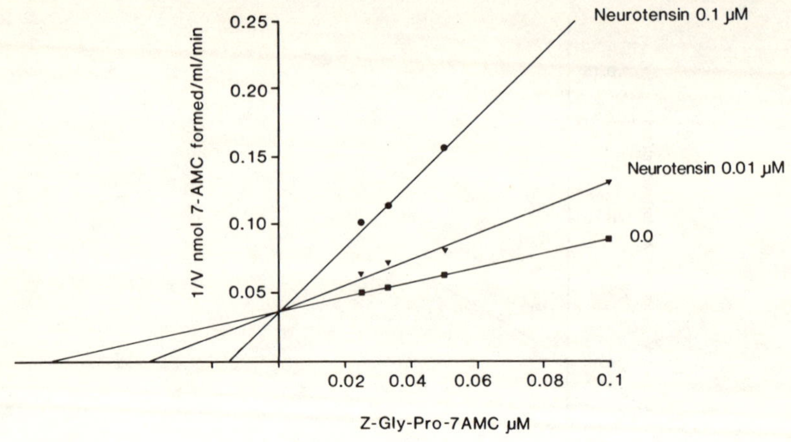

FIGURE 10.   Lineweaver-Burk plot illustrating the effects of low concentrations of neurotensin on the Triton-X-100 solubilized proline endopeptidase activity.

proline endopeptidase is localized within the synaptic cleft where it might contribute to the inactivation of released neurotensin. However, with the availability

TABLE 5

K_i VALUES FOR THE EFFECTS OF PEPTIDES ON THE
PROLINE ENDOPEPTIDASE ACTIVITY (FRACTION I)
OF RAT BRAIN MEMBRANES

Peptide	Concentration (μM)
Neurotensin	0.04
Angiotensin III	0.09
Bradykinin	0.13
SQ 20881 (teprotide)	0.90
Substance P	1.27
Substance P_{1-8}	1.45

TABLE 6

DISTRIBUTION OF PROLINE ENDOPEPTIDASE AND NEUROTENSIN-DEGRADING ACTIVITY
IN SYNAPTIC MEMBRANE FRACTIONS FROM RAT BRAIN

Fraction	Proline Endopeptidase Activity (nmol/mg protein/hour)	Neurotensin-degrading Activity (nmol/mg protein/hour*)
Myelin	0.9	0.42
Light synaptic membranes	30.7	3.52
Synaptic membranes	23.2	6.74
Mitochondria	1.30	<0.1

*Neurotensin-degrading activity as determined by the COOH-terminal specific radioimmunoassay.

of the antibody against rat proline endopeptidase, immunohistochemical studies can be undertaken to localize the enzyme. Further work will concentrate on purification of fractions I and II, and we hope also to purify the dipeptidylcarboxypeptidase activity present in fraction I. Taken together all these various enzyme activities should rapidly inactivate released neurotensin.

ACKNOWLEDGMENTS

We are grateful to Dr. L. L. Iversen for reading and criticizing the text, to Drs. P. Andrews and J. E. Dixon for proline endopeptidase antiserum, and to Drs. J. I. Nagy and P. D. Marley for help in the anterior pituitary study. M. Goedert is a research scholar of Trinity College, Cambridge. M. Ninkovic is supported by an MRC Research Studentship.

REFERENCES

1. CARRAWAY, R. E. & S. E. LEEMAN. 1973. The isolation of a new hypotensive peptide, neurotensin, from bovine hypothalami. J. Biol. Chem. **248:** 6854–6861.
2. KITABGI, P., R. CARRAWAY & S. E. LEEMAN. 1976. Isolation of a tridecapeptide from bovine intestinal tissue and its partial characterisation as neurotensin. J. Biol. Chem. **251:** 7053–7058.
3. HAMMER, R. A., S. E. LEEMAN, R. CARRAWAY & R. H. WILLIAMS. 1980. Isolation of human intestinal neurotensin. J. Biol. Chem. **255:** 2476—2480.
4. KOBAYASHI, R. M., M. BROWN & W. VALE. 1977. Regional distribution of neurotensin and somatostatin in rat brain. Brain Res. **126:** 584–588.
5. UHL, G. & S. SNYDER. 1976. Regional and subcellular distribution of brain neurotensin. Life Sci. **19:** 1827, 1832.
6. EMSON, P. C., M. GOEDERT, P. HORSFIELD, F. RIOUX & S. ST. PIERRE. 1982. The regional distribution and chromatographic characterisation of neurotensin-like immunoreactivity in the rat central nervous system. J. Neurochem. **38:** 992–999.
7. UHL, G., M. KUHAR & S. SNYDER. 1977. Neurotensin: Immunohistochemical localization in rat central nervous system. Proc. Natl. Acad. Sci. USA **74:** 4059–4063.
8. GOEDERT, M., S. L. LIGHTMAN, J. I. NAGY, P. D. MARLEY & P. C. EMSON. 1982. Neurotensin in the anterior pituitary gland. Nature **298:** 163–165.
9. SUNDLER, F., R. HAKANSON, R. A. HAMMER, J. ALUMETS, R. CARRAWAY, S. E. LEEMAN & E. A. ZIMMERMAN. 1977. Immunohistochemical localisation of neurotensin in endocrine cells of the gut. Cell Tissue Res. **178:** 313–321.
10. SCHULTZBERG, M., T. HOKFELT, G. NILSSON, L. TERENIUS, J. F. REHFELD, M. BROWN, R. ELDE, M. GOLDSTEIN & S. SAID. 1980. Distribution of peptide- and catecholamine-containing neurons in the gastrointestinal tract of rat and guinea pig: Immunohistochemical studies with antisera to substance P, vasoactive intestinal polypeptide, enkephalin, somatostatin, gastrin/cholecystokinin neurotensin and dopamine ~hydroxylase. Neuroscience **5:** 689–744.
11. SUNDLER, E., J. ALUMETS, R. HAKANSON, R. CARRAWAY & S. E. LEEMAN. 1977. Ultrastructure of the gut neurotensin cells. Histochemistry **53:** 25–34.
12. EMSON, P. C., M. GOEDERT, H. BENTON, S. ST. PIERRE & F. RIOUX. 1982. The regional distribution and chromatographic characterization of neurotensin-like immunoreactivity in the rat. *In* Advances in Biochemical Psychopharmacology. M. Trabucchi & E. Costa, Eds. **33:** 477–487.
13. IVERSEN, L. L., S. D. IVERSEN, F. BLOOM, C. DOUGHAS, M. BROWN & W. VALE. 1978. Calcium-dependent release of somatostatin and neurotensin from rat brain *"in vitro."* Nature (Lond) **273:** 161–163.

14. MAEDA, K. & L. A. FROHMAN. 1981. Neurotensin release by rat hypothalamic fragments *in vitro*. Brain Res. **210**: 261–269.

15. YOUNG, W. S. III & M. J. KUHAR. 1981. Neurotensin receptor localization by light microscopic autoradiography in rat brain. Brain Res. **206**: 273–285.

16. UHL, G. R. & SNYDER. 1977. Neurotensin receptor binding, regional and subcellular distribution favour transmitter role. Eur. J. Pharmacol. **41**: 89–91.

17. KITABGI, P., R. CARRAWAY, J. VAN RIETSCHOTEN, C. GRANIER, J. L. MARGOT, A. Menez, S. E. LEEMAN & P. FREYCHET. 1977. Neurotensin: specific binding to synaptic membranes from rat brain. Proc. Natl. Acad. Sci. USA **74**: 1846–1849.

18. KITABGI, P., C. POUSTIS, C. GRANIER, J. VAN RIETSCHOTEN, J. RIVIER, J. L. MARGOT & P. FREYCHET. 1980. Neurotensin binding to extraneural and neural receptors: comparison with biological activity and structure–activity relationships. Mol. Pharmacol. **18**: 11–19.

19. YOUNG, W. S. III, G. R. UHL & M. J. KUHAR. 1978. Iontophoresis of neurotensin in the area of the locus coeruleus. Brain Res. **150**: 431–435.

20. MILETIC, V. & M. RANDIC. 1979. Neurotensin excites cat spinal neurones in laminae I–III. Brain Res. **169**: 600–604.

21. UHL, G. R. & S. H. SNYDER. 1980. Neurotensin. *In* The Role of Peptides in Neuronal Function. J. L. Barker & T. G. Smith Jr., Eds. Marcel-Dekker. New York.

22. ANDERSSON, S., D. CHANG, K. FOLKERS & S. ROSELL. 1977. Inhibitions of gastric acid secretion in dogs by neurotensin. Life Sci. **19**: 367–370.

23. ANDERSSON, S., S. ROSELL, U. HJELMQUIST, D. CHANG & K. FOLKERS. 1977. Inhibition of gastric and intestinal motor activity in dogs by (Gln^4)-neurotensin. Acta Physiol. Scand. **100**: 231–235.

24. CARRAWAY, R. & S. E. LEEMAN. 1976. Radioimmunoassay for neurotensin, a hypothalamic peptide. J. Biool. Chem. **251**: 7035–7044.

25. WEIHE, E. & M. REINECKE. 1981. Peptidergic innervation of the mammalian sinus nodes: Vasoactive intestinal polypeptide, neurotensin, substance P. Neurosci. Lett. **26**: 283–288.

26. KAHN, D., G. ABRAMS, E. A. ZIMMERMAN, R. CARRAWAY & S. E. LEEMAN. 1981 Neurotensin neurons in the rat hypothalamus: An immunocytochemical study. Endocrinology **107**: 47–54.

27. CLINESCHMIDT, B. V. & J. McGRIFFIN. 1977. Neurotensin administered intraceisternally inhibits responsiveness of mice to noxious stimuli. Eur. J. Pharmacol. **46**: 395–396.

28. NEMEROFF, C. B., A. J. OSBAHR III, P. J. MANBERG, G. N. ERVIN & A. J. PRANGE JR. 1979. Alterations in nociception and body temperature after intracisternal administration of neurotensin, β-endorphin, other endogenous peptides and morphine. Proc. Natl. Acad. Sci. USA **76**: 5368–5376.

29. MARTIN, G. E. & S. T. NARUSE. 1982. Differences in the pharmacological actions of intrathecally administered neurotensin and morphine. Regulatory Peptides **3**: 97–103.

30. HUNT, S. P., J. S. KELLY, P. C. EMSON, J. R. KIMMELL, R. J. MILLER & J. Y. WU. 1981. An immunohistochemical study of neuronal populations containing neuropeptides and γ-aminobutyric acid within the superficial layers of the dorsal horn. Neuroscience **6**: 1883–1888.

31. NINKOVIC, M., S. P. HUNT & J. S. KELLY. 1981. Effect of dorsal rhizotomy on the autoradiographic distribution of opiate and neurotensin receptors and neurotensin-like immunoreactivity within the rat spinal cord. Brain Res. **230**: 111–119.

32. HUNT, S. P., J. S. KELLY & P. C. EMSON. 1980. The electron microscope localization of methionine-enkephalin within the superficial layers (I and II) of the spinal cord. Neuroscience **5**: 1871–1890.

33. DUPONT, A. & Y. MERAND. 1978. Enzyme inactivation of neurotensin by hypothalamic and brain extracts of the rat. Life Sci. **22**: 1623–1630.

34. ANDREWS, P. C., C. M. HINES & J. E. DIXON. 1979. Characterisation of proline endopeptidase from rat brain. Biochemistry **19**: 5494–5500.

35. KATO, T., T. NAKANO, K. KOJIMA, T. NAGATSIN & S. SAKAKIBARA. 1980. Changes in

prolyl endopeptidase during maturation of rat brain and hydrolysis of substance P by purified enzyme. J. Neurochem. **35:** 527–535.

36. LEE, C. M., P. C. EMSON & L. L. IVERSEN. 1980. The development and application of a novel NH_2-terminal-directed substance P antiserum. Life Sci. **27:** 535–543.

37. LEE, C. M., B. E. B. SANDBERG, M. R. HANLEY & L. L. IVERSEN. 1981. Purification and characterisation of a membrane-bound substance P degrading enzyme from human brain. Eur. J. Biochem. **114:** 315–327.

38. BLUMBERG. S-. V. I. TEICHBERG, J. L. CHARLIE, L. B. HERSH & J. F. MCKELVY. 1980. Cleavage of substance P to an NH_2-terminal tetrapeptide and a COOH-terminal heptapeptide by a post-proline cleaving enzyme from bovine brain. Brain Res. **192:** 477–486.

39. FOSTER, A. C., E. E. MENA, G. E. FAGG & C. W. COTMAN. 1981. Glutamate and aspartate binding sites are enriched in synaptic junctions isolated from rat brain. J. Neurosci. **1:** 620–625.

40. ANDREWS, P. C., C. D. MINTH & J. E. DIXON. 1982. Immunochemical characterisation of a proline endopeptidase from rat brain: Its relationship to proline endopeptidase from other tissue and from other species. Biochemistry. In press.

———————◆———————

DISCUSSION OF THE PAPER

P. KITABGI, (*INSERM, Nice, France*): We have done rather similar studies trying to elucidate the degradation of neurotensin by brain synaptic membranes. We have not tried to purify the enzymes involved. When we incubate neurotensin in this preparation, we see, as the major breakdown product, neurotensin 1–8 suggesting that there would be a cleavage between Arg-8 and Arg-9. Do you have any evidence for such an enzyme?

P. C. EMSON (*MRC Centre, Cambridge, England*): We had a quick look at the possible presence of a trypsin-like enzyme. There certainly is a small amount of trypsin-like activity. It is not one of the major, as I have called them, NT-degrading activities. In fact there is only low activity if you use a fluorescent substrate to monitor the tryptic-like activity in the membrane.

R. CHIPKIN (*Schering Corp., Bloomfield, NJ*): Do you know much about the dipeptidylcarboxypeptidase? Does it have any relationship to enkephalinase, which is another dipeptidylcarboxypeptidase?

EMSON: We have not looked at that. However, with the availability of the antibody to proline endopeptidase, we can go back and look at the other enzymes in fraction I.

J. W. HOLADAY, (*Walter Reed Army Medical Center, Washington, DC*): I was very interested in your slide that showed the presence of neurotensin receptors in the adrenal medulla. Yesterday we heard that some of the pressor responses to neurotensin are possibly mediated by adrenergic systems. Are you aware of any literature that links an effect of neurotensin to alteration of adrenal medullary secretion? Do you have any data on peptidase activity in adrenal medullary tissue?

EMSON: We presently have no information that pertains to either question.

ELECTROPHYSIOLOGICAL STUDIES ON THE NEUROACTIVE PROPERTIES OF NEUROTENSIN *

James L. Henry

*Departments of Research in Anaesthesia,
Physiology, and Psychiatry
McGill University
Montreal, Quebec, Canada H3G 1Y6*

INTRODUCTION

Neurotensin has been implicated as a chemical mediator of synaptic transmission in the central nervous system (CNS) on the basis of its presence and uneven distribution there,[1] its localization within the synaptosomal fraction of brain homogenates,[2] the existence of a receptor substrate for neurotensin,[3] its inactivation, presumably by enzymatic degradation, in nervous tissue,[4, 5] the calcium-dependent release of neurotensin from CNS tissue by depolarizing concentrations of K^+,[6, 7] and effects observed upon its intraventricular or intracerebral administration, which include changes in body temperature,[8] in locomotion,[9] in pituitary output,[10, 11] in pain threshold,[13, 14] and in arterial pressure.[15]

In addition to these important criteria, neurotensin has been found to have an effect on central neurons particularly in regions where it is found in abundance, and in some cases this effect has resembled the effect of the natural transmitter.

This manuscript is directed at these last two criteria, and examines specifically the electrophysiological evidence on the effects of neurotensin on neurons in the central nervous system. To do this, experiments done by the author will be described in which neurotensin was applied by microiontophoresis locally into the vicinity of single neurons *in vivo* in the dorsal and lateral horns of the spinal cord. The evidence obtained will then be discussed within the broader context of electrophysiological evidence from other laboratories on the neuroactive properties of neurotensin on spinal and supraspinal neurons.

A preliminary report of some of the present evidence appears elsewhere.[16]

METHODS

Experiments were done on cats anesthetized with alpha chloralose (60 mg/kg i.v.) after induction with halothane-oxygen.

Two series of experiments were done, one on sensory neurons in the dorsal horn of the lumbar spinal cord, the other on sympathetic preganglionic neurons in the intermediolateral nucleus of the upper thoracic cord. In experiments on sensory neurons, spinal segments L_5–L_7 were exposed for recording and covered with warm mineral oil to prevent drying. The cords were transected at L_1 to eliminate the influence of supraspinal structures on the excitability of neurons studied. In experiments on sympathetic preganglionic neurons, segments T_1–T_3

*This work was supported by grants from the Canadian Medical Research Council, the Dysautonomia Foundation, and the Quebec Heart Foundation. The author is a "Chercheur-Boursier" of the Conseil de la Recherche en Santé du Québec.

216

were exposed and covered with warm mineral oil. The sympathetic chain on the right side was exposed at the level of the stellate ganglion for activation of preganglionic axons. To reduce movements of the spinal cord, particularly those associated with breathing, the cats were paralyzed with pancuronium bromide (Pavulon, Organon) and ventilated artificially after bilateral pneumothorax.

Core temperature was maintained at 38°C. Arterial pressure was monitored on a Grass polygraph, and mean arterial pressure was always above 90 mm Hg. Endtidal CO_2 concentration was monitored using a Beckman LB-2 Medical Gas Analyzer and was always between 3 and 5%. Spinal circulation was checked periodically throughout the experiment using a Carl Zeiss zoom dissecting stereomicroscope.

Recording and Iontophoresis

Extracellular single-unit spikes were recorded using multibarrelled micropipettes with overall tip diameters of 5–10 μm. The central recording barrel was filled with 2.7 M NaCl. Each remaining barrel was filled with one of the following solutions: Na-L-glutamate (1 M, pH 8.0, Sigma), neurotensin (1 mM in 165 mM NaCl at pH 5.5, Peninsula). At least one additional barrel contained a control solution or another peptide to identify artifactual changes in unit activity due to local changes in current or pH at the electrode tip.

Classification of Units

All sensory neurons were tested thoroughly for their responses to cutaneous stimulation. Stimuli used were blowing with an air stream, brushing with a camel hair brush, gentle-to-heavy mechanical pressure with a blunt, glass probe, pinching with serrated forceps, heating likely to produce a burning sensation (43–50°C) with an infrared bulb and passive movements of the limb. In some cases, units were also tested for their responses to electrical stimulation of the skin: bipolar needle electrodes were inserted into the receptive field identified using the natural stimuli described above, and rectangular pulses of 20–500 μA and width 0.5–2.0 msec were delivered. From the latencies of the evoked responses obtained, approximate conduction velocities of the afferent fibers could be calculated.

Sympathetic preganglionic neurons were identified by their antidromic activation by spikes generated by bipolar electrical stimulation of sympathetic preganglionic fibers. Classification as a sympathetic preganglionic neuron was based on the collision test in which spikes traveling orthodromically and antidromically will cancel each other when they meet, and on the identification of discrete initial-segment versus soma-dendritic components of the evoked spike.

Identification of Sites of Recording

Sites of recording were identified in histological sections. One barrel of each electrode was filled with a 2% solution of Pontamine Sky Blue 6BX (Gurr, London) in 0.5 M sodium acetate. Marking was done by passing a negative current of 10 μA through this barrel for 10 min. Further details of the method of identification of sites of recording may be found elsewhere.[17]

RESULTS

Recordings were made from 22 sensory neurons and from nine sympathetic preganglionic neurons. Of these, 16 sensory neurons and seven sympathetic preganglionic neurons are included in this study because they meet the following criteria: (1) they yielded clear results with neurotensin; (2) classification of neuron type was complete; and (3) they showed a drug-induced response, either to neurotensin or to glutamate.

An effect of neurotensin was considered to be a genuine response if it was reversible, reproducible, and not mimicked by a current control.

Sensory Neurons in Lumbar Dorsal Horn

Of the 16 neurons from which reliable results were obtained, neurotensin depressed eight, it excited one, and it had no effect on the remaining seven. These results are summarized in TABLE 1.

Character of Depression

In all eight cases where depression occurred, it was typically slow in time course. Two typical responses are shown in FIGURES 1 and 2. The depression was

TABLE 1

EFFECT OF NEUROTENSIN ON NUMBERS OF SENSORY NEURONS IN LUMBAR
DORSAL HORN CLASSIFIED ACCORDING TO THE ADEQUATE STIMULI OF THESE NEURONS

	Depression	Excitation	No Effect
Nociceptive	4	1	4
Non-nociceptive	2	0	2
Unidentifiable	2	0	1

slow and prolonged, a character shared by most responses to peptides. Comparison of FIGURES 1 and 2 demonstrates, however, the wide variation in the time course of this depression, that in FIGURE 1 being relatively faster, although a component of the initial part of the depression must be attributed to a current effect because of the quickly reversible depressions of glutamate-induced excitation by iontophoretic application of a potent benzodiazepine (B) and of a peptide (P).

FIGURE 1 illustrates the depression of glutamate-induced excitation by neurotensin without an effect on basal firing rate, except a possible minor increase in this rate. Of five neurons tested in this way, four showed a decrease in the response to glutamate. Typically, the effect was observed during the application of neurotensin, and recovered only slowly during 2–5 min following application.

FIGURE 2 illustrates the depression of the excitatory response to noxious stimulation of the cutaneous receptive field and of the basal discharge rate, seen as the activity just preceding each excitatory response. In each case, the time course of this depression was strikingly different: its onset was considerably slower, and recovery occurred either very slowly, over 10–20 min, or not at all. Of four neurons thus tested, three were depressed.

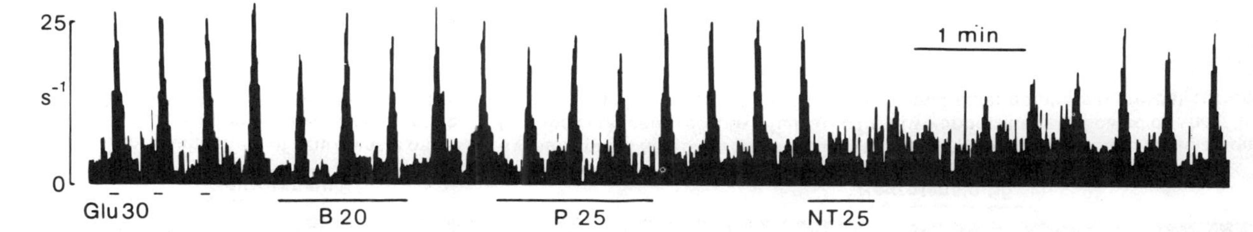

FIGURE 1.   Depressant effect on neurotensin (NT) cn a wide dynamic range neuron in laminae IV. Pen recorder tracing shows the output from a gating ratemeter counting the activity and displaying it as spikes per 5 as the ordinate. This unit was responding reproducibly to automatically controlled periodic applications of glutamate (Glu) by iontophoresis (the first three applications indicated by short horizontal bars below the records). Iontophoretic application of a potent benzodiazepine (B) and of an endogenous peptide (P) failed to alter the excitatory response to glutamate or basal discharge rate, seen as the activity between responses to glutamate; because of its time course the minor decrease in amplitude of the response to glutamate is probably a current artifact. Neurotensin, on the other hand, abolished the response to glutamate, and this effect returned only gradually over a period of 2–3 min. Long horizontal bars below the records indicate periods of drug application. Numbers indicate current in nA.

FIGURE 2. Depressant effect of neurotensin on a wide dynamic range neuron in lamina I. This unit was responding reproducibly to automatically controlled periodic applications of noxious radiant heat (H) to the cutaneous receptive field (the first two applications indicated by short horizontal bars below the records). Neurotensin was applied twice, during periods indicated by the long horizontal bars; numbers represent current in nA.

Character of Excitation

Neurotensin had an excitatory effect on only one neuron. This effect is illustrated in FIGURE 3, where 30 nA of current was used to apply the neurotensin. This effect was also slow in time course, beginning about 30 sec after the onset of the application of neurotensin and outlasting it by more than five min. Interestingly, the excitation consisted only of an increase in basal firing rate without any noticeable change in the excitatory response to noxious radiant stimulation. A similar excitatory effect of neurotensin was observed about 20 min later, but in this case 85 nA were required.

Correlation of Neurotensin Effects with Depth in Gray Matter

Neurons were tested in laminae I–IV. Depression was observed in laminae I, III, and IV; the only neuron excited was in lamina IV. The three neurons in lamina II were all unaffected by neurotensin.

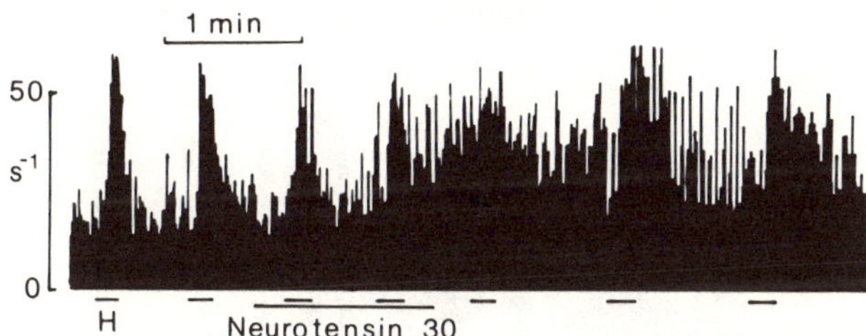

FIGURE 3.   Excitatory effect of neurotensin on a wide dynamic range neuron in lamina IV. This unit was responding reproducibly to automatically controlled periodic applications of noxious radiant heat (H) to the cutaneous receptive field (periods of application are indicated by short horizontal bars between the records). Neurotensin was applied with 30 nA of current during the period indicated by the long horizontal bar.

Correlation of Neurotensin Effects with Modality

TABLE 1 presents a comparison of the effects of neurotensin with the type of adequate stimuli to which the respective neurons responded. The results fail to indicate that the effects of neurotensin are associated with one modality rather than another; thus, neurotensin depressed about half of each of nociceptive and non-nociceptive neurons. The one case of excitation occurred with a nociceptive neuron, but in this case the nociceptive response itself remained unaltered.

Sympathetic Preganglionic Neurons in Thoracic Lateral Horn

Of the seven neurons from which reliable results were obtained, three were excited, two were depressed and two were unaffected. A typical excitatory effect is illustrated in FIGURE 4. As with dorsal horn neurons, the excitation and depression induced by neurotensin were slow in onset and outlasted the period of application by several minutes.

DISCUSSION

Results of the present study show variable effects on sensory neurons in the dorsal horn of the lumbar spinal cord as well as sympathetic preganglionic neurons of the upper thoracic spinal cord.

Depression and excitation were both observed, with the former predominating with sensory neurons and the latter with sympathetic motor neurons. In the dorsal horn, the effects failed to show a correlation with either functional type of neuron or with depth in the gray matter, with the possible exception that no effects were observed on the three neurons tested in lamina II.

The depression of glutamate-induced excitation would suggest that the effects of neurotensin are postsynaptic, except for the fact that there was no comcomitant effect on basal firing rate. One possible explanation of this differential depressant effect might be that this depression may be through an interference with the interaction between glutamate and its receptor substrate.

It is noteworthy also that the time course of the depression of glutamate-induced excitation, although slow, is considerably different from that of noxious heat-evoked excitation, which recovers much less slowly and which is associated

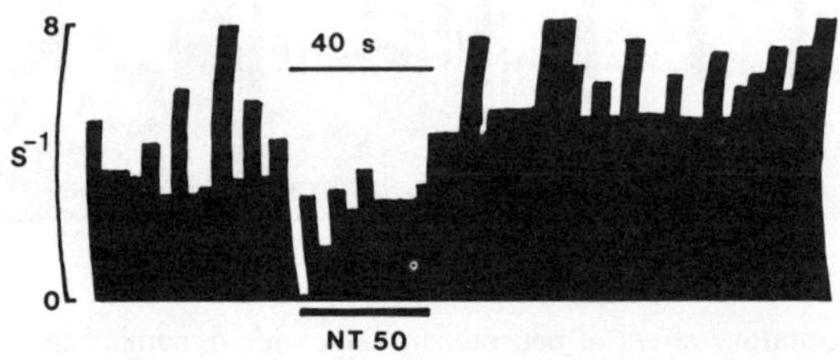

FIGURE 4. Excitatory effect of neurotensin on a sympathetic preganglionic neuron in the intermediolateral nucleus of the upper thoracic spinal cord. Application of neurotensin with 50 nA current caused an abrupt arrest of activity, probably due to current effects. The period of application was followed by a marked increase in the rate of discharge of this unit.

with a similar, prolonged depression of on-going activity. Such marked differences in time course raise the possibility that different mechanisms are involved in the mediation of the two types of depression.

In fact, in one case, an excitatory effect on a wide dynamic range neuron was observed. In this case, however, the response was manifested only as an increase in on-going activity, with no effect on the naturally elicited response.

Taken together, these results support the possibility that neurotensin is involved as a chemical mediator of synaptic transmission in the dorsal horn, but they also suggest that this is not a simple one, and may in fact consist of several unique roles, depending upon properties of neurons beyond those selected for study in the present experiments.

A similar conclusion can be reached from the sample of sympathetic preganglionic neurons studied in the upper thoracic spinal cord. These are a more homogeneous population of neurons,[18] yet, again, neurotensin had inconsistent effects.

A survey of the literature substantiates this point because while some studies have reported a predominantly depressant or excitatory effect in specific sites, there has been no consistent effect on all neurons, as is seen with, for example, enkephalins,[19] but conflicting reports exist on the effects of neurotensin within the same CNS site.

For example, Miletić and Randić[20] have reported an exclusively excitatory effect on dorsal horn neurons, while in the present study the effects were predominantly depressant. The only consistency of effects of neurotensin in the spinal cord were those from *in vitro* studies on amphibian[21, 22] and rat[23] spinal cord. In all cases, depolarization of motoneurons was observed, and this effect was also observed on dorsal root terminals.[22] The reduction of the depolarization of both dorsal and ventral roots with TTX, which blocks synaptic transmission, suggests an action on intrinsic spinal neurons, a conclusion that is consistent with the depression of glutamate-evoked excitation in the present study; this, incidentally, is also consistent with a postsynaptic site of action in the dorsal horn.

In any case, it must be pointed out that the results of the present study are not consistent with those of Miletić and Randić.[20] They reported that neurotensin had an exclusively excitatory effect in laminae I–III; in the present experiments, the ratio of excitation to depression was only 1:8, and the caution is expressed against the belief that in the spinal dorsal horn, neurotensin produces only excitation. In fact, the records shown in Figure 2 of Miletić and Randić[20] do not convincingly represent an excitatory effect: halfway through the application of neurotensin, the rate of firing falls to zero, hardly what one would expect to see during an excitatory response. Also, during the two minutes immediately following application (the best period to look for effects as current artifacts are not likely to manifest themselves, yet a lingering response should be observable, especially as the authors state that the response outlasted the period of application by up to three minutes) the activity was clearly depressed below that during the period immediately preceding current application. Another point of difference between the present results and those of Miletić and Randić[20] is with regard to the effects of neurotensin in lamina IV. They report that neurotensin did not modify the spontaneous firing rate of any of 10 units in laminae IV–VII. In the present study, however, of the four units in lamina IV yielding clear results, two showed responses (see, e.g., Figs. 1 and 3). A third, and minor point of difference is that Miletić and Randić describe the excitatory effects as slight to moderate, while in the present study, Figure 3 shows an excitation that is neither slight nor moderate.

Thus, rather than describing the effects of neurotensin in the dorsal horn as excitatory, the author visualizes them as being variable in nature: some neurons are excited while others are depressed. Variable effects of neurotensin have been described in other parts of the CNS. Thus, while Guyenet and Aghajanian[24] report neurotensin to be without effect in the locus ceruleus (although admittedly only a small number of neurons were tested), Young, Uhl and Kuhar[25] presented clear evidence of fast depressant effects of neurotensin with low (ca. 15 nA) ejection currents. Cerebellar cortical neurons were unaffected in one study[26, 27] yet were depressed in another,[28] the discrepancy being attributable perhaps to technique.[28] Contradictory or variable effects have been reported also for the following: the hippocampus,[27, 29] the stria terminalis,[27, 30, 31] the thalamus,[27, 29] the preoptic area,[27] and the septum.[27] Surprisingly, the hypothalamus, where one might expect to see effects (see paper by Frohman *et al.*, this volume), was without effects at the single neuron level.[27] Depression has been reported in the nucleus accumbens[26] and excitation in the substantia nigra,[32] but in each case only one study has been reported. The only region of the CNS where consistent effects have been reported is the cerebral cortex, where neurons are excited.[27, 29, 33]

The point of citing these other reports of the effects of neurotensin in different structures of the CNS is to draw attention to the varied nature of the responses. While the initial reaction to this may be disappointment, because neuroscientists prefer to see organization within neural systems rather than variability, the fact is, nevertheless, that neurotensin is without consistent effects in the CNS.

However, this lack of consistency of effects must have its own meaning, and in an attempt to establish just what that meaning is, a number of possible interpretations are offered. One is that while some of the responses to neurotensin were due to an action directly upon the neurons whose spikes were being recorded, others were indirect, being due to an action on neighboring neurons that in turn projected to the one being studied. Another possibility is that neurotensin may be causing the release of another agent that leads to the responses observed. Such an agent may be, for example, noradrenaline which, in earlier experiments of the author, had inconsistent effects in the spinal dorsal horn[17] as well as in the brain stem.[34] A third possibility is that a number of different receptor types exist to which neurotensin has affinity and through which neurotensin can induce changes in membrane ionic permeability. This may even be the case within most CNS regions where neurotensin is found; in fact, in view of the possibility that neurotensin may consist of a family of peptides with neurotensin-like activity (see Discussions this volume) the specificity of effect may lie in the juxtapositions of specific member peptides of this family with their respective type of receptor.

ACKNOWLEDGMENTS

The author acknowledges the participation of Steven B. Backman in experiments on sympathetic preganglionic neurons. The following gifts are gratefully acknowledged: halothane (Somnothane) from Hoechst Pharmaceuticals, Montreal, and pancuronium bromide (Pavulon) from Organon, West Hill, Ontario.

REFERENCES

1. UHL, G. R. & S. H. SNYDER. 1981. Neurotensin. *In* Neurosecretion and Brain Peptides Advances in Biochemical Psychopharmacology. J. B. Martin, S. Reichlin & K. L. Bick, Eds. Raven Press. New York.
2. UHL, G. R. & S. H. SNYDER. 1976. Regional and subcellular distribution of brain neurotensin. Life Sci. **19:** 1837–1832.
3. YOUNG, W. S. & M. J. KUHAR. 1981. Neurotensin receptor localization by light microscopic autoradiography in rat brain. Brain Res. **206:** 273–285.
4. DUPONT, A. & Y. MERAND. 1978. Enzymic inactivation of neurotensin by hypothalamic and brain extracts of rat. Life Sci. **22:** 1623–1630.
5. UHL, G. R, J. P. BENNETT, JR. & S. H. SNYDER. 1977. Neurotensin, a central nervous system peptide: Apparent receptor binding in brain membranes. Brain Res. **130:** 299–313.
6. IVERSEN, L. L., S. D. IVERSEN, F. BLOOM, C. DOUGLAS, M. BROWN & W. VALE. 1978. Calcium-dependent release of somatostatin and neurotensin from rat brain *in vitro*. Nature. **273:** 161–163.
7. MAEDA, K. & L. A. FROHMAN. 1981. Neurotensin release by rat hypothalamic fragments *in vitro*. Brain Res. **210:** 261–269.
8. NEMEROFF, C. B., G. BISSETTE, A. J. PRANGE, JR., P. T. LOOSEN, T. S. BARLOW & M. A. LIPTON. 1977. Neurotensin: Central nervous system effects of a hypothalamic peptide. Brain Res. **128:** 485–496.

9. Ervin, G. N., L. S. Birkemo, C. B. Nemeroff & A. J. Prange. 1981. Neurotensin blocks certain amphetamine-induced behaviours. Nature **291**: 73–76.

10. Vijayan, E. & S. McCann. 1979. *In vivo* and *in vitro* effects of substance P and neurotensin on gonadotropin and prolactin release. Endocrinology **105**: 64–68.

11. Maeda, K. & L. A. Frohman. 1978. Dissociation of systemic and central effects of neurotensin on the secretion of growth hormone, prolactin, and thyrotropin. Endocrinology **103**: 1903–1909.

12. Clineschmidt, B. V. & J. C. McGuffin. 1977. Neurotensin administered intracisternally inhibits responsiveness of mice to noxious stimuli. Eur. J. Pharmacol. **46**: 395–396.

13. Clineschmidt, B. V., J. C. McGuffin & P. B. Bunting. 1979. Neurotensin: Antinocisponsive action in rodents. Eur. J. Pharmacol. **54**: 129–139.

14. Nemeroff, C., A. Osbahr, P. Mauberg, G. Ervin & A. Prange. 1979. Alterations in nociception and body temperature after intracisternal administration of neurotensin, β-endorphin, other endogenous peptides, and morphine. Proc. Natl. Acad. Sci. USA **76**: 5368–5371.

15. Rioux, F., R. Quirion, S. St-Pierre, D. Regoli, F. B. Jolicoeur, F. Belanger & A. Barbeau. 1981. The hypotensive effect of centrally administered neurotensin in rats. Eur. J. Pharmacol. **69**: 241–247.

16. Henry, J. L. 1982. Effects of neurotensin on single units in cat spinal dorsal horn. Proc. Can. Fed. Biol. Soc. **25**: In press.

17. Henry, J. L. 1976. Effects of substance P on functionally identified units in cat spinal cord. Brain Res. **114**: 439–451.

18. Henry, J. L. & F. R. Calaresu. 1972. Topography and numerical distribution of neurons of the thoraco-lumbar intermediolateral nucleus in the cat. J. Comp. Neurol. **144**: 205–214.

19. Zieglgänsberger, W. & I. F. Tulloch. 1979. The effects of methionine-and leucine-enkephalin on spinal neurones of the cat. Brain Res. **167**: 53–64.

20. Miletić, V. & M. Randić. 1979. Neurotensin excites cat spinal neurones located in laminae I–III. Brain Res. **169**: 600–604.

21. Nicoll, R. A. 1978. The action of thyrotropin-releasing hormone, substance P, and related peptides on frog spinal motoneurons. J. Pharmacol. Exp. Ther. **207**: 814–824.

22. Phillis, J. W. & J. R. Kirkpatrick. 1979. Actions of various gastrointestinal peptides on the isolated amphibian spinal cord. Can. J. Physiol. Pharmacol. **57**: 887–899.

23. Suzue, T., N. Yanaihara & M. Otsuka. 1981. Actions of vasopressin, gastrin releasing peptide, and other peptides on neurons of newborn rat spinal cord *in vitro*. Neurosci. Lett. **26**: 137–142.

24. Guyenet, P. G. & G. K. Aghajanian. 1977. Excitation of neurons in the nucleus locus coeruleus by substance P and related peptides. Brain Res. **136**: 178–184.

25. Young, W. S., G. R. Uhl & M. J. Kuhar. 1978. Iontophoresis of neurotensin in the area of the locus coeruleus. Brain Res. **150**: 431–435.

26. McCarthy, P. S., R. J. Walker, H. Yajima, K. Kitagawa & G. N. Woodruff. 1979. The action of neurotensin on neurons in the nucleus accumbens and cerebellum of the rat. Gen. Pharmacol. **10**: 331–333.

27. Dao, W. P. C., H. Yajima, K. Kitagawa & R. J. Walker. 1981. The effect of neurotensin on neurons in the ventral basal complex of the thalamus and other regions of the rat central nervous system. *In* Advances in Physiological Sciences. E. Stark, G. B. Makara, B. Halasz & G. Y. Rappay, Eds. Vol. **14**: 249–254.

28. Marwaha, J., B. Hoffer & R. Freedman. 1980. Electrophysiological actions of neurotensin in rat cerebellum. Regulatory Peptides **1**: 115–125.

29. Zieglgänsberger, W., G. Siggins, M. Brown, W. Vale & F. Bloom. 1978. Actions of neurotensin upon single neurone activity in different regions of the rat brain. Proc. 7th Int. Congr. Pharmacol., Paris, P. 126.

30. Sawada, S., S. Takada & C. Yamamoto. 1980. Electrical activity recorded from thin sections of the bed nucleus of the stria terminalis, and the effects of neurotensin. Brain Res. **188**: 578–581.

31. SAWADA, S. & C. YAMAMOTO. 1980. Effects of neurotensin, enkephalin, and dop-
     amine on electrical activity of the interstitial nucleus of the stria terminalis. Neurosci.
     Lett. 4(Suppl): 58.
32. ANDRADE, R. & G. K. AGHAJANIAN. 1981. Neurotensin selectively activates dopami-
     nergic neurons of the substantia nigra. Soc. Neurosci. Abstr. 7: 573.
33. PHILLIS, J. W. & J. R. KIRKPATRICK. 1980. The actions of motilin, luteinizing hor-
     mone releasing hormone, cholecystokinin, somatostatin, vasoactive intestinal pep-
     tide, and other peptides on rat cerebral cortical neurons. Can. J. Physiol. Pharmacol.
     58: 612–623.
34. HENRY, J. L. 1976. Correlation of unit responses to narcotic drugs and to some
     putative transmitters in cat brain stem. Neurosci. Abstr. 2: 785.

◄─────────────►

DISCUSSION OF THE PAPER

A. J. PRANGE (*University of North Carolina, Chapel Hill*) Tomorrow Dr.
Clineschmidt is going to talk about antinocisponsive effects of NT in the whole
animal.

We have studied that phenomenon as well. When we look at neurotensin in a
model that is dependent upon the modification of deep pain—visceral pain as
opposed to superficial pain—we find that it seems to emerge as more potent in that
setting. It appears to be more active on deep pain than on superficial pain. If this
is the case, are the different laminae of the substantia gelatinosa excited differentially according to the type of pain? Can you sort that out?

J. L. HENRY (*McGill University, Montreal, Que., Canada*): I am not arguing
with the analgesic effects of NT at all, nor about the specificity of neurotensin or
preferential effect of neurotensin on one type of noxious stimulus versus another.
I am only raising the doubt that this is mediated at the spinal level and it might be
more likely to act at a supraspinal level where neurotensin is also distributed.

In terms of deep pain I do not have the means of studying that because to obtain
a reliable result to consider it a genuine effect, it should be repeatable and reversible. If one studies a single episode of deep pain, it may take 15 minutes to
dissipate and one is not sure that other conditions have not changed to alter
responsiveness to neurotensin. Therefore we are limited to the study of abrupt and
brief painful stimuli.

G. E. MARTIN (*Merck Sharp & Dohme, West Point, PA*): Data that Dr. Henry
described essentially agree with out observation that intrathecally applied neurotensin does not cause any antinociception whereas it does produce changes in
body temperature. We would share the same general conclusion that you have
reached, i.e. that antinociceptive action of NT is probably not spinally mediated
but supraspinal.

T. L. YAKSH (*Mayo Clinic, Rochester, MN*): How certain are you that you are
delivering a given amount of neurotensin with repeated applications through the
same pipette? It is well known that a depolarizing block may occur at high concentrations of an agonist. Therefore, it is difficult to know, under certain conditions,
whether you might be driving the cell to depolarization block or conversely to
hyperpolarization.

HENRY: Like any other technique, iontophoresis is actually full of pitfalls.
Thinking of the first record that I showed of the glutamato-evoked excitation, if
there had been depolarization block, the ongoing activity would also have been

depressed. This was not in that case. Another point is that depolarization block usually reverses fairly quickly.

D. J. Braitman (*Armed Forces Research Institute, Bethesda, MD*): I think that it is very interesting that you were able to suppress the glutamate excitation with NT without finding any effect on background activity. In fact, if anything you increased it, which would indicate that you had a disassociation of activity. One of my particular interests is the interaction of peptide transmitters with other transmitters. Perhaps what we are seeing here is an action of neurotensin interacting with other transmitters that you are not aware of, but in fact you can examine very readily with multibarrel pipettes by looking at the individual actions of transmitters applied individually and then together. We have performed some work in our laboratory with TRH and acetylcholine which indicates that, in fact, peptides can have very specific actions on other transmitters. TRH potentiates the action of acetylcholine without having any effect on glutamate. Furthermore, I think it is very important that you, in fact, localized where you are recording from. Because as Dr. Prange pointed out, different levels of spinal cord may be involved with different functional activities and again in our hands TRH specifically has effects on pyramidal tract neurons and not on nonpyramidal tract neurons.

Henry: Yes, I was interested in the fact that the three lamina II neurons that I studied were clearly unaltered by high concentrations of neurotensin. Concerning my records with the glutamate, I was pleased to observe such a specific effect of NT on glutamate excitation; it is not very common.

Martin: One of the reasons why we also think that neurotensin is not exerting its analgesic action via the spinal cord is that in comparison with morphine in the spinal cord, it is impotent. I am curious as to whether you have tested enkephalins or morphine in your system, and seen more of a consistent response than you see with neurotensin?

Henry: Yes, absolutely consistent, modality specific, but not lamina specific. I have seen it with the iontophoretic application of morphine, enkephalin, D-Ala^2-enkephalinamide, and β-endorphin. With the peptides and the opiates, I see naloxone-reversible depression that is specific to nociceptive neurons. And that is totally different from what I see with neurotensin.

PHARMACOLOGICAL STUDIES ON THE APPLICATION, DISPOSITION, AND RELEASE OF NEUROTENSIN IN THE SPINAL CORD*

T. L. Yaksh,† C. Schmauss,† P. E. Micevych,†
E. O. Abay,† and V. L. W. Go‡

*Departments of †Neurosurgical Research and ‡Gastroenterology
Mayo Clinic
Rochester, Minnesota 55905*

Introduction

Current interest has been directed at the 13-amino-acid peptide neurotensin. Following its original discovery, and subsequent isolation in brain and characterization,[1-3] it has been shown that this peptide has numerous biological effects on both central and peripheral tissues. Pursuant to our interests in the spinal cord and the pharmacology of sensory transmission, we developed an interest in neurotensin because of three observations: (1) The report by Clineschmidt and colleagues[4] (in this volume) that intracisternally administered neurotensin was analgesic; (2) that iontophoretically administered neurotensin produced a significant depolarization of dorsal horn neurons;[5] and (3) that neurotensin was highest in the dorsal horns and was contained in axons traveling along the longitudinal axis of the spinal cord in the nucleus proprius.[6] In the present series of experiments, we present some of our recent work dealing with the spinal localization of neurotensin; the effects of intrathecal neurotensin on nociceptive responding; and preliminary studies on the physiological factors governing the release of neurotensin from spinal cord.

Methods

Radioimmunoassay for Neurotensin

The radioimmunoassay for neurotensin is derived using antiserum that was obtained by making multiple subcutaneous injections of neurotensin coupled to bovine serum albumin 1-ethyl-3-(3-dimethylaminopropyl)-carbodiimide. Neurotensin was iodinated using chloramine-T solution to which was added sodium metabisulfite. Purification was obtained by passing the material over a cellulose CF-1 column and washing with $0.2\,M$ glycine. The column was eluted with acetic acid plasma. The antiserum (N398) was employed at a dilution, in these assays, of $1:110,000$. Dilution with buffer, $0.05\,M$ KH_2PO_4; o0.15 M NaCl; 0.1% gelatin; at pH = 7.4, was employed to obtain 10,000 cpm/100 μl solution. The assay consisted of incubating antibody (500 μl) with 100 μl of sample. All tubes were preincubated for 24 hr at 4°C at which time, 100 μl of neurotensin-I^{125} was added to all tubes, and a further incubation of 24 hr at 4°C was carried out. At this time, 1 ml of charcoal-dextran solution was added to each tube and centrifuged. The

* This work was supported in part by funds from the Mayo Foundation, DA 02110 and NS 16549.

228

supernatant was aspirated and the count of the free fraction bound to the charcoal-dextran was carried out. In each assay run, standard curves were run with the known standards, samples and blank tubes in which the sample and the neurotensin buffer (i.e., no antiserum) were added. All assays were performed in duplicate and the means taken. The absolute sensitivity per tube was 3 picogram.

To assay tissue, brains were taken from anesthetized animals and immediately frozen on dry ice. For assay, dissected pieces are brought to a concentration of 0.1 g of tissue per ml with 0.1 normal HCl. Solutions were homogenized with a Teflon tissue grinder and boiled for 5 min.

CSF (cerebrospinal fluid) samples were collected on ice and immediately frozen and lyophilized before assay. They were then reconstituted in assay buffer. Blanks were run with equal volumes of unperfused artificial CSF, which was subjected to freezing, lyophilization, and reconstitution as the perfused CSF samples.

Immunohistochemistry for Neurotensin

Cats were anesthetized with pentobarbital (30 mg/kg) and perfused transcardially with iced, 4% paraformaldehyde in a 0.1 M cacodylate buffer with 10 g/l zinc salicilate.[7] The spinal cord was removed and placed into the perfusion medium overnight at 4°C. The spinal cord was then transferred to a 10% sucrose solution for 24 hr at 4°C. Frozen sections (30 μm) were obtained and incubated with antineurotensin serum (1/500 or 1/1000, ImmunoNuclear Corp., Stillwater, MN) at 4°C with constant agitation, then processed using the Vectastain ABC kit (Vector Laboratories Inc., Burlingame, CA). The tissues are then mounted on glass slides, counterstained with toluidine blue, dehydrated, and coverslipped. Absorption controls were accomplished by preabsorbing 0.5 ml of the diluted anti-neurotensin serum with 100 μg of authentic neurotensin (Sigma, St. Louis, MO) on cat and rat spinal cord or by substitution of immune serum with normal rabbit serum.

To reveal neurotensin-like immunoreactivity in cell bodies, cats were injected through an intrathecal catheter whose tip lay at the lumbar enlargement, with 500 μg/0.3 ml colchicine (Sigma) and allowed to survive 24 or 48 hr before perfusion. These cats were processed identically to the noncholchicinized cats.

Surgical Lesions

To determine the distribution of neurotensin-containing fibers in spinal cord, cats were subjected to either hemisection of the spinal cord at the C2 level or dorsal rhizotomy extending from the L2–S1 root. Lesions were carried out with the animal under pentobarbital anesthesia and pretreated with Decadron. Postoperatively, the animals received penicillin and were allowed to survive for 7–10 days. At this time, the animals were deeply anesthetized with sodium pentobarbital and the lumbar L4–L6 segments were removed and immediately placed on dry ice. As soon as they had reached a firm consistency, the cord was divided into the dorsal and ventral quadrants on the left and right side.

INTRATHECAL INJECTION

To intrathecally administer agents in the spinal space of the unanesthetized animal, rats are implanted with indwelling intrathecal catheters (PE-10), placed subdurally to the level of the lumbar enlargement through the cisterna magna. Catheter tips are externalized on the tops of the heads subcutaneously. To perform an intrathecal injection, drugs are dissolved in a concentration such that the final dose is given in a volume of 10 μl of saline. This is followed subsequently by 10 μl of saline to flush the catheter. Drugs used were neurotensin (Sigma), naloxone (Endo), picrotoxin (Sigma), phentolamine-HCl (Sterling-Winthrop), and methysergide bimaleate (Sigma). Solutions were mixed fresh daily except neurotensin, which was diluted to 10 mg/ml, aliquoted, and frozen at $-20\,^\circ$C until used.

Measurement of Nociceptive Thresholds

Three procedures were employed in these experiments to examine the effects of intrathecal neurotensin on the nociceptive threshold: tail flick; hot plate; writhing.

Tail Flick

The animal is placed in a semi-circular restraining cage and the tail placed over a hot lamp. The time between the onset of the thermal stimulus and the abrupt flip of the tail is counted as the response latency. The mean response latency for 40 rats on this tail flick system is 2.1 ± 0.3 sec. In the absence of any response, the test is terminated at 6 sec and that value assigned.

Hot plate

The animal is placed upon a metal surface measuring 15 by 20 cm and surrounded by 40 cm plexiglass walls. The surface is maintained at $52.5\,^\circ$C. The time between the placing of the animal on the hot surface and the endpoint, that is, lifting of the hindpaws or a leaping response, is counted as the response latency. In the absence of any response, the test is terminated at 60 sec and that value assigned. The mean of 40 untreated rats in this test was 14.3 ± 2.1 sec.

Writhing Response

Groups of 12 animals with intrathecal catheters are placed in individual plastic compartments measuring $10 \times 15 \times 20$ cm. The animals are divided into two randomly assigned groups, 6 animals receiving drug vehicle (saline) and the other 6 receiving a drug at time 0. Ten min later, the animals receive an intraperitoneal injection of acetic acid (1:10; 0.5 ml/300 g). Ten min after the injection of the dilute acetic acid, an observer, blinded as to the intrathecal drug treatment received by each animal, assigned scores 0 through 3 to each animal at 5 min intervals for the succeeding 60 min. The writhing scores are assigned such that:

- 0 = Animal sitting in normal four-square posture with all paws flat on floor or displaying normal sniffing and exploratory behavior.
- 1 = Animal leaning so as to favor either left or right side.
- 2 = Animal on side with one or both hind paws extended caudally with the abdomen stretched.
- 3 = Writhing response where the animal rubs the belly on the floor of the cage and shows caudal extension of the hind paws.

To determine dose–response curves, the percent of the control writhing score for the drug group was estimated, the sum of the observations for each rat was obtained and the mean ± SE for each group was obtained. This was divided by the mean of the saline group. The normalized value was then plotted versus dose as the percent of the control writhing score.

Intrathecal Perfusion Procedures

Animals are prepared for intrathecal perfusion as described elsewhere.[8] Briefly, animals are anesthetized with Brevital anesthesia, prepared with a tracheal tube and jugular and carotid catheters. The basal artery is then closed by bipolar coagulation and the two carotid arteries are ligated. The animal, now rendered anemically decerebrated, is placed in the stereotaxic head holder and a spinal perfusion catheter (PE-10) is inserted 35 cm down the spinal subarachnoid space through the cisterna magna. Artificial CSF is infused by a peristaltic pump (100 μl/min) through the system and collected at the midthoracic level by a second concentric catheter through which fluid is withdrawn by a second channel of the roller pump. To provide somatic stimulation, the sciatic nerve on both sides are dissected free at the ischial notch and placed on tubular electrodes. Stimulation intensities are judged by either (1) the nature of the compound action potential generated on the distal sural nerve or (2) by the motor response generated before the application of muscle relaxant. In previous experiments, we have observed that Aδ/C fiber populations are evoked at intensities 7 to 10 times that required to evoke a threshold motor response.

RESULTS

Distribution of Neurotensin in Spinal Cord

Radioimmunoassay in Normal and Surgically Lesioned Cats

TABLE 1 presents the levels of neurotensin immunoreactivity in the dorsal horn, ventral horn, and dorsal root ganglia of control cats. As can be seen, the predominance of activity is to be found in the dorsal horn. About 50% lower levels of immunoreactivity are observed in the ventral horn and none at the sensitivity of this assay in dorsal root ganglia. Neither extensive unilateral dorsal rhizotomy (L2–S1) nor hemisection at the level of the C2 cord had any effect on the levels of neurotensin immunoreactivity in the lumbar spinal cord. The levels of substance P in cat spinal cord and the effects of these manipulations on its level are shown for comparison. Note that dorsal rhizotomy and, to a lesser extent, C2 hemisection resulted in a significant reduction in the levels of substance P on the lesioned side.

Radioimmunoassay in Normal and Neurotoxin-treated Rats

TABLE 2 presents the levels of neurotensin and substance P immunoreactivity in the whole rat lumbar spinal cord. Treatment of the animals with intrathecal capsaicin and subsequent sacrifice at 7 days indicated no effect on the levels of neurotensin immunoreactivity but produced a 45% reduction in the levels of substance P immunoreactivity in the same animal. Intrathecal kianic acid (0.7 μg) resulted in a significant reduction in the levels of both neurotensin and, to a lesser extent, substance P in rat spinal cord as compared to normal or vehicle-injected

TABLE 1

LEVELS OF NEUROTENSIN (NT) AND SUBSTANCE P (sP) IMMUNOREACTIVITY IN CAT SPINAL CORD IN CONTROL
AND AFTER RHIZOTOMY OR HEMISECTION

| | Control* | | Dorsal Rhizotomy | | C-2 Hemisection | |
	(n = 6) ng/gr[†]		(n = 6) % control[†]		(n = 3) % control[†]	
DH	2.9 ± 0.9	67.2 ± 6.0	101 ± 28	46 ± 7[‡]	103 ± 15	60 ± 7[‡]
VH	1.4 ± 0.8	19.5 ± 2.8	94 ± 22	76 ± 8[‡]	112 ± 14	32 ± 4[‡]
DRG	—	8.3 ± 2.2	—	122 ± 26	—	102 ± 12

*n = number of animals.
[†] mean $\pm$ S.D.
[‡] $p < 0.05$

controls. It should be noted that the hot-plate and tail-flick latencies were significantly elevated by the capsaicin, but not by the kianic acid.

Immunohistochemical Localization of Spinal Cord Neurotensin in Cat and Rat

Both the 1/500 and 1/1000 dilutions of anti-neurotensin serum revealed specific fiber/terminal structures in the dorsal and ventral horns of the cat spinal cord. Preabsorbing the antiserum with authentic neurotensin totally abolished immunoreactivity in both the colchicine- and noncolchicine-treated cats. Similarly, the use of normal rabbit serum did not reveal any neurotensin-like immunoreactivity. The most striking immunoreactivity was localized in a band within the superficial layers of the cat dorsal horn (FIG. 1). The neurotensin immunoreactivity appeared more concentrated in the outer portion of lamina II and lamina I (Rexed) with immunoreactivity lamina II. This immunoreactivity appeared to reveal neuro-

TABLE 2

LEVELS OF NEUROTENSIN (NT) AND SUBSTANCE P (sP) IMMUNOREACTIVITY IN RAT
SPINAL CORD AND TAIL FLICK RESPONSE LATENCY AFTER INTRATHECAL
INJECTION OF NEUROTOXINS*

| Intrathecal Treatment | N | Levels (ng/gr)[†] | | Tail Flick (sec)[†] |
		(NT)	(sP)	
Control[‡]	5	24 ± 9	163 ± 66	1.9 ± 0.2
Capsaicin (30 μg)	7	26 ± 9	90 ± 14[§]	5.8 ± 0.1[§]
Kianic acid (0.3 μg)	6	11 ± 4[§]	68 ± 22[§]	2.0 ± 0.2

* Animals sacrificed 7 days after injection.
[†] Mean ± SD
[‡] Control animals had intrathecal catheters and received intrathecal injection of capsaicin vehicle.
[§] $p < 0.05$ as compared to control.

tensin-like staining of fibers and/or terminals. In colchicine-treated cats, immunoreactivity was localized in cells in the substantia gelatinosa whose distribution resembled the distribution of fibers. These neurons had a long axis oriented in the longitudinal plane of the spinal cord (FIG. 1).

Immunohistochemistry of the rat spinal cord revealed a similar distribution of neurotensin-immunoreactivity in the rat dorsal horn. TABLE 3 summarizes the effects of intrathecal treatments on neurotensin and substance P immunoreactivity. Intrathecal capsaicin (70 μg) had no effect on the levels of neurotensin but resulted in a significant reduction in substance P-like immunoreactivity in adjacent sections. In contrast, intrathecal kianic acid (0.7 μg) resulted in a total abolition of neurotensin with no detectable effect on substance P immunoreactivity.

Pharmacology of Spinal Neurotensin on the Nociceptive Threshold

The intrathecal administration of neurotensin evoked no signs of agitation at any dose (1-100 μg). Such treatment did, however, result in an increase in the

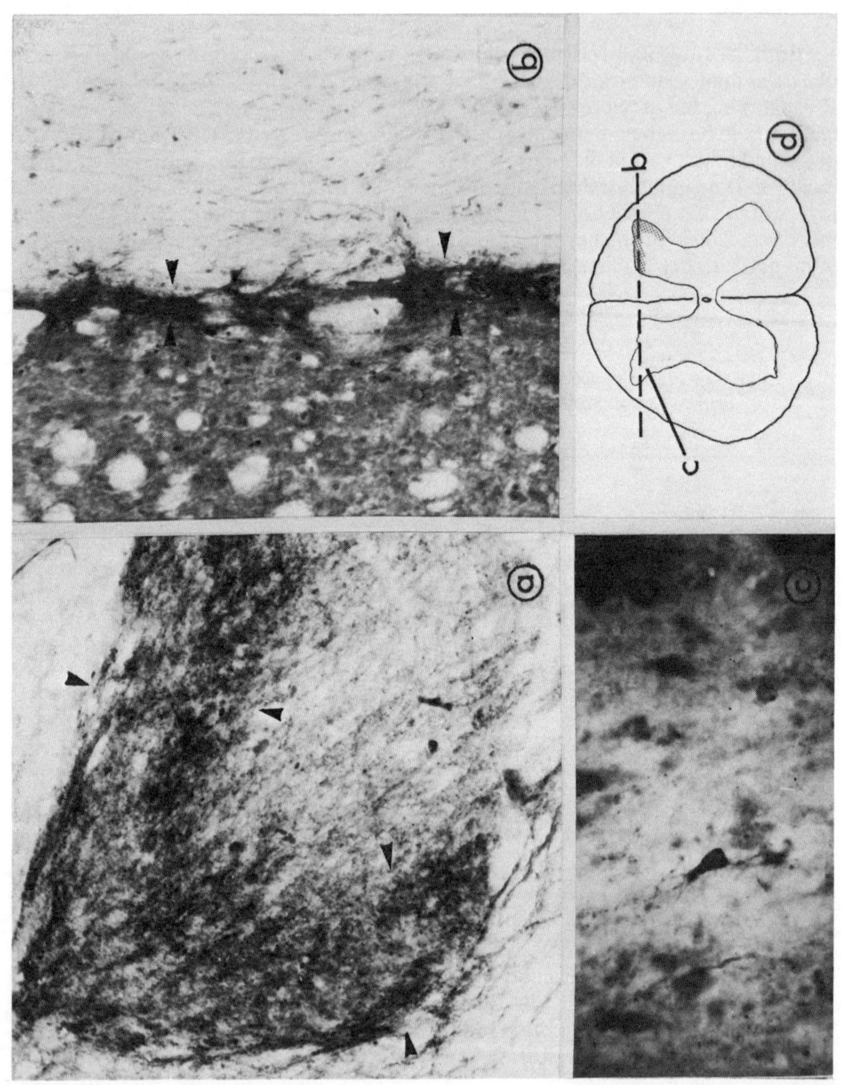

FIGURE 1. Photomicrographs of cat spinal cord incubated with antiserum raised against neurotensin and counterstained with toluidine blue. (a) The cross-section of the cat spinal cord at 1.5 shows the immunoreactivity distributed through laminae I and II. This fiber and terminal immunoreactive staining comprises a plexus in lamina I (a). This plexus is illustrated in the longitudinal section in (b). The neurotensin immuno-reactive plexus is in lamina I and the lower portions of lamina II (*arrows*, a). The neurotensin immunoreactive cell bodies were observed in the same distribution as the fiber and terminal staining. The neurotensin-containing cell in (c) is localized in deep lamina II with its long axis in a rostrocaudal direction, as seen in this longitudinal section of a colchicinized cat spinal cord at L6. Figure (d) is a schematic of the cat lumbar spinal cord for orientation. The plane of section (d) is indicated by the dashed line designated b and the location of the cell in (c) is indicated by the solid line designated c. The distribution of neurotensin immunoreactivity in the substantia gelatinosa is represented on the right side. Original magnification: (a) = ×145; (b) = ×174; (c) = ×245. All reduced to 60%.

TABLE 3

EFFECTS OF INTRATHECALLY ADMINISTERED DRUG ON THE IMMUNOREACTIVITY OF
SUBSTANCE P (sP) AND NEUROTENSIN (NT) IN THE SUBSTANTIA
GELATINOSA OF THE RAT

		Intensity of Staining*	
Drug[†]	Antisera:	NT	sP
DMSO (Control)		+++	+++
Capsaicin		+++	+
Kainate		0	+++

* +++: Normal immunoreactivity of the particular peptide.
  ++ : Slightly reduced immunoreactivity.
  +   : Profoundly reduced immunoreactivity.
  0   : No immunoreactivity detected.
[†] All drugs administered 7 days before sacrifice by transcardial perfusion with buffered paraformaldehyde (DMSO—50%, capsaicin—70 μg and kainate—0.7 μg).

nociceptive threshold which was dependent upon dose and test. FIGURE 2 presents the time-effect curves for intrathecal neurotensin on the hot-plate and tail-flick response latency. As can be seen on the hot plate and tail flick, maximum amplitude response were observed within 5 min of the intrathecal administration. The duration of the response as measured by the hot plate, was dose dependent with the animal returning to baseline latencies in intervals between 30 min and 2 hr at these doses. The effects on tail flick response were not impressive, though the maximum effect was observed 5 min after the intrathecal injection. FIGURE 3 presents a log dose–response curve for intrathecal neurotensin on the hot-plate and tail-flick test. The increase above baseline latencies reached statistical significance at doses of 4 μg of intrathecal neurotensin. With regard to the tail flick, the slope of the dose–response curve was not statistically greater than zero ($p > 0.05$), though effect produced by the highest dose (40 μg) was statistically greater than baseline.

The intrathecal action of neurotensin, exerted a clear and powerful inhibitory effect on the writhing response. FIGURE 4 (top) presents the effects of intrathecally administered vehicle (saline) and intrathecally administered neurotensin (10 μg) on the writhing score measured over the 60-min period after the intraperitoneal injection of the diluted acetic acid. FIGURE 4 (bottom) presents the dose–response relationship that was observed with intrathecal neurotensin on the percent control writhing score. As can be seen, a dose-dependent inhibition of the writhing response was observed at doses of 1 and 10 μg of intrathecal neurotensin. The measured effect of neurotensin on the writhing test was not, however, monatomically related to dose. At doses of 30 and 100 μg, an increased writhing score was observed.

In preliminary experiments designed to investigate the mechanism through which intrathecal neurotensin was exerting its modulatory effect on the hot-plate response, groups of animals received intrathecal injections of saline, naloxone (10 and 40 μg), picrotoxin (5 μg) or a combination of methysergide (15 μg) and phentolamine (15 μg), followed by an intrathecal administration of 10 μg of neurotensin. FIGURE 5 presents the results of these experiments. As can be seen, the ability of neurotensin to produce a statistically significant increase above the hot-plate response latency, was not affected by any treatment except naloxone. In this series of experiments, the inhibition produced by intrathecal naloxone was observed to be dose dependent. It is important to note that in none of these experiments did the pretreatment have any statistically significant effect on the hot-plate response latency before the administration of the intrathecal neurotensin. This ability of naloxone to at least partially antagonize the effects of intrathecal neurotensin were also observed on the writhing response. As shown in FIGURE 4, the administration of naloxone in both the parallel control (intrathecal saline) and neurotensin (intrathecal neurotensin, 10 μg) resulted in a total antagonism of the anti-writing effects of intrathecal neurotensin.

Release of Neurotensin from Spinal Cord

Superfusion of the spinal cord in the decerebrate and decerebrate-spinal transected cat, revealed the presence of resting levels of neurotensin-like immunoreactivity. FIGURE 6 presents the results of one such experiment showing the blood pressure and the levels of neurotensin immunoreactivity in the spinal superfusate (picograms/30 min sample) before and after C2 spinal transection. As shown, stimulation of sciatic nerve in the third sample resulted in a marked increase in neurotensin levels, which peaked during the second sample after the stimulus. Spinal transection, resulted in a spiking increase in blood pressure and a large increase in neurotensin immunoreactivity. Following spinal transection, sciatic nerve stimulation continued to produce measurable increases in the levels of neurotensin. Thus, release of spinal neurotensin into the extracellular space secondary to lumbar somatic stimulation did not require a supraspinal loop. In five

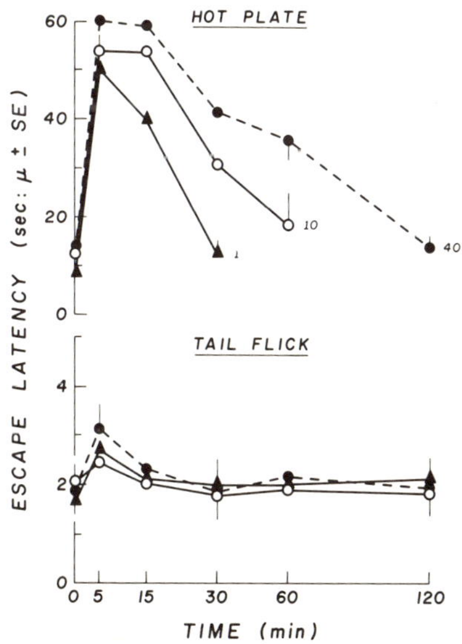

FIGURE 2.   Effects of intrathecally administered neurotensin (1, 10, and 40 μg) on the hot-plate (*top*) and tail-flick (*bottom*) response latencies as a function of time after the intrathecal administration. Each time–effect curve represents the mean of 6–10 rats.

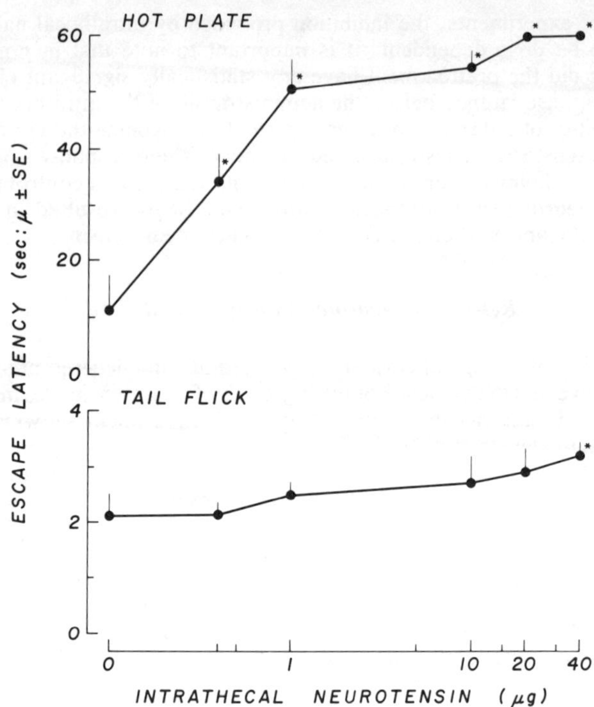

FIGURE 3.   Log dose–response curves for the effects of intrathecal neurotensin on the hot plate (*top*) and tail flick (*bottom*) response latencies. Each point represents the mean and standard error of 4–6 animals. *: $p < 0.05$.

decerebrate-spinal cats, the mean resting level was observed to be 4.1 ± 2.3 picograms/30 min sample. High-intensity stimulation of the sciatic nerves in the decerebrate and decerebrate-spinal cat, resulted in a 1.8 ± 0.3-fold increase ($p < 0.05$) in the levels of neurotensin immunoreactivity in the spinal superfusate.

DISCUSSION

The distribution of neurotensin as measured by radioimmunoassay is largely restricted to systems intrinsic to the cat dorsal horn. Neither hemisections nor rhizotomies affected its presence in the lumbar cord. These results are confirmed by immunohistochemistry carried out in both the cat and rat. Cell bodies were noted to exist largely within the ventral half of Rexed lamina II and in lamina III. Fibers, in general, appeared to run largely along a longitudinal axis and distribute in lamina I, and in the ventral substantia gelatinosa. The distribution of these pathways appears to parallel the dorsal intracornual tract previously discussed by Kerr[9] as a potential intersegmental control system for dorsal horn processing.

These observations of an intrinsic system are confirmed by the effects of intrathecal kianic acid, a glutamate-like neurotoxin thought to destroy cell bodies having glutamate receptors,[10] and the lack of effect of capsaicin, a neurotoxin thought to have significant effects on certain populations of primary afferents.[11] The significant effect on dorsal horn substance P and a correlation of this effect with thermal analgesia has been previously noted.[12]

The present observations indicate that the spinal action of neurotensin results in a statistically significant, dose-dependent increase in the nociceptive threshold as measured by the hot plate and response latency in the acetic-acid-induced writhing response. The lack of effect of neurotensin at the doses employed on general motor function, as well as the failure to alter tail-flick latencies leads us to discount the possibility that the elevated escape latencies or loss of writhing resulted from an inability of the animal to make the appropriate response.

We cannot at present explain the failure of neurotensin to affect the tail flick or the U-shaped dose–response curves observed on the writhing. It is possible that the tail-flick stimulus might be sufficiently intense to preclude block by the

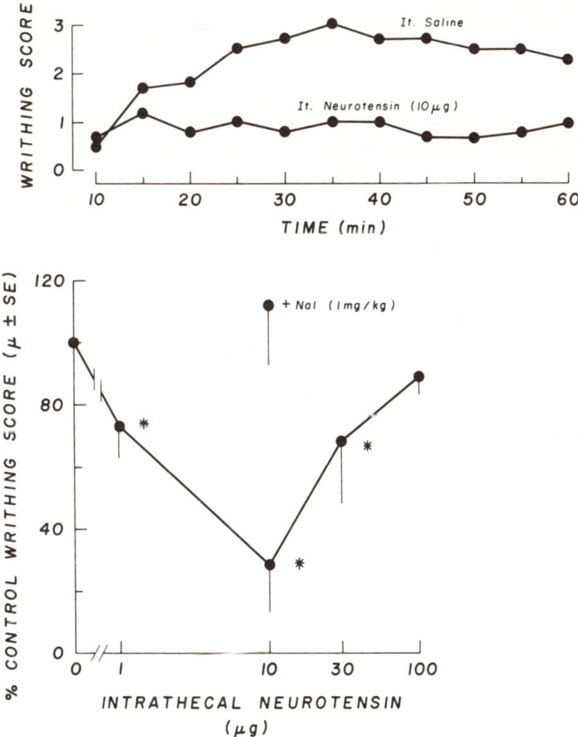

FIGURE 4. (*Top*) Time–effect curve showing the mean writhing score after the intrathecal administration of saline or neurotensin (10 μg). Each curve represents the observations obtained from 6 animals each. (*Bottom*) Log dose–response curve for the effects of intrathecal neurotensin on the percent control writhing score. Pretreatment with naloxone (Nal, 1 mg/kg, s.c.) completely reversed the antiwrithing effects produced by 10 μg of neurotensin administered intrathecally. Each point represents the mean and standard error of 6 animals. Details of the writhing test and analysis are as described in the text. *: $p < 0.05$.

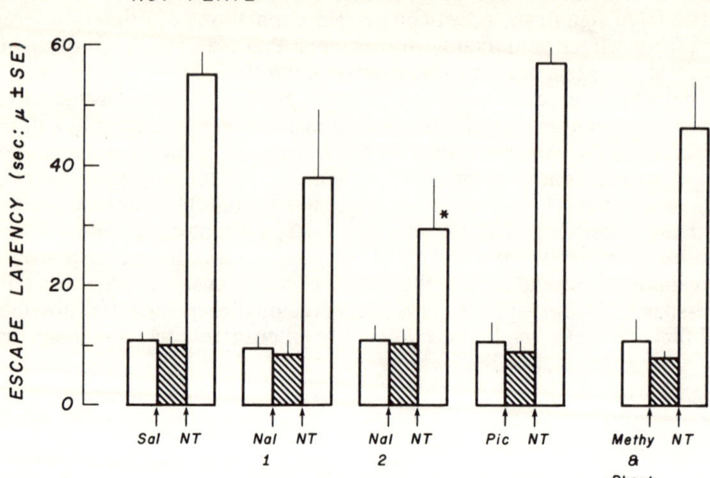

FIGURE 5. Effects of intrathecally pretreating rats with saline (Sal), naloxone (10, 15 μg), picrotoxin (5 μg), or a combination of methysergide and phentolamine (15 μg) each on the effects of intrathecal neurotensin (10 μg). Each set of 3 bars, presents the predrug-response latency, the second bar indicates the latency observed after the intrathecal administration of the pretreatment, and the third bar indicates the effects of intrathecal neurotensin. Each set of data presents the mean and standard error of 6–10 rats. *: $p < 0.05$ as compared to the neurotensin response following intrathecal saline.

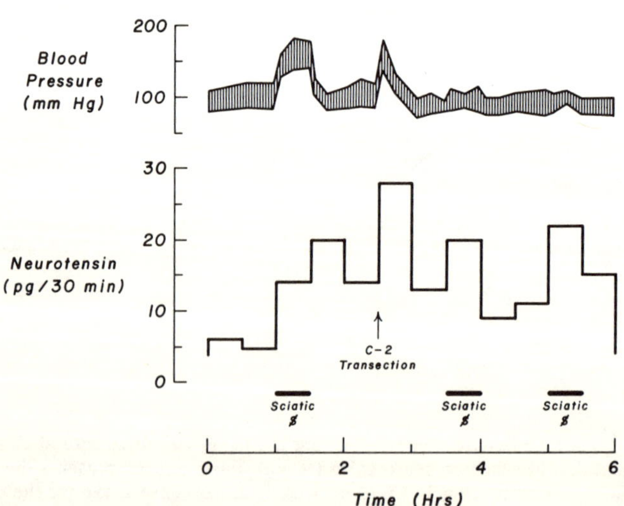

FIGURE 6. Presents the blood pressure and neurotensin levels in spinal superfusate during sciatic nerve stimulation before and after transection of the spinal cord at C2. Other details are as described in the text.

systems through which the neurotensin works. It is known that intrathecal, partial opiate agonists block writhing but not tail flick.[12]

The failure of a variety of agent known to antagonize GABA (γ-aminobutiric acid), serotonin, and norepinephrine or to affect the analgesia produced by neurotensin, would appear to exclude, within the limitations of the present experiments, an action of neurotensin mediated by these spinal receptor systems. It is significant that a dose-dependent effect of naloxone was observed to at least partially antagonize the antinociceptive effects of the intrathecal neurotensin. The apparent antagonism does not appear to result from a simple facilitation of the measured response as baselines were not altered by the antagonists alone. Such antagonism as observed in the present experimental series has not been found by other investigators following intracerebral or intracisternal injection (see this volume) of the neurotensin and systemic administration of the naloxone. This surprising effect, in conjunction with the failure of neurotensin to interact with opiate binding sites, suggests that neurotensin may be acting in part through the release of an endogenous agent whose effects are on a receptor sensitive to naloxone. The fact that the effect of naloxone on the neurotensin analgesia was observed on both hot-plate and writhing responses, suggests that the interaction between naloxone and neurotensin is not test dependent.

The effects of kianic acid strongly suggests that neurotensin-containing cell bodies may have glutamate receptors. This plus the observation that spinal neurotensin effects are antagonized by naloxone suggests that excitatory input (perhaps from primary afferents) may drive neurotensin-containing cells that serve to activate an endogenous opioid interneuron, resulting in a modulation by somatic input of dorsal horn processing. It is important to note that neurotensin cell bodies are located in close proximity to the high concentration of enkephalin-containing cells in the substantia gelatinosa,[14] that neurotensin is excitatory,[5] and that somatic input will in fact release enkephalin.[8] One may therefore speculate as to whether neurotensin will also drive enkephalin release from spinal cord.

ACKNOWLEDGMENTS

We would like to thank Ms. Gail Harty for her technical assistance in these experiments and Ms. Ann Rockafellow for preparing the manuscript.

REFERENCES

1. CARRAWAY, R. & S. E. LEEMAN. 1975. J. Biol. Chem. **250:** 1907–1911.
2. CARRAWAY, R. & S. E. LEEMAN. 1975. J. Biol. Chem. **250:** 1912–1918.
3. CARRAWAY, R. & S. E. LEEMAN. 1976. J. Biol. Chem. **251:** 7045–7052.
4. CLINESCHMIDT, B. R., J. C. McGUFFIN & P. B. BUNTING. 1979. Eur. J. Pharmacol. **54:** 129–139.
5. MILETIC, V. & M. RANDIC. 1979. Brain Res. **169:** 600–604.
6. UHL, G. R., M. J. KUHAR & S. H. SNYDER. 1977. **74:** 4059–4063.
7. MUGNAINI, E. University of Connecticut. Personal communication.
8. YAKSH, T. L. & R. P. ELDE. 1981. J. Neurophysiol. **46:** 1056–1075.
9. KERR, F. W. L. 1975. Pain **1:** 325–356.
10. HERUDEN, R. M. & J. T. COYLE. 1977. Science **198:** 71–72.
11. YAKSH, T. L., P. E. MICEVYCH, R. P. ELDE & V. L. W. GO. 1981. Adv. Pain Res. Ther. In press.
12. YAKSH, T. L., D. FARB, S. LEEMAN & T. JESSELL. 1979. Science **206:** 481–483.

13. YAKSH, T. L. 1981. Pain **11:** 293–346.
14. HÖKFELT, T., R. ELDE, O. JOHANSSON, R. LUFT & A. ARIMURA. 1976. Neuroscience **1:** 131–136.

DISCUSSION OF THE PAPER

R. CHIPKIN (*Schering Corp., Bloomfield, NJ*): In your system you have the ability to both perfuse and to sample the perfusate. Have you ever examined your perfusate for enkephalin after neurotensin administration?

T. L. YAKSH (*Mayo Clinic, Rochester, MN*): No, but we shall endeavor to do it.

CHIPKIN: We have looked at enkephalinase inhibitors and examined their effects on neurotensin-induced analgesia. We can get analgesia in the mouse tailflick test after intraventricular neurotensin administration, but we cannot get any blockade of this effect by naloxone and we do not have potentiation with enkephalinase inhibitors such as thiorophan.

C. B. NEMEROFF (*University of North Carolina, Chapel Hill*): Dr. Chipkin, I might suggest to you that one way to clarify this issue would be to give the enkephalinase inhibitor and then intrathecal neurotensin to determine whether, indeed, the spinal cord system is opiate sensitive as regards neurotensin, whereas the higher CNS centers may not be.

G. E. MARTIN (*Merck Sharp & Dohme, West Point, PA*): I am somewhat perplexed by your results on the intrathecal injections of NT, because in our lab we have tested as much as 80 μg and seen no effect.

One other point that I would like to make in that regard, especially in regard to opiate-induced mechanisms is the fact that the two major physiological events that have been found to occur after neurotensin administration are hypothermia and analgesia. After intrathecal injections of NT, we have analgesic and body temperature. After intrathecal injections of neurotensin, we observed a fall in body temperature. However as you and Rudy originally reported, with intrathecal morphine a rise in body temperature occurs. Have you measured body temperature after intrathecal NT?

YAKSH: No.

MARTIN: That is another way of addressing the question of whether the NT effect is opiate-mediated. Because you have done so much work with morphine, I wonder about the relative potency of neurotensin to morphine in the spinal cord.

YAKSH: On a μg basis it is slightly more potent in the tailflick test than morphine, but obviously morphine blocks the hotplate. On the writhing, it is clearly less potent.

D. GMEREK (*Temple University School of Medicine, Philadelphia, PA*): To add to the controversy, I just thought that I would point out that in doses from 1–100 μg into the lateral cerebral ventricle of rats, we could not show any activity of neurotensin in the tail pressure test or in the hypertonic saline test.

MARTIN: Did you look at other tests as well?

GMEREK: No, those are the only two that we have attempted at this point, except that we did show that neurotensin antagonizes the wet dog shakes that one observes after TRH administration. If you use that as an analgesic test, then NT is effective and in a naloxone-irreversible manner.

F. B. JOLICOEUR (*McGill University, Montreal, Que., Canada*): In our lab neurotensin administered into the lateral ventricle does not affect latency to jump off a hotplate.

YAKSH: The real problem is that we do not have a NT antagonist.

G. R. UHL (*Johns Hopkins University, Baltimore, MD*): Speaking as an anatomically oriented person I think that there could very well be neurotensin receptors on dorsal horn enkephalinergic interneurons which themselves are known to synapse on primary afferent terminals.

THE EFFECTS OF CENTRALLY ADMINISTERED
ANTISERA TO NEUROTENSIN AND RELATED
PEPTIDES UPON NOCICEPTION AND
RELATED BEHAVIORS*

R. J. Bodnar and M. M. Wallace

Department of Psychology
Queens College, City University of New York
Flushing, New York 11367

G. Nilaver and E. A. Zimmerman

Department of Neurology
Columbia University College of Physicians & Surgeons
New York, New York 10032

The investigation into the role of peptides in antinociceptive processes was
encouraged by the isolation of the endogenous opioid peptides including the
enkephalins[1] and β-endorphin,[2] and the subsequent observations that they increase pain thresholds following central administration.[3-6] However, activation of
pain-inhibitory systems by either electrical stimulation of the brain[7-9] or acute
exposure to stressful events[10-12] indicated that effects upon the endogenous opioid
system could not fully account for all analgesic processes, suggesting that some
nonopioid peptides or transmitters might also modulate pain perception. Therefore, the findings that a number of other neuropeptides could induce analgesia
following central injections corroborated the existence of possible nonopioid
pain-inhibitory systems. Two potential candidates are the neuropeptides
neurotensin[13,14] and substance P.[15,16] Neurotensin has been shown to possess
antinociceptive properties on the tail-immersion, acetic-acid-writhing, and hot-
plate tests,[17-19] effects that are abolished by thyrotropin-releasing hormone,[20] but
not naloxone.[18-20] Substance P induces naloxone-reversible analgesia following
intracerebroventricular and intracerebral administration.[21-24] This effect appears
to be sensitive to dose, basal responsivity, and site of administration. While low
substance P doses produce analgesia, higher doses elicit a hyperalgesic effect.[25]
Moreover, substance P elicits analgesia in rats with low pain thresholds and
hyperalgesia in rats with high pain thresholds.[26] While intraperitoneal injections of
substance P have been reported to produce analgesia in some, but not all,[24,27] pain
paradigms, both intracerebral[28] and intrathecal[29] injections of substance P produce
biting and scratching responses that are inhibited by morphine and disinhibited
by naloxone.[28]

To assess fully whether a given endogenous substance plays a role in nociceptive processes, two other approaches should be employed in addition to the above-
cited exogenous administration paradigm. The first alternative approach involves
the administration of exogenous amounts of the substance to an animal in which

*This research was supported in part by NIH Grants 14449 and AM 20337, PSC/CUNY
Grant 13493, and NIH GRSG 5-SO5-RR-07064.

244

that substance has been depleted. This approach has not been pursued for neurotensin and substance P since selective agents that deplete the peptides are either generally not available or are relatively nonspecific. A case in point involves the use of capsaicin as a specific depleter of substance P. While initial reports indicated that capsaicin depleted substance P from primary sensory neurons in the superficial dorsal horn[30,31] and produced an insensitivity to noxious stimuli following systemic[32-34] and intrathecal[35] administration, recent work has shown that capsaicin alters other neuropeptide systems[36-38] and produces hyperalgesia as well as impaired opiate responses following intracerebroventricular injection.[39]

Moreover, the central sites of nociceptive action of neurotensin and substance P are not precisely known. For instance, neurotensin cell bodies localized in the medial-basal hypothalamus send projections to the periaqueductal gray, locus ceruleus, reticular formation, and the substantia gelatinosa, the latter possessing neurotensin cells.[40-42] Substance P is located within primary sensory neurons involved in the transmission of nociceptive information[43-45] and also coexists with serotonin in a descending bulbospinal system.[44,46,47] Yet, the antinociceptive property of either peptide has not been delimited, precluding depletion studies.

The second alternative approach involves administration of selective antagonists or inactivators of neurotensin or substance P to an intact animal. While considerable effort has been expended in producing effective antagonists, assessment of nociceptive reactivity has not been carried out, and it was our laboratory's decision to employ antisera raised against particular peptides as a means of inactivating the endogenous stores of a given peptide while leaving intact other systems.[48] The following studies were designed to: (a) examine the respective effects of antisera raised against neurotensin and substance P upon reactivity to various levels of thermal stimuli as measured by the tail-flick test; (b) determine whether any observed antisera effects were due to the antibody reaction itself by investigating whether normal rabbit serum or preimmune serum were without effect; (c) replicate any antisera effects upon reactivity to shock as measured by the flinch-jump test; and (d) ascertain whether these antisera produced concomitant changes in thermoregulatory and activity measures.

EXPERIMENT 1: ANTISERA AND TAIL-FLICK LATENCIES

Method

Surgery

Seventy-two male albino Sprague-Dawley rats weighing between 250 and 350 g were anesthetized with Pentothal (50 mg/ml sterile water/kg body weight, i.p.) and were stereotaxically implanted with one stainless steel 22-gauge guide cannula (Plastic Products), the tip of which was aimed 0.3 mm above the left lateral ventricle. With the incisor bar set at +5 mm, lateral ventricle coordinates were 0.5 mm anterior to the bregma suture, 1.3 mm lateral to the sagittal suture, and 3.6 mm from the top of the skull. The guide cannula was secured to three stainless steel anchor screws with dental acrylic. Ten days were allowed for recovery from surgery. Animals were maintained on a 12-hr light : 12-hr dark cycle in individual polyethylene cages with food and water provided *ad libitum*.

Tail-Flick Test

Before surgery and testing, the rats were adapted to mild restraint in a plexiglass cylinder (IITC Bootstrap Holder) for 5 min a day for 3 days in order to minimize movement and stress reactions during the experimental procedure. A radiant heat source (IITC Analgesia Meter), which served as the thermal stimulus, was mounted 8 cm above the dorsum of the tail of the restrained animal 6 cm proximal to its tip.[49] Three levels of radiant heat were employed with each differentially approaching the noxious range (tail temperature: 49°–53°C).[50] These levels were confirmed in a pilot experiment in which a 26-gauge microprobe (Bailey Instruments) was inserted subdermally into the dorsal surface of the tail of 5 lightly anesthetized rats. Tail temperatures (±0.1°C) were continuously monitored on a digital thermometer (Bailey Instruments) immediately before and 3, 6, 9, 12, and 15 sec following exposure to each of the three beam settings. Significant temporal differences were observed among the beam settings to reach the 46°C–53°C criterion temperatures: high (2.6–4.6 sec), medium (3.6–7.0 sec), and low (5.4–12.0 sec). These rise times into the nociceptive range correlate well with the mean postoperative baseline tail-flick latencies of 3.5, 4.0, and 5.5 sec achieved for each of these beams respectively.

After restraint adaptation, baseline latencies were determined over a three-trial session separated by a 30-sec intertrial interval. Mean baseline latencies for each beam setting were based on four preoperative and four postoperative sessions. In order to avoid tissue damage, cut-off values of 6.0, 8.0 and 10.0 sec were employed for the three respective beam settings.[51]

Antisera

The neurotensin antiserum (ANT) was purchased from Immuno-Nuclear (Stillwater, MN) and appeared not to cross-react with other peptide antisera in independent immunocytochemical techniques. The substance P antiserum (ASP) was generated by Drs. Nilaver and Zimmerman in New Zealand albino rabbits using synthetic substance P (Beckman) conjugated to bovine thyroglobulin by the carbodiimide method.[52] Conjugated immunogen (2 mg/2 ml buffered saline) was emulsified with killed tubercle bacilli (Difco) in complete Freunds adjuvant (10 mg/2 ml) and injected into multiple intradermal sites along the rabbit's back.[53] Conjugated immunogen (1 mg) in complete Freunds adjuvant was administered by the same route every three weeks following the initial immunization. Blood was collected from the dorsal ear vein two weeks after each booster immunization and the presence of antibodies was evaluated by immunocytochemical staining of rat brain sections following incubation in serial dilutions of the antiserum. The optimal antibody titer was then determined. Booster immunizations were continued for two months until no further rise in the peak antibody titer of 1:1000 was obtained. The rabbit was then anesthetized and the blood collected by transcardiac puncture. The staining specificity was demonstrated elsewhere.[39]

Procedure

The 72 animals were subdivided into six groups of 12 rats each, matched on the basis of baseline postoperative tail-flick latencies. The first, second, and third groups underwent the following protocol receiving ANT microinjections as well as vehicle (VEH) normal saline controls at antiserum concentrations of 100%, 50%, and 10%, respectively. The fourth, fifth, and sixth groups underwent the following protocol receiving ASP and VEH microinjections at identical antiserum concentrations. The microinjection procedure was prefaced by preadapting all rats to handling and minimal bodily restraint in a towel. Microinjections were administered through a 5 μl Hamilton syringe that was connected to a 28-gauge stainless steel internal cannula (Plastic Products) by polyethylene tubing. The internal cannula was inserted into and extended 0.2 mm beyond the tip of the guide cannula. Injectate volume (5 μl) was held constant across antiserum concentrations and was infused at a rate of 1 μl every 15 sec to avoid tissue and ventricular damage.[54] A minimum of 48 hr separated successive microinjections during which time a dummy cannula insured patency of the guide cannula.

Each animal in each group received six microinjections in this protocol with a pair of injections (antiserum and VEH) delivered at each beam level. Half of the animals were tested in a descending series of high to low levels and the remainder were tested in an ascending series. The injection sequence within a given group at a given beam level was determined by an incomplete counterbalanced design. Following every microinjection, each rat was tested for tail-flick latencies 15, 30, 60, and 90 min thereafter. Each data point was based upon the mean of three trials.

Histology

Following completion of experimental testing, rats were overdosed with pentobarbitol and perfused via a transcardiac puncture of 0.9% normal saline followed by a 10% formalin solution. The brains were removed, fixed, blocked, frozen and cut into 40 μm sections. Coronal sections through the lateral ventricle were mounted and stained with cresyl violet for cell body visualization. The cannula tip localizations were verified using a light microscope by an observer uninformed of the behavioral results. The cannula tips of all animals reported in the present study were located in the lateral ventricle.

Results

ANT and Tail-Flick Latencies

FIGURE 1 displays the mean ANT-induced alterations in tail-flick latencies relative to VEH across thermal stimulus levels, across concentrations and across test times. A split-plot analysis of variance revealed signifiant differences in tail-flick latencies across the thermal stimulus levels [F (2, 66) = 155.41, $p < .001$], across test times [F (3, 99) = 9.26, $p < .001$] and for the interaction between thermal stimulus level and injection condition [F (2, 66) = 4.75, $p < .012$]. Since the main purpose of the study was to determine whether antiserum administration altered tail-flick latencies relative to VEH at a particular concentration, thermal level, and postinjection time, Tukey comparisons were employed to analyze each paired ANT-VEH observation. A transient, though significant, increase in tail-flick latencies [t (11) = 2.37, $p < .05$] was observed 60 min following administration of the 50% concentration at the high thermal stimulus level. This thermal level was unaltered by ANT at the 100% and 10% concentrations. In contrast, tail-flick latencies induced by the moderate thermal stimulus level were decreased following all ANT concentrations with significant effects observed 15 min [t (11) = 3.48, $p < .01$], 30 min [t (11) = 2.73, $p < .05$], and 60 min [t (11) = 2.46, $p < .05$] following the 10% concentration. Finally, tail-flick latencies induced by the low thermal stimulus were significantly increased 30 min [t (11) = 2.20, p < .05], and 60 min [t (11) = 3.70, p < .01] following undiluted ANT, and unaltered by the lower concentrations. These data indicate that the level of the thermal stimulus, the postinjection time course, and the antiserum concentration are all integral variables in assessing ANT effects.

ASP and Tail-Flick Latencies

FIGURE 2 displays ASP-induced alterations in tail-flick latencies relative to VEH across thermal stimulus levels, concentrations, and test times. A split-plot analysis of variance revealed significant differences in tail-flick latencies across the thermal stimulus levels [F (2, 66) = 112.78, p < .001], between ASP and VEH injections [F (1, 33) = 4.46, p < .05], across test times [F (3, 99) = 17.36, p < .001] and for the interactions between thermal level and concentration [F (4, 66) = p < .002] and between thermal level and test time [F (6, 198) = 2.57, p < .02]. Tukey comparisons revealed that tail-flick latencies induced by the high thermal level were significantly decreased by undiluted ASP 60 min following injection [t (11) = 2.22, p < .05]. This effect was not observed at lower ASP concentrations. In contrast, tail-flick latencies induced by the moderate thermal level were significantly increased by undiluted ASP across the entire postinjection time course: 15 min [t (11) = 2.93, p < .05), 30 min [t (11) = 2.38, p < .05], 60 min [t (11) = 2.66, p < .05] and 90 min [t(11) = 2.16, p < .05]. This effect was not observed at lower ASP concentrations. Finally, tail-flick latencies induced by the low thermal level were significantly increased by ASP in a transient manner at 90 min following the 50% concentration [t (11) = 2.15, p < .05] and at 60 min following the 10%

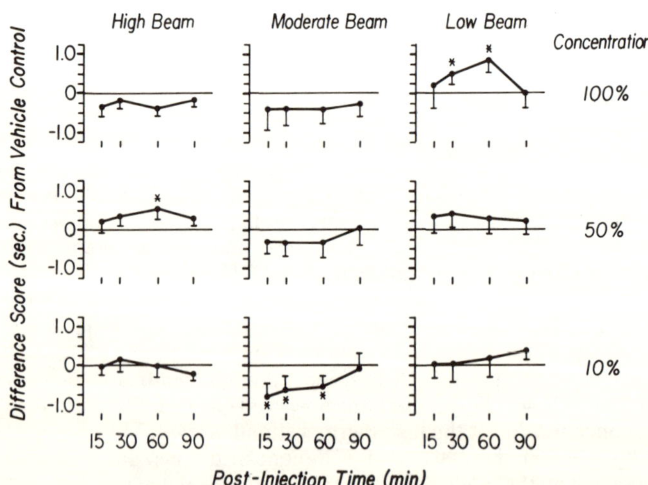

FIGURE 1. Alterations in tail-flick latencies (±SEM) relative to vehicle control values following neurotensin antiserum as functions of the thermal stimulus level, the concentration, and the postinjection test time. *p < .05.

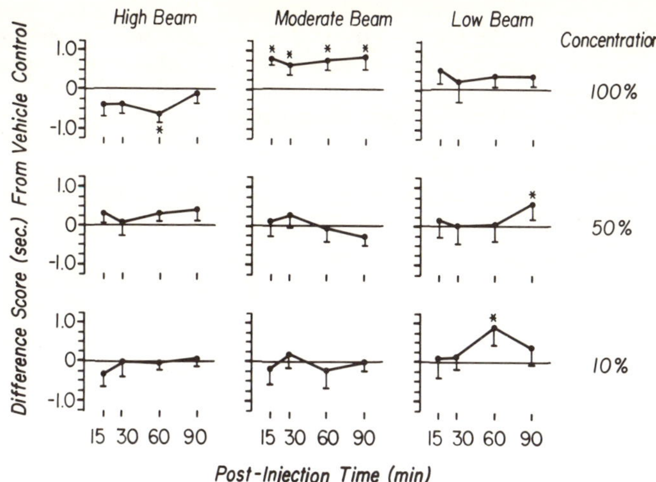

FIGURE 2.   Alterations in tail-flick latencies (±SEM) relative to vehicle control values following substance P antiserum as functions of the thermal stimulus level, the concentration and the postinjection test time. *p < .05.

concentration [t (11) = 3.29, p < .01]. While undiluted ASP increased latencies at this level, they failed to achieve statistical significance. Again, the variables of stimulus intensity, concentration, and time course were important in assessing ASP effects.

EXPERIMENT 2: SERUM CONTROLS AND TAIL-FLICK LATENCIES

Method

Six male albino Sprague-Dawley rats were surgically prepared, maintained, and assessed for baseline tail-flick latencies as in Experiment 1. Each animal received nine 5 μl microinjections with three at each thermal stimulus level. The three treatments consisted of VEH, normal rabbit serum (NORM), and pre-immune rabbit serum (PREIM) to investigate whether the observed antisera effects were due to nonspecific factors. Administration of NORM controls for such factors as: (a) the presence of inorganic anions, inorganic cations, and plasma proteins; (b) the changes in osmotic pressure; and (c) some, but not all, of the nonspecific serum constituents. To control for the effects of nonspecific antibody constituents, the PREIM serum for substance P was employed because it consisted of serum in which ASP was present, but was then removed. The PREIM serum control contained those constituents used in the generation of ASP and thus represents a true blank. The injection sequence within a given beam level was determined by an incomplete counterbalanced design. Following every microinjection, each rat was tested for tail-flick latencies 15, 30, 60, and 90 min thereafter. Each data point was based upon the mean of three trials. Histological verification proceeded as described previously.

Results

TABLE 1 indicates the lack of effect of the serum control conditions across thermal stimulus levels and time. While a split-plot analysis of variance revealed significant differences in tail-flick latencies across thermal stimulus levels [F (2, 10) = 69.89, p < .001] and across postinjection test times [F (3, 15) = 7.56, p < .003], tail-flick latencies failed to differ significantly among the VEH, NORM, and PREIM injection conditions [F (2, 10) = 0.54] and for the interactions between injection condition and thermal stimulus level [F (4, 20) = 0.18], between injection condition and postinjection time [F (6, 30) = 0.57] and among the intensity, time, and injection variables [F (12, 60) = 0.81].

TABLE 1

TAIL-FLICK LATENCIES* FOLLOWING NORMAL (NORM) RABBIT SERUM, PREIMMUNE (PREIM) RABBIT SERUM OR VEHICLE (VEH) INJECTIONS

Thermal Stimulus Level	Condition	Postinjection Time†			
		15	30	60	90
High	VEH	3.31	3.09	3.04	2.83
	NORM	3.09	3.01	3.09	2.88
	PREIM	3.21	3.05	2.82	3.01
Moderate	VEH	4.69	4.53	4.11	4.15
	NORM	4.61	4.18	4.09	4.13
	PREIM	4.43	3.91	3.86	4.02
Low	VEH	5.13	4.96	5.58	5.28
	NORM	5.37	5.03	5.22	5.36
	PREIM	5.38	5.16	5.28	5.01

*Given in seconds.
†Given in minutes.

EXPERIMENT 3: ANTISERA AND FLINCH-JUMP THRESHOLDS

Method

Twenty-four male albino Sprague-Dawley rats were surgically prepared and maintained as in Experiment 1. They were assessed for flinch-jump thresholds according to a modified ascending method of limits procedure[55,56] in which electric shocks were delivered through a 30-cm by 24-cm floor composed of 16 grids by a 60-Hz constant current shock generator (BRS/LVE) and grid scrambler (Campden Instruments). The flinch threshold was defined in mA as the lowest intensity that elicited a withdrawal of a single paw from the grids. The jump threshold was defined as the lowest of two consecutive intensities that elicited simultaneous withdrawal of both hindpaws from the grids. Each trial began with the animal receiving a 300-msec foot shock at a current intensity of 0.1 mA. Subsequent shocks occurred at 10-sec intervals and were increased in equal 0.05 mA steps until each nociceptive threshold was determined. After each trial, the current intensity was reset to 0.1 mA for the next trial until 6 trials were completed. Daily flinch and jump thresholds were each computed as the mean of these 6 trials. Stable flinch-jump thresholds were determined over four days.

Matched grops of 12 rats each received two microinjections consisting of either undiluted ANT and VEH or undiluted ASP and VEH, respectively, with injection sequences counterbalanced across animals. Following each microinjection, each rat was tested for flinch-jump thresholds 15, 30, 60, and 90 min thereafter. Histological verification proceeded as described previously.

Results

ANT and Flinch-Jump Thresholds

TABLE 2 displays the ANT-induced alterations in jump thresholds relative to VEH across the postinjection period. A split-plot analysis of variance revealed that while significant differences in jump thresholds failed to occur between VEH and ANT injections [$F(1, 11) = 0.83$], significant alterations were observed across test times [$F(3, 33) = 3.10$, $p < 0.04$] and for the interaction between injection condition and test time [$F(3, 33) = 4.45$, $p < .009$]. However, Tukey comparisons failed to indicate that ANT significantly alters jump thresholds following injection as compared to appropriate VEH values. In contrast, flinch thresholds failed to differ significantly between injection conditions [$F(1, 11) = 2.01$], across test times [$F(3, 33) = 0.94$], and for the interaction between injection condition and test time [$F(3, 33) = 0.91$].

ASP and Flinch-Jump Thresholds

TABLE 2 summarizes the inability of ASP to significantly alter jump thresholds. Thresholds failed to differ significantly between injection conditions [jump: $F(1, 11) = 0.14$; flinch: $F(1, 11) = 0.05$], across test times [jump: $F(3, 33) = 0.36$; flinch: $F(3, 33) = 0.39$], and for the interaction between injection condition and test time [jump: $F(3, 33) = 0.30$; flinch: $F(3, 33) = 0.16$].

EXPERIMENT 4: ANTISERA, THERMOREGULATION, AND LOCOMOTOR ACTIVITY

Method

Thirty-two male albino Sprague-Dawley rats were surgically prepared and maintained as in Experiment 1. The core body temperatures of 16 of the animals were determined using a rectal probe that allowed a reading, stable to 0.1°C, on a digital thermometer (Bailey Instruments). Two groups of eight rats each received two microinjections consisting of either undiluted ANT and VEH or undiluted ASP and VEH, respectively, with injection sequences counterbalanced across animals. Core body temperature, at ambient temperatures between 22–24°C, was monitored immediately before and 15, 30, 60, and 90 min following injection.

The activity levels of the remaining 16 animals were determined in a test cage (45 × 24 × 21 cm) that contained the sawdust of their home cage to minimize novelty. The test cage was situated on an activity meter (Omicron Electronics), the sensitivity of which was adjusted to a setting that recorded an animal's horizontal and vertical (rearing) displacements, but failed to record grooming movements and autonomic activity such as heart rate and respiration. Testing was conducted in a

TABLE 2

JUMP THRESHOLDS* FOLLOWING ADMINISTRATION OF ANTISERA RAISED
AGAINST NEUROTENSIN (ANT) AND SUBSTANCE P (ASP) OR VEHICLE (VEH)

	Postinjection Time†			
Condition	15	30	60	90
ANT	.354	.398	.421	.421
VEH	.365	.411	.379	.412
ASP	.386	.406	.408	.400
VEH	.381	.409	.412	.413

*Given in mA.
†Given in minutes.

quiet, secluded room that minimized reactivity to noise. Two groups of eight rats each received two microinjections consisting of either undiluted ANT and VEH or undiluted ASP and VEH, respectively, with injection sequences counterbalanced across animals. Following each injection, activity counts were monitored in 15-min blocks over a 90-min observation period. Histological verification of all animals proceeded as described previously.

Results

ANT and Thermoregulation

TABLE 3 summarizes ANT effects upon core body temperatures. While significant differences across test times [$F_{(4, 28)} = 3.74$, $p < .015$] were observed, temperatures failed to differ between ANT and VEH injection conditions [$F_{(1, 7)} = 0.58$] and for the interaction between injection condition and test time [$F_{(4, 28)} = 0.81$]. Tukey comparisons indicated that core body temperature was significantly increased 15 min following the VEH condition [$t_{(7)} = 2.46$, $p < .05$] and 60 min following the ANT condition [$t_{(7)} = 3.19$, $p < .05$]. ANT failed to alter thermoregulation relative to VEH values.

ASP and Thermoregulation

TABLE 3 displays ASP effects upon core body temperatures. While significant differences across test time [$F_{(4, 28)} = 4.22$, $p < .009$] were observed, temperatures failed to differ between ASP and VEH injection conditions [$F_{(1, 7)} = 0.75$] and for the interaction between injection condition and test time [$F_{(4, 28)} = 0.16$]. Tukey comparisons indicated that core body temperature was significantly increased 30 min following the VEH condition [$t_{(7)} = 2.45$, $p < .05$] and 15 min following the ASP condition [$t_{(7)} = 2.38$, $p < .05$). ASP failed to alter thermoregulation relative to VEH values.

ANT and Locomotor Activity

TABLE 4 summarizes ANT effects upon spontaneous locomotor activity. While activity levels predictably declined over the post-injection test interval [$F_{(5, 35)} = 16.06$, $p < .001$], locomotion failed to differ between ANT and VEH-injection conditions [$F(1, 7) = 0.33$] and for the interaction between injection condition and test time [$F(5, 35) = 0.99$].

ASP and Locomotor Activity

TABLE 4 summarizes ASP effects upon spontaneous locomotor activity. Again, activity significantly declined over time [$F(5, 35) = 7.20$, $p < .001$] without differing significantly either between injection conditions [$F(1, 7 = 4.06$] or for the interaction between injection condition and test time [$F(5, 35) = 0.41$].

DISCUSSION

The present set of experiments demonstrate that both ANT and ASP are capable of modulating pain perception in rats as a function of the pain test employed, the magnitude of the nociceptive stimulus, the concentration of the antiserum, and the postinjection testing interval. Both ANT- and ASP-mediated effects are differentially selective and appear not to be due either to administration of serum per se or to secondary changes in thermoregulatory or locomotor responses.

Neurotensin produces analgesic responses on the tail immersion, acetic-acid-writhing, and hot-plate tests[17-19] that are abolished by thyrotropin-releasing hormone,[20] and not naloxone.[18,20] The antinociceptive response is, on a molar basis, greater than that observed for Met-enkephalin, Leu-enkephalin, and morphine yet less than that observed for β-endorphin.[19] ANT produced biphasic effects on the tail-flick test with the temperature of the thermal stimulus a crucial variable. ANT transiently increases tail-flick latencies at tail temperatures that typically elicit baseline latencies around 3.5 sec. It should be noted that this latency is a "standard" in most antinociceptive tests. However, more dramatic effects are observed if the thermal stimulus is diminished, thereby lengthening the baseline latencies. If baseline latencies are preset to values between 4 and 5 sec, ANT administration will produce a hyperalgesic response at 100%, 50%, and 10% concentrations. Furthermore, the effect occurs quickly (within 15 min) and lasts between 60 and 90 min following injection. By contrast, if baseline latencies are lengthened further to values over 6 sec by reducing the thermal stimulus, ANT elicits an analgesic effect that is concentration dependent, that is, effects are observed at undiluted, but not lower, concentrations. The time course was more gradual with significant

TABLE 3

CORE BODY TEMPERATURE (°C) FOLLOWING ADMINISTRATION OF ANTISERA
RAISED AGAINST NEUROTENSIN (ANT) AND SUBSTANCE P
(ASP) OR VEHICLE (VEH)

| Condition | PRE | Postinjection Time* | | | |
		15	30	60	90
ANT	37.8	38.0	38.0	38.1	38.1
VEH	37.9	38.2	38.2	38.0	38.0
ASP	37.8	38.0	38.1	37.8	37.8
VEH	37.5	37.9	37.9	37.6	37.5

*Given in minutes.

effects observed 30 and 60 min following injection. The capability of ANT to alter nociceptive reactivity is further complicated by the observation that jump thresholds were unaltered following undiluted ANT. Since neurotensin is also a potent hypothermic,[57] and hypoactive,[58] possibilities exist that its analgesic properties may merely be a covariant of its hypothermic, though not poikilothermic, capabilities or its hypoactive capabilities. The present data fail to support this notion since ANT induces selective alterations in pain perception without affecting thermoregulation at ambient temperatures or altering spontaneous locomotor activity.

Substance P, though generally reported as an analgesic peptide,[21-24] can exert biphasic effects with dose,[25] route of administration,[24,27-29] and basal responsivity of the animal[26] as important variables in determining the magnitude and direction of effect. Like ANT, ASP produced biphasic effects on the tail-flick test with the temperature of the thermal stimulus a crucial variable. However, ASP differed from ANT in terms of the configuration of effects, the effective concentrations, and the magnitude. Undiluted ASP significantly reduced tail-flick latencies at intense tail temperatures that elicited baseline latencies around 3.5 sec. In this

TABLE 4

ACTIVITY LEVELS FOLLOWING ADMINISTRATION OF ANTISERA RAISED AGAINST NEUROTENSIN (ANT) AND SUBSTANCE P (ASP) OR VEHICLE (VEH)

Condition	Postinjection Time*					
	15	30	45	60	75	90
ANT	829.0	604.5	397.1	228.8	160.9	168.9
VEH	691.2	535.2	478.6	345.1	227.4	283.8
ASP	619.4	471.0	400.8	318.1	377.0	222.9
VEH	608.4	362.4	344.8	184.7	274.0	126.8

*Given in minutes.

regard, the ASP effect was the reciprocal of observed analgesic substance P effects,[21-24] especially confirming the observation that animals with low basal thresholds display analgesia following substance P.[26] By contrast, undiluted ASP significantly increased tail-flick latencies at moderate tail temperatures that elicited baseline latencies around 5 sec, reciprocally confirming the observation that animals with higher basal thresholds display hyperalgesia following substance P.[26] Both the analgesic and hyperalgesic effects of ASP were observed at the highest, but not at lower, concentrations with the former showing a gradual peak effect at 60 min postinjection and the latter showing a quickly acting and prolonged 90-min effect. In addition to differential ASP effects within a given nociceptive modality, differential ASP effects were observed across nociceptive modalities since tail-flick latencies, but not flinch-jump thresholds, were affected. In contrast to neurotensin-induced hypothermia and hypoactivity,[57,58] substance P produces hyperactivity[59-62] at low doses and hyperthermia.[57] Yet the induced alterations in pain perception do not appear to be covariants of these properties of the peptide since ASP failed to alter either thermoregulatory responses at ambient temperatures or spontaneous locomotor activity.

That the effects observed for ANT and ASP were specific to the antibodies involved are supported by the following points. First, a given antiserum had concentration-dependent and time-dependent effects as a function of a stimulus level within a modality as well as across modalities. Second, though substance P and neurotensin have some similar properties as a function of characterization,[13-16] the antisera employed in the present study indicate differential neuroanatomical distributions[39-47] as well as differential effects on the flinch-jump and tail-flick tests. If nonspecific factors within the serum medium or in the antibody complex were responsible for the behavioral alterations, then the effects of the respective antisera should have been similar, if not identical. Third, the question of nonspecificity was controlled further by the failure of the NORM and PREIM serum controls to alter tail-flick latencies in any manner. This is in striking contrast to the significant selective alterations described for ANT and ASP as well as for differential effects observed respectively for antisera raised against vasopressin, oxytocin, Met-enkephalin and β-endorphin.[48]

If ASP and ANT are producing selective effects upon pain perception, what then can be stated about the roles of these peptides in nociceptive processes. It appears that the effects are due to changes within the central nervous system given the intracerebroventricular route of administration. However, beyond this rather general statement, one can only speculate. If the antibodies are exerting their effects through neural mechanisms, it would be necessary for these molecules of relatively large size to permeate through the ependymal tissue and act presynaptically or postsynaptically. That centrally administered antisera can indeed enter neural tissue has been confirmed in pilot work by our laboratories, but further studies must be carried out to ascertain the extent of diffusion. It should be noted that this diffusion need not be over great distances since many relevent structures involved in pain inhibition are located in the periventricular or periaqueductal gray. Alternatively, the antisera may be binding to and thereby inactivating relevent peptides within the cerebrospinal medium. Whether neurotensin or substance P is found in cerebrospinal fluid has not been determined conclusively, but other peptides may possess functional significance within the cerebrospinal fluid. In this regard, analgesic stimulation of the midbrain central gray results in increased levels of both β-endorphin[63] and Met-enkephalin[64] in the cerebrospinal fluid. Further questions involving site of action both within the central nervous system and on the peptide chains need to be established. Yet, these results indicate that the use of antisera in behavioral paradigms can provide useful, specific information into the roles of neuropeptides upon behavior.

ACKNOWLEDGMENTS

We thank M. Korber for manuscript preparation and M. Tucker for histological assistance.

REFERENCES

1. HUGHES, J., T. W. SMITH, B. A. MORGAN & L. A. FOTHERGILL. 1975. Life Sci. **16:** 1753–1758.
2. LI, C. H. & D. CHUNG. 1976. Proc. Natl. Acad. Sci. USA **73:** 1145–1148.
3. BELLUZZI, J. D., N. GRANT, V. GARSKY, D. SARANTAKIS, C. D. WISE & L. STEIN. 1976. Nature **260:** 625–626.
4. GRAF, L., J. I. SZEKELY, A. Z. RONAI, Z. DUNAI-KOVACS & S. BAJUSZ. 1976. Nature **263:** 240–241.
5. BLOOM, F. E., D. SEGAL, N. LING & R. GUILLEMIN. 1976. Science **194:** 630–632.
6. JACQUET, Y. & N. MARKS. 1976. Science **194:** 632–635.

7. AKIL, H., D. J. MAYER & J. C. LIEBESKIND. 1976. Science **191**: 961–963.
8. PERT, A. & M. WALTER. 1976. Life Sci. **19**: 1023–1032.
9. YAKSH, T. L., J. C. YEUNG & T. A. RUDY. 1976. Life Sci. **18**: 1193–1198.
10. BODNAR, R. J., D. D. KELLY, M. BRUTUS & M. GLUSMAN. 1980. Neurosci. Biobehav. Rev. **4**: 87–100.
11. LEWIS, J. W., J. T. CANNON & J. C. LIEBESKIND. 1980. Science **208**: 623–625.
12. GRAU, J. W., R. L. HYSON, S. F. MAIER, J. MADDEN & J. D. BARCHAS. 1981. Science **213**: 1409–1411.
13. CARRAWAY, R. & S. E. LEEMAN. 1973. J. Biol. Chem. **248**: 6854–6861.
14. CARRAWAY, R. & S. E. LEEMAN. 1975. J. Biol. Chem. **250**: 1907–1911.
15. VONEULER, U. S. & J. H. GADDUM. 1931. J. Physiol. (Lond.) **72**: 74–87.
16. CHANG, M. M., S. E. LEEMAN & H. D. NIALL. 1971. Nature New Biol. **232**: 86–87.
17. CLINESCHMIDT, B. V. & J. C. McGUFFIN. 1977. Eur. J. Pharmacol. **46**: 395–396.
18. CLINESCHMIDT, B. V., J. C. McGUFFIN & P. B. BUNTING. 1979. Eur. J. Pharmacol. **54**: 129–139.
19. NEMEROFF, C. B., A. J. OSBAHR, P. J. MANBERG, G. N. ERVIN & A. J. PRANGE. 1979. Proc. Natl. Acad. Sci. USA **76**: 5368–5371.
20. OSBAHR, A. J., C. B. NEMEROFF, D. LUTTINGER, G. A. MASON & A. J. PRANGE. 1981. J. Pharmacol. Exp. Ther. **217**: 645–651.
21. STEWART, J. M., C. J. GETTO, K. NELDNER, E. BASIL-REEVE, W. A. KRIVOY & E. ZIMMERMAN. 1976. Nature **262**: 784–785.
22. MALICK, J. B. & J. M. GOLDSTEIN. 1978. Life Sci. **23**: 835–844.
23. STARR, M. S., T. A. JAMES & D. GAYTTEN. 1978. Eur. J. Pharmacol. **48**: 203–212.
24. MOHRLAND, J. S. & G. F. GEBHART. 1979. Brain Res. **171**: 556–559.
25. FREDERICKSON, R. C. A., V. BURGIS, C. E. HARRELL & J. D. EDWARDS. 1978. Science **199**: 1359–1362.
26. OEHME, P., H. HILSE, E. MORGENSTERN & E. GORES. 1980. Science **208**: 305–307.
27. GOLDSTEIN, J. M. & J. B. MALICK. 1979. Life Sci. **25**: 431–436.
28. SHARE, N. N. & A. RACKHAM. 1981. Brain Res. **211**: 379–386.
29. HYLDEN, J. L. K. & G. L. WILCOX. 1981. Brain Res. **217**: 212–215.
30. JESSELL, T. M., L. L. IVERSEN & A. C. CUELLO. 1978. Brain Res. **152**: 183–188.
31. GAMSE, R., A. MOLNAR & F. LEMBECK. 1979. Life Sci. **25**: 629–636.
32. JANSCO, N., A. JANSCO-GABOR & J. SZOLCSANYI. 1968. Br. J. Pharmacol. **33**: 32–41.
33. HAYES, A. G. & M. B. TYERS. 1980. Brain Res. **189**: 561–563.
34. NAGY, J. I., S. R. VINCENT, W. A. STAINES, H. C. FIBIGER, T. D. REISINE & H. I. YAMAMURA. 1980. Brain Res. **186**: 435–444.
35. YAKSH, T. L., D. H. FARB, S. E. LEEMAN & T. M. JESSELL. 1979. Science **206**: 481–483.
36. NAGY, J. I., P. C. EMSON & L. L. IVERSEN. 1981. Brain Res. **211**: 497–502.
37. NAGY, J. I., S. P. HUNT, L. L. IVERSEN & P. C. EMSON. 1981. Neuroscience. **6**: 1923–1934.
38. AINSWORTH, A., P. HALL, P. D. WALL, G. ALLT, M. L. MacKENZIE, S. GIBSON & J. M. POLAK. 1981. Pain **11**: 379–388.
39. BODNAR, R. J., A. KIRCHGESSNER, G. NILAVER, J. MULHERN & E. A. ZIMMERMAN. 1982. Neuroscience **7**: 631–638.
40. UHL, G., M. J. KUHAR & S. H. SNYDER. 1977. Proc. Natl. Acad. Sci. USA **74**: 4059–4063.
41. UHL, G., R. R. GOODMAN & S. H. SNYDER. 1979. Brain Res. **167**: 77–94.
42. KAHN, D., G. M. ABRAMS, E. A. ZIMMERMAN, R. CARRAWAY & S. E. LEEMAN. 1980. Endocrinology **107**: 47–54.
43. CUELLO, A. C. & I. KANAZAWA. 1978. J. Comp. Neurol. **178**: 129–156.
44. HOKFELT, T., O. JOHANSSON, A. LJUNGDAHL, J. M. LUNDBERG & M. SCHULTZBERG. 1980. Nature (London) **284**: 515–521.
45. HOKFELT, T., J. O. KELLERTH, G. NILSSON & B. PERNOW. 1975. Science **190**: 889–890.
46. HOKFELT, T., A. LJUNGDAHL, H. STEINBUSCH, A. VERHOFSTAD, G. NILSSON, E. BRODIN, B. PERNOW & M. GOLDSTEIN. 1978. Neuroscience **3**: 517–538.

47. Ljungdahl, A., T. Hokfelt & G. Nilsson. 1978. Neuroscience **3**: 861–943.
48. Wallace, M. M., R. J. Bodnar, D. Badillo-Martinez, G. Nilaver & E. A. Zimmerman. 1981. Neurosci. Soc. Abstr. **7**: 167.
49. D'Amour, F. E. & D. L. Smith. 1941. J. Pharmacol. Exp. Ther. **72**: 74–79.
50. Price, D. D. & R. Dubner. 1977. Pain **3**: 307–338.
51. Levine, J. D., D. T. Murphy, D. Seidenwurm, A. Cortez & H. L. Fields. 1980. Brain Res. **201**: 129–141.
52. Odell, W., R. Skowsky, G. Abraham, M. Hescox, D. Fisher & P. K. Grover. 1972. Biol. Reprod. **6**: 427–442.
53. Vaitukaitas, J., J. B. Robbins, E. Nieschlag & G. T. Ross. 1971. J. Clin. Endocrinol. **33**: 988–991.
54. Yaksh, T. L. & T. A. Rudy. 1978. Pain **4**: 299–359.
55. Evans, W. O. 1961. Psychopharmacologia **2**: 318–325.
56. Bodnar, R. J., D. D. Kelly, S. S. Steiner & M. Glusman. 1978. Pharmacol. Biochem. Behav. **8**: 661–666.
57. Nemeroff, C. B., G. Bissette, A. J. Prange, P. T. Loosen & M. A. Lipton. 1977. Brain Res. **128**: 485–496.
58. Nemeroff, C. B. 1980. Biol. Psychiatry **15**: 283–302.
59. Stinus, L., A. E. Kelly & S. D. Iversen. 1978. Nature **276**: 616–618.
60. Kelley, A. E. & S. D. Iversen. 1978. Brain Res. **158**: 474–478.
61. Rondeau, D. B., F. B. Jolicoeur, F. Belanger & A. Barbeau. 1978. Pharmacol. Biochem. Behav. **9**: 769–775.
62. Kelley, A. E., L. Stinus & S. D. Iversen. 1979. Neurosci. Lett. **11**: 335–339.
63. Hosobuchi, Y., J. Rossier, F. E. Bloom & R. Guillemin. 1979. Science **203**: 279–281.
64. Akil, H., D. E. Richardson, J. Hughes & J. D. Barchas. 1978. Science **201**: 463–465.

Discussion of the Paper

J. L. Henry (*McGill University, Montreal, Que., Canada*): How do you know that 15 minutes is the right time to begin your sampling?

R. J. Bodnar (*Queens College, CUNY, Flushing, NY*): Most of the time courses that I chose were really from the time courses of the peptide effects themselves.

Henry: I have been doing some intrathecal injection studies with substance P and the principal effect that I see begins at 15 seconds and is over within 5 to 10 minutes. Following the effect there is a rebound; so if I were to begin testing at 15 minutes, I would actually be measuring a reflex change. I raise it as an important issue: when to begin your testing and when to finish it.

Bodnar: Let me refer to the people who do direct neurotensin injections.

C. B. Nemeroff (*University of North Carolina, Chapel Hill*): I think you are exactly right in saying that when you are mapping out a new phenomenon, you do need to look very closely at its time course. In fact, when we first discovered neurotensin-induced hypothermia, we wondered whether it was just such a secondary effect. The appropriate thing to do, as you suggest, is to measure the variable of interest at 30 seconds, 1 minute, etc. and once you have defined that and you know that 15 minutes is appropriate, then you can use that to set up your paradigm.

My question to Dr. Bodnar is, did you ever give neurotensin antiserum intraventricularly and then give neurotensin intraventricularly to see whether the exogenous antiserum actually prevented the effects of exogenous neurotensin?

BODNAR: Not yet.

J. W. HOLADAY (*Walter Reed Army Medical Center, Washington, D.C.*): Since there is less controversy involved in the possible role of endogenous opiates than there might be with regard to neurotensin on nociceptive latencies, it would have been a nice positive control to see what effect antisera against opoids produced.

BODNAR: β-Endorphin antiserum produces, at the high beam, a hyperalgesia at 15 and 30 minutes, but not beyond. The Met-enkephalin antiserum failed to show any effects.

V. SEYBOLD (*University of Minnesota, Minneapolis*): Are you injecting whole serum, diluted serum, or purified IGgS?

BODNAR: Neurotensin antiserum was whole serum.

R. E. CARRAWAY (*University of Massachusetts, Worcester*): I have a few comments concerning the use of antisera in this manner and a few cautions as well. First, in regard to the apparent nonspecificity of the effect that you saw, i.e., that you saw very similar action for all of these antisera, would suggest to me that there might be antibodies in the preparation that are directed towards other substances, for example, the carrier molecule that might have been employed.

I would also caution you in regards to the content of the peptide in these antisera. That is, these antibodies are circulating and they pick up the peptide to which they are directed. Therefore you are injecting not only the antibody, but the neurotensin bound to the antibody. We have noticed this in our laboratory, and in fact antisera toward a substance have very elevated levels of that particular substance. I am talking about 50- to 100-fold higher levels. I think it would be very useful for you to purify your antisera by affinity chromatography, for example. And, also be sure that you don't have any ligand attached to the antibody. That might have something to do with this dual response you are seeing.

EFFECTS OF NEUROTENSIN ON THE ACTIONS OF BARBITURATES AND ETHANOL*

Daniel Luttinger,†‡ Gerald D. Frye,†§ and Garth Bissette†‡

†*Biological Sciences Research Center, ‡Neurobiology Program,*
and the §Center for Alcohol Studies
University of North Carolina School of Medicine
Chapel Hill, North Carolina 27514

Several studies have shown that peptides can modify the sedative effect of barbiturates and other central nervous system (CNS) depressants.[1-3] Many peptides reduced the duration of barbiturate-induced sleep, while others exerted no effect.[1] Of the active peptides, thyrotropin-releasing hormone (TRH) was studied most extensively to determine the mechanism of its analeptic action, after intracisternal (i.c.) or intraperitoneal (i.p.) administration. The reduction in the duration of pentobarbital-induced sleep by TRH was independent of its pituitary effect (i.e., TRH-induced release of thyroid-stimulating hormone and prolactin).[4] Furthermore, the analeptic action of TRH, following intracerebral administration, was found to be mediated by only a few discrete regions within the CNS.[5]

In a study designed to evaluate the effects of several peptides on the duration of pentobarbital-induced sleep in mice, only neurotensin (NT), administered i.c., was found to *increase* the duration of sleep.[1] In these experiments the duration of sleep was defined as the time from the loss of the righting reflex until it was regained (three rightings within one minute). Mice were injected i.p. with pentobarbital (55 mg/kg) and 10 minutes later with an i.c. injection of NT (46.1 μg, 27.6 nmoles). The duration of pentobarbital-induced sleep was increased by 76% in NT-treated mice when compared with vehicle-injected controls (mean sleeping time = 91.6 ± 8.8 min). None of the doses of NT (0.02–100 μg, 0.01–59.9 nmoles) tested caused loss of the righting reflex when injected alone. The potentiation of the barbiturate effect appeared to be mediated within the CNS, because peripherally administered NT (23.5 mg/kg i.p.) did not alter the duration of pentobarbital-induced sleep.

Additional experiments indicated that i.c. NT doses as small as 300 ng significantly enhance the duration of pentobarbital-induced sleep.[6] Interestingly, an extremely low dose of NT (10 ng, i.c.) significantly *shortened* the duration of pentobarbital-induced sleep. Figure 1 illustrates that the effect of NT to increase the duration of pentobarbital-induced sleep was dose related. In addition, NT administered i.c. produced dose-related lethality when administered in combination with a nonlethal dose of pentobarbital (50 mg/kg, i.p.). The LD_{50} for NT administered with this fixed dose of pentobarbital was calculated to be 5.23 μg (3.1 nmoles).

Nemeroff *et al.*[6] investigated the possibility that NT might alter the disposition of pentobarbital, an effect that could account for the potentiation of pentobarbital-induced sleep. They found that NT (30 μg, 17.9 nmoles, i.c.) decreased the rate

*This work was supported by grants from NIMH MH-22536, MH-32316, MH-33127, MH-34121, MH-14277, NIAAA AA-02334, NICHHD HD-03110 and the North Carolina Alcoholism Research Authority (7908, 7920, and 8010).

Table 1

THE EFFECT OF INTRACISTERNAL NEUROTENSIN (30 μg, 17.9 nmoles) ON THE DISPOSITION OF [³H] PENTOBARBITAL

	Time after Pentobarbital Injection (min)				
	10	15	30	60	90
1. Plasma†					
Saline	4.61 ± 0.19	4.46 ± 0.12	4.84 ± 0.17	3.39 ± 1.31	2.27 ± 0.30
Neurotensin	4.34 ± 0.19	4.03 ± 0.16	3.64 ± 0.33**	3.53 ± 0.09	3.83 ± 0.20††
2. Liver‡					
Saline	88.72 ± 3.59	78.59 ± 2.31	75.34 ± 2.54	41.49 ± 7.12	32.26 ± 5.90
Neurotensin	81.41 ± 3.28	71.46 ± 3.47	66.50 ± 3.42	67.62 ± 3.35**	64.93 ± 4.84††
3. Liver§					
Saline	31.20 ± 3.82	23.61 ± 1.80	14.12 ± 1.55	5.28 ± 1.86	4.45 ± 1.16
Neurotensin	34.68 ± 4.41	31.08 ± 2.55#	‖	16.11 ± 4.61	11.50 ± 1.72**
4. Liver¶					
Saline	55.18 ± 7.36	55.35 ± 2.57	56.42 ± 4.85	38.44 ± 7.33	27.45 ± 7.37
Neurotensin	49.17 ± 8.88	41.79 ± 4.66#	46.09 ± 5.42	45.40 ± 8.03	47.04 ± 2.24#
5. Brain‡					
Saline	11.45 ± 1.95	7.77 ± 0.48	6.14 ± 0.15	4.86 ± 0.43	4.40 ± 0.29
Neurotensin	11.33 ± 1.16	13.49 ± 2.90	6.78 ± 0.21	8.43 ± 1.09**	5.38 ± 0.47
6. Brain§					
Saline	8.10 ± 1.51	6.04 ± 1.46	3.04 ± 0.38	0.97 ± 0.24	0.73 ± 0.19
Neurotensin	11.07 ± 2.37	11.81 ± 2.98	6.46 ± 1.19**	5.30 ± 1.20††	1.85 ± 0.32**
7. Brain¶					
Saline	3.74 ± 0.52	3.37 ± 0.51	4.66 ± 0.30	4.75 ± 0.07	4.71 ± 0.12
Neurotensin	2.51 ± 0.16	3.22 ± 0.38	‖	4.24 ± 0.23	5.08 ± 0.21

*Adult male mice were injected intraperitoneally with 50 mg/kg pentobarbital plus [³H] pentobarbital (specific activity 1 μCi/mg) and treated 10 min later with 30 μg neurotensin or vehicle (0.9% NaCl, pH 7.5) intracisternally. There were 5 mice/group/time point. Animals were sacrificed 10, 15, 30, 60, and 90 min after receiving the barbiturate. Probability values were estimated using Student's t-test (two-tailed). Adapted from data in Nemeroff et al.[6] †Total radioactivity, disint./min × 10³/ml ± SEM. ‡Total radioactivity, disint./min × 10³/g tissue ± SEM. § Extracted [³H] pentobarbital, disint./min × 10³/g tissue ± SEM. ¶[³H] pentobarbital metabolites, disint./min × 10³/g tissue ± SEM. ‖Data not obtained because of equipment failure. #p < 0.05; **p < 0.02; ††p < 0.01.

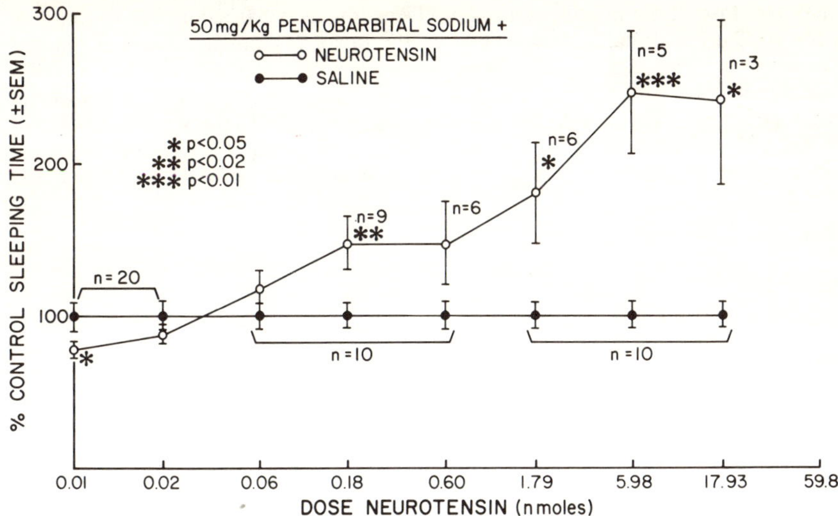

FIGURE 1. The effect of intracisternally administered neurotensin on the sleeping time of mice pretreated with 50 mg/kg pentobarbital (i.p.). Data are expressed as percent control sleeping time ± SEM. The pentobarbital- and vehicle-treated mice slept for 85 ± 5 minutes (n = 40). p values refer to the Mann-Whitney U-test. Adapted from Nemeroff *et al.*[6]

of metabolism of [³H] pentobarbital in both brain and liver of mice (TABLE 1). The effect on pentobarbital metabolism was most striking for [³H] pentobarbital extracted from brain; 60 min postinjection there was five times as much pentobarbital in brains of NT-treated animals as in vehicle-treated controls. Thus, it appeared that the potentiation of pentobarbital-induced sleep by NT may be secondary to NT-induced decrease in pentobarbital metabolism.

To further characterize the effects of NT on the actions of CNS depressants we studied the interactions of NT and ethanol. Ethanol was of particular interest since it is well established that different enzymes are involved in the metabolism of pentobarbital and ethanol and thus the effects of NT on ethanol-induced sleep might be less likely to be complicated by altered drug metabolism. As with pentobarbital-treated mice, NT (i.c.) potentiated the duration of sleep in ethanol-treated mice.[7] The dose–response curve for NT-induced potentiation of ethanol-induced sleep (FIG. 2) was almost identical to that obtained with pentobarbital (FIG. 1). The smallest effective dose of NT that potentiated sleep due either to ethanol or to pentobarbital was the same (300 ng, 0.18 nmoles). The effects of NT in ethanol-treated mice were probably not due to changes in drug metabolism because brain and blood ethanol concentrations were not altered by NT (TABLE 2), unlike the concentrations of pentobarbital (see above). This finding suggests that NT can potentiate the actions of some CNS depressants independent of any action on drug metabolism.

Although NT potentiates sleeping time in both pentobarbital- and ethanol-treated animals, certain distinctions should be noted. The first was emphasized above: NT delays the removal of pentobarbital but not the removal of ethanol. Moreover, NT never shortened the duration of ethanol-induced sleep, however small the dose of the peptide. Doses of NT as low as 0.0001 μg i.c. (0.06 pmoles) were ineffective. Finally, however large the dose, NT never increased ethanol lethality.

Concomitant with the prolongation of ethanol-induced sleep by NT was the potentiation of ethanol-induced hypothermia (FIG. 2). There was a significant (p < 0.001), inverse linear correlation between the duration of sleep and colonic temperature. Because of this correlation and because i.c. NT, when injected

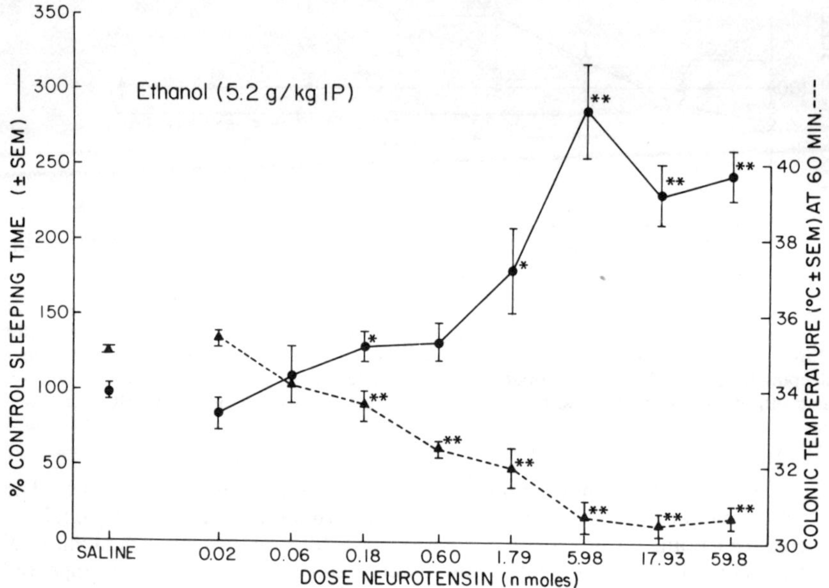

FIGURE 2. Effect of intracisternally administered neurotensin on the duration of sleep and colonic temperature after an i.p. injection of ethanol (5.2 g/kg). The ethanol- and vehicle-treated mice slept for 75 ± 5 minutes (n = 96). Experimental values were compared with those of controls by Dunnett's test for multiple comparisons (two-tailed). Each point is the mean of 8 or more mice.
————represents the effect of neurotensin on the duration of ethanol-induced sleep
– – – –represents the effect of neurotensin on ethanol-induced hypothermia
*p < 0.05; **p < 0.01 when compared with ethanol plus saline treatment. Adapted from Luttinger et al.[7]

alone, produces a decrease in colonic temperature,[6] we sought to determine whether a possible summation of the hypothermic effects of ethanol and NT were sufficient to increase the duration of sleep. This does not appear to be the case.[8] As shown in TABLE 3, at an ambient temperature of 24°C, ethanol reduced colonic temperature by about 3.5°C. At an ambient temperature of 16°C, ethanol reduced colonic temperature by about 7.1°C. Thus, at an ambient temperature of 16°C, ethanol produced decreases in colonic temperature comparable to those observed after ethanol plus NT at an ambient temperature of 24°C (TABLE 3). However, in mice placed in a 16°C environment, the duration of ethanol-induced sleep was no greater than that of mice in a 24°C environment. This suggested that the hypo-

TABLE 2

THE EFFECT OF INTRACISTERNALLY ADMINISTERED NEUROTENSIN
ON BLOOD AND BRAIN ETHANOL CONCENTRATIONS IN MICE*

| Treatment | N | Ethanol Concentrations | | |
		Blood†	N	Brain‡
Ethanol + vehicle	8	5.62 ± 0.20	7	3.09 ± 0.13
Ethanol + neurotensin	8	5.88 ± 0.13	8	3.08 ± 0.15

*Mice were treated with ethanol (5.2 g/kg, i.p.). Ten minutes later, they were injected intracisternally with vehicle (0.9% NaCl) or peptide (10 μg, 5.98 nmoles). Sixty minutes after ethanol treatment, mice were decapitated and brain and blood ethanol concentrations assayed. No significant differences were found when ethanol levels were compared between ethanol-vehicle and ethanol-neurotensin groups. Adapted from data in Luttinger *et al.*[7]

†mg ethanol/ml
‡mg ethanol/g tissue

thermic interaction of NT with ethanol was not responsible for the potentiation of ethanol-induced sleep.

The tripeptide TRH, when administered peripherally or centrally, has been found to antagonize several of the CNS effects of NT.[9] In addition, the effects of NT on pentobarbital-induced sleep and ethanol-induced sleep and hypothermia are essentially the opposite of TRH, which under certain conditions can antagonize these depressant actions. Thus, we sought to determine how the depressant actions of ethanol would be altered by the administration of both of these peptides. Neurotensin potentiation of ethanol-induced sleep and hypothermia were both antagonized by a dose of TRH (30 μg, 82.6 nmoles, i.c.). This dose, when administered to mice treated with ethanol (5.2 g/kg, i.p.) but not with NT, had no effect on sleeping time. This finding was especially intriguing when additional studies demonstrated that TRH (30 μg, i.c.) attenuated the relatively short duration of sleep induced by 4.7 g/kg of ethanol but not the longer duration of sleep induced by 5.2 g/kg of ethanol (data not shown).[8]

To further characterize the interaction of NT and ethanol, we chose to study

TABLE 3

THE EFFECT OF AMBIENT TEMPERATURE AND NEUROTENSIN ON ETHANOL-INDUCED
HYPOTHERMIA AND DURATION OF SLEEP*

Treatment	Ambient Temperature (°C)	N	Colonic Temperature (°C ± SEM)	Duration of Sedation (% Control ± SEM)
Ethanol + vehicle	24°	17	35.5 ± 0.2	100 ± 8
Ethanol + vehicle	16°	15	31.9 ± 0.2†	96 ± 8
Ethanol + neurotensin	24°	8	30.8 ± 0.5†	286 ± 32†

*Mice were injected with ethanol (5.2 g/kg, i.p.) and 10 minutes later were injected with peptide (10 μg, 5.98 nmoles, i.c.). Colonic temperature was measured 60 minutes after ethanol injection. Ethanol + vehicle-injected mice slept for 65 ± 5 minutes at an ambient temperature of 24°C.

†$p < 0.01$ when compared to ethanol + vehicle-injected mice at an ambient temperature of 24°C (Dunnett's test for multiple comparisons).

the effects of NT on ethanol-induced impairment of the aerial righting reflex in mice. This simple test was used to assess motor incoordination. Ethanol (2.0–4.0 g/kg, i.p.) produced a dose-dependent impairment in the ability of mice to right themselves when dropped from an inverted position 60 min after treatment.[10] Neurotensin, administered i.c. after ethanol injection, further increased the height necessary for the mice to right themselves (TABLE 4). As with the potentiation of depressant-induced sleep, this was clearly not due to a summation of effects of ethanol and NT because NT did not alter the animals' aerial righting reflex when administered i.c. to non-ethanol-treated mice (TABLE 4). Furthermore, when NT was administered i.c. with a dose of ethanol (3.5g/kg, i.p.) that alone did not cause sleep, the result was loss of the aerial righting reflex (i.e., sleep) 30 minutes postinjection.

Despite the interactions of ethanol and NT following acute administration, no clear interactions were observed following chronic administration of either compound. In two complementary studies, it was found that induction of tolerance to either NT or ethanol did not cause cross tolerance in the hypothermic response to the other compound.[11,12] In rats, Wenger *et al.*[11] found that 14 hours of a constant intraventricular infusion of NT produced tolerance to its hypothermic effect. However, following this treatment, there was no change in the magnitude of hypothermia induced by an i.p. injection of ethanol (3.0 g/kg). Thus, with regard to hypothermia, there does not appear to be cross tolerance between NT and ethanol. Further, these findings suggest that ethanol-induced hypothermia is probably not dependent on endogenous NT.[11] Recently we found that tolerance to the hypothermic action of ethanol produced by feeding mice an ethanol containing liquid diet did not alter the magnitude of NT-induced hypothermia.[12] The hypothesis that NT and ethanol-induced hypothermia may involve separate mechanisms is supported by other evidence. Neurotensin can induce hypothermia at high ambient temperatures,[13] while ethanol fails to lower body temperature under these conditions.[14]

TABLE 4

EFFECT OF NEUROTENSIN ON ETHANOL-INDUCED IMPAIRMENT OF THE
AERIAL RIGHTING REFLEX IN MICE*

Treatment		N	Sleep (%)†	Height of Aerial
i.c.	i.p.			Righting‡ (cm ± SEM)
Vehicle	+ vehicle	29	0	4 ± 1¶
Neurotensin	+ ”	16	0	6 ± 1
Vehicle	+ ethanol (3.5 g/kg)	34	3	19 ± 2
Neurotensin	+ ”	16	69¶	29 ± 3§

*Mice were injected i.p. with vehicle or ethanol (3.5 g/kg) and at the same time, an i.c. injection of vehicle or neurotensin (30 μg, 17.93 nmoles) was administered. The percentage of mice asleep was determined 30 minutes postinjection, and the aerial righting reflex was assessed 60 minutes postinjection.

†Statistical analysis of the frequency of peptide and ethanol-induced sleep. Frequency was calculated by Chi-square analysis.

‡Changes in the height of aerial righting were evaluated statistically by the Mann-Whitney U-test.

§$p < 0.05$.

¶$p < 0.01$ when compared to vehicle-ethanol treatment.

TABLE 5

EFFECT OF NEUROTENSIN ON ETHANOL WITHDRAWAL TREMOR
AND AUDIOGENIC SEIZURE SUSCEPTIBILITY IN RATS*

Treatment (nmoles)	N	Tremor Score	Percent Exhibiting Seizures		
			Wild Running	Clonic Seizure	Tonic Seizure
Neurotensin					
0	12	2.2 ± 0.2	83	75	58
5.98	11	2.2 ± 0.2	73	73	45
17.93	12	2.3 ± 0.2	83	83	58
59.8	12	2.3 ± 0.2	75	75	25†

*Rats were fed an ethanol-containing liquid diet for 12 days before ethanol withdrawal at 0800 hours. Seven hours and twenty minutes after ethanol withdrawal, saline or neurotensin dissolved in saline were administered by intracisternal injection. Withdrawal signs were evaluated 10 min later. Significance of frequency data was evaluated by Chi-square analysis.

†$p < 0.05$ when compared with saline treatment.

‡$p < 0.01$ when compared with saline treatment.

Consistent with the studies of tolerance described above, studies of withdrawal in rats physically dependent on ethanol have not revealed evidence for an action of NT. The effects of i.c. NT on ethanol withdrawal tremors and audiogenic seizures were not striking (TABLE 5). Only the highest dose of NT tested (100 μg 59.8 nmoles) had any effect, and this was to reduce the frequency of tonic seizures, the most severe component of the withdrawal seizure. However, there was no effect on other less severe components of ethanol withdrawal seizures and no changes in tremors.

Two other peptides have been found to share many of the properties of NT as regards interactions with ethanol. These two peptides are bombesin and β-endorphin. All of the acute effects and lack of cross tolerance effects observed with NT plus ethanol also occur with bombesin or β-endorphin and ethanol.[7,8,11] However, the effects of bombesin and β-endorphin on ethanol withdrawal signs are different from those of NT. Thus, bombesin reduced all three signs of seizures measured (i.e., wild running, clonic and tonic seizures), and β-endorphin reduced tremor score and tonic seizures (data not shown). Hence, the effects of NT on ethanol are not unique because some of the features are shared by two other neuropeptides. Further, the conclusions one can draw from the NT-ethanol interactions may apply equally well to bombesin– or β-endorphin–ethanol interactions.

The studies described in this report, taken together, suggest that NT may be involved in some of the organism's responses to ethanol, pentobarbital, and possibly other CNS depressants. Perhaps the most significant findings are those involving the effects of NT on ethanol-induced sleep. Neurotensin administration not only potentiated ethanol-induced sleep without affecting ethanol concentrations in blood or brain, but also was able to induce sleep following injection of a non-sleep-inducing dose of ethanol. This occurred even though injection of the peptide alone does not induce sleep. One of the many intriguing possibilities suggested by these studies is that different sensitivities to ethanol could be due in part to differences in endogenous NT activity.

ACKNOWLEDGMENTS

The authors express their appreciation to Robert Considine, Barbara Gau, Ossie Hatley, Bill Hudson, and David Knight for their excellent technical assistance and to Faygele ben Miriam for his skillful assistance in the preparation of this manuscript. We are grateful for the helpful comments of Drs. Arthur Prange, Jr. and Charles Nemeroff.

REFERENCES

1. BISSETTE, G., C. B. NEMEROFF, P. T. LOOSEN, G. R. BREESE, G. B. BURNETT, M. A. LIPTON & A. J. PRANGE, JR. 1978. Modification of pentobarbital-induced sedation by natural and synthetic peptides. Neuropharmacology 17: 229–237.
2. BISSETTE, G., C. B. NEMEROFF, P. T. LOOSEN, A. J. PRANGE, JR. & M. A. LIPTON. 1977. Comparison of the analeptic potency of TRH, ACTH$^{4-10}$, LHRH, and related peptides. Pharmacol. Biochem. Behav. 5 (Suppl. 1): 135–138.
3. COTT, J. M., G. R. BREESE, B. R. COOPER, T. S. BARLOW & A. J. PRANGE, JR. 1976. Investigations into the mechanism of reduction of ethanol sleep by thyrotropin-releasing hormone. J. Pharmacol. Exp. Ther. 196: 594–604.
4. PRANGE, A. J., JR., G. R. BREESE, G. D. JAHNKE, B. P. MARTIN, B. R. COOPER, J. M. COTT, I. C. WILSON, L. B. ALLTOP, M. A. LIPTON, G. BISSETTE, C. B. NEMEROFF & P. T. LOOSEN. 1975. Modification of pentobarbital effects by natural and synthetic polypeptides: Dissociation of brain and pituitary effects. Life Sci. 16: 1907–1914.
5. KALIVAS, P. W. & A. HORITA. 1979. Thyrotropin releasing hormone, central site of action in antagonism of pentobarbital narcosis. Nature (London) 278: 461–463.
6. NEMEROFF, C. B., G. BISSETTE, A. J. PRANGE, JR., P. T. LOOSEN, T. S. BARLOW & M. A. LIPTON. 1977. Neurotensin: Central nervous system effects of a hypothalamic peptide. Brain Res. 128: 485–496.
7. LUTTINGER, D., C. B. NEMEROFF, G. A. MASON, G. D. FRYE, G. R. BREESE & A. J. PRANGE, JR. 1981. Enhancement of ethanol-induced sedation and hypothermia by centrally administered neurotensin, β-endorphin, and bombesin. Neuropharmacology 20: 305–309.
8. LUTTINGER, D., G. D. FRYE, C. B. NEMEROFF & A. J. PRANGE, JR. 1982. The effects of neurotensin, β-endorphin, and bombesin on ethanol-induced behaviors in mice and rats. Psychopharmacology. In press.
9. BURGESS, S. K., D. LUTTINGER, D. E. HERNANDEZ, P. W. KALIVAS, C. B. NEMEROFF & A. J. PRANGE, JR. 1982. Antagonism of the central nervous system effects of neurotensin by thyrotropin-releasing hormone. Thyrotropin-releasing hormone (TRH) satellite symposium of the 8th ISN meeting, Sept. 1981. Raven Press. In press.
10. FRYE, G. D., D. LUTTINGER, C. B. NEMEROFF, R. A. VOGEL, A. J. PRANGE, JR. & G. R. BREESE. 1981. Modification of the actions of ethanol by centrally active peptides. Peptides 2 (Suppl. 1): 99–106.
11. WENGER, J., J. YUSIM, F. BLOOM & M. BROWN. 1981. Rapid induction of tolerance to the hypothermic effect of β-endorphin (βE), bombesin (BN), and neurotensin (NT) and lack cross-tolerance to the hypothermic effect of ethanol (E). Neurosci. Soc. Abst. 7: 505.
12. PRANGE, A. J., JR., C. B. NEMEROFF, P. B. CHAPPELL, P. W. KALIVAS, D. LUTTINGER, D. A. STANLEY & G. D. FRYE. 1982. Lack of tolerance to neurotensin-induced hypothermia in the ethanol-tolerant mouse. Alcoholism, Clinical and Experimental Research. In press.
13. MASON, G. A., C. B. NEMEROFF, D. LUTTINGER, O. L. HATLEY & A. J. PRANGE, JR. 1980. Neurotensin and bombesin: Differential effects on body temperature of mice after intracisternal administration. Regulatory Peptides 1: 53–60.
14. MYERS, R. D. 1981. Alcohol's effect on body temperature: Hypothermia, hyperthermia, or poikilothermia? Brain Res. Bull. 7: 209–220.

DISCUSSION OF THE PAPER

J. CRAWLEY (*DuPont, Neurobiology Central Research, Glenolden, PA*): Your data are really interesting, especially since we are thinking of peptides as modulators of other transmitter systems. It is interesting that you showed neurotensin's effects on ethanol in doses in which neurotensin had no sedating effects in itself. My question is whether you have analyzed your data in terms of whether the neurotensin-alcohol interaction is additive or synergistic?

D. LUTTINGER (*University of North Carolina School of Medicine, Chapel Hill*): They are synergistic and not additive since the peptide does not induce sleep. When given with ethanol, NT induces sleep in non-sleep mice treated with an inducing dose of ETOH.

D. J. BRAITMAN (*AFRRI, Bethesda, MD*): I was intrigued by the last piece of data that you presented where neurotensin did not affect ethanol-induced seizure withdrawal. How do you see the difference in this test versus all the other tests, and does that indicate anything about the mechanism of the neurotensin action?

LUTTINGER: Not directly. However in animals rendered tolerant to ethanol, there was no alteration in response to neurotensin, at least as regards hypothermia.

All of our other studies were done with acute ethanol treatment.

E. A. ZIMMERMAN (*Columbia University, New York*): It was interesting that the interperitoneal administration of NT had no effect and I was thinking about this problem of getting peptides across the blood-brain barrier. Have you ever tried putting it in the carotid, though this would be difficult in the mouse?

LUTTINGER: We have not tried that.

NEUROTENSIN AND THERMOREGULATION*

Garth Bissette, Daniel Luttinger, George A. Mason,
Daniel E. Hernandez, and Peter T. Loosen

*Biological Sciences Research Center
University of North Carolina School of Medicine
Chapel, Hill, North Carolina 27514*

The physiological roles of the tridecapeptide neurotensin (NT) have not yet been clearly defined. Its heterogeneous distribution in brain and gut and the manifold effects observed after central nervous system (CNS) or systemic injection allow preliminary formulation (see Prange and Nemeroff, this volume)[1-3] but not closure.

Our group first reported the hypothermic effect of NT in 1976.[4] This study demonstrated that NT injected intracisternally (i.c.) caused a dose-dependent decrease in the colonic temperature of mice in a 4°C environment. The lowest effective dose was 30 ng (FIG. 1). At an ambient temperature of 25°C, animals also responded to i.c. NT with loss of body temperature, though the effect was less pronounced than at 4°C. Somatostatin (SRIF), substance P, melanocyte-stimulating hormone inhibiting factor (MIF-I), luteinizing-hormone releasing hormone (LHRH) and thyrotropin-releasing hormone (TRH) were not effective in inducing hypothermia at 4°C after i.c. injection of doses equimolar to 30 μg of NT. These findings have been confirmed in several laboratories.[5-8] It is noteworthy that NT given peripherally appears to lack a hypothermic effect; intravenous (i.v.) administration, at whatever ambient temperature, was ineffective in doses as large as 100 μg/kg^{-1}.

Using NT fragments and D-amino-acid-substituted analogs of NT, synthesized and generously supplied by Dr. Jean Rivier of the Salk Institute, we next sought to identify the part of the NT molecule[9] responsible for hypothermic activity. D-Tyr^{11}-NT, NT-NHCH^3, and bombesin, a peptide derived from frog skin, were found to be more potent hypothermic agents in mice at 4°C than NT when injected i.c. in doses equimolar to 30 μg NT. The carboxyl terminus and, specifically, the two adjacent arginine residues were shown to be crucial for hypothermia (TABLE 1). These structure–activity findings are largely supported by Rivier *et al.*[10] The hypothermic effect of bombesin confirms the original report by Brown and Vale.[5]

The impetus for the examination of the thermoregulatory effects of NT came from the use of a pharmacological screening test for CNS activity of peptides.[11] The interactions of NT with CNS depressants is discussed in another chapter of these proceedings, but the profound potentiation by i.c. NT of a narcotic dose of pentobarbital deserves mention. This potentiation was shown to be due to a marked decrease in the metabolism of [^{3}H] pentobarbital in brain, blood, and liver of rats.[12] As seen in TABLE 2, the hypothermia elicited by 50 mg/kg pentobarbital is suppressed at 33 minutes and augmented after 66 minutes after i.c. NT in mice at 25°C, although no clear dose–response relationship is apparent. This effect of

*Supported by NIMH MH-32316, MH-33127, MH-34121, MH-14277 and NICHHD HD-03110.

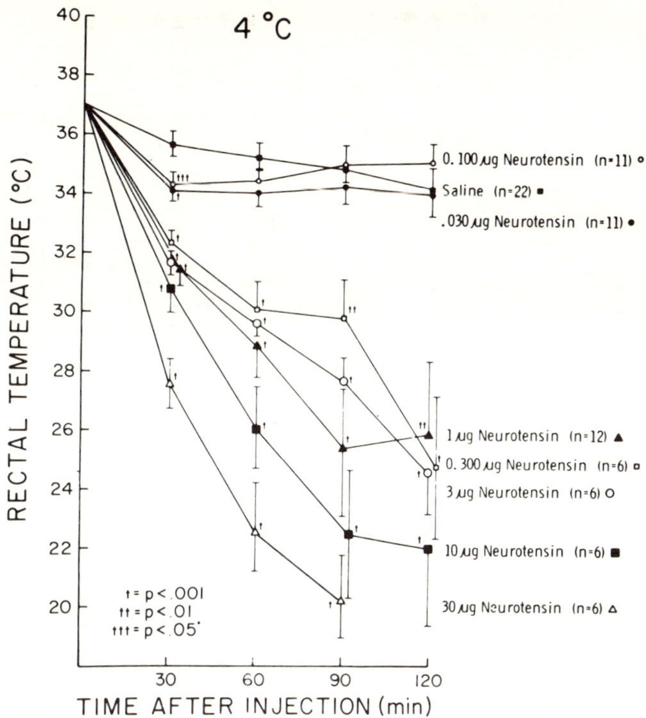

FIGURE 1. Effect of i.c. neurotensin on the core temperature of mice subjected to a cold environment (4°C). Data are expressed as °C ± SEM. Experimental values were compared with those of controls by Student's t-test (two-tailed). Small open circles, 0.1 µg of neurotensin (n = 11); small closed squares, saline (n = 22); closed circles, 0.030 µg of neurotensin (n = 11); closed triangles, 1 µg of neurotensin (n = 12); small open squares, 0.3 µg of neurotensin (n = 6); large open circles, 3 µg of neurotensin (n = 6); large closed squares, 10 µg of neurotensin (n = 6); open triangles, 30 µg of neurotensin (n = 6). † = p < 0.001; †† = p < 0.01; ††† = p < 0.05. (Reproduced from Bissette, *et al.*,[4] with permission.)

NT on pentobarbital sedation was seen only after i.c. injection, indicating that a CNS interaction was responsible for these changes. Consistent with these findings, Chandra *et al.*[29] found NT hypothermia at ambient temperatures of 8° and 22° to be associated with decreased metabolic heat production rather than by increased heat loss from skin.

We next sought to determine the phylogenetic responses of different vertebrates to the hypothermic effect of i.c. NT.[13,14] A representative fish, amphibian, and reptile were selected as poikilotherms and a bird, several species of rodent (both hibernators and nonhibernators), and one species of primate were chosen as the homeotherms to be tested. As seen in TABLE 3, the poikilotherms, the obligatory hibernators, and the rabbit showed no response to i.c. NT at 4° or 23°C, whereas the guinea pig and golden hamster exhibited NT-induced hypothermia at 4°C only. The mouse, rat, gerbil, and monkey exhibited the hypothermic response to NT at both temperatures tested. The rabbit's lack of hypothermia after i.c. injection was puzzling, but this effect was not observed after as much as 300 µg

TABLE 1

THE EFFECT OF NEUROTENSIN, NEUROTENSIN ANALOGUES, BOMBESIN, AND PHYSALEMIN ON BODY TEMPERATURE OF RATS AT 4°C*

Treatment	Dose, i.c. (μg)	N	Rectal temperature (°C ± SEM)				
			0 min	30 min	60 min	90 min	120 min
Saline		69	37.8 ± 0.06	36.9 ± 0.11	37.3 ± 0.08	37.4 ± 0.08	37.5 ± 0.08
Neurotensin	30	64	37.8 ± 0.06	34.8 ± 0.12†	34.3 ± 0.15†	34.7 ± 0.18†	35.3 ± 0.21†
[D-Lys[6]]-NT	10	6	38.0 ± 0.16	36.3 ± 0.38	36.9 ± 0.50	37.5 ± 0.33	37.5 ± 0.32
[D-Lys[6]]-NT	30	6	37.7 ± 0.15	35.2 ± 0.27†	36.0 ± 0.26	36.7 ± 0.21†	37.1 ± 0.24
[D-Pro[7]]-NT	10	5	37.5 ± 0.07	36.2 ± 0.21	36.5 ± 0.32	36.9 ± 0.36†	37.6 ± 0.18
[D-Pro[7]]-NT	30	6	37.3 ± 0.08	35.1 ± 0.30†	34.5 ± 0.44†	34.6 ± 0.64†	35.4 ± 0.79†
[D-Arg[8]]-NT	10	6	37.1 ± 0.20	34.4 ± 0.32†	33.7 ± 0.49†	34.5 ± 0.49†	35.1 ± 0.45†
[D-Arg[8]]-NT	30	6	37.9 ± 0.12	34.3 ± 0.25†	33.7 ± 0.29†	33.8 ± 0.26†	34.3 ± 0.17
[D-Arg[9]]-NT	10	6	37.6 ± 0.15	36.3 ± 0.17	36.6 ± 0.27	37.4 ± 0.23	37.7 ± 0.49
[D-Arg[9]]-NT	30	6	37.7 ± 0.21	36.0 ± 0.28	36.2 ± 0.49	36.8 ± 0.43	37.0 ± 0.35
[D-Pro[10]]-NT	10	6	38.2 ± 0.21	35.8 ± 0.54†	36.4 ± 0.36	36.7 ± 0.33†	37.1 ± 0.36
[D-Pro[10]]-NT	30	6	38.6 ± 0.09	35.1 ± 0.40†	35.9 ± 0.36	36.7 ± 0.32†	37.1 ± 0.27
[D-Tyr[11]]-NT	10	6	37.5 ± 0.15	34.0 ± 0.40†	32.8 ± 0.34†	32.3 ± 0.47†	32.3 ± 0.41†
[D-Tyr[11]]-NT	30	6	37.5 ± 0.11	33.8 ± 0.40†	32.3 ± 0.59†	31.9 ± 0.63†	32.1 ± 0.72†
[Phe[11]]-NT	10	6	37.7 ± 0.16	36.1 ± 0.27	36.7 ± 0.30	37.4 ± 0.24	37.9 ± 0.18
[Phe[11]]-NT	30	6	37.8 ± 0.18	34.8 ± 0.20†	35.5 ± 0.25	36.6 ± 0.29†	36.9 ± 0.29
NT-NHMe	10	6	38.0 ± 0.08	36.0 ± 0.44†	36.4 ± 0.43	37.0 ± 0.33	37.1 ± 0.29
NT-NHMe	30	5	38.0 ± 0.14	33.3 ± 0.56	34.2 ± 0.96†	34.9 ± 0.88†	35.9 ± 0.59
NT[9-13]	10	6	38.0 ± 0.14	36.5 ± 0.48	36.9 ± 0.45	37.2 ± 0.35	37.6 ± 0.41
NT[9-13]	30	5	38.1 ± 0.24	35.8 ± 0.27	37.0 ± 0.18	37.2 ± 0.05	37.4 ± 0.18
NT[8-13]	10	6	37.2 ± 0.17	36.6 ± 0.33	37.0 ± 0.34	37.2 ± 0.19	37.4 ± 0.30
NT[8-13]	30	6	37.3 ± 0.07	34.9 ± 0.50†	35.9 ± 0.50	36.7 ± 0.44†	37.0 ± 0.46
Bombesin	10	6	37.4 ± 0.09	33.5 ± 0.31†	32.0 ± 0.41†	31.5 ± 0.42†	31.4 ± 0.47†
Bombesin	30	6	37.6 ± 0.19	34.0 ± 0.23†	32.3 ± 0.40	31.1 ± 0.45†	30.6 ± 0.49†
Physalemin	10	6	37.4 ± 0.09	37.8 ± 0.3	37.8 ± 0.14	37.6 ± 0.12	37.6 ± 0.13
Physalemin	30	6	37.9 ± 0.20	37.1 ± 0.3	37.3 ± 0.28	37.6 ± 0.16	37.6 ± 0.16

* Rats (N ≥ 5/group) were lightly anesthetized with ether, injected intracisternally with vehicle (0.9% NaCl, pH 7.5), neurotensin (30 μg), or a neurotensin analogue (equimolar to 10 or 30 μg of neurotensin) and placed in plastic cages at 4°C. Rectal temperature was recorded immediately after injection (0 min) and at 30-min intervals over a 2-hr period. Data are expressed as °C ± SEM. (Reproduced from Loosen et al.,[9] with permission.) † $p < 0.05$ one-way ANOVA, Neuman-Keul's test.

TABLE 2

THE EFFECT OF INTRACISTERNALLY ADMINISTERED NEUROTENSIN ON
PENTOBARBITAL-INDUCED NARCOSIS AND HYPOTHERMIA *

Temperature	N	Sleeping Time (min ± SEM)		Body Temperature (°C ± SEM) 33 min		66 min	
Saline	40	84.8	4.76	31.77	0.300	32.32	0.476
Neurotensin (μg)							
0.01	30	61.8	3.83†	33.12	0.311‡	33.46	0.456
0.03	29	68.9	5.72	33.28	0.383§	33.37	0.469
0.10	9	92.3	11.68	31.47	0.316	31.09	0.695
0.3	9	117.3	12.66	31.34	0.464	30.38	0.809
1.0	6	116.9	21.0	30.71	0.288	29.80	0.522§
3.0	6	162.8	29.22	32.04	0.430	30.37	0.610
10.0	5	222.1	37.23§	32.10	0.186	30.92	0.666
30	3	218.2	49.23‡	32.19	0.275‡	30.36	0.354

* All male mice (both control and experimental groups) received 50 mg/kg pentobarbital i.p. and were injected with vehicle (10 μl 0.9% NaCl, pH 7.5) or neurotensin. The small N at the higher doses of neurotensin utilized is due to the mortality observed in these experimental groups. P was assessed with the Mann-Whitney U-test. (Reproduced from Nemeroff et al.,[12] with permission.)

† = p < 0.05
‡ = p < 0.02
§ = p < 0.0002

of NT at 4°C. Retrospective analysis showed that only those species whose basal metabolic rate is stimulated at 4°C were responsive to i.c. NT at this temperature.

At the 4th International Symposium on the Pharmacology of Thermoregulation in Oxford, England, 1979, several contributions were made to the understanding of NT-induced hypothermia. Metcalf confirmed the lack of effect of i.c. NT (1–100 μg) in inducing hypothermia in the rabbit.[15] Our group reported the partial antagonism of NT-induced hypothermia in rats by the simultaneous i.c. injection of TRH[14] (TABLE 4). Although the TRH antagonism was not dose related, L-triiodothyronine (T3), MIF, atropine, propranolol, phenoxybenzamine, naloxone, or apomorphine had no attenuating effect on NT-induced hypothermia in rats at room temperature. Haloperidol pretreatment significantly potentiated the hypothermic effect of i.c. NT. Brown and Vale[7] demonstrated TRH antagonism of the hypothermia induced by central administration of NT, bombesin or β-endorphin. A SRIF fragment was also shown to reverse NT and bombesin-induced hypothermia, but to have no effect on β-endorphin-induced hypothermia. In a paper of related interest, Stanton et al.[16] reported that TRH infusion into the ventricles of ground squirrels could rouse them from hibernation. This species is not hypothermic after i.c. NT, but does exhibit a thermoregulatory response to TRH.

Differences in NT- and bombesin-induced hypothermia have been clearly demonstrated. Tache et al.[17] showed bombesin to be a poikilothermic agent, causing *hyper*thermia at 36°C ambient temperature, some *hypo*thermia at 24°C, and marked *hypo*thermia at 4°C. Bombesin showed no effect on temperature at 33°C ambient temperature. Mason et al.[18] have shown NT to be an effective hypothermic agent at ambient temperatures of 4°C, 23°C, and 26°C in mice. NT showed no tendency toward hyperthermia at higher ambient temperatures, while

TABLE 3

THE EFFECT OF INTRACISTERNAL NEUROTENSIN ON RECTAL TEMPERATURE IN WARM (23°C) AND COLD (4°C) AMBIENT TEMPERATURES*

Species	Sex	Body Wt. (kg approx.)	Brain Wt. (kg approx.)	Dose (μg)	Volume (μl)	Warm (23°C)			Cold (4°C)		
						No. times N	sig.	NT effect	No. times N	sig.	NT effect
Poikilotherms											
Nonmammals											
Nonhibernators											
Fish											
Bluegill	M,F	0.06	0.1	30	10	—	—	?	6	0	No
Amphibian											
Frog	M,F	0.06	0.2	30	10	8	0	No	8	0	No
Reptile											
Lizard	M,F	0.004	0.05	15	5	16	0	No	7	0	No
Homeotherms											
Nonmammals											
Nonhibernators											
Bird											
Pigeon	M,F	0.55	2.5	100	10	6	0	No	6	0	No

Mammals										
Obligatory hibernators										
Rodents										
Ground Squirrel	M,F	0.10	3.5	30	10	7	0	No	5	0	No
Woodchuck	M,F	5.5	15.0	100	20	3	0	No	4	0	No
Permissive hibernators										
Rodent										
Golden hamster	M	0.14	1.0	30	10	6	0	No	5	3	Yes
Nonhibernators										
Rodents										
Rabbit	M	3.5	8.0	300	30	5	0	No	5	0	No
Guinea pig	M	0.35	4.0	30	10	6	0	No	6	3	Yes
Gerbil	M,F	0.05	1.0	10	10	6	4	Yes	7	3	Yes
Mouse	M	0.03	0.5	10	10	7	4	Yes	6	4	Yes
Rat	M	0.30	2.0	30	10	6	4	Yes	6	4	Yes
Primate										
Monkey	M	4.0	80.0	150	10	4	3	Yes	—	—	?

*Groups of animals were injected intracisternally with neurotensin or vehicle (0.9% NaCl, pH 7.5). Thereafter they were individually housed at ambient temperature (23°C) or, in other experiments, in a cold room (4°C). Rectal temperature was assessed immediately after injection (0 time) and at 30-minute intervals for two hours. "Cold" experiments in lizards and frogs were conducted at 15°C because of their intolerance to colder temperatures. Bluegills were subjected to gradually cooling water (see reference 13). Monkeys were studied only in a warm environment; bluegills only in a cold environment. In each experiment, significant differences occurred at most times (3–4) or at none. Student's t-test (two-tailed interpretation) was used to assess statistical significance. When for a given species, a positive effect of neurotensin is shown, the table lists the minimal effective dose; when a null effect is shown, the largest dose employed is listed. (Reproduced from Prange *et al.*,[13] with permission.)

TABLE 4

EFFECTS OF PHARMACOLOGICAL AGENTS ON NT-INDUCED HYPOTHERMIA IN THE RAT AT 4°C*

Drug	N	Dose	Pretreatment Interval (min)	No Effect	Antagonism	Potentiation
Atropine	6	4 mg kg^{-1} i.p.	15	+		
Propranolol	6	12 mg kg^{-1} i.p.	30	+		
Phenoxybenzamine	6	5 mg kg^{-1} i.p.	30	+		
Naloxone	6	2 mg kg^{-1} i.p.	30	+		
Apomorphine	6	0.3 mg kg^{-1} i.p.	30	+		
Haloperidol	6	2 mg kg^{-} i.p.	30			+
MIF (Pro-Leu-Gly-NH$_2$)	6	30 μg i.c.	0	+		
TRH (pGlu-His-Pro-NH2)	5	10 μg i.c.	0		+	
L-Triiodothyronine (T3)	6	20 μg s.c.	240	+		

* Reproduced from Nemeroff et al.,[14] with permission.

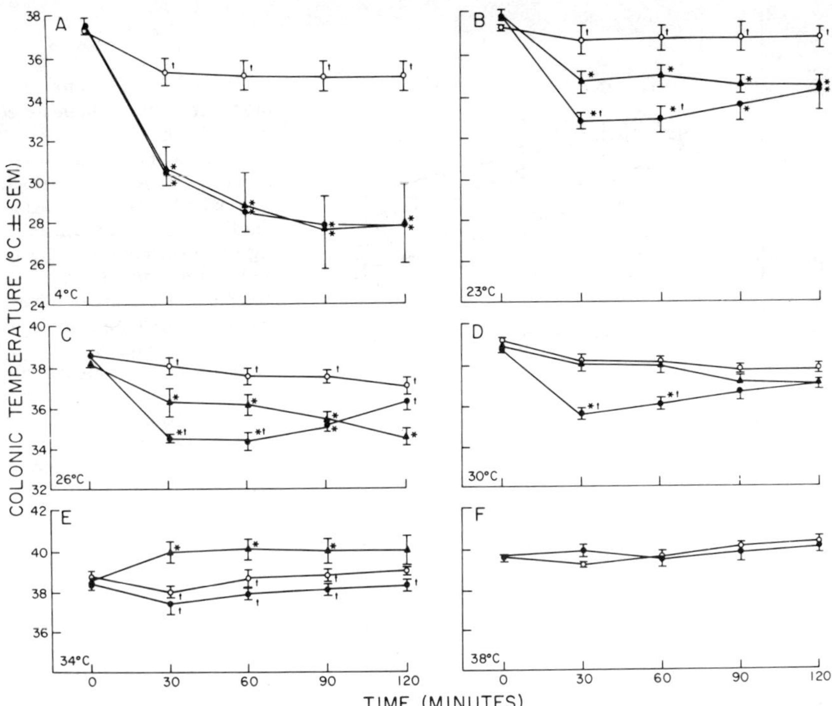

FIGURE 2. The effects of intracisternally administered neurotensin 0.6 nmol (closed circles) (O——O); bombesin, 0.6 nmol (closed triangles) (△——△), or vehicle (O——O) on colonic temperatures of mice at ambient temperatures of 4°C (A), 23°C (B), 26°C (C), 30°C (D), 34°C (E), and 38°C (F). See reference 18 for methods and statistical analysis used. * = p < 0.05 when compared to saline-treated controls, † = p < 0.05 when compared to bombesin-treated mice. (Reproduced from Mason *et al.*,[18] with permission.)

equimolar doses of bombesin were clearly hyperthermic (FIG. 2). Naloxone has been shown to partially block bombesin-induced hypothermia,[19] while NT-induced hypothermia is not blocked by this opiate antagonist.[14] Atropine has also been reported to block bombesin hypothermia (M. Brown, personal communication) while this cholinergic receptor blocker is without effect on NT-induced hypothermia.[14] Thus the two peptides act in different ways and through apparently different mechanisms.

Prostaglandin E2 (PGE2) has been shown to inhibit NT-induced hypothermia in mice[21] (FIG. 3). The hypothermic effect of i.c. NT was not antagonized by the inhibitors of prostaglandin synthesis, indomethacin, or aspirin, but blockade of the hypothermic effect of NT by TRH was prevented by indomethacin pretreatment. Interestingly, TRH blockade of NT-induced hypothermia was only antagonized by indomethacin at 4°C: at 28°C the effect was not seen. These data indicate that an intact prostaglandin system is necessary for TRH antagonism of NT hypothermia at 4°C. The hypothermic effect of bombesin is also blocked by PGE2.[19]

A comparison of the potencies of several peptides in their effects on thermo-

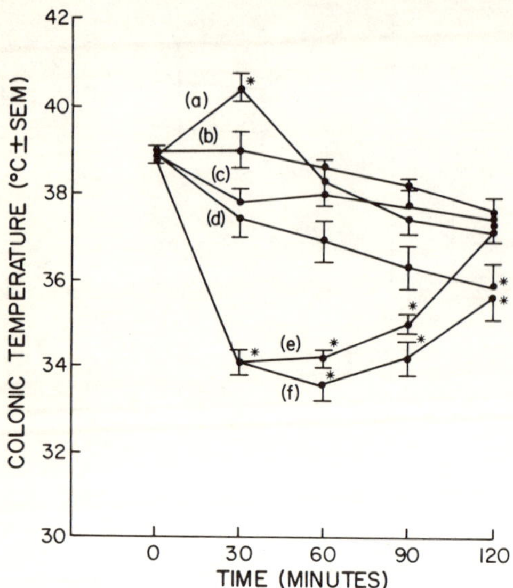

FIGURE 3. The effects of intracisternally administered prostaglandin E2, 4 μg (a); prostaglandin E2 (4 μg) + neurotensin, 1 μg (b); vehicle (normal saline) (c); prostaglandin E2, 4 μg + neurotensin, 10 μg (d); neurotensin (1 μg (e) and neurotensin, 10 μg (f) on colonic temperature of mice at an ambient temperature of 26°C. See reference 21 for methods and statistical analysis used. *$p < 0.05$ when compared to saline-treated controls. (Reproduced from Mason et al.,[21] with permission.)

regulation and nociception after i.c. injection was undertaken by our group.[20] We found 1 μg of NT in mice at 25°C to be a more potent hypothermic agent than equimolar amounts of β-endorphin or bombesin (FIG. 4). PGE2 and SRIF, two agents capable of reversing NT or bombesin hypothermia, caused hyperthermia when injected i.c. by themselves. Thus, PGE2, TRH, and SRIF are all capable of reversing NT- and bombesin-induced hypothermia.

The ontogeny of the hypothermic effect of NT in rats has been outlined by Hernandez et al.[22] (TABLE 5). Rats 5, 10, 20, 30, and 60 days old were shown to be responsive with hypothermia after i.c. NT. Ten-day-old rat pups, selected for further study, showed significant hypothermia after i.c. NT, even in the presence of their mothers. Thus the hypothermic effect of i.c. NT is a mechanism that is established early in the animal's development.

As previously emphasized, NT give peripherally appears to be utterly devoid of hypothermic properties, while NT given centrally, even in very small amounts, has been proven to be a potent hypothermic agent. Dorsa et al.[23] applied this site dependency of NT-induced hypothermia to demonstrate retrograde transport from the pituitary to the brain. They injected NT into the anterior pituitary gland and observed a hypothermic response. After pituitary stalk section, such injection was without effect. Thus, the hypothermia elicited by anterior pituitary injection of NT indicates retrograde transport of this peptide into the CNS.

Our pharmacological approach to exploring the mechanism of NT's hypothermic effect was addressed to possible interactions with DA systems, earlier work by others having shown that a variety of DA-active agents affect body temperature.[33-35] We demonstrated that the DA-receptor blocker, haloperidol, and the direct DA agonist, apomorphine, potentiate the effects of NT, but at high doses, of course, these agents produced hypothermia by themselves.[27] Selective DA depletion with 6-hydroxydopamine in rats also potentiated the hypothermic effects of i.c. NT (TABLE 5). In this experiment, desmethylimipramine pretreatment spared the norepinephrine levels of rats treated with 6-hydroxydopamine, while brain DA levels were lowered 85%. Depletion of serotonin by parachloro-phenylalanine had no effect on NT-induced hypothermia. Thyroidectomized (THX) rats at 23°C showed an increased hypothermic response to i.c. NT, but also showed lower baseline temperatures than sham-operated controls. Circulating thyrotropin (TSH) levels of THX rats were characteristically high, but were signif-icantly depressed by i.c. NT one hour after injection, while rising again after two hours' duration. Thus, the thyroid axis is not necessary for the elaboration of NT-induced hypothermia, although NT can affect TSH levels.

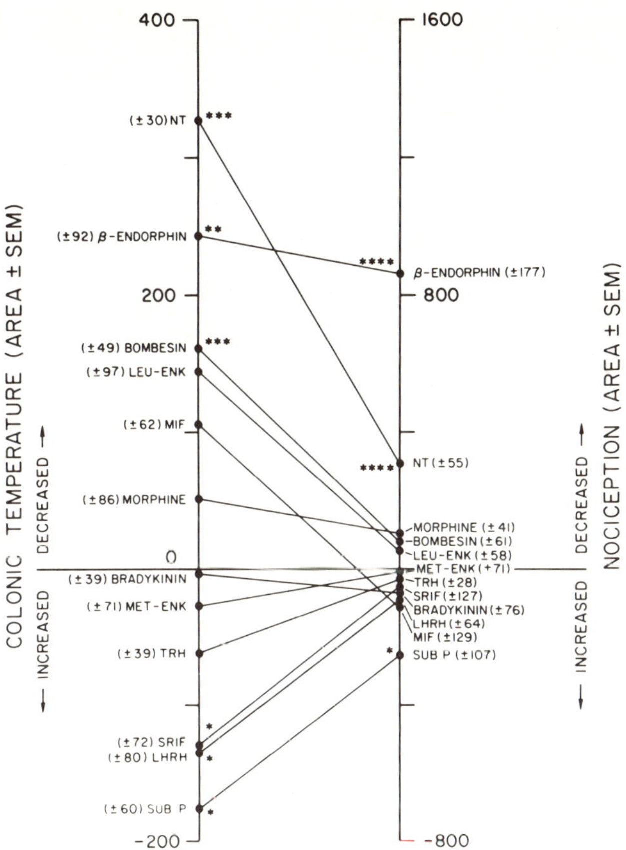

FIGURE 4.   Effects of 11 endogenous peptides and morphine on response latency to a noxious stimulus and on alterations of body temperature after intracisternal administration to mice. NT was administered in a dose of 1 μg; all other substances were administered in an equimolar dose to allow direct comparison of potency. SEM is shown in parentheses. By Dunnett's t-test for multiple comparisons, NT was the most potent hypothermic substance. Asterisks indicate significant antinociceptive, hyperalgesic, hypothermic, or hyperthermic responses when compared to the controls (Student's t-test, two-tailed, * = $p < 0.05$; ** = $p < 0.02$; *** = $p < 0.01$; **** = $p < 0.001$). LEU-ENK, Leu-enkephalin; MIF, melanostatin, MET-ENK Met-enkephalin; TRH, thyroliberin; SRIF, somatostatin; LHRH, luliberin; SUB P, substance P. (Reproduced from Nemeroff, *et al.*,[20] with permission.)

TABLE 5

EFFECT OF SELECTIVE DEPLETION OF BRAIN NE OR DA ON NT-INDUCED HYPOTHERMIA IN RATS AT 4°C

Treatment	N	Rectal temperature (°C ± SEM)				
		0 min	30 min	60 min	90 min	120 min
Controls (saline i.c.)	16	38.2 ± 0.1	37.4 ± 0.2	37.7 ± 0.2	37.7 ± 0.2	37.9 ± 0.2§
NT (30 μg i.c.)	17	38.3 ± 0.1	35.8 ± 0.2§	35.2 ± 0.2§	35.5 ± 0.2§	35.8 ± 0.3
↓ NE + saline i.c.	18	37.8 ± 0.1†	37.1 ± 0.2	37.4 ± 0.3	37.7 ± 0.2	38.0 ± 0.2
↓ NE + NT (30 μg i.c.)	16	37.8 ± 0.1†‖	36.1 ± 0.2§	35.6 ± 0.3§	35.9 ± 0.3§	36.6 ± 0.3§
↓ DA + saline i.c.	11	37.6 ± 0.2†	37.4 ± 0.3	37.7 ± 0.3	37.6 ± 0.4	37.6 ± 0.4
↓ DA + NT (30 μg i.c.)	11	38.0 ± 0.2	35.2 ± 0.2§‖	34.6 ± 0.3§	34.7 ± 0.3§‖	34.8 ± 0.4§¶

* Levels of brain NE and DA are shown in TABLE 2. See reference 27 for description of experimental methods utilized. ↓ DA = 6-OHDA (240 μg i.c.) + DMI (40 mg/kg i.p.) ↓ NE = 6-OHDA (3 × 25 μg i.c.). Reproduced from Nemeroff *et al.*,[27] with permission.

†p < 0.05 when compared with controls (saline)

‡p < 0.02

§p < 0.001

¶p < 0.05 when compared with NT 30 μg

‖p < 0.01

Recently, we have examined the effects of DA agonists on the hypothermic effect of i.c. NT.[28] Indirect DA agonists such as *d*-amphetamine, methylphenidate, and cocaine were effective in blocking the hypothermic action of NT at 25°C in mice. Apomorphine treatment alone produced hypothermia at 25°C, but i.c. NT in apomorphine-treated animals did not exhibit additive effects in decreasing the colonic temperature of rats. Interestingly, the effective DA agonists produced significant *hyper*thermia at 30 minutes when blocking NT, an effect not seen in animals receiving the indirect DA agonist alone at any time point.

The question remains whether the hypothermic effects of centrally administered NT reflect a physiological role for this peptide in temperature regulation. A growing body of evidence supports this possibility. NT is found in areas known to be involved in the control of temperature and specific injection of NT into these regions causes a decline in core temperature. NT has been microinjected into discrete brain loci in order to map regions of the brain that mediate the hypothermic response. Martin *et al.*[25] found 63 sites in rat brain that evoked significant hypothermia after infusion of 2.5 μg of NT. The medial preoptic nucleus and periaqueductal gray were the areas of highest density for NT-active cells, and these areas have high levels of endogenous NT. The diagonal band of Broca, the ventral thalamus and dorsomedial hypothalamus also contained NT-active sites. No hyperthermia to NT was seen at any site tested. Kalivas *et al.*[26] mapped both thermoregulatory and nociceptive sites in rat brain that responded to bilateral 2.5 μg NT infusion. The location of NT-thermoreactive sites rarely overlaps the sites where NT produces nociception. Lateral ventricular infusion of NT or injection of NT into cells on the floor of the fourth ventricle produced hypothermia. These cells may be the target site for i.c. or i.c.v. NT-induced hypothermia. The caudal medial preoptic area was shown to contain a discrete colony of active sites that did not extend beyond 1 μm of the midline or above the anterior commisure. No NT-thermoactive sites were seen in the periaqueductal gray area of the mesencephalon, but several sites were seen in the anterior hypothalamus, the ventral tegmental area, and the spinal tract of the trigeminal nerve. In contradiction to these data, Taube *et al.*[30] have reported a *hyper*thermic effect of NT injected into the preoptic area of rats. These investigators are alone in reporting such an effect of NT at this location in rodent brain. Further evidence for a physiological role for NT in thermoregulation is provided by Wallace *et al.*[24] They have demonstrated that antisera to NT cause *hyper*thermia when infused into rat lateral ventricles. This hyperthermia is presumably due to the removal of endogenous NT through binding to the antisera, although penetration of antisera to NT-containing sites in brain has not yet been demonstrated.

The NT hypothermic response is developed by day five in rats and is seen in a variety of mammalian species, but is not seen in poikilotherms and obligatory hibernators. The hypothermic effect seems to be mediated by receptors to NT that recognize the COOH-terminal and basic portion of the molecule. The system mediating NT-induced hypothermia is complicated as it is antagonized by TRH, an analog of SRIF, PGE2, and indirect DA agonists. The antagonism of NT by TRH may deserve special consideration. It is not dependent upon TRH activation of the pituitary-thyroid axis, thus indicating a CNS interaction of these peptides, possibly through a prostaglandin-mediated mechanism. Brown and Vale,[3] in a recent review, hypothesized that bombesin exerts its hypothermic effect by preventing TRH from reaching and activating pituitary thyrotrophs after cold exposure. This hypothesis is not adequate to explain the hypothermic effects of NT, however. NT can cause decreases in metabolism and thus would be able to limit heat production.[32] This concept is supported by the lack of hypothermic action of i.c. NT at temperatures not requiring heat production for the maintenance of core temperature. Neurotensin is one of several endogenous peptides that appear to contribute to thermoregulatory homeostasis by exerting a hypothermic effect, after direct CNS administration either through interactions with peptides that produce hyperthermic effects or by counterbalancing their effects on some common effector systems.

REFERENCES

1. NEMEROFF, C. B., D. LUTTINGER & A. J. PRANGE, JR. 1982. *In* The Handbook of Psychopharmacology. L. L. Iversen, S. D. Iversen & S. H. Snyder, Eds. Vol. 16: 363–467. Plenum Press. New York.
2. NEMEROFF, C. B., D. LUTTINGER & A. J. PRANGE, JR. 1980. Trends in Neurosciences. September pp. 212–215.
3. BISSETTE, G., P. J. MANBERG, C. B. NEMEROFF & A. J. PRANGE, JR. 1978. Life Sci. **23:** 2173–2182.
4. BISSETTE, G., C. B. NEMEROFF, P. T. LOOSEN, A. J. PRANGE, JR. & M. A. LIPTON. 1976. Nature **262:** 607–609.
5. BROWN, M. & W. VALE. 1977. Science **196:** 998–1000.
6. CLINESCHMIDT, B. V. & J. C. McGUFFIN. 1977. Eur. J. Pharmacol. **464:** 395–396.
7. BROWN, M. & W. VALE. 1980. *In* Thermoregulatory Mechanisms and Their Therapeutic Implications. B. Cox, P. Lomax, A. S. Milton & E. Schonbaum, Eds. S. Karger. Basel. pp. 186–194.
8. JOLICOEUR, F. B., A. BARBEAU, F. RIOUX, R. QUIRION & S. ST. PIERRE. 1980. Peptides **2:** 171–175.
9. LOOSEN, P. T., C. B. NEMEROFF, G. BISSETTE, G. B. BURNETT, A. J. PRANGE, JR. & M. A. LIPTON. 1978. Neuropharmacology **17:** 109–113.
10. RIVIER, C., M. BROWN & W. VALE. 1977. Endocrinology **102:** 519–522.
11. BISSETTE, G., C. B. NEMEROFF, P. T. LOOSEN, G. R. BREESE, G. B. BURNETT, M. A. LIPTON & A. J. PRANGE, JR. 1978. Neuropharmacology **18:** 229–237.
12. NEMEROFF, C. B., G. BISSETTE, A. J. PRANGE, JR., P. T. LOOSEN, T. S. BARLOW & M. A. LIPTON. 1977. Brain Res. **128:** 485–496.
13. PRANGE, A. J., JR., C. B. NEMEROFF, G. BISSETTE, P. J. MANBERG, A. J. OSBAHR III, G. B. BURNETT, P. T. LOOSEN & G. W. KRAEMER. Pharmacol. Biochem. Behav. **II:** 473–477.
14. NEMEROFF, C. B., G. BISSETTE, P. J. MANBERG, S. I. MOORE, G. N. ERVIN, A. J. OSBAHR III, A. J. PRANGE, JR. 1979. *In* Thermoregulatory Mechanisms and Their Therapeutic Implications. B. Cox, P. Lomax, A. S. Milton & E. Schonbaum, Eds. Karger. Basel. pp. 180–185.
15. METCALF, G., P. W. DETTMAR & T. WATSON. 1979. *In* Thermoregulatory Mechanisms and Their Therapeutic Implications. B. Cox, P. Lomax, A. S. Milton & E. Schonbaum, Eds. Karger. Basel. pp. 175–179.
16. STANTON, T. L., A. L. BECKMAN & A. WINOKUR. 1979. *In* Thermoregulatory Mechanisms and Their Therapeutic Implications. B. Cox, P. Lomax, A. S. Milton & E. Schonbaum, Eds. Karger. Basel. pp. 195–197.
17. TACHE, Y., Q. PITTMAN & M. BROWN. 1980. Brain Res. **188:** 525–530.
18. MASON, G. A., C. B. NEMEROFF, D. LUTTINGER, O. L. HATLEY & A. J. PRANGE. 1980. Regulatory Peptides **1:** 53–60.
19. BROWN, M., J. R. RIVIER & W. VALE. 1977. Life Sci. **20:** 1681–1687.
20. NEMEROFF, C. B., A. J. OSBAHR III, P. J. MANBERG, G. N. ERVIN & A. J. PRANGE, JR. 1979. Proc. Natl. Acad. Sci. USA **76:** 5368–5371.
21. MASON, G. A., D. E. HERNANDEZ, C. B. NEMEROFF, J. W. ADCOCK, O. L. HATLEY & A. J. PRANGE, JR. 1982. Regulatory Peptides. In press.
22. HERNANDEZ, D. E., C. B. NEMEROFF & A. J. PRANGE, JR. 1982. Dev. Brain Res. **3:** 497–501.

23. DORSA, D. M., E. R. DEKLOET, E. MEZEY & D. DE WIED. 1979. Endocrinology **104:** 1663–1666.
24. WALLACE, M. M., R. J. BODNAR, D. BADILLO-MARTINEZ, G. NILAVER & E. A. ZIMMERMAN. 1981. Soc. Neurosci. Abst. **7.** p. 167.
25. MARTIN, G. E., C. B. BACINO & N. L. PAPP. 1980. Peptides **1:** 333–339.
26. KALIVAS, P., C. B. NEMEROFF & A. J. PRANGE, JR. 1982. Ann. N.Y. Acad. Sci. **400:** 307–318. This volume.
27. NEMEROFF, C. B., G. BISSETTE, P. J. MANBERG, A. J. OSBAHR III, G. R. BREESE & A. J. PRANGE, JR. 1980. Brain Res. **195:** 69–84.
28. BISSETTE, G., D. LUTTINGER, C. B. NEMEROFF & A. J. PRANGE, JR. 1981. Soc. Neurosci. Abst. **8,** p. 584.
29. CHANDRA, A., H. C. CHOU & M. T. LIN. 1981. Neuropharmacology **20:** 715–718.
30. TAUBE, D., M. COHN & M. L. COHN. 1980. Fed Proc. **39:** Abst. No. 4252.
31. BROWN, M., Y. TACHE, J. RIVIER & Q. PITTMAN. 1980. Proc. NINCDS Workshop. In press.
32. STITT, J. T. 1978. *In* Current Studies of the Hypothalamic Function. W. L. Veale & K. Lederis, Eds. Karger. Basel. vol. 2: 44–54.
33. YEHUDA, S. & R. J. WURTMAN. 1972. Nature **240:** 477–478.
34. YEHUDA, S. & R. J. WURTMAN. 1975. Eur. J. Pharmacol. **30:** 154–158.
35. MYERS, R. D. 1974. Handbook of Drug and Chemical Stimulation of the Brain. Reinhold. New York.

DISCUSSION OF THE PAPER

D. KAHN (*Columbia University, New York*): I am under the impression that the dopamine-blocking agents such as haloperidol are themselves poikilothermic agents and I wondered if that is correct? Did you look at neurotensin's ability to cause hypothermia or poikilothermia in an animal that had been pretreated with a dopamine-blocking agent?

G. BISSETTE (*University of North Carolina School of Medicine, Chapel Hill*): We have only looked at the interaction of NT and haloperidol at 4°C in rats and potentiation of hypothermia was observed. Higher environmental temperatures have not been studied.

R. ANDRADE (*Yale University, New Haven, CT*): I would like to know whether you ever tried reversing the effects of neurotensin with apomorphine or replacing the endogenous dopamine that you destroyed with the 6-hydroxydopamine with apomorphine?

BISSETTE: Yes, we have tried apormophine. The problem is that apomorphine is a hypothermic agent in its own right; it does not block neurotensin-induced hypothermia.

C. FERRIS (*University of Massachusetts, Worcester*): Where do you think the neurotensin is acting and how do you think it is inducing the hypothermia?

BISSETTE: There is some indication that neurotensin is decreasing metabolic activity in general. The data shown by Dr. Luttinger on the disposition of ^{3}H-pentobarbital show that the metabolism of this compound is slowed by i.c. neurotensin. Other workers have suggested that the hypothermia seen after neurotensin is not due to an increase in heat loss through the skin, but, rather to a decrease in heat production.

FERRIS: Does the hypothermic response last for 2 hours?

BISSETTE: Yes.

FERRIS: How do you explain that? The neurotensin must certainly be gone by then.

BISSETTE: This is not uncommon in neuropeptide research. The peptide probably binds to its receptor setting off a train of events that persists long after the peptide is degraded.

R. J. BODNAR (*Columbia University, New York*): Have you looked at the possible effect of NT on brain temperatures?

BISSETTE: No we have not done that, and as far as I know, no other group has studied that question.

ANTINOCISPONSIVE EFFECTS OF NEUROTENSIN AND NEUROTENSIN-RELATED PEPTIDES

Bradley V. Clineschmidt, Gregory E. Martin,
and Daniel F. Veber

Merck Sharp & Dohme Research Laboratories
West Point, Pennsylvania 19486

INTRODUCTION

A naloxone-insensitive antinocisponsive action[1,2] and a decrease in body temperature[3] are two prominent effects in rodents of neurotensin (NT) given by the intracerebroventricular or intracisternal (i.c.) route of administration. Various experimental findings described herein indicate that these two actions are separable and independent, that is, the antinocisponsive effect of NT does not result secondarily as a consequence of the substance's hypothermic action. This report also summarizes previous[4-6] and recently obtained results dealing with the sites in the central nervous system responsible for the hypothermic and antinocisponsive effects of NT. Finally, we describe results on the initial biological testing of recently synthesized cyclic octapeptides related structurally to NT. As shown below along with the equivalent cyclic form of NT^{6-13}, the other two substances have a similar make-up, with D-Lysine in place of L-Lysine at position 6. They differ only in the amino acids at positions 8 and 9.

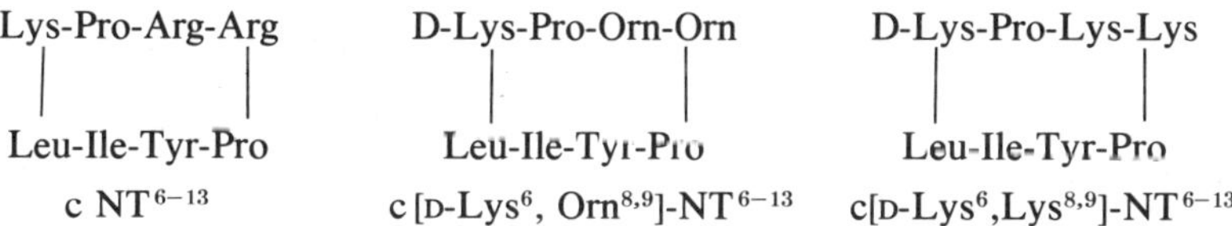

To the extent examined so far, $c[\text{D-Lys}^6,\text{Orn}^{8,9}]\text{-NT}^{6-13}$ and $c[\text{D-Lys}^6,\text{Lys}^{8,9}]\text{-}NT^{6-13}$ have exerted NT-like actions both *in vivo* (i.c. injection) and *in vitro*. However, neither analog administered intravenously was superior to NT itself with regard to producing hypothermia or antinocisponsive effects.

METHODS

Intracisternal and Intravenous Administration

Sterile saline (0.9%) or a peptide dissolved in saline was injected intracisternally (i.c.) or intravenously (i.v.) in mice (female Carworth CF_1) weighing 18–22 g or rats (female Charles River CRCD Cobs) weighing 60–80 g. The volume for i.c. injection was 5 μl in mice and 10 μl in rats, while 0.2 ml was used in both species for i.v. administration.

283

Antinocisponsive Tests

Mice were evaluated for responsiveness to noxious stimuli, using the well-known methods of the hot-plate (60°C) and acetic-acid-induced writhing. These methods were the same as described previously in detail,[1,2] except that in some studies the dose administered i.p. of acetic acid was 0.2 ml per 20 g of body weight of a 0.6% (v/v) solution instead of the previously employed volume of 0.4 ml/20 g. For studies with rats, we used the familiar hot-plate (55°C) and tail-flick procedures.[2] The endpoint for the hot-plate test in both rats and mice was licking of the front paws, hind paws, or both. The intensity of the light focused on the rat's tail was adjusted in preliminary experiments to give a reaction (flick of the tail) time of about 4 seconds in control animals.

Body Temperature

Mice were grasped by the skin at the back of the neck and held upside down while the thermistor probe was inserted 4 cm into the rectum and during the time (about 5 seconds) required to obtain a stable reading of temperature. Room temperature varied day-to-day from 21–23°C.

Data Analysis

Dunnett's Test[7] was used for comparison of several treatment groups with a control group and the Least Significant Difference Test[8] for comparison of the means of any two treatment groups.

Intracerebral and Intrathecal Administration

A single guide cannula (24-gauge thin wall or 26-gauge regular wall) was implanted using standard stereotaxic procedures in each of 67 male Sprague-Dawley rats (Blue Spruce Farms, Altamont, NY) weighing 320–400 g. Using coordinates described previously,[5] the cannulae were positioned above sites in the preoptic anterior hypothalamus (POAH) or the periaqueductal gray (PAG) region. In 50 additional rats, a catheter (PE-10 tubing) was implanted within the spinal subarachnoid space as described by Yaksh and Rudy.[9] The tip of the cannula was positioned to terminate at the level of the lumbar enlargement. Substances injected intracerebrally or intrathecally were dissolved in an artificial cerebrospinal fluid (CSF.)[10] The volume injected intracerebrally was 0.5 μl given over a period of 30 seconds through a 30- or 34-gauge injector needle inserted *via* the guide cannula. The volume of CSF-containing drug injected into the subarachnoid space was 15 μl administered over a period of 3–4 seconds. This was followed by a 10 μl injection of artificial CSF to flush the cannula. The sites of the intracerebral microinjections were determined using standard histological procedures[11] with the atlas of Konig and Klippel[12] for reference.

Body Temperature and Hot-Plate Reaction Time

Each rat served as its own control for studies where NT was delivered by intracerebral microinjection, whereas a separate group of animals treated with the CSF vehicle served as the control in experiments employing delivery by the intrathecal route. For the intracerebral microinjection studies, baseline (control) temperature was taken as the average of three determinations made at 60 minutes, 30 minutes, and just before the injection. A thermistor probe was inserted 6 cm beyond the anus for 15 seconds to record rectal temperature. Ambient temperature varied between 20–23°C. A single baseline determination of reaction time upon exposure to the hot plate (55°C) was taken just before the intracerebral microinjection. The criterion in these studies was latency to lick a hind paw. Reaction time and body temperature were recorded again at 30 and 60 minutes following the injection. The protocol was similar for studies employing intrathecal administration, except the rats were tested at only one time point, 60 minutes, after injection. Reaction time was determined immediately after body temperature in all experiments.

Data Analysis

At either time, 30 or 60 minutes, following intracerebral microinjection of NT, a hypothermic effect was arbitrarily defined as a decrease in temperature of 0.8°C or greater. An antinocisponsive effect was arbitrarily defined as being an increase in latency equal to or more than 30% of the maximum possible effect (MPE), using a cut-off time of 30 seconds in the hot-plate test. MPE was calculated as follows:

$$\text{MPE} = \frac{\text{latency postinjection-baseline latency}}{\text{30 seconds-baseline latency}} \times 100.$$

The possible correspondence between sites producing hypothermia and those eliciting an antinocisponsive effect was examined with the Spearman Rank Correlation Test.[13] Data from studies with intrathecal injection were analyzed with Dunnett's Test.[7]

Rat Fundus

Stomachs were excised from Sprague-Dawley rats (King Animal Laboratories, Oregon, Wisconsin) of either sex weighing 200–250 g, and a strip of the fundus was prepared as described by Vane.[14] The tissue was mounted in a 10-ml organ bath containing Krebs-Henseleit solution (NaCl, 7.8; KCl, 0.3; $CaCl_2$, 0.42; NaH_2PO_4, 0.165; $NaHCO_3$, 1.37; dextrose, 1.4 g/liter) maintained at 30°C and continuously bubbled with 95% O_2 and 5% CO_2. The strip of fundus was placed under 1 g of tension, with repeated adjustment of the tension back to 1 g during a period of 30–45 minutes of equilibration. The tissue was then exposed to NT, 2.5 nM, and the contraction recorded on a Grass polygraph. Tissues failing to respond with a brisk contraction were discarded. Following exposure to NT, the tissue was washed with fresh solution every 5 minutes for 30 minutes. The concentration–response studies were then carried out.

Four tissues were used for the comparison of NT with another NT-related peptide. A cross-over design was employed, that is, the concentration–response to NT was determined first with two tissues, whereas the other two tissues were tested first with the NT-related peptide. The two tissues tested first with NT were subsequently tested with the NT-related peptide and vice versa. The range of concentrations examined was 0.02 to 312.5 nM, proceeding from the lowest to the highest concentration. Following the maximum contractile response at each concentration, the tissue was washed repeatedly for 15 minutes before proceeding to test the next higher concentration of peptide. After completion of the concentration–response studies with one peptide, at least 30 minutes elapsed before beginning studies with the second substance.

For each tissue, the contraction induced by the NT-related peptide was expressed as a per cent of the maximum contraction obtained with NT in that same tissue. The results from the four tissues were then combined and normalized. EC_{50} values were estimated graphically; the EC_{50} value for the NT-related peptide was expressed as that concentration producing 50% of the maximal contraction caused by NT.

Preliminary studies were conducted with the rat fundus to ascertain if the contractile effect of NT was "direct" as opposed to being secondary to the release of another substance. The response to NT, 2.5 nM, after 5 minutes exposure to a possible antagonist was compared to the average of two control responses obtained with NT, one obtained before adding the antagonist to the bath and the other after washing the antagonist out of the tissue bath. The procedure outlined above was changed slightly in the cases of atropine and clozapine. The exposure time was 15 rather than 5 minutes, allowing enough time for the modest contraction induced by these two antagonists to subside before challenging the tissue with NT. The average % changes in the response to NT for four tissues (unless otherwise noted) in the presence of the potential antagonist were as follows: saline (control) = +5.2 (N = 8); atropine, 0.1 μM = −19.1, 1 μM = −6.4; clozapine, 0.1 μM = −18.8, 1 μM = +12.9; mecamylamine, 1 μM = −7.0; chlorpheniramine, 1 μM = −1.3; Pyribenzamine, 1 μM = +13.1; cimetidine, 1 μM = +14.2; methysergide, 1 μM = +1.6; methergoline, 1 μM = +1.2; propranolol, 1 μM = −1.4; HEAT (an α-adrenoceptor blocker), 1 μM = −1.5; theophylline, 1 μM = −2.8; indomethacin, 1 μM = −9.3; tetrodotoxin, 1 μM = −19.7 (N = 8); naloxone, 1 μM = +8.4; haloperidol, 1 μM = −9.9; desipramine, 1 μM = +5.0; baclofen, 1 μM = −22.

RESULTS

Antinocisponsive Action of Neurotensin

Intracisternal Injection

The results from dose–response studies in mice and rats tested on the hot plate are summarized in TABLE 1. With either species, a significant increase in reaction time occurred at a dose of 15 pmoles/animal administered i.c. At 100 times this dose, that is, 1,500 pmoles/animal given i.v., NT did not change the reaction time (TABLE 1). NT administered i.c. at doses of 15 and 1,500 pmoles/rat was without effect on latency to tail-flick (TABLE 1). Since NT given to mice in doses as low as 2 pmoles i.c. will reduce body temperature (see below), it was of interest to determine if a decrease in body temperature *per se* was a sufficient condition to increase hot-plate reaction time. As shown by the results in TABLE 2, this was not the case. Mice treated with D,L-5-hydroxytryptophan, oxotremorine, or chlorpromazine had a markedly lowered body temperature, but their response times in the hot-plate test were not different compared to vehicle-treated animals.

NT given i.c. also produced an antinocisponsive effect in the writhing test. The dose–response data are given in TABLE 3. The minimum dose, 0.15 pmoles/mouse, shown to be effective as an inhibitor of writhing was 100 times less than

TABLE 1

ANTINOCISPONSIVE ACTION OF NEUROTENSIN IN THE HOT-PLATE AND TAIL-FLICK TESTS

Treatment* (pmoles/animal)	Reaction Time (Seconds) †		Tail-Flick
	Hot Plate		
	Mouse ‡	Rat §	Rat ¶
I. Intracisternal			
Saline ‖	6.0 ± 1.6	8.5 ± 3.0	4.0 ± 0.5
Neurotensin (0.015)	6.3 ± 2.0	—	—
(0.15)	6.4 ± 2.3	10.7 ± 3.6	—
(1.5)	8.9 ± 5.9	—	—
(15)	8.8 ± 3.2 #	12.3 ± 6.0 #	4.0 ± 0.5
(150)	12.3 ± 8.2 #		—
(1,500)	13.0 ± 5.4 #	13.9 ± 6.1 #	4.6 ± 0.8
II. Intravenous			
Saline (200 µl)	5.1 ± 1.7	6.8 ± 1.9	—
Neurotensin (1,500)	4.8 ± 1.2	8.1 ± 2.2	—

*Administered 30 minutes before testing.

† All values are mean ± SD.

‡ N = 20 mice/treatment group, except for the intracisternal saline group, which was comprised of 40 animals.

§ N = 36 (saline) or 37 rats per group for the intracisternal treatments, whereas N = 10 rats/group for the intravenous injection of saline or neurotensin.

¶ N = 7 rats for all treatment groups.

‖ The volume was 5 µl in mice and 10 µl in rats.

# p < 0.05 (Dunnett's test).

TABLE 2

EFFECTS OF D,L-5-HYDROXYTRYPTOPHAN, OXOTREMORINE, AND CHLORPROMAZINE ON BODY TEMPERATURE AND REACTION TIME IN THE HOT-PLATE TEST

Treatment*	Body Temperature (°C) †	Reaction Time (Seconds) †
Vehicle	37.5 ± 0.7	5.0 ± 1.4
D,L-5-Hydroxytryptophan (300 mg/kg i.p.)	‡ 32.7 ± 0.9	4.7 ± 1.5
Oxotremorine (0.1 mg/kg s.c.)	‡ 32.0 ± 1.6	5.4 ± 1.9
Chlorpromazine (4 mg/kg i.p.)	‡ 34.3 ± 1.5	5.1 ± 1.3

*The drugs (10 mice/treatment group) were injected 30 minutes before determination of temperature; the animals were tested on the hot plate immediately after recording their temperatures.

† All values are mean ± SD.

‡ p < 0.05 (Dunnett's test).

TABLE 3

ANTINOCISPONSIVE ACTION OF NEUROTENSIN IN THE WRITHING TEST

Treatment * (pmoles/mouse)	Writhes†/10 Minutes
I. Intracisternal	
Saline (5 µl)	25.3 ± 9.3 ‡
Neurotensin (0.015)	19.5 ± 12.8
(0.15)	10.4 ± 3.4 §
(1.5)	8.3 ± 4.0 §
(15)	5.0 ± 7.5 §
(150)	2.7 ± 4.0 §
(1,500)	2.1 ± 2.6 §
II. Intravenous	
Saline (200 µl)	28.4 ± 6.9
Neurotensin (1,500)	35.5 ± 7.3

*Administered 25 minutes before an i.p. injection of acetic acid (0.4 ml/20 g of body weight; 0.6% (v/v) solution).

†Writhes were counted for 10 minutes beginning 5 minutes after injecting acetic acid.

‡All values are mean ± SD for 8 mice/treatment group, except for the intracisternal saline group where the number of animals was 16.

§ p < 0.05 (Dunnett's test).

the minimum effective dose in the hot-plate procedure. NT, 1,500 pmoles/mouse injected i.v., did not inhibit writhing induced by acetic acid (TABLE 3).

The duration of the antinocisponsive action of NT given i.c. in mice was previously demonstrated to be approximately 2 hours following doses of 150 or 1,500 pmoles.[2] Previous studies also showed that the antinocisponsive effect of NT was not antagonized by anticholinergic drugs, serotonin antagonists, antihistaminics, α- or β-adrenoceptor blockers, antagonists of dopamine or an inhibitor of prostaglandin synthesis (indomethacin).[2]

Lack of Interaction with Naloxone

Treatment with naloxone, 1 mg/kg s.c., did not affect the antinocisponsive action of NT (TABLE 4). This was true in both rats and mice, using the hot-plate test. Note that both a "low" and "high" dose of NT were examined in mice. Naloxone was likewise ineffective in offsetting NT in the writhing test, regardless of whether the dose of NT was 1.5 or 15 pmoles/mouse (TABLE 4).

Intracerebral Microinjection of NT

Preoptic Anterior Hypothalamus (POAH)

Of all regions examined previously, the medial POAH demonstrated the highest density of sites where microinjection of 1.5 nmoles of NT caused a decrease in body temperature.[5] Illustrated in FIGURE 1 are 14 sites where NT, 1.5 nmoles, elicited a decrease in body temperature ($\geq$0.8°C). The hot-plate response of these

14 rats was unaffected by the microinjection of NT, as their change in reaction time was only 0.2 ± 4.1 seconds ($\overline{X} \pm$ SD). The correlation coefficient (r_s) between the two parameters was $+0.4$ ($p > 0.05$). Moreover, none of the 14 animals reached the criterion of demonstrating an antinocisponsive effect ($\geq 30\%$ of MPE), as defined in METHODS.

Periaqueductal Gray Region (PAG)

An increase in reaction time following exposure to the hot plate was observed in 20 of 53 rats receiving a microinjection of 1.5 nmoles of NT in mesencephalic sites in the region of the PAG (FIG. 2). A decrease in rectal temperature also occurred in 14 of these 20 rats. There was, however, no significant correlation ($r_s = -0.05$) between these two parameters, analyzing the results from all 20 animals exhibiting an antinocisponsive effect. Also depicted in FIGURE 2 are 7 sites at which NT elicited hypothermia without causing an accompanying increase in latency to respond in the hot-plate test.

TABLE 4

FAILURE OF NALOXONE TO ANTAGONIZE THE ANTINOCISPONSIVE
EFFECT OF NEUROTENSIN IN MICE AND RATS

Treatment I * (pmoles/animal)	Treatment II † (mg/kg s.c.)	Hot-Plate Reaction Time (Seconds)		Acetic-Acid- Induced Writhing (writhes/10 min)
		Mouse	Rat	
Saline	Vehicle	4.3 ± 1.2 ‡	—	—
Saline	Naloxone (1)	4.3 ± 1.2	—	—
Neurotensin (15)	Vehicle	6.4 ± 1.9 ‖	—	—
Neurotensin (15)	Naloxone (1)	6.7 ± 2.7 ‖	—	—
Saline	Vehicle	4.4 ± 1.1 ‡	—	—
Saline	Naloxone (1)	4.0 ± 1.2	—	—
Neurotensin (60)	Vehicle	7.9 ± 2.7 ‖	—	—
Neurotensin (60)	Naloxone (1)	7.2 ± 3.7 ‖	—	—
Saline	Vehicle	—	6.9 ± 1.7 §	—
Saline	Naloxone (1)	—	8.7 ± 3.4	—
Neurotensin (15)	Vehicle	—	13.1 ± 5.5 ‖	—
Neurotensin (15)	Naloxone (1)	—	13.7 ± 4.9 ‖	—
Saline	Vehicle	—	—	44.4 ± 11.4 ¶
Saline	Naloxone (1)	—	—	48.3 ± 10.8
Neurotensin (1.5)	Vehicle	—	—	9.6 ± 8.9 ‖
Neurotensin (1.5)	Naloxone (1)	—	—	5.8 ± 3.1 ‖
Neurotensin (15)	Vehicle	—	—	10.4 ± 9.8 ‖
Neurotensin (15)	Naloxone (1)	—	—	9.6 ± 7.9 ‖

* Administered i.c. thirty minutes before test.
† Twenty minutes before test.
‡ Mean $\pm$ SD with 20 mice in each treatment group.
§ Mean $\pm$ SD with 20 rats in each treatment group.
¶ Mean $\pm$ SD with 8 mice in each treatment group.
‖ $p < 0.05$ (Dunnett's test).

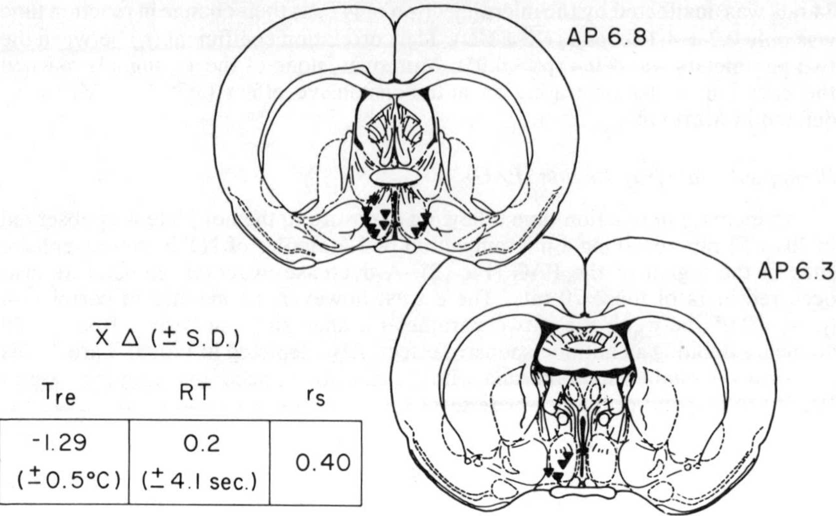

FIGURE 1. Coronal sections of the rat brain depicting the location of 14 sites where injection of 1.5 nmoles of NT produced a decrease of rectal temperature (T_{re}) of 0.8°C or more. The inset shows the mean changes in T_{re} and hot-plate reaction time (RT) for these 14 rats, along with the Spearman's rank order correlation coefficient (r_s) for these two parameters.

Intrathecal Administration of NT

The effects of NT and morphine were quite distinct after injection intrathecally in rats (TABLE 5). At doses ranging from 0.6 to 48 nmoles, NT did not significantly change reaction time in the hot-plate test. NT, at the two highest dose levels (24 and 48 nmoles), did cause a significant decrease in rectal temperature. Morphine, on the other hand, significantly increased response latency at doses of 120 nmoles or larger, but, depending upon the dose, rectal temperature was either elevated or unchanged following doses from 3.8 to 240 nmoles (TABLE 5).

NT-related Cyclic Octapeptides

Since NT itself administered systemically (i.v.) in large doses does not mimic the CNS effects produced by i.c. administration of the peptide, we have synthesized and tested related peptides in the hope of finding a systemically active analog. The most interesting compounds examined to date are c[D-Lys^6,Orn^{8,9}]-NT$^{6-13}$ and c[D-Lys^6,Lys^{8,9}]-NT$^{6-13}$.

Rat Fundus

As illustrated in FIGURES 3 and 4, both cyclic octapeptides caused contraction of fundic strips. The compound with lysine at positions 8 and 9 was more potent than the one with ornithine at 8 and 9 (compare FIGS. 3 and 4).

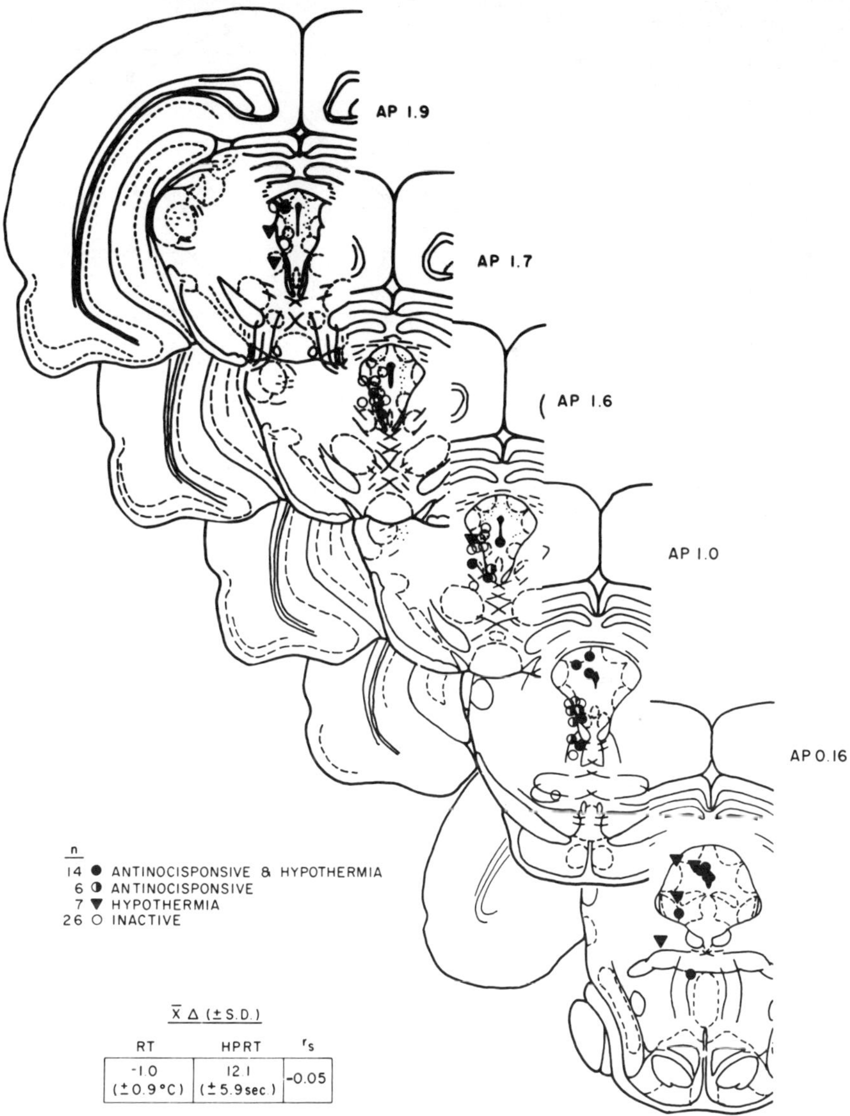

FIGURE 2. Coronal sections of rat brain showing 53 sites injected with NT, 1.5 nmoles. At 7 sites (▼), only hypothermia ($-0.8°C$ or more) was obtained, whereas at 6 sites (◑), only an antinocisponsive action ($\geq 30\%$ of maximum possible) in the hot-plate test occurred. Both effects followed injection of NT at 14 sites (●), and neither effect was found at 26 sites (○). Shown in the inset are the mean changes in hot-plate reaction time (HPRT) and rectal temperature (RT) at the time (30 or 60 minutes) of peak antinocisponsive effect in 20 animals (◑ + ●). r_s is Spearman's rank order correlation coefficient.

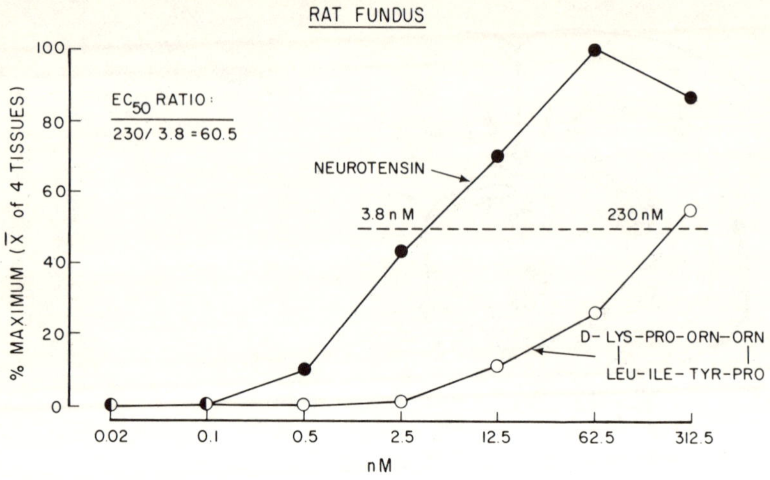

FIGURE 3.   Concentration–response curves for contractile effects of neurotensin and c[D-Lys^6,Orn^{8,9}]-NT$^{6-13}$ on the isolated rat fundus.

In Vivo

Like NT, both cyclic octapeptides produced hypothermia and increased the hot-plate latency following their injection i.c. in mice (FIGS. 5 and 6). c[D-Lys^6,Orn^{8,9}]-NT$^{6-13}$ was less potent than NT, with the dose–response lines for both reaction time and body temperature lying to the right of the lines for NT (FIG. 5). It may be noted in FIGURE 5, that the two dose–response lines for reaction

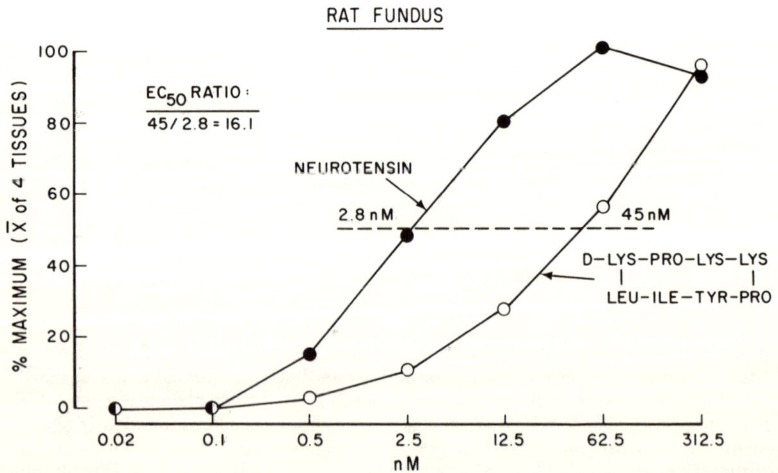

FIGURE 4.   Concentration–response curves for neurotensin and c[D-Lys^6,Lys^{8,9}]-NT$^{6-13}$ on the fundus.

time lie closer together than those for body temperature. Results obtained with $c[\text{D-Lys}^6,\text{Lys}^{8,9}]\text{-NT}^{6-13}$ were quite similar to those for $c[\text{D-Lys}^6,\text{Orn}^{8,9}]\text{-NT}^{6-13}$. As illustrated in FIGURE 6, $c[\text{D-Lys}^6,\text{Lys}^{8,9}]\text{-NT}^{6-13}$ was less potent than NT and the dose–response lines for reaction time were closer together than those for body temperature.

Data from the linear portions of the dose–response lines shown in FIGURES 5 and 6 were analyzed by regression analysis, using three dose levels of each substance. These results are summarized in TABLE 6, along with the findings obtained *in vitro* with isolated fundic strips. Because the dose–response relationship for NT

TABLE 5

CHANGES IN RECTAL TEMPERATURE AND HOT-PLATE RESPONSE
TIME ONE HOUR AFTER THE INJECTION OF NEUROTENSIN OR MORPHINE
INTO THE SPINAL SUBARACHNOID SPACE IN RATS

Treatment (nmoles)	Δ Rectal Temperature § (°C)	Δ Response Time § (sec)
Neurotensin *		
(48)	-1.1 ± 1.1 ¶	3.2 ± 2.3
(24)	-1.1 ± 0.7 ¶	1.7 ± 3.0
(12)	-0.8 ± 0.9	3.1 ± 5.1
(6)	-0.6 ± 1.3	0.6 ± 2.9
(3)	-0.1 ± 0.5	4.6 ± 10.7
(0.6)	0.2 ± 0.7	2.5 ± 6.2
Morphine †		
(240)	0.9 ± 0.8	10.2 ± 8.1 ¶
(120)	1.0 ± 0.9	14.3 ± 9.1 ¶
(60)	1.3 ± 0.4 ¶	8.0 ± 6.4
(30)	1.3 ± 0.8 ¶	5.4 ± 5.3
(15)	1.0 ± 0.3	-1.9 ± 3.5
(7.5)	0.9 ± 0.7	1.0 ± 8.1
(3.8)	0.7 ± 0.3	
Artificial CSF ‡ (25 μl)	0.1 ± 0.1	0.2 ± 1.9

* N = 8/treatment group.
† N = 5–8/treatment group.
‡ N = 8 (response time) or 14 (rectal temperature).
§ All values are mean ± SD.
¶ Significantly different from the CSF control group (p < 0.05, Dunnett's test).

and the cyclic octapeptides was shallow, the 95% confidence limits for the estimated ED values in TABLE 6 were rather wide. This makes for some equivocation in interpretation. Nevertheless, it would appear that $c[\text{D-Lys}^6,\text{Lys}^{8,9}]\text{-NT}^{6-13}$ and $c[\text{D-Lys}^6,\text{Orn}^{8,9}]\text{-NT}^{6-13}$ were about equipotent *in vivo*, despite the greater potency of the former compound on the rat fundus *in vitro* (TABLE 6).

Comparison with Related Compounds

Data obtained with substances related to $c[\text{D-Lys}^6,\text{Orn}^{8,9}]\text{-NT}^{6-13}$ and $c[\text{D-Lys}^6,\text{Lys}^{8,9}]\text{-NT}^{6-13}$ are summarized in TABLE 7. The linear octapeptide, NT^{6-13},

TABLE 6

COMPARISON *in Vitro* AND *in Vivo* OF TWO CYCLIC OCTAPEPTIDES WITH NT[1-13]

Test Substance	Rat Fundus		Body Temperature		Hot Plate Reaction Time	
	EC_{50} (nm)	Ratio*	$ED_{34.5°C}$ † (pmoles/mouse)	Ratio*	ED_{13} seconds † (pmoles/mouse)	Ratio*
c[D-Lys^6,Orn^{8,9}]-NT$^{6-13}$	230		489(257–929) §		849(287–2,513)	
NT[1-13]	3.8	60.5	28.4(13.1–61.9)	17.2	193(55.0–677)	4.40
c[D-Lys^6,Lys^{8,9}]-NT$^{6-13}$	45		272(174–427)		352(157–789)	
NT[1-13]	2.8	16.1	9.75(4.21–22.6)	27.9	110(36.7–331)	3.19

* Cyclic octapeptide/NT[1-13].
† Dose (i.c.) to reduce body temperature to 34.5°C.
‡ Dose (i.c.) to increase reaction time to 13 seconds.
§ The 95% confidence limits are given in parentheses.

TABLE 7

CYCLIC OCTAPEPTIDES WITH NT-LIKE ACTIVITY *in Vitro* AND *in Vivo*

NT-Related Peptide	Rat Fundus EC$_{50}$ (nM)		Hot Plate Reaction Time (sec)‡ (pmoles/mouse i.c.)			Body Temperature (°C)‡ (pmoles/mouse i.c.)		
	Peptide	NT[1-13]	Saline	20	2,000	Saline	20	2,000
NT[6-13]	0.35	2.2	5.5 ± 1.3†	7.0 ± 2.0	§11.3 ± 2.9	36.1 ± 1.8†	§33.7 ± 0.8	§32.0 ± 1.0
cNT[6-13]	>312.5(2.4)*	4.9	5.9 ± 2.1	7.5 ± 2.7	9.4 ± 5.5	36.5 ± 0.9	36.6 ± 0.8	§34.7 ± 0.7
c[D-Lys[6]]-NT[6-13]	>312.5(45.3)*	3.8	5.9 ± 2.1	6.5 ± 1.9	§12.2 ± 5.7	36.5 ± 0.9	36.2 ± 1.7	§33.1 ± 0.8
	>312.5(32.3)*	1.6						
c[D-Lys[6],Orn[ε,9]]-NT[6-13]	230	3.8	5.9 ± 1.7	9.8 ± 3.7	§15.7 ± 4.6	36.8 ± 1.6	§34.7 ± 1.5	§33.3 ± 0.5
c[D-Lys[6],Lys[ε,9]]-NT[6-13]	45	2.8	5.3 ± 1.9	6.4 ± 1.9	§17.3 ± 8.0	37.2 ± 1.2	36.3 ± 0.8	§31.3 ± 0.7

* Contraction at 312.5 nM as % of maximum response to NT[1-13].
† All values are mean ± SD (N = 10/group).
‡ The mice were tested at 30 minutes following the i.c. injection of saline or one of the peptides.
§ $p < 0.05$ (Dunnett's test).

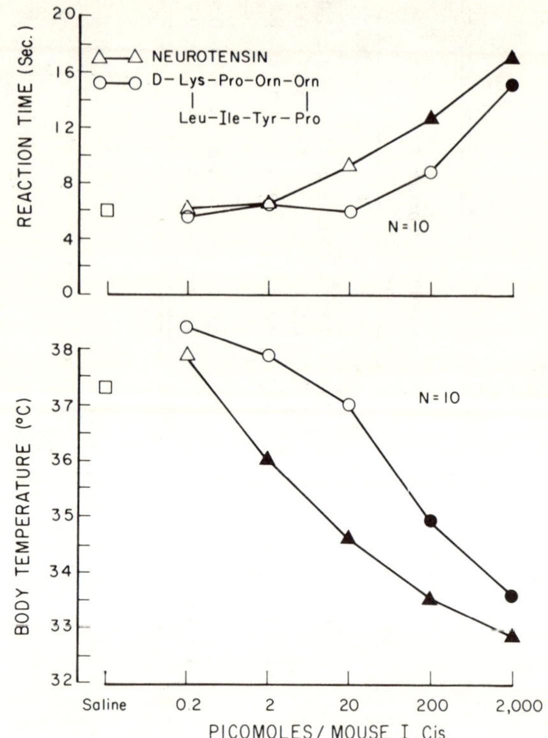

FIGURE 5. Dose-response relationship for the antinocisponsive (hot-plate test) and hypothermic actions of neurotensin and c [D-Lys⁶,Orn⁸,⁹]-NT⁶⁻¹³ in mice at 30 minutes after i.c. injection. A separate group of mice was used for each different dose of each peptide. Filled symbols indicate a significant difference compared to the saline-treated mice (Dunnett's test).

was very potent on the fundus, the EC_{50} being less than that obtained for NT itself. This fragment of NT was also effective *in vivo*, causing a decrease in body temperature at a dose of 20 pmoles i.c. and an increase in reaction time at 2,000 pmoles (TABLE 7). The cyclic octapeptide, cNT^{6-13}, was considerably weaker *in vitro* and *in vivo* than the linear equivalent, NT^{6-13} (TABLE 7). Substitution of D-lysine for L-lysine at position 6 of the cyclic octapeptide enhanced potency both *in vitro* and *in vivo*. For example, c [D-Lys⁶]-NT⁶⁻¹³ at 2,000 pmoles i.c. significantly increased reaction time, whereas cNT^{6-13} did not (TABLE 7). Potency was further increased by replacing the arginines at positions 8 and 9 with lysine or ornithine.

Systemic (i.v.) Administration

At a dose of 8 nmoles injected i.v. in mice 15 minutes before test, neither the cyclic octapeptides nor NT caused a significant change in body temperature (FIGURE 7). Reaction time in the hot-plate test and writhing elicited by acetic acid were likewise unaffected, as also shown in FIGURE 7.

Given i.v. at 80 nmoles/mouse, all three substances produced hypothermia, with c[D-Lys^6,Orn^{8,9}]-NT$^{6-13}$ and c[D-Lys^6,Lys^{8,9}]-NT$^{6-13}$ having a duration of action similar to that of NT (TABLE 8). Similar results were obtained in the hot-plate test, except for the longer duration of action exhibited by c[D-Lys^6,Orn^{8,9}]-NT$^{6-13}$ (TABLE 9). All three substances at 80 nmoles/mouse i.v. inhibited writhing induced by acetic acid, this inhibition being short lived in all cases (TABLE 10).

DISCUSSION

The potent antinocisponsive action of neurotensin (NT) in rodents described herein and first reported elsewhere[1,2] has been confirmed by others.[15,16] Besides the confirmatory studies using the hot-plate and writhing tests, Osbahr *et al.*[16] also examined the antinocisponsive effect of i.c. administered NT in mice using the tail-immersion procedure. Following exposure to the noxious stimulus, hot water, mice treated with NT (180 pmoles or more) showed a significant delay in the time to a tail-flick response. The tail-immersion test in mice would seem to be quite similar to the tail-flick procedure in rats, since both rely on heat applied to the tail and the endpoint in both procedures is a flick of the tail. However, NT at doses

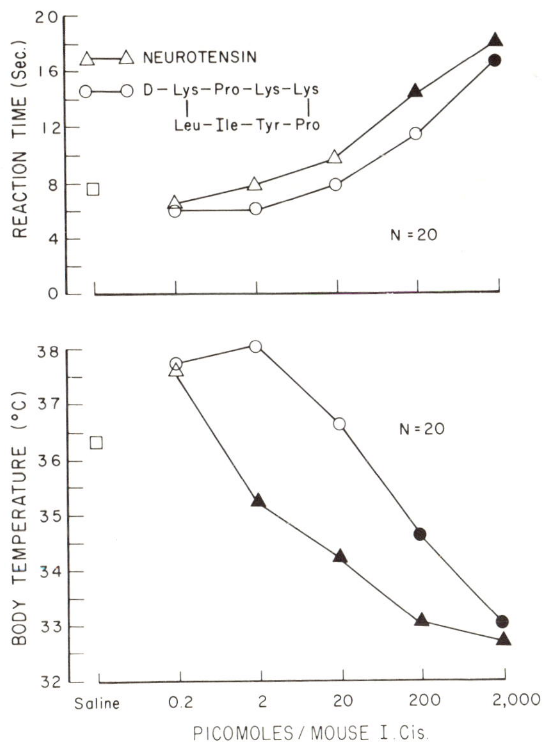

FIGURE 6. c[D-Lys^6,Lys^{8,9}]-NT$^{6-13}$ compared *in vivo* with neurotensin. See legend to FIGURE 5 for details.

of 15 or 1,500 pmoles i.c. was, in our hands, not antinocisponsive in the tail-flick test in rats. This apparent discrepancy is probably attributable to the different species or the methodological differences.

The original observation[1,2] that naloxone did not antagonize the antinocisponsive action of NT was recently confirmed.[15,16] This indicates that the mode of action of NT is independent of receptors for opiates. Besides being insensitive to naloxone, NT differs from opiate-like antinocisponsive agents in several other

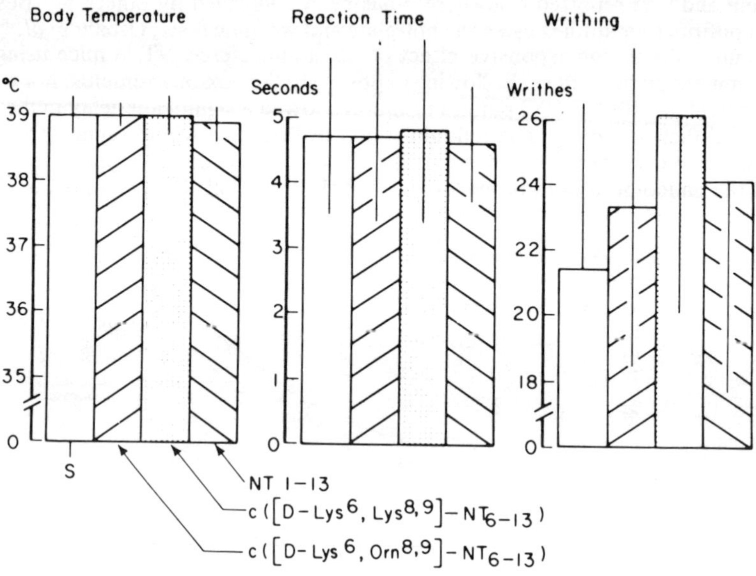

FIGURE 7.   Failure of neurotensin and the cyclic octapeptides at a dose of 8 nmoles given i.v. to exert hypothermic or antinocisponsive actions in mice. The top of each bar shows the mean measurement, with the vertical lines indicating the standard deviation. In each of the three tests, four groups of 10 mice were treated with either saline or one of the three peptides. Data were collected at 15 minutes after intravenous injection of one of the four treatments. Neither neurotensin (NT[1−13]) nor cyclic octapeptide significantly affected body temperature, reaction time (hot-plate test), or the number of writhes evoked by acetic acid.

ways. For example, the depression of motor activity in mice by NT,[2] rather than stimulation, distinguishes this peptide from the opiates and enkephalins. Moreover, NT did not exhibit an antinocisponsive effect in the tail-flick test in rats, whereas it is well known that opiates do so. Finally, the effects of NT and morphine administered intrathecally were dissimilar. NT reduced rectal temperature in rats, whereas morphine increased it. Morphine increased response latency in the hot-plate test, while NT in doses as high as 48 nmoles had no significant effect.

A relationship between the antinocisponsive and hypothermic effects of NT was explored, since some association between these two actions seemed possible,

TABLE 8

HYPOTHERMIA PRODUCED IN MICE FOLLOWING I.V. ADMINISTRATION
OF CYCLIC OCTAPEPTIDES OR NT$^{1-13}$

| Treatment | Body Temperature (°C) | | |
| | Time (minutes) after Injection | | |
(nmoles/mouse i.v.)	15	30	60
I.*			
Saline (200 μl)	38.7 ± 0.3‡	39.0 ± 0.3	38.6 ± 0.4
c[D-Lys^6,Orn^{8,9}]-NT$^{6-13}$ (80)	37.1 ± 0.7 §	37.6 ± 1.1 §	38.2 ± 0.5
II.†			
Saline (200 μl)	38.6 ± 0.4	38.3 ± 0.4	37.9 ± 0.5
c[D-Lys^6,Lys^{8,9}]-NT$^{6-13}$ (80)	35.9 ± 1.1 §	36.6 ± 1.3 §	37.6 ± 0.5
III.†			
Saline (200 μl)	39.1 ± 0.1	38.8 ± 0.3	38.5 ± 0.6
NT$^{1-13}$	36.5 ± 0.9 §	37.7 ± 0.9 §	38.2 ± 0.5

*N = 10/group.
† N = 20/group.
‡ All values are mean ± SD.
§ p < 0.05 (Dunnett's test).

TABLE 9

INCREASE IN REACTION TIME (HOT-PLATE TEST) FOLLOWING
I.V. ADMINISTRATION OF CYCLIC OCTAPEPTIDES OR NT$^{1-13}$

| Treatment | Reaction Time (seconds) | | | |
| | Time (minutes) after Injection | | | |
(nmoles/mouse i.v.)	15	30	60	120
I*				
Saline (200 μl)	5.0 ± 1.3‡	4.8 ± 1.1	4.8 ± 1.1	4.7 ± 0.8
c[D-Lys^6,Orn^{8,9}]-NT$^{6-13}$ (80)	6.5 ± 2.5 §	6.1 ± 1.8 §	6.2 ± 2.1 §	5.1 ± 1.1
II†				
Saline (200 μl)	4.7 ± 0.7	4.5 ± 1.0	4.9 ± 1.7	—
c[D-Lys^6,Lys^{8,9}]-NT$^{6-13}$ (80)	6.3 ± 2.5 §	4.9 ± 1.7	5.4 ± 1.5	—
III†				
Saline (200 μl)	4.5 ± 1.0	5.0 ± 1.8	5.4 ± 1.2	—
NT$^{1-13}$ (80)	5.6 ± 1.3 §	5.5 ± 1.6	4.9 ± 1.1	—

*N = 30/group except at 120 minutes N = 10.
† N = 20/group.
‡ All values are mean ± SD.
§ p < 0.05 (Dunnett's test).

TABLE 10

INHIBITION OF WRITHING BY CYCLIC OCTAPEPTIDES
OR NT[1–13] ADMINISTERED INTRAVENOUSLY

Treatment (nmoles/mouse i.v.)	Writhes/10 Minutes		
	15–25 †	30–40 ‡	60–70 §
I. ¶			
Saline (200 μl)	28.4 ± 7.6 **	23.1 ± 6.9	26.9 ± 9.2
c[D-Lys^6,Orn^{8,9}]-NT$^{6-13}$ (80)	16.5 ± 7.4 ††	20.3 ± 7.8	24.0 ± 10.5
II. ‖			
Saline (200 μl)	27.6 ± 12.1	28.5 ± 7.8	20.8 ± 4.9
c[D-Lys^6,Lys^{8,9}]-NT$^{6-13}$ (80)	17.7 ± 8.1 ††	24.6 ± 8.0	25.3 ± 7.8
III. #			
Saline (200 μl)	24.0 ± 4.9	26.1 ± 8.4	—
NT$^{1-13}$ (80)	16.2 ± 9.7 ††	21.7 ± 8.3	—

* Administered at 15, 30, or 60 minutes before counting number of writhes/10 minutes.
† Acetic acid, 0.2 ml of 0.6% (v/v), was injected i.p. at 10 minutes following treatment.
‡ Acetic acid, 0.2 ml of 0.6% (v/v), was injected i.p. at 25 minutes following treatment.
§ Acetic acid, 0.2 ml of 0.6% (v/v), was injected i.p. at 55 minutes following treatment.
¶ N = 20/group.
‖ N = 10 (15 and 60 minutes) or 20 (30 minutes)/group.
N = 10/group.
** All values are mean ± SD.
†† p < 0.05 (Dunnett's test).

especially in the hot-plate test.[17] Several observations indicate, however, that these two effects of NT are independent of each other. A reduction in body temperature is not a sufficient condition *per se* to increase hot-plate reaction time in mice. Mice rendered hypothermic by administration of chlorpromazine, 5-hydroxytryptophan or oxotremorine exhibited an unaltered response latency in the hot-plate test. The microinjection studies in rats also point to the independent nature of these two actions of NT. Fourteen rats with a decrease of 0.8°C or more in rectal temperature caused by administration of NT into the POAH region showed, as a group, no significant change in their hot-plate reaction time. Furthermore, at the time of the individual peak decreases in temperature, no significant correlation (r_s = +0.40) was found between the magnitude of the changes in rectal temperature and reaction time. Results obtained following microinjection of NT into the region of the PAG are likewise relevant to the question of the separability of these two actions of NT. Active sites ($\geq$30% of the maximum possible increase in hot-plate latency) were identified in 20 rats, and 14 of these animals also had a decrease in rectal temperature of 0.8°C or greater. At the time of peak change in hot-plate reaction latency, the magnitude of the changes in response latency and temperature for the 20 rats were, however, not significantly correlated (r_s = −0.05). In the 14 animals showing both an increase in response time and a decrease in rectal temperature, the site of microinjection in 4 of these impinged on the aqueduct, raising the possibility of diffusion to other sites of action via the CSF. Finally, it should be noted that there were 7 sites in or around the PAG at which hypothermia occurred after injection of NT, without an accompanying increase in response latency. Although somewhat less clear than the findings with microinjection of NT into the POAH, the results obtained with NT given into the PAG region also suggest that the antinocisponsive and hypothermic effects of NT are not closely linked. The dose–response results generated with the two cyclic octapeptides compared with NT also support this suggestion. The dose–response lines in the hot-plate test obtained following i.c. injection of c[D-Lys^6,Orn^{8,9}]-NT$^{6-13}$ or NT in mice were closer together than the dose–response lines for the decrease in rectal temperature. The same relationship was found when c[D-Lys^6,Lys^{8,9}]-NT$^{6-13}$ was compared with NT. A straightforward interpretation of these data is that NT and the cyclic octapeptides act at distinct sites in the CNS to produce these two effects. As a final point, it is not readily apparent *a priori* why the hypothermia caused by NT should necessarily result in an inhibition of writhing induced by acetic acid. All of the foregoing considerations lead us to conclude that the antinocisponsive and hypothermic effects of NT are two distinct pharmacological actions of the peptide.

Like many other substances affecting body temperature via an action in the CNS, NT (1.5 nmoles) was active in a high percentage of sites following microinjection into the region of the POAH,[5] an area containing substantial amounts of the peptide.[18,19] However, NT also reduced rectal temperature after microinjection into several other sites in the rat's brain such as the PAG, ventral thalamus, the dorsomedial hypothalamus, and the diagonal band of Broca.[5] NT administered intrathecally in the rat (TABLE 5 and references 4 and 6) decreased rectal temperature, but only at doses greater than 12 nmoles, suggesting this is not a principal site of action.

The PAG is a popular site for microinjecting antinocisponsive substances, and Sullivan and Pert[20] have indeed reported that NT elicits an antinocisponsive effect in rats following injection into this area. In our studies, 53 mesencephalic sites in and around the PAG were explored. NT, 1.5 nmoles, applied at 20 of these caused an antinocisponsive effect in rats tested on the hot plate. Excluding the four sites in very close proximity to the aqueduct, 16 of 49, or 32%, of the remaining sites were positive, that is, response latency was increased to 30% or more of the maximum possible. In similar unpublished experiments with morphine, microinjection of 13.3 nmoles at comparable, but overall somewhat more caudal, PAG sites resulted in a positive hot-plate effect in 46% of the rats (17 of 37 sites). The region of the PAG ventral and lateral to the aqueduct is a favorable area with regard to demonstrating an antinocisponsive action for morphine,[21] and most (13) of the positive sites for NT were located here as well. If the PAG region is an important site for the antinocisponsive action of NT, as our results would indicate, this could be of physiological consequence. Antinociception follows electrical stimulation of the PAG,[22-24] an effect antagonized partially by naloxone. NT is a viable candidate for mediating the naloxone-resistant component, since NT has been shown immunohistochemically[19] to be present in the PAG. The lack of a potent and selective antagonist for NT precludes for the time being a direct test of this hypothesis.

c[D-Lys^6,Orn^{8,9}]-NT$^{6-13}$ and c[D-Lys^6,Lys^{8,9}]-NT$^{6-13}$ mimicked the contractile action of NT on the rat fundic strip and caused hypothermia and an increase in response latency (hot plate) in mice following i.c. administration. These cyclic octapeptides are, therefore, considered to act as agonists on smooth muscle and neuronal receptors for NT. The rat fundus was selected over other tissues because NT probably acts directly on specific receptors for NT in the muscle of the fundus,[25-27] as opposed, for example, to acting indirectly by releasing some other substance. Our data (see METHODS) support this view, since the contractile effect of NT on the fundus was not substantially blocked by a wide variety of pharmacological agents, including tetrodoxotin, anticholinergics, antihistaminics, antisero-toninergics, etc. The contractile effect of NT on the fundus was, however, modestly (-22%) antagonized by baclofen, a compound known to act as an antagonist of substance P.[28] This observation is possibly of interest, because in the isolated guinea-pig ileum the contractile effect of NT is mainly indirect and mediated at least partly by acetylcholine and substance P.[29–32] In contrast to the guinea-pig ileum,[31] rat stomach strips desensitized to substance P maintain their sensitivity and responsiveness to NT.[27] In view of this latter finding, the modest antagonism of NT by baclofen at 1 μM in fundic strips is probably attributable to a "nonspecific" inhibitory effect, a proposition that will be tested by examining other substances that contract the fundus, for example, 5-hydroxytryptamine.

The cyclic octapeptide having ornithines substituted for arginines at positions 8 and 9 was approximately one-fourth as potent on the rat fundus as the corresponding compound with lysine at positions 8 and 9. No difference in potency between these two compounds was demonstrable *in vivo*, perhaps because the slopes of the dose–response curves were so shallow. A difference in potency of only 4-fold would be difficult at best to show *in vivo*. Nevertheless, it is clear that c[D-Lys^6,Orn^{8,9}]-NT$^{6-13}$ and c[D-Lys^6,Lys^{8,9}]-NT$^{6-13}$ were both more potent than cNT$^{6-13}$ or c[D-Lys^6]-NT$^{6-13}$ *in vivo* as well as *in vitro*.

Studies of analogs of somatostatin have shown that analog design based on conformational constraint can be used to reduce the number of amino acids required for activity and to lock in the "bioactive" conformation. Cyclic analogs of somatostatin have been designed which show a high degree of metabolic stability and long duration of action.[33,34] In a similar manner, we hoped to limit the metabolism of neurotensin through conformational constraint and thereby allow its entry into the CNS after systemic injection. Despite a reduction in size, cyclization, substitution of D- for L-Lysine at position 6, and replacement of two arginines with ornithine or lysine, these analogs show little or no improvement of potency after i.v. administration compared to neurotensin. The high intrinsic potency of these analogs suggests a folded rather than extended conformation for the bioactive form of neurotensin.

ACKNOWLEDGMENT

We thank Drs. M. Christy, K. Shepard, S. Varga, and Mr. R. Strachan for supplying the synthetic peptides.

REFERENCES

1. CLINESCHMIDT, B. V. & J. C. McGUFFIN. 1977. Neurotensin administered intracisternally inhibits responsiveness of mice to noxious stimuli. Eur. J. Pharmacol. **46:** 395–396.
2. CLINESCHMIDT, B. V., J. C. McGUFFIN & P. B. BUNTING. 1979. Neurotensin: Antinocisponsive action in rodents. Eur. J. Pharmacol. **54:** 129–139.
3. NEMEROFF, C. B., G. BISSETTE, A. J. PRANGE, P. T. LOOSEN, T. S. BARLOW & M. A. LIPTON. 1977. Neurotensin: Central nervous system effects of a hypothalamic peptide. Brain Res. **128:** 485–486.
4. MARTIN, G. E. & T. NARUSE. 1982. Differences in the pharmacological actions of intrathecally administered neurotensin and morphine. Regulatory Peptides 3: 97–103.
5. MARTIN, G. E., C. B. BACINO & N. L. PAPP. 1980. Hypothermia elicited by the intracerebral microinjection of neurotensin. Peptides 1: 333–339.

6. MARTIN, G. E., T. NARUSE & N. L. PAPP. 1981. Antinociceptive and hypothermic actions of neurotensin administered centrally in the rat. Neuropeptides 1: 447–454.
7. DUNNETT, C. W. 1964. New tables for multiple comparisons with a control. Biometrics 20: 482–491.
8. SNEDECOR, G. W. & W. G. COCHRAN. 1967. Statistical Methods. 6th edit. The Iowa State University Press. Ames, Iowa. pp. 272.
9. YAKSH, T. L. & T. A. RUDY. 1976. Chronic catherization of the spinal subarachnoid space. Physiol. Behav. 17: 1031–1036.
10. MYERS, R. D. 1971. General laboratory procedures. *In* Methods in Psychobiology. R. D. Myers, Ed. Vol. 1: 27–65. Academic Press. London.
11. WOLF, G. 1971. Elementary histology for neuropsychologists. *In* Methods in Psychobiology. Vol. 1: 281–299. Academic Press. London.
12. KONIG, J. F. R. & R. A. KLIPPEL. 1963. The Rat Brain. Williams and Wilkins. Baltimore, Maryland.
13. SIEGEL, S. 1956. Nonparametric Statistics. McGraw-Hill. New York, N.Y. pp. 202–213.
14. VANE, J. R. 1957. A sensitive method for the assay of 5-hydroxytryptamine. Br. J. Pharmacol. Chemother. 12: 344–349.
15. NEMEROFF, C. B., A. J. OSBAHR, P. J. MANBERG, G. N. ERVIN & A. J. PRANGE. 1979. Alterations in nociception and body temperature after intracisternal administration of neurotensin, β-endorphin, other endogenous peptides, and morphine. Proc. Natl. Acad. Sci. USA 76: 5368–5371.
16. OSBAHR, A. J., C. B. NEMEROFF, D. LUTTINGER, G. A. MASON & A. PRANGE. 1981. Neurotensin-induced antinociception in mice: Antagonism by thyrotropin-releasing hormone. J. Pharmacol. Exp. Ther. 217: 645–651.
17. WINTER, C. A. & L. FLATAKER. 1953. The relation between skin temperature and the effect of morphine upon the response to thermal stimuli in the albino rat and dog. J. Pharmacol. Exp. Ther. 109: 183–188.
18. KOBAYASHI, R. M., M. BROWN & W. VALE. 1977. Regional distribution of neurotensin and somatostatin in rat brain. Brain Res. 126: 584–588.
19. UHL, G. R., R. R. GOODMAN & S. H. SNYDER. 1979. Neurotensin-containing cell bodies, fibers, and nerve terminals in the brain stem of the rat: Immunohistochemical mapping. Brain Res. 167: 77–91.
20. SULLIVAN, T. L. & A. PERT. 1981. Analgesic activity of nonopiate neuropeptides following injections into the rat periaqueductal gray matter. Soc. Neurosci. Abst. 7: 504.
21. YAKSH, T. L., J. C. YEUNG & T. A. RUDY. 1976. Systematic examination of the rat brain sites sensitive to the direct application of morphine: Observation of differential effects within the periaqueductal gray. Brain Res. 114: 83–103.
22. AKIL, H., D. L. MAYER & J. C. LIEBESKIND. 1976. Antagonism of stimulation produced analgesia by naloxone, a narcotic antagonist. Science (Washington, D.C.) 191: 961–962.
23. GIESLER, G. J. & J. C. LIBESKIND. 1976. Inhibition of visceral pain by electrical stimulation of periaqueductal gray matter. Pain 2: 43–48.
24. HOSOBUCHI, Y., J. F. ADAMS & R. LINCHITZ. 1977. Pain relief by electrical stimulation of the central gray matter in humans and its reversal by naloxone. Science (Washington, D.C.) 197: 183–186.
25. ROKAEUS, A., E. BURCHER, D. CHANG, K. FOLKERS & S. ROSELL. 1977. Actions of neurotensin and (Gln⁴)-neurotensin on isolated tissues. Acta Pharmacol. Toxicol. 41: 141–147.
26. KATAOKA, K., A. TANIGUCHI, H. SHIMIZU, K. SODA, S. OKUNO, H. YAMIMA & K. KITAGAWA. 1978. Biological activity of neurotensin and its COOH-terminal partial sequences. Brain Res. Bull. 3: 555–557.
27. QUIRION, R., D. REGOLI, F. RIOUX & ST-PIERRE. 1980. The stimulating effects of neurotensin and related peptides in rat stomach strips and guinea-pig atria. Br. J. Pharmacol. 68: 83–91.
28. SAITO, K., S. KONISHI & M. OTSUKA. 1975. Antagonism between Lioresal and substance P in rat spinal cord. Brain Res. 97: 177–180.

29. KITABGI, P. & P. FREYCHET. 1979. Neurotensin contracts the guinea-pig longitudinal ileal smooth muscle by inducing acetylcholine release. Eur. J. Pharmacol. **56:** 403–406.
30. ZETTER, G. 1980. Antagonism of the gut-contracting effects of bombesin and neurotensin by opioid peptides, morphine, atropine, or tetrodotoxin. Pharmacology **21:** 348–354.
31. MONIER, S. & P. KITABGI. 1980. Substance P-induced autodesensitization inhibits atropine-resistant, neurotensin-stimulated contractions in guinea-pig ileum. Eur. J. Pharmacol. **65:** 461–462.
32. KITABGI, P. & P. FREYCHET. 1978. Effects of neurotensin on isolated intestinal smooth muscles. Eur. J. Pharmacol. **50:** 349–357.
33. VEBER, D. F., F. W. HOLLY, R. F. NUTT, S. J. BERGSTRAND, S. F. BRADY, R. HIRSCHMANN, M. S. GLITZER & R. SAPERSTEIN. 1979. Highly active cyclic and bicyclic somatostatin analogs of reduced ring size. Nature **280:** 512–514.
34. VEBER, D. F., R. M. FREIDINGER, D. S. PERLOW, W. J. PALEVED JR., F. W. HOLLY, R. G. STRACHAN, R. F. NUTT, B. H. ARISON, C. HOMNICK, W. C. RANDALL, M. S. GLITZER, R. SAPERSTEIN & R. HIRSCHMANN. 1981. A potent cyclic hexapeptide analogue of somatostatin. Nature **292:** 55–58.

DISCUSSION OF THE PAPER

C. B. NEMEROFF (*University of North Carolina, Chapel Hill*): In view of the difference in the dose-response curves you obtained when studying neurotensin-induced hypothermia and analgesia, do you think there is any evidence of distinct neurotensin receptor subtypes?

B. V. CLINESCHMIDT (*Merck Sharp & Dohme Research Laboratories, West Point, PA*): Well, I would speculate that 10–20 years from now we will be talking about distinct receptors for neurotensin. We need to learn more about the different agonists related to neurotensin. We need some highly potent and selective antagonists to work with.

I would interpret our data as, probably, indicating NT distribution to different sites of action.

R. CHIPKIN (*Schering Corporation, Bloomfield, NJ*): I have a few technical questions. What temperature hot plate were you using in these experiments?

CLINESCHMIDT: Our studies were done at 60°C for mice, 55°C for rats. I failed to mention that we tested neurotensin in doses as high as 1500 picomoles intracisternally in rats and we find no change in tail flick reaction time.

CHIPKIN: What cutoff time were you using?

CLINESCHMIDT: If an animal showed no response by 30 seconds, the test was ended.

CHIPKIN: It did not seem as if the NT response was that robust. What sort of effects do you see with maximal doses of morphine?

CLINESCHMIDT: With morphine, if you give a high dose, most animals will reach the cutoff. With neurotensin, if you inject 20 mice with 150 picomoles, there may be two or three that will reach the cutoff. The maximum effect that you get with NT is less than that of an opiate.

CHIPKIN: You said you were testing at 30 minutes. Did you measure earlier times?

CLINESCHMIDT: Yes. If you measure earlier times, the dose response does not change very much. It may shift a bit to the left. If you give 150 picomoles, the effect of neurotensin will last from 1 to 2 hours. If you give 1500 picomoles, it will last beyond 2 hours. Thus, our time course is very consistent with Dr. Bissette's data.

When we first did these studies, we thought that the neurotensin effect would be extremely transient. However, it is not.

CHIPKIN: Have you tried any other blockers to see whether or not they will antagonize the effects of NT? TRH, for example, has been suggested to be a neurotensin antagonist.

CLINESCHMIDT: The group at the University of North Carolina has done that experiment, and if I recall correctly TRH does antagonize neurotensin induced analgesia.

We have looked at anticholinergic antihistaminic, antiserotonin, anti-dopamine antagonists as well as prostaglandin synthetase inhibitors and we have not been able to block the antinocisponsive effects of neurotensin.

P. KITABGI (*INSERM Nice, France*): Regarding the comment of Dr. Nemeroff, as I said in my presentation it appears that the neurotensin receptors both in the periphery and in the central nervous system seem to be similar with respect to structural requirements. However, there are some bits of evidence that there might be different NT receptors. For instance, recently we have found that Trp^{11}-NT is as potent as neurotensin in the rat brain wth regard to binding studies, whereas it is 10–20 times as potent as neurotensin in guinea pig brains. So, there might be some species variations in neurotensin receptors.

We have also found recently using a ligand for neurotensin with very high specific activity that there were two populations of binding sites for neurotensin in brain, a high affinity binding site of affinity of 0.1 nanomolar and a lower affinity binding site of 2 nanomolars. We do not know as yet if there is any physiological relevance to this finding.

There are also indications from the Sherbrooke group that some analogs behave completely differently in brain and maybe Dr. St-Pierre or Dr. Jolicoeur could comment on that.

F. B. JOLICOEUR (*Sherbrooke University, Que., Canada*): As my poster presentation showed some analogs in which we substitute the tryosine[11] with moieties such as tryptophan, D-tyrosine, or D-phenylalanine—there is a series of such compounds—we definitely find differential behavioral effects.

One of the most fascinating results, we think, is that some of these analogs have a completely opposite effect on locomotor activity. They increase locomotor activity, whereas neurotensin decreases it. Whether or not this suggests the existence of different receptors remains to be seen.

CLINESCHMIDT: What we all would like to have is a very good antagonist of neurotensin. I think it is very interesting that you converted the locomotor depressant effect of neurotensin to stimulation with NT analogs, but as you indicated, the findings are subject to at least several interpretations.

JOLICOEUR: Absolutely. Could I ask you a question on methodology? In the rats, did you ever inject NT into the lateral ventricles? As I said yesterday, we were not able to get any analgesic effect using that route of administration.

CLINESCHMIDT: Dr. Martin, do you want to answer that?

G. MARTIN (*Merck Sharp & Dohme Research Laboratories, West Point, PA*): We have looked at threshold doses given intraventricularly, intracisternally, and intrathecally, and the results were published in *Neuropeptides* this past year. We found that you have to give about 20 μg in the ventricle to get an antinocisponsive effect and 15 μg to get a hypothermic effect.

R. J. BODNAR (*Queens College, CUNY, Flushing*): With morphine analgesia you get decreases with nucleus raphae magnus lesions, dorsal raphae lesions or spinal dorsolateral funiculus (DLF) cuts. Have you tried to eliminate neurotensin analgesia following any of those procedures?

CLINESCHMIDT: We have not.

S. ST-PIERRE (*National Institute of Mental Health, Bethesda, MD*): With the number of analogs of neurotensin you prepared, did you enter all these into the Merck modeling system? I would like to ask you or Dr. Nutt what sort of response you had from the computer concerning conformation of the active part of neurotensin?

R. F. NUTT (*Merck Sharp & Dohme Research Laboratories, West Point, PA*): We have not examined these analogs thoroughly by the computer system.

ST-PIERRE: Have you tried any other analogs, with other forms of cyclization, for example, through cystine bridges. Have you synthesized another series of analogs with free carboxyls at the COOH terminal?

NUTT: Yes, but we have not obtained any increased activity.

C. FERRIS (*University of Massachusetts, Worcester*): What were the volumes of your microinjections?

CLINESCHMIDT: $0.5\ \mu$l given over 30 seconds.

NEUROANATOMICAL SITES OF ACTION
OF NEUROTENSIN*

Peter W. Kalivas†, Charles B. Nemeroff,
and Arthur J. Prange, Jr.

Biological Sciences Research Center
Department of Psychiatry and the Neurobiology Program
University of North Carolina School of Medicine

INTRODUCTION

The administration of nanomolar quantities of neurotensin (NT) into the cerebroventricular system of rodents produces a variety of behavioral and neurochemical alterations. These include an increase in nociceptive threshold,[1,2] a decrease in body temperature,[3,4] and an increase in dopamine (DA) turnover in dopaminergic terminal fields.[5,6] While all of these NT-induced alterations are mediated centrally, injection into the cerebrospinal fluid (CSF) does not indicate where in the brain NT is acting. In an attempt to define the neuroanatomical site(s) mediating these effects of NT, a series of experiments was designed in which NT was microinjected into numerous neuroanatomical loci.

METHODS

Male Sprague-Dawley rats were used in all experiments. Rats were individually housed with a 12-hr light–dark cycle. After attaining a body weight between 300–450 g, rats were stereotaxically implanted with bilateral cannulae as previously described.[7] The neuroanatomical locus of implantation varied depending upon the experiment. To examine the central site of action for NT-induced antinociception and hypothermia, cannulae were placed at 0.5–1.5 mm intervals in all three stereotaxic planes.[8] When NT-induced effects were observed at an injection site, that site was then studied in more animals. To examine the effect of NT on DA systems, cannulae were placed in the ventral tegmental area (VTA; a region containing a high density of DA perikarya[9]) and the nucleus accumbens (NA; a terminal region for neuronal perikarya located in the VTA[10])

Microinjections were made with a 33-gauge injection needle inserted exactly 1 mm below the tip of the chronic guide cannula (26 gauge) into the desired injection locus. Infusion of NT or saline vehicle was made in a volume of 0.5 μl/side over 30 min using a Sage infusion pump. When more than one treatment was given per animal (maximum of 3 injections) an interval of 72 hr was used.

Behavioral testing was begun one week after surgery. A hot plate (IITC, Inc., Model 35-D) was used to assess nociceptive threshold and was maintained at 55.0 ± 0.5°C. The latency for the animal to lick its hind or forepaw or jump off the

*Supported by NIMH MH-08935, MH-34121, MH-32316, MH-33127, MH-22536, and NICHHD HD-03110.

†Correspondence may be sent to: Peter W. Kalivas, Ph.D., Biological Sciences Research Center (220-H), University of North Carolina, Chapel Hill, NC 27514. Telephone: (919) 966-1480.

hot plate was used as an endpoint, and was assessed by a trained observer ignorant of the treatment regimen employed. As described previously,[11] no significant difference in baseline measurements of response latency was observed between the lick versus jump endpoints. Two or three baseline measurements of hot-plate response latency were obtained before microinjection of NT (2.5 μg/side), and at 10, 30, and 60 min following microinjection. Immediately before obtaining each hot-plate observation, colonic temperature was measured with a YSI-402 temperature probe and a YSI Tele-thermometer.

Spontaneous motor activity was the behavioral endpoint measured after injection of NT into the VTA or NA. The injection parameters for the VTA were identical to those described above, but for the NA, 1.0 μl was infused over 30 sec. Motor activity was monitored with either a Stoelting electromagnetic detector,[12] or photocell apparatus capable of differentiating between locomotion and rearing. Ten rectangular photocell cages, 18 × 90 × 50 cm, were used, each containing a photocell at the midline 4.5 cm above the cage floor for measuring locomotion and a photocell across the long axis of the cage, 13 cm above the cage floor, for measuring rearing.

Levels of DA and its metabolites, 3,4-dihydroxyphenylacetic acid (DOPAC) and homovanillic acid (HVA) were measured with high-pressure liquid chromatography (HPLC) and electrochemical detection as described by Kilts *et al.*[13] Neurotensin or saline was microinjected into the VTA as described above, and animals sacrificed by decapitation 60 min following microinjection. The NA was dissected from each brain on an ice-cooled glass plate.

All animals were examined histologically for cannula placement. When animals were not used for neurochemical measurements, each animal received an intracardiac injection with 50 ml of 10% formalin, and with a microtome cryostat, 100 μm sections were obtained of the cannula tracts. The brain slices were then mounted on slides, stained with cresyl violet, and examined with a dissecting scope by an investigator ignorant of the animal's behavioral response. When brains were used for neurochemical measurements, the brain was not perfused, but after removal of brain regions for chemical analysis, it was surface washed and kept in 10% formalin for at least seven days. Tissue prepared in this way was found suitable for cutting and staining as described above. When HPLC was performed on animals implanted in the NA, cannulae placements were considered correct if the cannula tracts were observed to penetrate the dorsal, but not ventral, surface of the dissected nucleus accumbens.

RESULTS AND DISCUSSION

Antinociception and Hypothermia

TABLE 1 lists the brain regions where greater than 50% of the animals demonstrated an increase in hot-plate response latency following NT administration. Active sites included the central amygdaloid nucleus,[11] the medial preoptic area, mesencephalic periaqueductal gray, a region in the diencephalon ventral and lateral to the reuniens nucleus of the thalamus and medial to the medial lemniscus and zona incerta, and a region at the pontine-medullary border, including the dorsal nucleus raphe magnus and the medial aspect of the nucleus reticularis pontis caudalis.

TABLE 2 lists brain regions where greater than 50% of NT injections produced

TABLE 1

BRAIN REGIONS WHERE > 50% OF INJECTIONS WITH NT (2.5 μg/side)
PRODUCED AN INCREASE IN HOT-PLATE LATENCY

Brain Region	No. Injected*	No. Responding	% Responding
Medial thalamus †	3	3	100
Medial preoptic area	5	4	80
Periaqueductal gray	14	10	71
Central amygdaloid nucleus	6	4	67
Pontine reticular formation †	6	4	67

* All animals received injections with saline vehicle and NT at different trials. An antinociceptive response was defined as > 75% increase in hot-plate response latency by NT as compared to saline. Changes in response latency were determined by adding the three measurements obtained after injection and dividing this sum by the average baseline measurement. This value for NT was then divided by the corresponding saline value.

† The medial thalamus includes tissue ventral to the reuniens nucleus of the thalamus and medial to the medial lemniscus and zona incerta. The pontine reticular formation includes injections made into the medial nucleus reticularis pontis caudalis, nucleus raphe pontis and nucleus raphe magnus.

a decrease in colonic temperature. Active sites included the medial preoptic area, anterior hypothalamus, ventral tegmental area, floor of the 4th ventricle, and tissue in and adjacent to the spinal tract of the trigeminal nerve. When the data from TABLE 1 and TABLE 2 are compared, it is clear that with the exception of the medial preoptic area, brain regions where NT produced antinociception are not the same regions sensitive to NT-induced hypothermia.

All the brain regions listed in TABLES 1 and 2 contain immunohistochemically identified NT [20, 21, 24] and NT receptors.[22] Especially striking is the high density of NT and NT receptors in the central amygdaloid nucleus. Uhl *et al.*[14] have provided evidence for a neurotensinergic pathway with cell bodies in the central amygdaloid nucleus and axonal projections through the stria terminalis. We recently demonstrated that bilateral transection of the stria terminalis blocks the antinociceptive effect of intra-amygdaloid injection with NT.[11] Thus, afferents to, or efferents

TABLE 2

BRAIN REGIONS WHERE > 50% OF INJECTIONS WITH NT (2.5 μg/side)
PRODUCED HYPOTHERMIA

Brain Region	No. Injected*	No. Responding	% Responding
Floor of the 4th ventricle	8	8	100
Spinal tract of the trigeminal nerve	3	3	100
Medial preoptic area	5	4	80
Anterior hypothalamus	8	5	63
Ventral tegmental area	12	7	58

* All animals received injections of saline vehicle and NT at different trials. A hypothermic response to NT was defined as a > 2.0°C decline in colonic temperature as compared to saline. This value was arrived at as follows: add the three measurements made after injection and divide by the average baseline measurement. Then, subtract the value obtained with NT treatment from the corresponding saline value.

from, the central amygdaloid nucleus are necessary for NT-induced antinociception after intra-amygdaloid injection.

Dopamine Systems

In earlier reports,[12, 15] we demonstrated that NT microinjection into the VTA produces an increase in spontaneous motor activity, which is associated with an increase in DA release in the NA. Ervin *et al.*[18] have reported that microinjection of NT into the NA blocks the increase in exploratory behaviors seen after peripheral administration of amphetamine. Since this behavioral hyperactivity produced by amphetamine is thought to involve increased DA release from mesolimbic terminals in the NA,[16] it seemed plausible that intra-NA injection of NT would attenuate the behavioral hyperactivity produced by intra-VTA injection of NT. FIGURE 1 shows that administration of NT into the NA blocked the behavioral hyperactivity after intra-VTA NT. However, while NT injection into the NA clearly blocked the behavioral response, it did not attenuate the increase in DA metabolites seen in the NA after intra-VTA injection of NT (see FIG. 2).

The increase in DA metabolites in the NA after intra-VTA injection of NT provides evidence that NT is acting in the VTA to stimulate DA perikarya, thereby producing enhanced DA release in the NA.[17] The observation that intra-NA NT blocks the behavioral, but not the neurochemical effect of NT administration in the

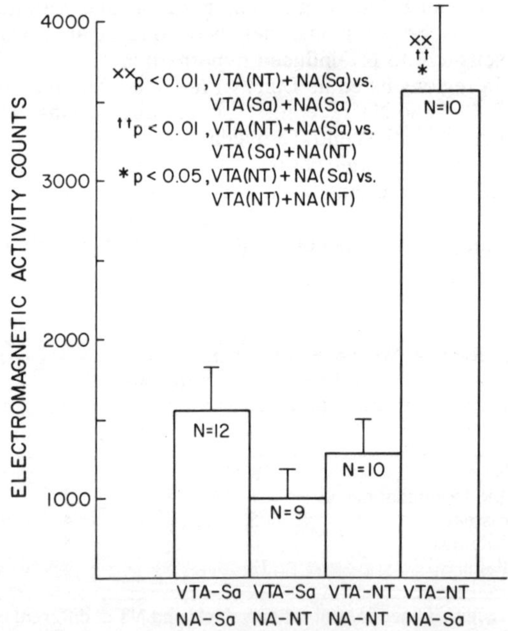

FIGURE 1. Blockade by intra-NA injection with NT (4.0 μg/side) of increased motor activity by intra-VTA injection of NT (2.5 μg/side). Statistical estimates were made with a one-way ANOVA followed by a Neumann-Keuls multiple comparison test.

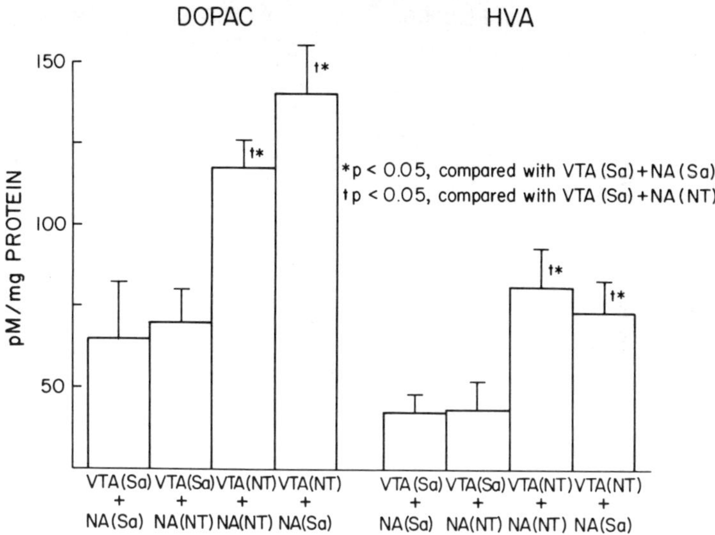

FIGURE 2. Lack of effect by intra-NA injection with NT (4.0 μg/side) on the increase in DA metabolites by intra-VTA injection of the NT (2.5 μg/side). intra-VTA saline + intra-NA saline (n = 5); intra-VTA saline + intra-NA NT (n = 6); intra-VTA NT + intra-NA NT (n = 6); intra-VTA NT + intra-NA saline (n = 6). Statistical estimates were made with a one-way ANOVA followed by a Neumann-Keuls multiple comparison test.

VTA indicates that in the NA, NT is not acting presynaptically to decrease DA release. Rather, NT may be acting postsynaptic to the DA terminal to diminish the consequence of mesolimbic DA release. To further examine this postulate, NT was microinjected into the NA simultaneously with DA. Dopamine administration into the NA is thought to produce behavioral hyperactivity via an action on postsynaptic receptors. Therefore, a blockade of DA-induced behavioral hyperactivity by NT would support the postulate that NT is acting postsynaptic to the DA terminal. As shown in FIGURE 3, NT injection simultaneously with DA into the NA antagonized the increase in both rearing and locomotion produced by DA alone.

In contrast to the data presented above, Reches *et al.*[6] have recently reported that NT acts on DA terminals in the NA. In these experiments, NT was given into the lateral cerebral ventricles to rats pretreated with γ-butyrolactone (GBL), and DOPA accumulation was measured in the NA after inhibition of DOPA decarboxylase. GBL is thought to produce a "reversible pharmacological lesion"[19] of central DA systems, allowing pharmacological manipulation of DA terminals without influence from drug effects on postsynaptic neurons or DA perikarya.[19] Reches *et al.*[6] found that NT produced an increase in DOPA accumulation in the NA. They concluded that NT is probably acting presynaptically on DA terminals in the NA to increase DOPA accumulation. To further examine this conclusion, we followed a similar protocol, but administered NT directly into the VTA or NA. As shown in FIGURE 4, NT injection into the VTA, but not the NA, produced a significant elevation of DA metabolites in the NA. Two conclusions can be made: (1) the "reversible pharmacological lesion"[19] of DA systems by GBL does not prevent pharmacological stimulation of mesolimbic DA perikarya by NT, and (2)

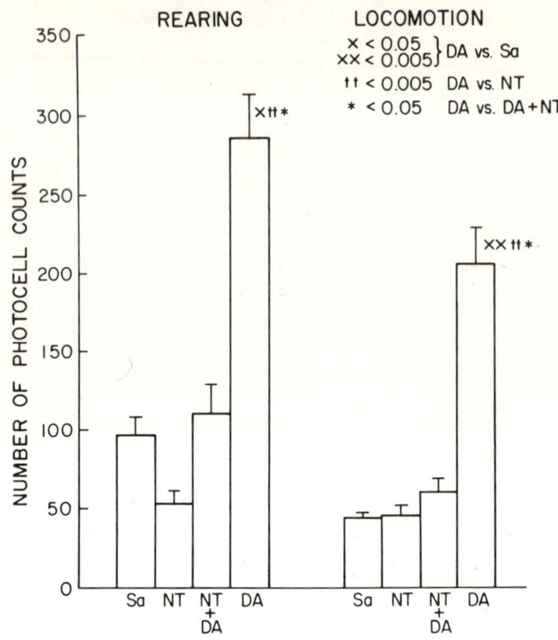

FIGURE 3. Blockade of the motor stimulant effects of intra-NA injection with DA (40.0 μg/side) by simultaneous intra-NA administration of NT (2.5 μg/side). Statistical estimates were made with a one-way ANOVA followed by a Neumann-Keuls multiple comparison test. Saline (n = 10); NT (n = 8); DA + NT (n = 8); DA (n = 8).

the effect of NT administration into the cerebroventricles observed by Reches et al.[6] is probably mediated by NT action on DA perikarya in the VTA, and not on DA terminals in the NA.

The existence of immunohistofluorescent NT[20, 21, 23] and NT receptors[22] in both the VTA and NA supports the possible physiological significance of an NT interaction with the mesolimbic DA system. While the VTA contains both NT perikarya and fibers, only NT fibers exist in the NA. These observations, in combination with the pharmacological data described in this report, are consistent with the model illustrated in FIGURE 5 showing NT modulation of the mesolimbic DA system. While the synaptic mechanism for NT action in the VTA is not clear, two reports support the possibility that NT acts directly on DA cell bodies or dendrites to excite the mesolimbic DA system. First, Andrade and Aghajanian[24] have reported that iontophoresis of NT onto DA neurons in the pars compacta of the substantia nigra and adjacent VTA (G.K. Aghajanian, personal communication) produces a marked short latency increase in firing frequency. Second, Palacios and Kuhar[25] recently reported that all NT receptors (identified by light microscopic autoradiographic localization) in the pars compacta are located on DA neurons. While DA neurons in the VTA were not examined, the similar response of DA neurons to iontophoretic NT in both the pars compacta and VTA supports the possibility that NT receptors in the VTA are also located on DA neurons.

In the NA, the synaptic mechanism by which NT acts is less clear than for the VTA. Data are presented that NT does not act presynaptically on DA terminals to alter DA release. Data have also been presented by Nemeroff *et al.*[26] that NT and DA do not compete for the same receptor. While these observations have been accommodated in FIGURE 5, our present ignorance regarding the action of NT on neurons in the NA prevents greater detail in the model. As presented, DA and NT release in the NA produce contrary effects, with DA release producing an increase in spontaneous motor activity and NT antagonizing this effect. However, the situation is undoubtedly more complex. For example, injection of NT alone (0.25–4.0 μg/side) does not alter spontaneous motor activity[35] (see FIG. 1). Therefore, a model depicting DA–NT synaptic interactions with neurons in the NA must accommodate the fact that NT is only effective in the presence of elevated DA release. Also, Govoni *et al.*[27] observed that chronic treatment with neuroleptics increases the level of immunoreactive NT in the NA. Thus, chronic blockade of

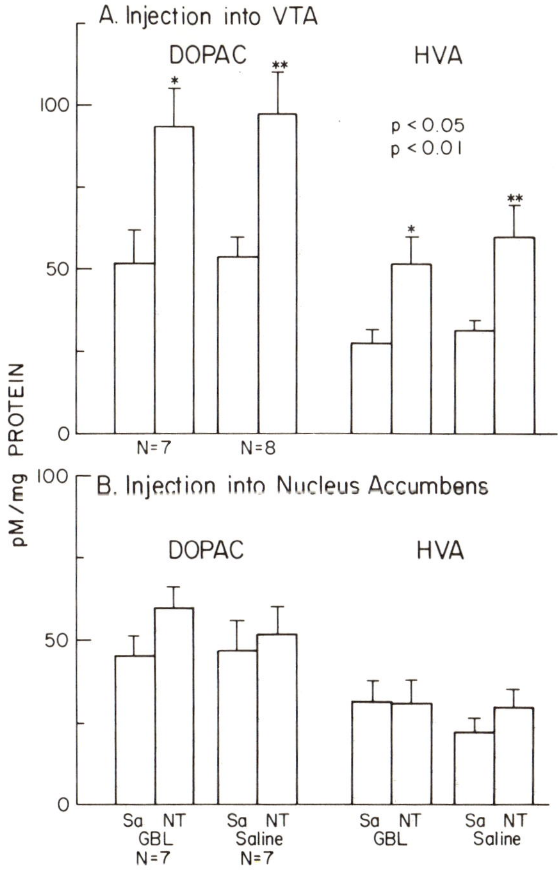

FIGURE 4.   Effect of intra-NA and intra-VTA injection with NT on DA metabolites in the NA after treatment with GBL. Rats were given GBL (750 mg/kg, i.p.) or saline. Five minutes before GBL or saline, each rat received an intra-VTA (A) or intra-NA (B) injection with NT (2.5 μg/side or 4.0 μg/side, respectively) or saline. All rats were sacrificed 35 min after GBL. NT was compared to saline with a two-tailed Student's t-test.

DA receptors presumably alters the neuronal activity of NT terminals in the NA. Finally, the iontophoresis of NT into the NA produces a decrease in firing frequency of DA-sensitive neurons.[28] All these observations provide clues that remain to be accommodated in a model of synaptic interaction between DA and NT systems in the NA.

Neurotensin may also be able to attenuate the behavioral consequence of increased mesolimbic DA function via action in brain sites outside the NA. Preliminary data indicate that NT injection into the lateral ventricles also antagonizes the behavioral hyperactivity produced by intra-VTA injection of NT (data not shown).

CONCLUSIONS

Data presented in this report demonstrate that NT injection into some brain loci, but not others, modifies some behavioral and neurochemical parameters in the rat. This includes alterations in nociceptive threshold, body temperature,

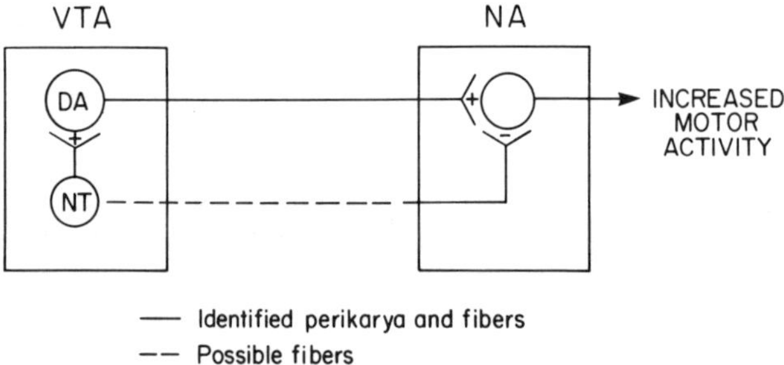

FIGURE 5. Tentative model illustrating NT modulation of the mesolimbic DA system. The dashed line indicates one possible source of NT terminals in the NA.

spontaneous motor activity, and DA release. The fact that moderate levels of immunohistochemical NT[20, 21, 24] and NT receptors[22] exist in all brain regions shown to be responsive to microinjected NT supports the possibility that these changes produced by exogenous NT reflect an underlying physiological function by endogenous NT. The brain areas where NT produces an increase in nociceptive threshold indicates that NT can act not only on classical nociceptive pathways in the brainstem (e.g., periaqueductal gray[29]) but may also interfere with more integrated behavioral and affective responses to painful stimuli (e.g., amygdala). Sensitivity to NT-induced hypothermia was observed in regions classically associated with temperature regulation (e.g., preoptic/anterior hypothalamus[30] and floor of the 4th ventricle[31]), and was also observed in and around the VTA, an area not frequently studied with regard to thermoregulation. Finally, data were presented that demonstrate that NT modulates mesolimbic DA activity and the behavioral consequences of such activity. When given in the VTA, NT has a stimulant action on the mesolimbic DA system, and in the NA, NT has a "neuroleptic-like"[32] action to inhibit the behavioral consequence of DA release. These site-specific modulations of the mesolimbic DA system have implications regarding mental disorders in humans. It has been proposed that a dysfunction of mesolimbic DA plays a causative role in schizophrenia.[33] In view of the present data that NT profoundly influences mesolimbic DA function, or the consequence thereof, it is plausible that a dysfunction of NT neuronal systems could also play a role in causing schizophrenia.

ACKNOWLEDGMENTS

The authors would like to thank Barbara Gau and Donald Stanley for their superb technical assistance, and Faygele ben Miriam for typing the manuscript.

REFERENCES

1. CLINESCHMIDT, B. V., J. C. McGUFFIN & P. B. BUNTING. 1979. Neurotensin: Antinociceptive action in rodents. Eur. J. Pharmacol. **54:** 129–139.
2. OSBAHR, A. J., C. B. NEMEROFF, D. LUTTINGER, G. A. MASON & A. J. PRANGE, JR. 1981. Neurotensin-induced antinociception in mice: Antagonism by thyrotropin-releasing hormone. J. Pharmacol. Exp. Ther. **217:** 645–651.
3. BISSETTE, G., C. B. NEMEROFF, P. T. LOOSEN, A. J. PRANGE, JR. & M. A. LIPTON. 1976. Hypothermia and cold tolerance induced by the intracisternal administration of the hypothalamic peptide neurotenisn. Nature (London) **262:** 607–609.
4. MARTIN, G. E., C. B. BACINO & N. L. PAPP. 1981. Hypothermia elicited by the intracerebral microinjection of neurotensin. Peptides 1: 333–339.
5. WIDERLOV, E., C. KILTS, R. MUELLER, R. MAILMAN, C. NEMEROFF, A. PRANGE & G. BREESE. 1981. Neurotensin increases dopamine metabolites in rat brain. Pharmacologist **23:** 139.
6. RECHES, A., R. E. BURKE, D.-H. JIANG, R. WAGNER & S. FAHN. The effect of neurotensin on dopaminergic neurons in rat brain. Ann. NY Acad. Sci. **400:** 420–421. This volume.
7. KALIVAS, P. W. & A. HORITA. 1980. Thyrotropin-releasing hormone: Neurogenesis of actions in the pentobarbital narcotized rat. J. Pharmacol. Exp. Ther. **212:** 203–210.
8. PELLIGRINO, L. K., A. S. PELLIGRINO & A. J. CUSHMAN. 1979. A Stereotaxic Atlas of the Rat Brain. Plenum Press. New York.
9. PHILLIPSON, O. T. 1979. The cytoarchitecture of the interfascicular nucleus and the ventral tegmental area of Tsai in the rat. J. Comp. Neurol. **187:** 85–98.
10. FALLON, J. H. & R. Y. MOORE. 1978. Catecholamine innervation of the basal forebrain. J. Comp. Neurol. **180:** 545–580.
11. KALIVAS, P. W., B. GAU, C. B. NEMEROFF & A. J. PRANGE, JR. 1982. Antinociception after microinjection of neurotensin into the central amygdaloid nucleus of the rat. Brain Res. **243:** 279–286.
12. KALIVAS, P. W., C. B. NEMEROFF & A. J. PRANGE, JR. 1981. Increase in spontaneous motor activity following infusion of neurotensin into the ventral tegmental area. Brain Res. **229:** 525–529.
13. KILTS, C. D., G. R. BREESE & R. B. MAILMAN. 1981. Simultaneous quantification of dopamine, 5-hydroxytryptamine and four metabolically related compounds by means of reverse-phase HPLC with electrochemical detection. J. Chromatogr. **225:** 347–357.
14. UHL, G. R., R. R. GOODMAN, M. J. KUHAR & S. H. SNYDER. 1978. Enkephalin and neurotensin: Immunohistochemical localization and identification of an amygdalofugal pathway. *In* Advances in Biochemical Psychopharmacology. E. Costa & M. Trabucchi, Eds. Vol **18:** 71–87. Raven Press. New York.
15. KALIVAS, P. W., S. K. BURGESS, C. B. NEMEROFF & A. J. PRANGE, JR. 1981. Stimulant effects of neurotensin microinjection into the ventral tegmental area. Soc. Neurosci. Abstr. **7:** 31.

16. BREESE, G. R., A. S. HOLLISTER & B. R. COOPER. 1976. Role of monoamine neural pathways in d-amphetamine and methylphenidate-induced locomotor activity. *In* Cocaine and Other Stimulants. E. H. Ellinwood, Jr. & M. M. Kilbey, Eds. Plenum Publishing Corp. New York. pp. 445–466.
17. ROTH, R. H., L. C. MURRIN & J. R. WALTERS. 1976. Central dopaminergic neurons: Effects of alterations in impulse flow on the accumulation of dihydroxyphenylacetic acid. Eur. J. Pharmacol. **36:** 163–171.
18. ERVIN, G. N., L. S. BIRKEMO, C. B. NEMEROFF & A. J. PRANGE, JR. 1981. Neurotensin blocks certain amphetamine-induced behaviors. Nature **291:** 73–76.
19. ROTH, R. H. 1979. Dopamine autoreceptors: Pharmacology, function and comparison with postsynaptic dopamine receptors. Commun. Psychopharmacol. **3:** 429–445.
20. UHL, G. R., M. J. KUHAR & S. H. SNYDER. 1977. Neurotensin: Immunohistochemical localization in rat central nervous system. Proc. Natl. Acad. Sci. USA **74:** 4059–4063.
21. UHL, G. R., R. R. GOODMAN & S. H. SNYDER. 1979. Neurotensin-containing cell bodies, fibers, and nerve terminals in the brain stem of the rat: Immunohistochemical mapping. Brain Res. **167:** 77–91.
22. YOUNG, W. S., III & M. J. KUHAR. 1981. Neurotensin receptor localization by light microscopic autoradiography in rat brain. Brain Res. **206:** 273–285.
23. JENNES, L., W. E. STUMPF, W. C. BECKMAN, JR., P. KALIVAS, C. B. NEMEROFF & A. J. PRANGE, JR. 1982. Topography and functional implications of neurotensin-containing structures in the rat brain. Ann. NY Acad. Sci. **400:** 403–404. This volume.
24. ANDRADE, R. & G. K. AGHAJANIAN. 1981. Neurotensin selectively activates dopaminergic neurons in the substantia nigra. Soc. Neurosci. Abstr. **7:** 573.
25. PALACIOS, J. M. & M. J. KUHAR. 1981. Neurotensin receptors are found on dopamine-containing neurons in rat brain: An autoradiographic study. Nature **294:** 587–589.
26. NEMEROFF, C. B., D. HERNANDEZ, D. LUTTINGER, P. W. KALIVAS & A. J. PRANGE, JR. Interactions of neurotensin with brain dopamine systems. Ann. NY Acad. Sci. **400:** 330–344. This volume.
27. GOVONI, S., J. S. HONG, H.-Y. T. YANG & E. COSTA. 1980. Increase of neurotensin content elicited by neuroleptics in nucleus accumbens. J. Pharmacol. Exp. Ther. **215:** 413–417.
28. MCCARTHY, P. S., R. J. WALKER, H. YAJIMA, K. KITAGAWA & G. N. WOODRUFF. 1979. The action of neurotensin on neurons in the nucleus accumbens and cerebellum of the rat. Gen. Pharmacol. **10:** 331–333.
29. FIELDS, H. L. & S. D. ANDERSON. 1978. Evidence that raphe spinal neurons mediate opiate and midbrain stimulation-produced analgesias. Pain **5:** 333–349.
30. MYERS, R. D. 1974. Handbook of Drug and Chemical Stimulation of the Brain. Van Nostrand Reinhold Co. New York. pp. 237–301.
31. LIPTON, J. M. 1973. Thermosensitivity of medulla oblongata in control of body temperature. Am. J. Physiol. **224:** 890–897.
32. NEMEROFF, C. B. 1980. Neurotensin: Perchance an endogenous neuroleptic? Biol. Psychiatry **15:** 283–302.
33. CROW, T. J., E. C. JOHNSTON, A. J. LONGDEN & F. OWEN. 1978. Dopaminergic mechanisms in schizophrenia: The antipsychotic effect and the disease process. Life Sci. **23:** 563–567.

DISCUSSION OF THE PAPER

B. V. CLINESCHMIDT (*Merck Sharp & Dohme Research Laboratories, West Point, PA*): Did you see whether there was any change in norepinephrine metabolites? Also could you block the effect of neurotensin with prazocin?

P. W. KALIVAS (*University of North Carolina School of Medicine, Chapel Hill*): We did not do that. We did look at serotonin and its metabolites, and there was no effect following intra-VTA injection or intra-accumbens injection of NT.

R. ANDRADE (*Yale University School of Medicine, New Haven, CT*): I was curious to see that you get hypothermic effects of NT after VTA injection. I was wondering whether you looked at the interaction between the animal's body temperature and the behavioral or biochemical effect you observed on dopaminergic turnover?

KALIVAS: Not the biochemical effect. However, I did put animals in the activity cage and measured colonic temperature. It appears that when animals are extremely hyperactive body temperature does not decrease. It is in the less hyperactive animals that I observed hypothermia after NT injection into the ventral tegmental area. In fact, the dose response is biphasic. At low NT doses you see no hypothermia. Between 0.5– and 1 μg, hypothermia is present. At higher doses of neurotensin, the hypothermic was not observed, presumably, because the rats are so hyperactive that they are not losing body heat.

D. J. BRAITMAN (*AFRRI, Bethesda, MD*): Do you say that neurotensin increases locomotor behavior.

KALIVAS: Only when given in the ventral tegmental area.

BRAITMAN: Does neurotensin also block the action of dopamine, which increases locomotor activity?

KALIVAS: That is right. Neurotensin in the nucleus accumbens blocks the effect of intra-accumbens injection of dopamine, as well as the effect of intra-VTA injection of neurotensin. So, both ways of producing hyperactivity, either intra-accumbens dopamine or intra-VTA neurotensin, are blocked by intra-accumbens neurotensin.

BRAITMAN: What does intra-accumbens NT alone do to locomotor activity?

KALIVAS: Nothing. The highest dose of NT that we studied was 4 μg per accumbens; we observed no significant effect on locomotor activity.

BRAITMAN: Did you try intra-VTA dopamine?

KALIVAS: No, I have not done that yet.

G. E. MARTIN (*Merck Sharp & Dohme Research Laboratories, West Point, PA*): Have you ever tried unilateral injections? Would you see a circling response?

KALIVAS: There is no circling response after unilateral injections of NT either in the VTA or the substantia nigra.

J. CRAWLEY (*Dupont, Central Research, Glenolden, PA*): In some ways your NT-induced hyperactivity in the VTA is similar to Susan Iversen's results with substance P. Would you care to contrast neurotensin with substance P?

KALIVAS: The effect that we observed with neurotensin occurred at a much lower dose than substance P. I have replicated Dr. Iversen's work and have been able to replicate it at the highest substance P dose that she has published. I believe it is 3 μg per VTA. However, at that dose I did not observe an increase in dopamine metabolites in the nucleus accumbens either at 30 minutes or 60 minutes. I think that if substance P activates the mesolimbic dopamine system, it is not as profound an activation as we see with neurotensin.

C. B. NEMEROFF (*University of North Carolina*): Because many people in the audience are not familiar with microinjection techniques in the CNS, it would be helpful if Dr. Kalivas would comment on the problems of diffusion, injection volume, etc.

KALIVAS: One explanation is that you need a high dose to create the same synaptic concentration that occurs after endogenous peptide release. Another explanation lies in the fact that the behavioral end points that we are looking at involve many brain systems. For example, you would not expect the periaqueductal gray to be the only brain area involved in antinociception. Therefore, we have to employ aphysiological concentrations of neurotensin in the periaqueductal gray to affect this system sufficiently thus producing an antinociceptive response that would involve many brain systems under physiological conditions. I think the same thing could hold true with temperature regulation. The preoptic area–anterior hypothalmus is thought to be very important in temperature regulation, but, clearly, the medulla is also very important.

Tj. B. VAN WIMERSMA GREIDANUS (*University of Utrecht, the Netherlands*): I think you may calculate the rate of spread of either peptides or antisera injected into the brain or into the ventricle at approximately 0.3 mm per hour.

KALIVAS: I might add that I have done some control experiments in which tritiated neurotensin was injected into some of the periventricular sites where we observed a behavioral response, e.g., the floor of the fourth ventricle, periaqueductal gray, and anterior hypothalamus. I measured the appearance of tritiated neurotensin in the cisterna magna. Following intra-periaqueductal gray injections or intrahypothalamic injections almost no radioactivity appeared in the cisterna magna as compared to direct ICV injection. In the floor of the fourth ventricle, the cannulae tips are half a millimeter below the ventricle and quite a bit of reflux was observed.

BEHAVIORAL EFFECTS OF NEUROTENSIN

Tj.B. van Wimersma Greidanus, M.C.G. van Praag,
R. Kalmann, G.J.E. Rinkel, G. Croiset, E.C. Hoeke,
M.A.H. van Egmond, and M. Fekete

Rudolf Magnus Institute for Pharmacology
University of Utrecht, Medical Faculty
3521 GD Utrecht, the Netherlands

Introduction

Numerous reports in literature deal with central nervous system effects of the tridecapeptide neurotensin.[1] Most of these reports concern the effects of this neuropeptide on body temperature,[2-4] on locomotor activity [2,3] and on responsiveness to a painful stimulus.[4-6] Relatively small quantities of neurotensin induce hypothermia,[2-4] diminish locomotor activity,[2,3] and potentiate some actions of barbiturates.[2,3] In addition, neurotensin increases reaction time of mice in a hotplate test.[4-6] Neurotensin exerts these effects after central, mostly intracerebroventricular, administration but not following systemic injection. This is probably due to the poor permeability of the blood–brain barrier to penetration by this neuropeptide.[1]

In order to further evaluate the influence of neurotensin on the central nervous system and to extend its behavioral profile, we examined the effects of this neuropeptide in a variety of behavioral tests, such as open field behavior, and active as well as passive avoidance behavior. Moreover, the effects of neurotensin on responsiveness to electric footshock were studied.

Material and Methods

Male rats of an inbred Wistar strain, weighing 120–160 g, were used. A polyethylene cannula was placed in the right lateral ventricle of the brain for intracerebroventricular (i.c.v.) administration of neurotensin or artificial cerebrospinal fluid (CSF), which was used as a placebo. The operation was performed 3 or 4 days before behavioral experimentation.

Locomotor activity of the animals was studied in a circular open field (diameter 75 cm) during 3-min sessions held on each of 3 consecutive days. The behavioral parameters used during the sessions were ambulation (frequency of crossings of lines drawn on the floor of the open field), rearing, grooming, and defecation.[7] Neurotensin was i.c.v. administered in doses of 3 ng, 30 ng, 300 ng, 3 μg, and 10 μg 1 hour before each session. In addition haloperidol was administered in a dose of 10 μg in a separate group of animals as well.

For studying the effect of neurotensin on amphetamine-induced increase in locomotor activity, a separate experiment was performed in which amphetamine was intraperitoneally (i.p.) injected 1 hour before a single behavioral observation in a dose of 3 mg/kg. Neurotensin was i.c.v. injected at the same time in a dose of 300 ng.

Responsiveness of rats to electric footshock (EFS) was determined by scoring

319

their percentages of jerk/run/jump reactions, flinches, and no responses to various shock levels, ranging from 0.031 mA to 0.187 mA.[8] Ten different shock levels were used and each shock level was presented twice. The 20 EFS's were presented in a randomly fixed order. The duration of each EFS presentation was 1 sec and the interval between presentations was 20 sec. The effect of neurotensin on responsiveness to EFS was studied in an acute experiment using graded doses (3 ng, 30 ng, 300 ng, and 3 μg) of the peptide i.c.v. injected 1 hour before the session. Also the influence of naltrexone and naloxone (1 μg, i.c.v.) on the neurotensin (300 ng, i.c.v.) induced change in responsiveness to EFS was investigated. In addition it was studied whether tolerance developed to the neurotensin-induced change in responsiveness to EFS by measuring the responsiveness to EFS of rats treated with 300 ng neurotensin (i.c.v.) for three consecutive days.

Active avoidance behavior was studied in a pole-jump, one-way, active avoidance situation. Rats were trained to jump onto a pole in order to avoid a scrambled EFS of .20 mA. The light of a bulb placed on top of the box was used as a conditioning stimulus (CS), which was presented 5 sec before the unconditioning stimulus (US) of EFS. As soon as the rat made an avoidance or an escape response, the presentation of the CS or of the CS/US combination was terminated immediately. Acquisition sessions of 10 trials/day with a mean intertrial interval of 60 sec were run on 4 consecutive days. On day 5, extinction trials were run during which only the CS was presented. During extinction, the CS presentation was terminated as soon as the animal made a response. In case an animal did not respond to CS presentation within 5 sec, the light was switched off after 5 sec. Three extinction sessions of 10 trials/session were performed with a 2-hr interval.[9] Neurotensin was administered in doses of 3 ng, 30 ng, 300 ng, and 3 μg immediately after the first extinction session.

Passive avoidance behavior was studied in a simple step-through situation.[10] The apparatus consisted of a dark compartment to which an illuminated elevated platform was attached. After adaptation to the situation, the initial latency of rats to enter the dark compartment was registered during 1 trial, on day 1 and during 3 trials on day 2. Immediately after entering the dark compartment during the third trial on day 2, the rats received a scrambled EFS (.25 mA) during 2 sec (learning trial). Avoidance latencies were registered during retention sessions performed at 24 hours and at 48 hours after the learning trial. A maximal observation time of 300 sec was used during the retention trials. Neurotensin was i.c.v. administered in doses of 30 ng, 300 ng, and 3 μg either immediately after the learning trial or 1 hr before the first retention session.

For statistical evaluation of the results, Student's t-test, Mann-Whitney's U-test or Dunnett's test were used as appropriate.

RESULTS

Locomotor Activity in the Open Field

Only the highest dose of neurotensin induced a significant effect on the behavior of rats in an open field. Ambulation as well as rearing, grooming, and defecation were significantly diminished after treatment with 10 μg neurotensin during the first day of observation. The other doses applied were all ineffective in this respect. On the second day, this effect of neurotensin was less pronounced and only a significant reduction of ambulation and of grooming was observed. On the third day, no effects of neurotensin were seen in this behavioral situation, except a reduction of defecation was noticed (TABLE 1). In contrast, the neuroleptic agent haloperidol induced a continuous reduction in ambulation during the three days of observation.

Neurotensin appeared to antagonize the stimulatory influence of amphetamine on ambulation and rearing in an open field. Amphetamine induced a marked increase in ambulation and rearing. This effect was not observed when amphetamine was administered in combination with neurotensin (TABLE 2).

Responsiveness to Electric Footshock

In the acute experiment, neurotensin exerts a dose-dependent decrease in responsiveness to electric footshock. This was demonstrated by the increased percentage of "no responses" and a decreased percentage of jerks/run/jumps following neurotensin administration (TABLE 3). The change in responsiveness to EFS was not accompanied by changes in the amount of vocalization. If naltrexone or naloxone (1 μg i.c.v.) was given at the same time as neurotensin (300 ng), the decreased responsiveness to EFS as induced by neurotensin was antagonized by the opiate antagonists (TABLE 4).

Observation of the neurotensin effect on responsiveness to EFS during three consecutive days revealed that not only on the first day of observation but also on the second and third day, neurotensin decreased responsiveness to EFS. This was illustrated by the increased percentage of no responses on days 2 and 3 and by the reduction of intense or weak motor reaction on days 2 and 3 respectively (TABLE 5). In addition, no differences in vocalization were found.

Active Avoidance Behavior

Neurotensin induced a dose-dependent inhibition of extinction of a conditioned avoidance response (CAR). Placebo-treated rats displayed percentages of 38% and 15% during the two extinction sessions performed at 2 and 4 hours after treatment. 30 ng of the neuropeptide resulted in a mean of 63% CAR's during the first extinction session after treatment and 29% CAR's during the second extinction session. These percentages amounted to 87% and 50% for the dose of 300 ng and to 83% and 84% for the dose of 3 μg (FIG. 1).

Passive Avoidance Behavior

Neurotensin increased passive avoidance latencies. Treatment of rats with 300 ng neurotensin immediately after the learning trial as well as before the first retention session induced an increase in the median latency score during the first retention session only. A dose of 3 μg induced an increased latency during the first and the second retention sessions (FIG. 2). A low dose (3 ng) of neurotensin, when administered after the learning trial, induced a weak but significant attenuation of passive avoidance behavior at the second retention session.

TABLE 1

THE EFFECT OF GRADED DOSES OF NEUROTENSIN (NT) AND OF HALOPERIDOL (HAL) ON BEHAVIOR OF RATS IN AN OPEN FIELD

	Number of Rats/Group	Ambulation	Rearing, Wall	Rearing, Free	Grooming	Defecation
Day 1						
CSF	(15)	94.6 ± 3.5	8.5 ± 0.7	4.0 ± 0.9	0.9 ± 0.3	4.2 ± 0.6
NT 3 ng	(6)	93.3 ± 5.8	9.2 ± 2.2	6.8 ± 1.4	0.7 ± 0.5	4.2 ± 0.8
30 ng	(6)	86.1 ± 4.1	10.4 ± 0.8	6.6 ± 1.7	0.6 ± 0.3	4.0 ± 1.1
300 ng	(6)	98.4 ± 9.6	6.7 ± 0.9	6.1 ± 1.7	0.7 ± 0.3	3.7 ± 1.2
3 μg	(6)	85.4 ± 7.3	9.6 ± 3.2	4.6 ± 1.8	0.7 ± 0.4	4.1 ± 1.4
10 μg	(6)	$63.7 \pm 5.7*$	$4.0 \pm 1.0*$	$2.4 \pm 1.1*$	$0.1 \pm 0.1*$	$1.7 \pm 0.9*$
HAL 10 μg	(5)	$64.9 \pm 5.7*$	$4.7 \pm 1.0*$	$2.3 \pm 0.9*$	0.7 ± 0.2	2.0 ± 1.0
Day 2						
CSF	(15)	58.5 ± 5.5	3.6 ± 0.9	0.9 ± 0.3	1.6 ± 0.4	3.3 ± 1.0
NT 3 ng	(6)	55.5 ± 10.9	2.7 ± 1.7	0.5 ± 0.5	1.0 ± 0.6	4.0 ± 1.0
30 ng	(6)	69.0 ± 14.5	4.4 ± 1.4	0.2 ± 0.2	1.3 ± 0.7	5.1 ± 0.3
300 ng	(6)	42.1 ± 6.6	1.7 ± 0.9	1.1 ± 1.0	0.9 ± 0.4	2.9 ± 1.2
3 μg	(6)	52.9 ± 9.4	3.9 ± 1.4	0.7 ± 0.5	$0.1 \pm 0.1*$	3.6 ± 1.1
10 μg	(6)	38.5 ± 8.8	2.7 ± 1.2	0.6 ± 0.4	$0.2 \pm 0.2*$	0.8 ± 0.8
HAL 10 μg	(6)	$36.3 \pm 7.5*$	$1.0 \pm 0.6*$	$0.1 \pm 0.1*$	0.7 ± 0.3	2.4 ± 1.2
Day 3						
CSF	(15)	39.0 ± 5.0	$1.0 \pm 0.5*$	$0.4 \pm 0.3*$	0.4 ± 0.1	3.5 ± 1.0
NT 300 ng	(6)	53.0 ± 7.9	3.6 ± 1.1	0.8 ± 0.5	1.0 ± 0.4	1.6 ± 0.8
3 μg	(6)	40.6 ± 9.3	2.8 ± 1.9	1.2 ± 1.0	0.1 ± 0.1	$0.1 \pm 0.1*$
10 μg	(6)	42.4 ± 6.3	1.8 ± 0.7	0.1 ± 0.1	1.2 ± 0.4	$0.3 \pm 0.3*$
HAL 10 μg	(5)	$19.6 \pm 6.2*$	$0.4 \pm 0.3*$	$0.0 \pm 0.0*$	0.9 ± 0.4	1.1 ± 0.9

*$p < .05$.

TABLE 2

THE EFFECT OF NEUROTENSIN* ON AMPHETAMINE†-INDUCED ACTIVITY OF RATS IN AN OPEN FIELD

i.c.v./i.p.	Number of Rats/Group	Ambulation	Rearing, Wall	Rearing, Free	Grooming	Defecation
CSF/saline	(8)	102.6 ± 5.7	10.3 ± 1.7	6.5 ± 2.4	0.2 ± 0.1	5.1 ± 0.8
CSF/Amph.	(7)	137.9 ± 11.1‡	19.3 ± 3.7‡	14.6 ± 2.6‡	0.1 ± 0.1	3.6 ± 1.0
NT/Amph.	(8)	84.8 ± 9.2	10.8 ± 2.8	8.5 ± 2.0	0.1 ± 0.1	3.4 ± 1.3
NT/saline	(15)	86.6 ± 6.1	7.5 ± 0.8	5.7 ± 1.0	0.3 ± 0.2	5.6 ± 0.6

*NT: 300 ng i.c.v.
†AMPH: 0.3 mg i.p.
‡$p < .05$ (vs. CSF/saline controls).

TABLE 3

THE EFFECT OF GRADED DOSES OF NEUROTENSIN (NT) ON RESPONSIVENESS TO
ELECTRIC FOOTSHOCK

	Number of Animals/Group	% No Responses	% Flinches	% Jerks/ Runs/Jumps	% Vocalization
CSF		12.9 ± 2.0	33.4 ± 3.0	53.6 ± 2.8	22.9 ± 4.3
NT 3 ng	(5)	21.3 ± 4.0	29.7 ± 3.0	49.0 ± 1.9	19.2 ± 9.0
30 ng	(5)	25.3 ± 3.0*	26.0 ± 3.5	48.7 ± 3.0	22.1 ± 6.6
300 ng	(5)	33.7 ± 3.0†	28.0 ± 2.5	38.3 ± 2.0*	20.0 ± 6.9
3 μg	(5)	27.6 ± 2.4†	31.2 ± 4.7	41.1 ± 3.0*	22.5 ± 5.7

*p < .05 (vs. CSF controls).
†p < .02 (vs. CSF controls).

DISCUSSION

Some effects of neurotensin have been claimed to be shared by neuroleptic agents.[3, 11] It has been reported that neurotensin exerts effects in various pharmacological and behavioral tests in which neuroleptic agents are active. It appeared that neurotensin releases prolactin[12] and that bilateral nucleus accumbens injections of neurotensin, like the neuroleptic drug haloperidol, significantly antagonized increased locomotor activity and rearing induced by amphetamine.[3, 11] However, unlike neuroleptic drugs, neurotensin seems not to inhibit the binding of ^{3}H-spiroperidol to dopamine receptors in rat brain membrane preparations from the nucleus caudatus or the nucleus accumbens.[3] Thus, neurotensin seems to share several, but not all, properties with neuroleptic agents.[3]

Generally the results obtained from the presently described experiments con-

TABLE 4

THE EFFECT OF NALTREXONE* AND OF NALOXONE† ON NEUROTENSIN‡-
INDUCED DECREASED RESPONSIVENESS TO ELECTRIC FOOTSHOCK

	No. of Animals/Group	% No Responses	% Flinches	% Jerks/ Runs/Jumps	% Vocalization
CSF	(5)	15.8 ± 2.0	30.8 ± 3.4	53.4 ± 4.0	22.9 ± 4.3
NALT	(5)	21.7 ± 3.1	34.2 ± 1.6	44.1 ± 2.1	18.3 ± 3.9
NT	(5)	44.2 ± 2.8‖	24.2 ± 3.3	31.6 ± 4.9§	20.0 ± 6.9
NT + NALT	(5)	24.1 ± 2.4	27.5 ± 2.5	48.4 ± 2.5	20.8 ± 5.3
CSF	(12)	17.0 ± 1.4	39.6 ± 1.4	43.4 ± 1.1	45.8 ± 4.0
NALO	(12)	17.4 ± 1.4	38.6 ± 1.3	44.4 ± 1.7	36.5 ± 3.0
NT	(9)	25.0 ± 1.2¶	39.8 ± 2.6	35.2 ± 2.2§	40.7 ± 2.2
NT + NALO	(10)	20.8 ± 1.6	37.9 ± 1.8	41.3 ± 2.3	41.7 ± 3.4

*NALT: 1 μg i.c.v.
†NALO: 1 μg i.c.v.
‡NT: 300 ng i.c.v.
§p < .05 (vs. CSF controls as well as vs. NT + NALT or NT + NALO).
¶p < .02 (vs. CSF controls as well as vs. NT + NALO).
‖p < .01 (vs. CSF controls as well as vs. NT + NALT).

TABLE 5

THE EFFECT OF NEUROTENSIN ON RESPONSIVENESS TO ELECTRIC
FOOTSHOCK DURING 3 CONSECUTIVE DAYS

		No. of Animals/Group	% No Responses	% Flinches	% Jerks/ Runs/Jumps	% Vocalization
Day 1	CSF	(12)	19.4 ± 2.5	33.1 ± 3.3	47.5 ± 3.0	26.6 ± 5.4
	NT	(12)	35.1 ± 3.5 †	37.9 ± 3.9	27.0 ± 3.5 †	23.8 ± 5.5
Day 2	CSF	(11)	25.0 ± 2.8	36.2 ± 4.0	38.8 ± 3.9	26.5 ± 6.0
	NT	(12)	38.2 ± 3.6 †	33.2 ± 4.0	28.4 ± 2.6 †	31.0 ± 7.7
Day 3	CSF	(11)	16.0 ± 1.6	50.8 ± 4.6	33.2 ± 3.5	32.9 ± 5.6
	NT	(12)	38.5 ± 4.6 †	32.0 ± 3.7 †	29.5 ± 3.1	30.7 ± 5.7

* NT: 300 ng i.c.v.
† p < .05 (vs. CSF controls).

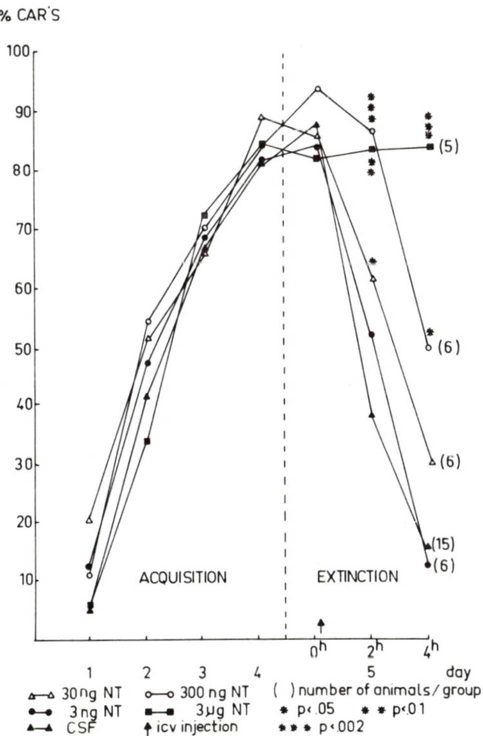

FIGURE 1.   Effect of neurotensin (NT) on extinction of a pole jump avoidance response.

firm earlier observations by others. The presently reported decreased locomotor activity and antagonism of amphetamine-induced increased locomotion has been reported previously.[3] Also the reported antinociceptive action of neurotensin [4-6, 11] has been confirmed in an EFS situation. The finding that neurotensin does not affect the amount of vocalization suggests that neurotensin has no analgesic action, but induces a decreased motor response to a painful stimulus. The antinociceptive effect of neurotensin as presently tested in the EFS was counteracted by

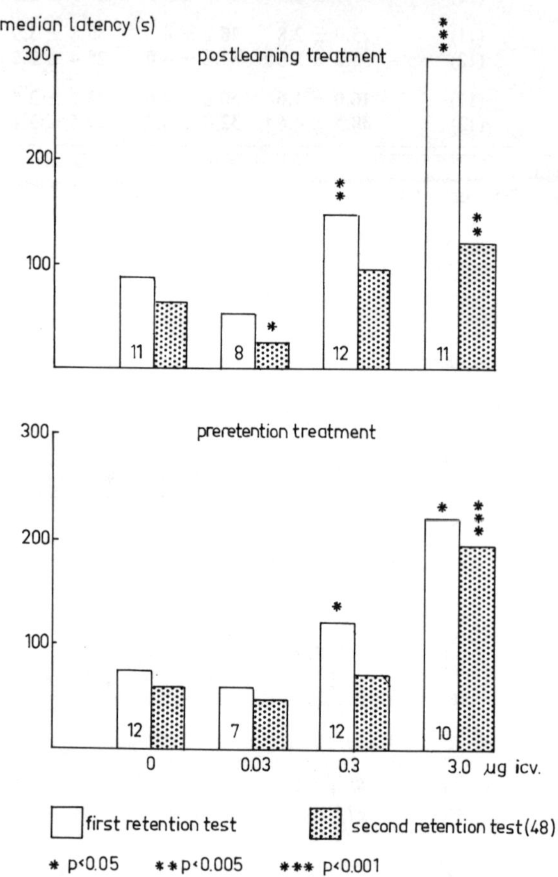

FIGURE 2. Effect of neurotensin on retention of a passive avoidance response.

the opiate antagonists naltrexone and naloxone. The latter antagonist has been reported[6] not to antagonize the antinociceptive action of neurotensin in the hotplate procedure. Thus, although opiate-receptors appear to be involved in the antinociceptive action of neurotensin, other systems concerning antinociception may play a role as well. Differences in behavioral test procedures, in doses of neurotensin as well as in those of the opiate-antagonists used, may explain the differences between the various studies. Interestingly, studies in which repeated testing and treatment with neurotensin were performed did not reveal a development of tolerance to the antinociceptive effect of neurotensin.

Studies performed to further evaluate the behavioral activity of neurotensin did not indicate that neurotensin has in general a neuroleptic-like action. Neuroleptics like haloperidol applied to rats submitted to active pole-jump avoidance behavior induces a facilitation of extinction of a conditioned avoidance response.[13] However neurotensin inhibits this extinction in a dose-dependent way, that is, opposite to what may be expected from treatment with a neuroleptic drug. In addition, neuroleptics attenuate passive avoidance behavior.[13] The results of the present experiments reveal that neurotensin administration during the acquisition as well as retention of the passive avoidance response generally results in increased latencies during retention sessions of this avoidance response. Thus, in fact neurotensin enhances passive avoidance behavior.

Whether or not the inhibitory effect of neurotensin on extinction of a CAR and the stimulatory effect of this neuropeptide on passive avoidance behavior are direct effects of neurotensin or mediated by a stimulation of the release of other neuropeptides (e.g., vasopressin and/or ACTH[14]) by neurotensin remains to be elucidated.

In order to determine the possible physiological role of endogenous neurotensin in behavior, a preliminary experiment was performed in which the amount of biologically available neurotensin was reduced by i.c.v. application of antiserum to this neuropeptide. Undiluted neurotensin antiserum (anti-NT)* was i.c.v. injected in an amount of 3 μl immediately after the first extinction session of the pole-jump avoidance response and the effect of this treatment was observed 4 hours later in a second extinction session. A tendency toward a facilitated extinction of conditioned avoidance behavior following administration of the antiserum was observed. This tendency is opposite to the marked effect found after treatment with neurotensin itself. Since binding capacity and binding specificity of the antiserum used are not known with respect to this kind of application, no conclusion can be drawn yet concerning the role of centrally available endogenous neurotensin in brain processes related to avoidance behavior.

In summary, it may be concluded that neurotensin shares some, but not all, effects with neuroleptic substances. In fact, it appears that neurotensin exerts some effects opposite to those induced by neuroleptics.

REFERENCES

1. NEMEROFF, C. B., D. LUTTINGER & A. J. PRANGE. 1980. Neurotensin: Central nervous system effects of a neuropeptide. TINS September: 212–215.
2. NEMEROFF, C. B., G. BISSETTE, A. J. PRANGE JR., P. T. LOOSEN, T. S. BARLOW & M. A. LIPTON. 1977. Neurotensin: Central nervous system effects of a hypothalamic peptide. Brain Res. **128:** 485–496.
3. NEMEROFF, C. B. 1980. Neurotensin: Perchance an endogenous neuroleptic? Biol. Psychiatry **15:** 283–302.
4. MARTIN, G. E., T. NARUSE & N. L. PAPP. 1981. Antinociceptive and hypothermic actions of neurotensin administered centrally in the rat. Neuropeptides **1:** 447–454.
5. CLINESCHMIDT, B. V. & J. C. McGUFFIN. 1977. Neurotensin administered intracisternally inhibits responsiveness of mice to noxious stimuli. Eur. J. Pharmacol. **46:** 395–396.

* Anti-NT was kindly supplied by Dr. R. E. Carraway, University of Massachusetts.

6. CLINESCHMIDT, B. V., J. C. McGUFFIN & P. B. BUNTING. 1979. Neurotensin: Antinociponsive action in rodents. Eur. J. Pharmacol. **54:** 129–139.
7. WEIJNEN, J. A. W. M. & J. L. SLANGEN. 1970. Effects of ACTH analogues on extinction of conditioned behavior. Prog. Brain Res. **32:** 221–235.
8. GISPEN, W. H., A VAN DER POEL & TJ. B. VAN WIMERSMA GREIDANUS. 1973. Pituitary-adrenal influences on behavior: Responses to test situations with or without electric footshock. Physiol. Behav. **10:** 345–350.
9. VAN WIMERSMA GREIDANUS, TJ. B. & D. DE WIED. 1971. Effects of systemic and intracerebral administratin of two opposite acting ACTH-related peptides on extinction of conditioned avoidance behavior. Neuroendocrinology **7:** 291–301.
10. ADER, R., J. A. W. M. WEIJNEN & P. MOLEMAN. 1972. Retention of a passive avoidance response as a function of the intensity and duration of electric shock. Psychon. Sci. **26:** 125–128.
11. ERVIN, G. N., L. S. BIRKEMO, C. B. NEMEROFF & A. J. PRANGE JR. 1981. Neurotensin blocks certain amphetamine-induced behaviors. Nature **291:** 73–76.
12. RIVIER, C., M. BROWN & W. VALE. 1977. Effect of neurotensin, substance P and morphine sulfate on the secretion of prolactin and growth hormone in the rat. Endocrinology **100:** 751–754.
13. KOVÁCS, G. L. & D. DE WIED. 1978. Effects of amphetamine and haloperidol on avoidance behavior and exploratory acitivity. Eur. J. Pharmacol. **53:** 103–107.
14. LEEMAN, S. E., E. A. MROZ & R. E. CARRAWAY. 1977. Substance P and neurotensin. *In* Peptides in Neurobiology. H. Gainer, Ed. Plenum Press. New York. pp. 99–144.

DISCUSSION OF THE PAPER

E. A. ZIMMERMAN (*Columbia University, New York*): As you well know one of the sources of the vasopressin for memory may be the paraventricular nucleus which projects to many areas. I am thinking about Dr. D. Kahn's demonstration of neurotensin cells next to the PVN. Could those neurotensin cells be regulating the vasopressin pathway?

TJ. B. VAN WIMERSMA GREIDANUS (*University of Utrecht, the Netherlands*): The only experiment related to your question we have done is to inject neurotensin ICV and measure at one single time point vasopressin levels in blood. We found a tendency towards an increase, but plasma levels of vasopressin may not really be related to changes in memory.

A. J. DUNN (*University of Florida, Gainesville*): Regarding extinction of pole jump avoidance, it seems to me that there are a number of peptides that seem to affect it one way or another: ACTH, vasopressin, endorphins and their derivatives, and neurotensin. Can you compare the potency of neurotensin with ACTH and vasopressin?

GREIDANUS: After i.c.v. injection neurotensin is 10–25 times less active than, for instance, vasopressin and the endorphins.

DUNN: Is it comparable to ACTH?

GREIDANUS: ACTH data showed were based on peripheral injections and generally we need 100 times less of a peptide after i.c.v. injection than after systemic injection.

D. KAHN (*Columbia University, New York*): To follow up on Dr. Zimmerman's comments, one suggestion would be to examine whether neurotensin's effects are mediated through vasopressin by looking at a vasopressin-deficient rat. Similarly if the NT effect is mediated through ACTH, one might look at an MSG-treated rat.

GREIDANUS: I agree.

A. J. Prange, Jr. (*University of North Carolina, Chapel Hill*): How would TRH compare with neurotensin in these two models?

Greidanus: I really do not know.

M. D. Hirsch (*Roche Institute of Molecular Biology, Nutley, NJ*): I recall some data that compared several peptides and TRH had about 50% of the activity of ACTH $_{1-10}$ in the pole jumping test.

Greidanus: Indeed, there was an effect of TRH, but it was only one single experiment in which the effects of TRH, LHRH and ACTH were compared.

S. Fielding (*Hoechst-Roussel, Somerville, NJ*): Was there an escape component in the pole jump avoidance response?

Greidanus: No, because we injected the peptide only during the extinction. An animal has no reason to escape because a shock is never applied during extinction. So, we never injected the peptide prior to the stimulus either in active or in passive avoidance behavior. The peptide was never present during the presentation of electric foot shock, either in active or in passive avoidance behavior.

Fielding: In evaluating neuroleptics in these kinds of procedures we usually look at a separation between the avoidance and the escape response. If there is wide separation between then it seems to point to a specific response for neuroleptics.

INTERACTIONS OF NEUROTENSIN WITH
BRAIN DOPAMINE SYSTEMS*

Charles B. Nemeroff, Daniel E. Hernandez, Daniel Luttinger,
Peter W. Kalivas, and Arthur J. Prange, Jr.

*Biological Sciences Research Center
Department of Psychiatry and the Neurobiology Program
University of North Carolina School of Medicine
Chapel Hill, North Carolina 27514*

INTRODUCTION

For over a decade, our research group has studied the role of peptides in the central nervous system (CNS). In particular, we have concentrated for the last six years on characterizing the pharmacological and behavioral effects of neurotensin (NT), the tridecapeptide about which this conference has been organized. In recent years, we have increasingly focused our attention on interactions of NT and brain dopamine (DA) systems and our findings, as well as those of others, form the theme of this review. Certain aspects of this subject are also discussed by Kalivas et al.[1] and by Prange and Nemeroff[2] in this volume.

As described by Luttinger et al.[3] in this volume, our early work with centrally administered NT was concerned with characterizing the effects of the peptide on the actions of two centrally acting depressants, pentobarbital and ethanol. The peptide significantly enhanced both pentobarbital- and ethanol-induced sleeping time and ethanol-induced hypothermia. In subsequent studies, reviewed by Bissette et al.[4] in this volume, our group demonstrated that injection of NT into cerebrospinal fluid (CSF) produced a potent hypothermia in a variety of mammals. In later studies it was shown that reduction in the functional activity of brain DA induced either by haloperidol (receptor blockade) or by 6-hydroxydopamine (DA neuronal destruction) results in a marked potentiation of NT-induced hypothermia in rats[5] (FIG 1). Subsequently, Bissette et al.[6] demonstrated that NT-induced hypothermia in mice is antagonized by the indirect DA agonists methylphenidate and *d*-amphetamine. While we were characterizing the effects of DA antagonists and agonists on NT-induced hypothermia, Rivier et al.[7] reported that intravenously administered NT produced a significant increase in plasma prolactin levels, a characteristic effect of DA antagonists. These findings, along with others (see refs. 8–10), illustrate the pharmacological properties that are shared by NT and neuroleptic drugs, all of which are DA receptor blockers. Moreover, brain NT and DA perikarya and terminals are found in close proximity in a variety of brain regions (e.g., nucleus accumbens, ventral tegmentum, and hypothalamus). These observations commanded a closer examination of NT-DA interactions.

*Supported by NIMH MH-34121, MH-32316, MH-33127, MH-22536, MH-08935, MH-14277, and NICHHD HD-03110.

Effects of Neurotensin in Neuroleptic Screening Tests

One research direction undertaken was a systematic comparison of the effects of centrally administered NT and neuroleptic drugs, and these data have been reviewed.[8–10] In an early study[11] we reported that intracerebroventricular (i.c.v.) administration of 30 µg of NT significantly reduced the locomotor activity of rats, as measured with electromagnetic activity cages. These findings have recently been confirmed.[12] Neuroleptic drugs also reduce locomotor activity in laboratory animals.

One commonly utilized screening test for neuroleptic activity is the Julou-Courvoisier muscle relaxation test. Mice are placed with their forepaws on a taut, horizontal rope mounted 30 cm high, and the latency in seconds for them to pull themselves up is measured by an observer, ignorant of the treatment regimen. In this test intracisternally (i.c.) administered NT (3–30 µg), like neuroleptics, produced significant muscle relaxation (TABLE 1).[13] The muscle relaxation, induced by

TABLE 1

THE ACTIVITY OF INTRACISTERNAL NEUROTENSIN IN THE
JULOU-COURVOISIER TRACTION TEST IN MICE *

Dose NT (μg i.c.)	N	No. Relaxed (%)	Rectal Temp. (°C) ($\bar{X}$ SEM) 45 min post-injection
0	23	0 (0)	38.2 ± 0.1
1	18	2 (11)	33.3 ± 0.4‡
3	21	13 (62)‡	33.1 ± 0.3‡
5	15	6 (40)‡	34.1 ± 0.4‡
10	21	7 (33)‡	34.1 ± 0.4‡
15	12	4 (33)†	34.2 ± 0.2‡
30	27	14 (52)‡	33.5 ± 0.1‡

* Mice were lightly anesthetized with ether and injected i.c. with NT or vehicle (10 µl) and placed with their forepaws on a taut rope strung horizontally. Mice that climbed up and touched the rope with their hindpaws in < 10 sec were considered nonrelaxed whereas those doing so in > 10 sec or not at all were considered relaxed. In all cases the hypothermic response to i.c. NT was confirmed. The X^2 test was applied to data pertaining to muscle relaxation, Students t-test to data pertaining to hypothermia. (From Osbahr *et al.*, 1979, with permission.)
† p < 0.002.
‡ p < 0.001.

both NT[13] and neuroleptics[14] is blocked by pretreatment with the tripeptide thyrotropin-releasing hormone (TRH).

Another pharmacological effect of neuroleptics often used in screening tests for new antipsychotic agents is inhibition of discrete-trial, conditioned avoidance responding. Recently Luttinger *et al.*[15] reported that NT, administered i.c.v. in doses greater than 1.0 µg, significantly inhibited the performance of rats in a "shuttle box" avoidance task. Like neuroleptics, NT diminished the number of avoidance responses (FIG. 2) without affecting the number of escape responses. The null effect on escape responding implies that the NT effect on avoidance responding is not simply secondary to motor impairment.

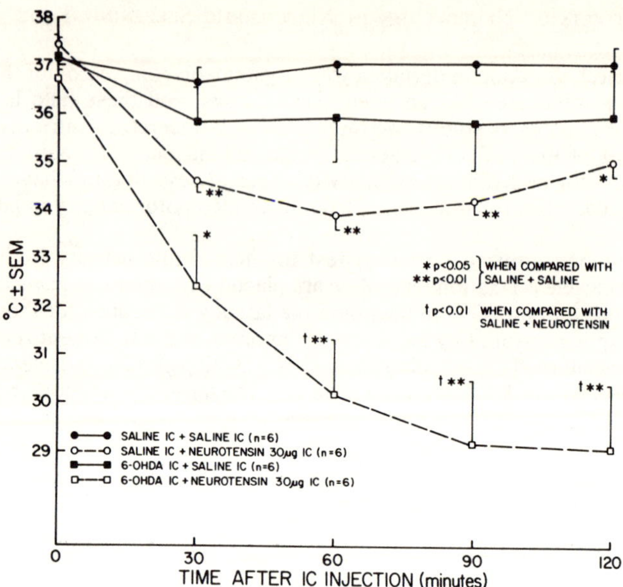

FIGURE 1. Effect of catecholamine depletion on NT-induced hypothermia. Adult rats were injected with 6-hydroxydopamine (240 μg i.c.) in order to destroy both brain norepinephrine and DA neurons. Rats were later injected i.c. with either NT (30 μg) or vehicle (10 μl 0.9% sterile saline) and the effects on rectal temperature in a cold (4°C) environment assessed. (Reproduced from Nemeroff et al.[5] with permission).

Acute neuroleptic treatment induces catalepsy in laboratory animals. As described recently by Snijders et al.[16] and in this volume by Dunn and Hurd,[17] NT administered i.c.v. in doses greater than 20 ng produces significant catalepsy in mice.

In recent studies in our laboratory, Luttinger and his associated (unpublished observations) have found that NT, administered i.c.v., reduces food consumption in food-deprived rats. This NT effect, which is shared by neuroleptic drugs, is not due to an aversive effect of the peptide because NT does not produce a conditioned taste aversion.

Another commonly utilized screen for detection of neuroleptic activity is antagonism of the behavioral effects of the indirect DA agonist d-amphetamine. This drug is believed to produce locomotor hyperactivity and stereotypic behaviors in rats by increasing DA release from dopaminergic nerve terminals in the mesolimbic and nigroneostriatal DA systems, respectively.[18] When administered peripherally, i.c.v., or directly into one of these DA terminal fields, neuroleptics block these d-amphetamine-induced behaviors.[18] We sought to determine whether NT interacts with these DA pathways. Accordingly we studied whether injection of NT into CSF (i.c.) blocks the behavioral effects of the indirect DA agonists, d-amphetamine, methylphenidate, and cocaine. Methylphenidate, like d-amphetamine, acts by releasing DA from nerve terminals (see above) whereas cocaine acts, at least partially, by blocking DA reuptake. Locomotor activity of rats was assessed with electromagnetic activity cages, of mice with photocell activity cages. Intracisternally administered NT, in both rodent species, significantly attenuated the behavioral hyperactivity observed after *d*-amphetamine, methylphenidate, and cocaine (FIG. 3)[19] and the effect is dose related. The effects of i.c. NT on the behavioral effects of two direct-acting DA agonists, apomorphine and lergotrile, were also investigated in both species. Centrally administered NT did not block the "hyperactivity" induced by either apomorphine or lergotrile in rats or mice. Unfortunately the doses of these drugs required to produce significant increases in locomotor activity also produced considerable stereotypic behavior. This was not a problem with *d*-amphetamine since low doses (1–2 mg/kg i.p.) produced locomotor hyperactivity whereas a higher dose (5 mg/kg i.p.) produced primarily stereotypic behavior. Stereotypic behavior will activate activity sensors in both our photocell and electromagnetic activity cages. Interpretation of these "locomotor activity" data with direct DA agonists is therefore rendered difficult. Intracisternal injection of haloperidol (5 μg), a butyrophenone neuroleptic, significantly attenuated the locomotor hyperactivity induced by both indirect and direct DA agonists.

To determine whether centrally administered NT acts on both the mesolimbic and nigroneostriatal DA systems, rats were implanted with bilateral cannulae either in the nucleus accumbens, a major terminal field of the mesolimbic DA system, or in the nucleus caudatus, a major terminal field of the nigroneostriatal DA system. NT or neuroleptics were administered into these sites afte receiving peripherally administered *d*-amphetamine.[20] Intra-accumbens injection of NT or haloperidol significantly blocked the locomotor hyperactivity induced by *d*-amphetamine (TABLE 2). This effect of intra-accumbens NT appears to be relatively specific because injection of another endogenous peptide of comparable molecular weight, luteinizing hormone releasing hormone (LHRH), was without effect. In contrast, intra-caudate injections of the peptide did not block *d*-amphetamine-induced stereotypic behaviors, though intra-caudate haloperidol did (TABLE 3).

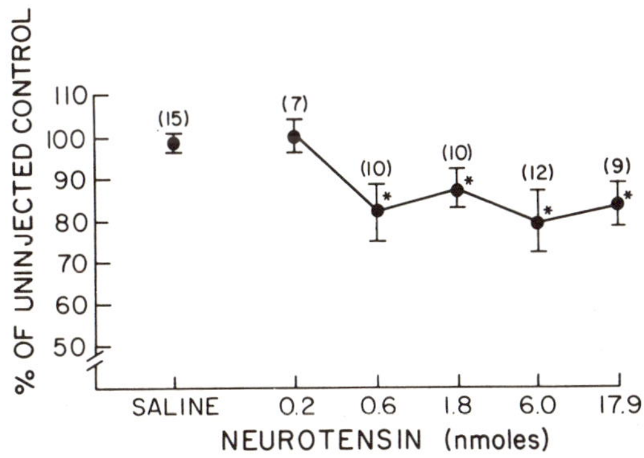

FIGURE 2. The effect of i.c.v. NT on discrete-trial, conditioned avoidance responding. Rats received NT or 0.9% NaCl (10 μl) injected into the lateral ventricle 30 min before behavioral testing. The number of animals from which a score was derived is enclosed in parentheses above each mean value (± SEM). Statistical comparisons were made between the effects of saline and the various doses of NT employed by use of the Mann-Whitney U-test (* p < 0.01). Reproduced from Luttinger *et al.*[15], with permission.

The effect of i.c. NT on apomorphine-induced climbing behavior in mice has been investigated. This climbing stereotypy has been reported to be mediated by DA receptors in the nucleus caudatus and blocked by neuroleptic treatment.[21] Intracisternal NT did not alter this apomorphine effect.[8]

Thus, when injected into DA terminal fields, NT blocks pharmacological activation of the mesolimbic DA system but not of the nigroneostriatal DA system.

In collaboration with King[22] we studied the effect of intra-accumbens NT injection on the rate of electrical self-stimulation from the A_{10} cell body region (ventral tegmental area), the site of origin of the mesolimbic DA system. In contrast to the d-amphetamine experiments, this paradigm permitted study of the effects of intra-accumbens NT on mesolimbic DA activation induced by discrete

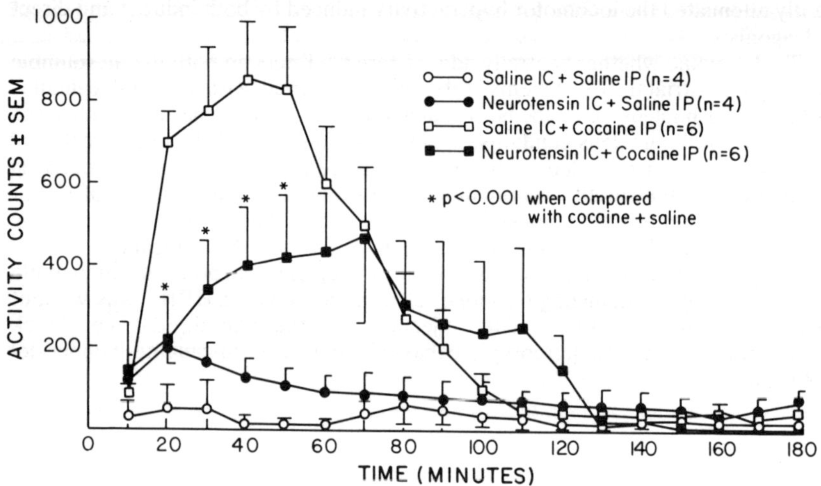

FIGURE 3. The effect of i.c. NT (30 μg) on the locomotor hyperactivity induced by cocaine (20 mg/kg i.p.) in rats. Spontaneous motor activity was assessed in Stoelting electromagnetic activity cages. Statistical analysis was performed with Dunnett's test for multiple comparisons.

electrical stimulation of DA perikarya from electrodes implanted in the VTA. The results obtained were congruent with those obtained in the d-amphetamine study; intra-accumbens injection of NT, as well as of the neuroleptic spiroperidol, significantly reduced the rate of electrical self-stimulation from the VTA (FIG. 4).

Considered together, these data support the hypothesis that centrally applied NT reduces activity of the mesolimbic, but not the nigroneostriatal, DA system. Two major research directions have been followed by our group. These include: (1) investigation of neurochemical correlates of NT–DA interaction and (2) evaluation of the behavioral and biochemical effects of exogenous NT injection into the ventral tegmental area. In addition, it is important to note that electrophysiological data have recently been obtained that support the notion of NT–DA interactions.[23] Iontophoretically applied NT inhibited DA-sensitive neurons in the nucleus accumbens.

TABLE 2

EFFECTS OF INJECTIONS INTO NUCLEUS ACCUMBENS ON SELECTED BEHAVIORAL RESPONSES TO AMPHETAMINE*

| | CSF | Neurotensin (μg) | | | | LHRH (3 μg) | Tartaric Acid | Haloperiodl (μg) | |
		0.1	0.3	1	5			2.5	5
Behavioral items (n)	30	10	9	11	15	8	10	10	10
Forward locomotion	15 (1–36)	8.5 (2–28)	6† (0–28)	8† (0–20)	5‡ (0–21)	14.5 (1–30)	14 (2–33)	3‡ (0–8)	1§ (0–30)
Rearing	4.5 (0–38)	2 (0–26)	1† (0–14)	0† (0–11)	0§ (0–11)	4.5 (0–13)	1 (0–4)	1 (0–3)	0 (0–11)
Sniffing	39.5 (10–53)	44 (18–50)	35 (27–40)	28 (11–46)	34 (12–54)	38.5 (19–50)	40 (30–51)	39.5 (26–48)	25† (0–49)
Sleeping	2.5 (0–18)	5 (0–15)	3 (0–15)	6 (0–18)	0 (0–15)	0 (0–12)	0 (0–12)	3 (0–21)	7.5 (0–22)

*Medians and ranges of behavioral scores over 3 hr after *d*-amphetamine (2 mg/kg i.p.) are shown. Immediately after amphetamine administration, all rats received injections into the nucleus accumbens through indwelling cannulae. Doses shown were delivered bilaterally in 1-μl injectate over 1 min. Animals were observed for 30 sec every 10 min for 3 hr and scored by a "blind" observer using a standard sampling technique. The maximum possible score for each item was 54. Peptide scores were compared with artificial CSF scores and haloperidol scores with tartaric acid scores, by the Mann-Whitney U-test. (From Ervin *et al.*[20] with permission.)

† p < 0.05.
‡ p < 0.0001.
§ p < 0.01.

TABLE 3

EFFECTS OF INJECTIONS INTO CAUDATE NUCLEUS ON SELECTED
BEHAVIORAL RESPONSES TO AMPHETAMINE*

		Neurotensin (µg)			
	CSF	3.0	5.0	Tartaric acid	Haloperidol (5.0 µg)
Behavioral items (n)	13	10	10	12	10
Stereotypes (consisting of sniffing stereotype, head bobbing, biting, and licking)	72 (46–98)	59.5 (32–92)	57 (14–98)	67 (45–99)	4.5† (0–39)
Forward locomotion	9 (0–53)	10 (1–33)	7 (0–16)	3 (0–26)	16† (1–52)
Rearing	2 (0–14)	3 (0–32)	3 (0–9)	1 (0–26)	7.5 (0–37)
Sniffing	11 (2–32)	10.5 (2–25)	20.5 (0–47)	13 (4–19)	38.5 (28–47)
Sleeping	0 (0)	0 (0)	0 (0)	0 (0)	0 (0–2)

*Medians and ranges of behavioral scores over 3 hr after d-amphetamine (5 mg/kg i.p.)
are shown. Immediately before amphetamine administration, all rats received injections into
the caudate nucleus through indwelling cannulae. Doses shown were delivered bilaterally in
1 µl injectate over 1 min. Animals were observed for 30 sec every 10 min for 3 hr, and scored
by a "blind" observer using a standard sampling technique. Peptide scores were compared
with artificial CSF scores, and haloperidol with tartaric acid scores, by the Mann-Whitney
U-test. (From Ervin et al.[20] with permission.)
†p < 0.01.

NEUROTENSIN–DOPAMINE INTERACTIONS: NEUROCHEMICAL CORRELATES

Because NT, given centrally, produces effects characteristic of neuroleptics,
and because neuroleptics block DA receptors, we examined the effect of NT, *in
vitro* on the binding of ³H-spiperone to the DA receptor in rat brain membranes of
nucleus accumbens and striatum. No effect of NT was observed.[8–10,13] In addition,
no effect on *in vitro* ³H-apomorphine binding to rat brain membranes was observed
after addition of NT (unpublished observations). We also studied the effects of i.c.
NT (30 µg) injection on subsequent *in vitro* ³H-spiperone binding to DA receptors
in rat brain membranes of nucleus accumbens and striatum. No effect of NT was
observed (Nemeroff *et al.*, in preparation). Furthermore, in collaboration with
Mailman, we recently observed no effect of NT on basal or DA-stimulated

adenylate cyclase activity in nucleus accumbens or striatal tissue. These two findings—the null NT effects on spiperone binding and on DA-stimulated adenylate cyclase activity—demonstrate that NT, unlike neuroleptics, does not interact directly with the DA receptor.

The above evidence notwithstanding, NT does alter the biochemical dynamics of DA systems. As described by Widerlöv and Breese in a short communication,[24] and in this volume,[25] as well as by others,[26] i.c.v. NT, like neuroleptic treatment, produces increases in the concentration of the metabolites, DOPAC and HVA, in the major DA terminal fields (nucleus accumbens, olfactory tubercle, hypothalamus and striatum). The NT effect on DA turnover in the striatum is problematical in view of the null effect of intra-caudate NT on nigroneostriatal-mediated behaviors.

Recently, Govoni *et al.*[27] reported that the content of immunoreactive NT in the nucleus accumbens is significantly increased after a single injection of haloperidol. Chronic neuroleptic treatment produced a gradual and progressive increase in NT levels in the nucleus accumbens (TABLE 4) and striatum; no such change was measured in the amygdala, septum, or hypothalamus. Other clinically efficacious neuroleptics including chlorpromazine and trifluoperazine exerted similar effects whereas two clinically inactive phenothiazines, promazine and promethazine, were without effect.

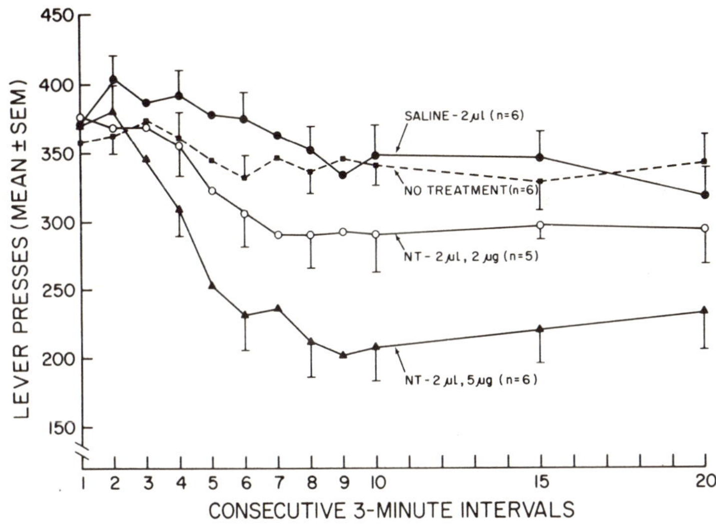

FIGURE 4. The effects of intra-accumbens neurotensin on bar pressing for electrical self-stimulation of the A-10 region in rats. Neurotensin (2 or 5 μg) was injected into the nucleus accumbens on the side ipsilateral to the site of electrical stimulation and rats were placed immediately into the experimental chamber. Rats received electrical stimulation following each bar press. The number of bar presses in successive three-minute periods is represented. Neurotensin produced a dose-dependent reduction in the number of lever presses.

TABLE 4

Time Course of Neurotensin-IR Increase in Rat Nucleus Accumbens
after Repeated Haloperidol Administration*

Treatment	Neurotensin-IR	% of Controls
	pmol/mg protein	
Saline	0.32 ± 0.03	
Haloperidol (2 mg/kg i.p.)		
7 days	0.57 ± 0.05†	174
14 days	0.62 ± 0.02†	191
21 days	0.68 ± 0.03†	211

*Values are expressed as mean ± SEM, N = 6 to 9. (From Govoni et al.,[27] with permission.)

† p < .01 when compared with saline-treated rats.

The neuroleptic-like action of NT in laboratory animals as well as the findings of Govoni et al.[26] led to our study of CSF NT levels in schizophrenics (Widerlöv et al., this volume).[28] A subgroup of drug-free schizophrenics had very low CSF levels of the peptide; those patients showed increases in CSF NT levels after neuroleptic treatment. The precise role of NT in the pathogenesis of schizophrenia and other mental disorders remains to be elucidated.

EFFECTS OF NEUROTENSIN INJECTION INTO THE
VENTRAL TEGMENTAL AREA (VTA)

Recently, we have investigated the behavioral and biochemical sequelae of intra-VTA NT injection,[29] and these data are described elsewhere in this volume.[1] In brief, bilateral injection of NT into this area of origin of DA cell bodies of the mesolimbic DA system produces activation of this system as evidenced by both large increases in locomotor activity (FIG. 5) and in the concentration of DA metabolites in the nucleus accumbens, a major termination site of the circuit. Bilateral injection of NT into the substantia nigra (Aq), site of origin of the nigro-neostriatal system, produces no overt behavioral effects. Further evidence that the increase in spontaneous locomotor activity after intra-VTA NT is due to activation of the mesolimbic DA system was provided by the observation that i.c.v. injection of haloperidol blocked this NT effect. Finally, direct evidence that NT modulates neuronal activity in both the VTA and the nucleus accumbens is provided by the observation that i.c.v. injection of haloperidol blocked this NT effect. Finally, direct evidence that NT modulates neuronal activity in both the VTA and the nucleus accumbens is provided by the recent observations of Kalivas et al.[30] that intra-accumbens injection of NT blocks the behavioral hyperactivity induced by intra-VTA injection of NT.

CONCLUSIONS

It appears clear from the data reviewed in this report that centrally administered NT modulates activity of the mesolimbic DA system, both at its origin

(VTA, A_{10}) and at its terminal fields (e.g., nucleus accumbens, olfactory tubercles). Injection of NT into the nucleus accumbens, like neuroleptic drugs, blocks activation of this system whether induced by *d*-amphetamine, electrical stimulation of the VTA, or intra-VTA NT. Intraventricularly or i.c. administered NT produces a variety of physiological and pharmacological alterations that are similar to the effects of neuroleptic drugs. The similarities and differences between the effects of centrally administered NT and neuroleptics are summarized in TABLE

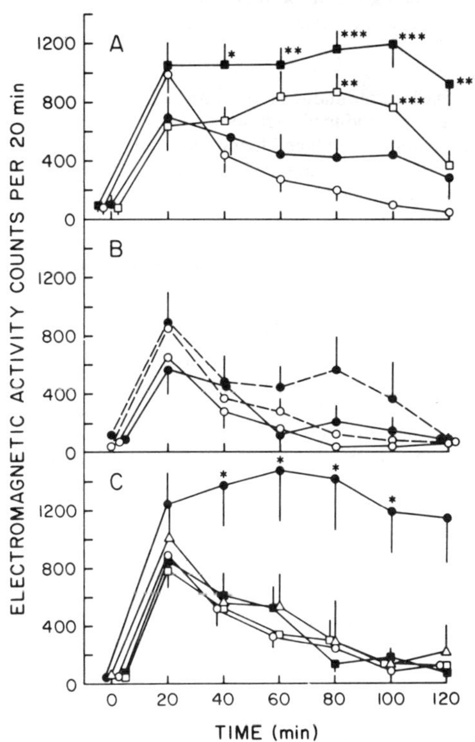

FIGURE 5. Electromagnetic motility scores following intra-VTA infusion of neurotensin and other peptides. Motor activity was integrated over 20 min intervals and data are shown as mean ± SEM. (In some instances SEM has been omitted for illustrative clarity.) Time zero is the mean number of counts in the 20 min just before injection. All doses are given as the amount of peptide injected/brain. A: ○, saline (n = 22); ●, 1 µg NT (n = 8); ■, 5 µg NT (n = 14). B: infusions were made bilaterally into the substantia nigra (----; n = 5; A/P 2.2 mm; M/L 2.0 mm; D/V --3.3 mm, relative to the interaural line) or unilaterally (0.6 µl/0.6 min) into the interpeduncular nucleus (---; n = 6; A/P 1.4 mm; M/L 0.0 mm; D/V--4.5 mm). Animals were given both saline (○) and 5 µg NT (●) on separate trials in randomized order. C: intra-VTA injection of saline (○; n = 18), luteinizing hormone releasing hormone (△; n = 5), thyrotropin-releasing hormone (□; n = 5), D-Pro[10]-NT (■; n = 9) or D-Tyr[11]-NT (●; n = 7). All peptides were injected at a dosage equimolar to 5 µg NT. * p < 0.05, using a two-tailed Student's paired t-test with probability values increased according to the Bonferroni method as described by Dunn. ** p < 0.01. *** p < 0.0001. (Reproduced from Kalivas *et al.*[29], with permission).

5. These findings have, of course, raised the issue as to whether NT may, at certain CNS loci, act as an endogenous neuroleptic-like substance. Although the peptide certainly does not interact directly with the DA receptor, it is clear that the peptide is capable of modulating DA neurotransmission. The question as to exactly where NT acts, at the synaptic level, to produce its effects remains unanswered. As noted

TABLE 5

COMPARISON OF THE PROPERTIES OF CENTRALLY ADMINISTERED
NEUROTENSIN AND NEUROLEPTIC DRUGS

I. Similarities
1. Hypothermia
2. Potentiation of barbiturate-induced sedation
3. Potentiation of ethanol-induced sedation
4. Blockade of amphetamine-induced locomotor activity and rearing
5. Blockade of methylphenidate-induced locomotor activity and rearing
6. Blockade of cocaine-induced locomotor activity
7. Muscle relaxation
8. Inhibition of a conditioned avoidance response
9. Reduction in the rate of rewarding electrical stimulation from the ventral tegmental area
10. Catalepsy
11. Reduction in food consumption in rats
12. Both neurotensin-induced and neuroleptic-induced hypothermia is antagonized by thyrotropin-releasing hormone (TRH)
13. Both neurotensin-induced and neuroleptic-induced muscle relaxation is antagonized by thyrotropin-releasing hormone (TRH)
14. Activation of the mesolimbic dopamine system by intra-VTA NT is blocked by both intra-accumbens NT and haloperidol

II. Differences
1. Neuroleptic drugs inhibit the *in vitro* binding of dopamine, dopamine agonists, and dopamine antagonists to brain membrane receptors. Neurotensin does not.
2. Certain neuroleptic drugs antagonize dopamine activation of adenylate cyclase activity in striatum. Neurotensin does not.
3. Neuroleptic drugs antagonize the behavioral effects of direct dopamine agonists (e.g. apomorphine) including stereotypy and increased locomotor activity. Neurotensin does not.
4. Neuroleptic drugs antagonize the stereotypic behavior induced by indirect dopamine agonists (e.g., d-amphetamine). Neurotensin does not.
5. Neurotensin produces a significant antinociceptive response. Neuroleptics do not.
6. Neuroleptics increase serum levels of prolactin. Centrally administered neurotensin decreases serum levels of prolactin.
7. Neuroleptic drugs shift the dose–response curve for amphetamine lethality to the right. Centrally administered neurotensin does not.
8. Neurotensin blocks the development of cold-restraint stress-induced gastric ulcers in rats. Neuroleptics do not.

above, i.c. NT blocks hyperactivity induced by indirect, but not direct, DA agonists. This would imply a presynaptic action of the peptide, but the behavioral apomorphine data were difficult to interpret (see above). In fact, Costall and Naylor[31] have recently accrued compelling evidence that the "routine use of apomorphine as a classical 'DA' agonist may require considerable caution." Recently Kalivas *et al.* have reported that intra-accumbens NT blocks the hyperactivity induced by intra-accumbens DA.[1] These results taken together with other available data discussed above indicate that NT probably acts distal to the DA receptor in the nucleus accumbens to reduce activity in the mesolimbic DA system. A partial presynaptic effect has not yet been ruled out and certain data are consistent with this notion.[24,32]

REFERENCES

1. KALIVAS, P. W., C. B. NEMEROFF & A. J. PRANGE JR. Neuroanatomical sites of action of neurotensin. Ann. N.Y. Acad. Sci. **400:** 307–318. This volume.
2. PRANGE, A. J. JR. & C. B. NEMEROFF. The manifold actions of neurotensin: A first synthesis. Ann. N.Y. Acad. Sci. **400:** 368–375. This volume.
3. LUTTINGER D., G. BISSETTE & G. D. FRYE. Effects of neurotensin on the actions of barbiturates and ethanol. Ann. N.Y. Acad. Sci. **400:** 259–267. This volume.
4. BISSETTE, G., D. LUTTINGER, G. A. MASON, D. E. HERNANDEZ & P. T. LOOSEN. Neurotensin and thermoregulation. Ann. N.Y. Acad. Sci. **400:** 268–282. This volume.
5. NEMEROFF, C. B., G. BISSETTE, P. J. MANBERG, A. J. OSBAHR III, G. R. BREESE & A. J. PRANGE, JR. 1980. Neurotensin-induced hypothermia: Evidence for interaction with dopaminergic systems and the hypothalamic-pituitary-thyroid axis. Brain Res. **195:** 69–84.
6. BISSETTE, G., D. LUTTINGER, C. B. NEMEROFF & A. J. PRANGE, JR. 1981. The effects of d-amphetamine, methylphenidate, and apomorphine on neurotensin-induced hypothermia in mice. Soc. Neurosci. Abstr. **7:** 32.
7. RIVIER, C., M. BROWN & W. VALE. 1977. Effect of neurotensin, substance P and morphine sulfate on the secretion of prolactin and growth hormone in the rat. Endocrinology **100:** 751–754.
8. NEMEROFF, C. B. 1980. Neurotensin: Perchance an endogenous neuroleptic? Biol. Psychiatry **15:** 283–302.
9. NEMEROFF, C. B., A. J. PRANGE, JR., D. LUTTINGER, D. E. HERNANDEZ, R. A. KING & S. K. BURGESS. 1981. Similarities and differences in the effects of centrally administered neurotensin and neuroleptics. Psychopharmacol. Bull. **17:** 145–147.
10. NEMEROFF, C. B., D. LUTTINGER, D. E. HERNANDEZ, P. W. KALIVAS, E. WIDERLOV, G. R. BREESE, & A. J. PRANGE, JR. 1982. Neurotensin, a peptide with certain neuroleptic-like properties. Proc. Third World Congress of Biol. Psychiatry. Elsevier, Amsterdam. pp. 363–367.
11. NEMEROFF, C. B., G. BISSETTE, A. J. PRANGE, JR., P. T. LOOSEN & M. A. LIPTON. 1977. Neurotensin: Central nervous system effects of a hypothalamic peptide. Brain Res. **128:** 485–496.
12. JOLICOEUR, F. B., A. BARBEAU, F. RIOUX, R. QUIRION & S. ST. PIERRE. 1981. Differential neurobehavioral effects of neurotensin and structural analogues. Peptides **2:** 171–175.
13. OSBAHR, A. J. III, C. B. NEMEROFF, P. J. MANBERG & A. J. PRANGE, JR. 1979. Centrally administered neurotensin: Activity in the Jolou-Courvoisier muscle relaxation test in mice. Eur. J. Pharmacol. **54:** 299–302.
14. KRUSE, H. 1975. Thyrotropin-releasing hormone: Interaction with chlorpromazine in mice, rats, and rabbits. J. Pharmacol. (Paris) **6:** 249–268.
15. LUTTINGER, D., C. B. NEMEROFF & A. J. PRANGE, JR. 1982. The effects of neuropeptides on discrete trial-conditioned avoidance responding. Brain Res. **237:** 183–192.
16. SNIJDERS, R., N. R. KRAMARCY, R. W. HURD, C. B. NEMEROFF & A. J. DUNN. 1982. Neurotensin induces catalepsy in mice. Neuropharmacology. **21:** 465–468.
17. DUNN, A. J. & R. HURD. Induction of catalepsy by central nervous system administrations of neurotensin and other neuropeptides. Ann. N.Y. Acad. Sci. **400:** 345–353. This volume.
18. PIJNENBERG, A. J., W. M. M. HONIG & J. M. VAN ROSSUM. 1975. Inhibition of d-amphetamine-induced locomotor activity by injection of haloperidol into the nucleus accumbens of the rat. Psychopharmacology **41:** 87–95.

19. NEMEROFF, C. B., D. LUTTINGER, A. J. OSBAHR III & A. J. PRANGE, JR. 1981. Neurotensin blocks behaviors induced by indirect but not direct dopamine agonists: Evidence for mediation at a mesolimbic dopamine termination site, the nucleus accumbens. Soc. Neurosci. Abstr. 6: 843.

20. ERVIN, G. N., L. S. BIRKEMO, C. B. NEMEROFF & A. J. PRANGE, JR. 1981. Neurotensin blocks certain amphetamine-induced behaviors. Nature 291: 73–76.

21. PROTAIS, P., J. COSTENIN & J. C. SCHWARTZ. 1976. Climbing behavior induced by apomorphine in mice: A simple test for the study of dopamine receptors in striatum. Psychopharmacology 50: 1–6.

22. LUTTINGER, D., C. B. NEMEROFF, R. A. KING, G. N. ERVIN & A. J. PRANGE, JR. 1981. The effect of injection of neurotensin into the nucleus accumbens on behaviors mediated by the mesolimbic dopamine system in the rat. *In* The Neurobiology of the Nucleus Accumbens. R. B. Chronister & J. F. de France, Eds. Haer Inst. Electrophysiol. Res. Sebasco, Maine. pp. 322–332.

23. MCCARTHY, P. S., R. J. WALKER, H. YAJIMA, K. KITAGAWA & G. N. WOODRUFF. 1979. The action of neurotensin on neurones in the nucleus accumbens and cerebellum of the rat. Gen. Pharmacol. 10: 331–333.

24. WIDERLÖV, E., C. KILTS, R. MUELLER, R. MAILMAN, C. B. NEMEROFF, A. J. PRANGE, JR. & G. R. BREESE. 1981. Neurotensin increases dopamine metabolites in rat brain. The Pharmacologist 23: 139.

25. WIDERLÖV, E. & G. R. BREESE. Actions of neurotensin on dopaminergic and serotoninergic pathways in rat brain. Ann. N.Y. Acad. Sci. 400: 428–430. This volume.

26. GARCIA-SEVILLA, J. A., T. MAGNUSSON, A. CARLSSON, J. LEBAN & K. FOLKERS. 1978. Neurotensin and its amide analogue (Gln⁴)-neurotensin: Effects on brain monoamine turnover. Naunyn-Schmiedebergs Arch. Pharmakol. 305: 213–218.

27. GOVONI, S., J. S. HONG, H.-Y. T. YANG & E. COSTA. 1980. Increase of neurotensin content elicited by neuroleptics in nucleus accumbens. J. Pharmacol. Exp. Ther. 215: 413–417.

28. WIDERLÖV, E., L. H. LINDSTROM, G. BESEV, P. J. MANBERG, C. B. NEMEROFF, G. R. BREESE, J. S. KIZER & A. J. PRANGE, JR. Subnormal CSF-levels of neurotensin in a subgroup of schizophrenics and normalization after neuroleptic treatment. Ann. N.Y. Acad. Sci. 400: 418–419. This volume.

29. KALIVAS, P. W., C. B. NEMEROFF & A. J. PRANGE, JR. 1981. Increase in spontaneous motor activity following infusions of neurotensin into the ventral tegmental area. Brain Res. 229: 525–529.

30. KALIVAS, P. W., C. B. NEMEROFF & A. J. PRANGE, JR. 1982. Neuroanatomical site-specific modulation of spontaneous motor activity by neurotensin. Eur. J. Pharmacol. 78: 471–474.

31. COSTALL, B. & R. J. NAYLOR. 1980. Behavioral aspects of dopamine agonists and antagonists. *In* The Neurobiology of Dopamine, A. S. Horn, J. Korf & B. H. C. Westerink, Eds. Academic Press. New York. pp. 555–576.

32. RECHES, A., R. E. BURKE, D. JIANG, R. H. WAGNER & S. FAHN. The effect of neurotensin on dopaminergic neurons in rat brain. Ann. N.Y. Acad. Sci. 400: 420–421. This volume.

DISCUSSION OF THE PAPER

C. F. FERRIS (*University of Massachusetts, Worcester, MA*): As I understand it the microinjection technique is a tool to find a site of action of a peptide. I do not think that the way it was employed in some of the studies presented earlier was effective in defining where NT acts in the central nervous system. One should consider three things: the dose, the volume, and the rapidity with which what you are giving is being cleared and metabolized. The doses of neurotensin that are being injected are 1000 times greater than all of the neurotensin in the central nervous system.

C. B. NEMEROFF (*University of North Carolina, Chapel Hill*): As far as I know, no one has ever measured the concentration of neurotensin in the synaptic cleft. Therefore, when we microinject NT into a specific brain area we do not know what concentration we should attempt to achieve.

FERRIS: I understand that reasoning certainly, but when one microinjects three micrograms into an area, one exceeds the endogenous level at that site by as much as a million times.

NEMEROFF: As you know, there are very avid peptidases present throughout the brain, so, we do not know what the time course of NT inactivation is.

What I wanted to say previously is that as Dr. L. Iversen has said, the amount of norepinephrine one needs to inject either intra-ventricularly or directly into a brain site to see a behavioral effect is many orders of magnitude greater than the endogenous concentration of that substance. Yet, few of us would argue that norepinephrine or acetylcholine, for example, are not involved in certain discrete behavioral and physiological functions.

I would also point out that certain of the injections use very small doses. The doses described by Dr. Greidanus this morning were very low. We see effects on analgesia with neurotensin in mice of 0.1–1 ng. Certain of Dr. Kalivas's doses of NT were 0.2 μg micrograms. There is also relative specificity in that when Dr. Kalivas injects LHRH into the VTA, even at high doses, he sees no effect.

FERRIS: Apart from dose, volume is another consideration. This is important because the area of diffusion where the peptide is going is actually a function of that volume. In an elegant study Dr. Swanson microinjected different labeled amino acids and looked at this diffusion. His microinjection volumes were 10–50 nl and significant diffusion was observed. We have microinjected 50 nl of iodinated neurotensin, and done serial frozen sections, and found diffusion as far as 200 microns.

NEMEROFF: Given the diffusion you have observed it is difficult to understand why Dr. Kalivas' data revealed inactive sites in close proximity to active sites.

Perhaps, Dr. Kalivas might comment on this.

P. W. KALIVAS (*University of North Carolina, Chapel Hill*): It is true that we can identify a responsive area and then circumscribe it with microinjections that prove ineffective.

FERRIS: Theoretically if you inject one μl it diffuses into a space of one cubic millimeter.

KALIVAS: You must assume a concentration gradient across the cubic millimeter, and the concentration at the center will be much higher than at the perimeter. Therefore the concentration one mm away from the site would not be high enough to induce the behavior.

NEMEROFF: Although there are surely problems of interpretation of neurotensin data, I think that it should be recalled that the technique has proven enormously useful in neurobiology. This has been amply demonstrated in the study of angiotensin and drinking behavior.

P. C. EMSON: (*MRC Centre, Medical School, Cambridge, England*): If you take one of the most abundant cortical peptides, cholecystokinin, in the rat brain and then you estimate that it is present in perhaps 1% of the total synapses, you obtain a rough estimate of the concentration of this peptide at the synaptic terminal in the millimolar range.

Now, with the classical transmitters you are talking about, perhaps 100 mM concentrations. I think you can argue that some of the concentrations that are microinjected are still physiologically relevant.

R. ANDRADE (*Yale University of Medicine, New Haven, CT*): I am curious about the lack of effects you observe on dopaminergic function in the striatum. As you know, Kuhar has located the receptors to dopaminergic neurons that project to the caudate. Dr. Agajamian and I have shown that actually neurotensin is just as effective in exciting dopamine neurons in substantia nigra as in the ventral tegmental area.

Do you see any effect if you microinject neurotensin into the caudate?

NEMEROFF: We did not see any effect of neurotensin injected into the caudate.

As was reported by Dr. Kalivas this morning, bilateral injection of NT into the pars compacta of the substantia nigra, does not produce any overt behavioral effects. Dr. A. Pert at NIMH has unilaterally injected it and has not seen rotation.

F. M. JOLICOEUR (*McGill University, Montreal, Que., Canada*): We have also been having a lot of problems with apomorphine. It is difficult to distinguish between apomorphine-induced stereotypy and locomotor activity. Recently we have been using *n*-propylapomorphine, a drug that increases locomotor activity. Neurotensin decreases this drug-induced hyperactivity, again, evidence for postsynaptic DA modulation.

INDUCTION OF CATALEPSY BY CENTRAL NERVOUS SYSTEM ADMINISTRATION OF NEUROTENSIN*

Adrian J. Dunn, Rolf Snijders,† Russell W. Hurd, and Neal R. Kramarcy

Department of Neuroscience
University of Florida College of Medicine
Gainesville, Florida 32601

INTRODUCTION

A number of reports have indicated that neurotensin (NT) has physiological and behavioral effects that resemble those of dopamine receptor antagonists.[1] These observations have led Nemeroff[2] to postulate that NT could be an endogenous neuroleptic. A common characteristic of neuroleptics is their ability to induce catalepsy. Yet in no previous reports has NT been reported to cause catalepsy, and in one recent study no cataleptic effects were found following intracerebroventricular (i.c.v.) injection of NT in rats.[3]

In these studies,[4] we sought to determine whether i.c.v. NT causes catalepsy in mice, and if so whether this effect was related to the hypothermia previously reported to follow this treatment.[5, 1]

METHODS AND MATERIALS

CD-1 male mice were obtained from Charles River (Wilmington, MA). Mice were implanted with plastic cannulae (Clay Adams, PE-50 bilaterally in each lateral ventricle as previously described.[6] After cannulation, mice were housed in individual cages on a 12–12 light cycle (lights on at 7 a.m.). Room temperature averaged 26°C. Injections were made with a 10 μl syringe (Unimetrics Anaheim, CA: 5010 RM) fitted with a 27-gauge needle extending 9 mm from a plastic (shrink wrap) spacer. One microliter of fluid was injected through each cannula.

Neurotensin was obtained from Bachem and TRH (thyrotropin-releasing hormone) from Abbott Laboratories. Peptides were dissolved in 0.9% saline. Oxotremorine was obtained from Nutritional Biochemicals. Naloxone was a gift from Endo Laboratories.

Catalepsy was tested using the bar test previously described.[7] A metal bar, 1.3 cm in diameter, was clamped horizontally 5 cm above the bench. Mice were placed with their forepaws on the bar and their hindpaws on the bench. Placing of the mice on the bar was important so as not to elevate baseline catalepsy scores (cf. also ref. 8). The mouse was placed in position holding it by the back of the neck taking care not to touch the tail or hindquarters in the process. The time taken from placing the mouse in position until both forepaws touched the bench was measured with a stopwatch to the nearest second. Uninjected or saline-injected mice typi-

*This research was supported in part by a grant from the National Institute of Mental Health (MH 25486).

†An exchange student under the University of Utrecht-University of Florida Neurobiology exchange program.

345

cally showed catalepsy scores of less than one second. Each mouse was tested 5 times at each time point and the median scores are reported.

After each catalepsy test, rectal temperature was measured with a thermister probe (Yellow Springs Instruments), and the mouse was returned to its home cage.

In any given experiment, half the mice were tested with NT on the first day and half with saline. On the following day, the treatments were reversed. Data presented are the mean scores of each group of animals. Because repeated testing of mice for catalepsy frequently revealed high scores in untreated animals, data are presented from animals tested only twice as described above.

RESULTS

In the first series of experiments, mice were injected with 0.5, 1.0, or 5.0 μg of NT or saline i.c.v. Catalepsy and body temperature was tested immediately before injection (0 min) and 10, 30, and 60 min and 24 hr afterwards. The mean rectal temperatures at each of these times are plotted in FIGURE 1A. As previously reported,[5, 1] NT caused a hypothermia of about 2°C. The hypothermia appeared in some animals at 10 min and was clearly evident in most at 30 and 60 min but had disappeared by 24 hours. The mean catalepsy scores for the same mice are shown in FIGURE 1B. There appeared to be a dose-dependent increase in catalepsy scores reaching a maximum at 60 min. Scores had not completely returned to baseline by 24 hr, but mice injected with saline vehicle also frequently showed apparent catalepsy at this time. This presumably reflects the "learned catalepsy" reported previously by Brown and Handley.[8] These results with NT confirmed the data obtained in a large number of preliminary studies in which catalepsy was observed with doses of NT as low as 20, 50, or 100 ng, but only in a minority of the mice tested.

To determine the duration of the catalepsy, a second experiment was performed in which catalepsy and body temperature were measured 1, 2, 3, and 24 hr following 1 μg of NT. The results showed maximum catalepsy scores 2 hours following i.c.v. NT; whereas maximum hypothermia occurred at one hour (FIG. 2).

While most mice injected with NT showed hypothermia, 9 out of 67 mice that showed hypothermia did not show catalepsy. Conversely, 8 of the 66 mice that showed catalepsy did not show hypothermia. Analyses of linear regression were performed between the fall in body temperature and the catalepsy score at both 30 and 60 min following NT. No significant correlations between the hypothermia and catalepsy were found.

As a further test of the relationship between hypothermia and catalepsy, we administered oxotremorine i.c.v. Oxotremorine (2 μg) induced pronounced hypothermia (2–3°C), while failing to induce catalepsy in any of the mice.

Because NT-induced hypothermia has been shown to be reversed by TRH,[8] we tested the effect of TRH on NT-induced catalepsy. In the first experiment, catalepsy and body temperatures were measured following i.c.v. injection of a mixture of 5 μg NT and 5 μg TRH. The results show that whereas TRH injected with the NT prevented the hypothermia, it only attenuated the catalepsy caused by NT (FIG. 3). In a second experiment, TRH (10 mg/kg) was injected intraperitoneally 5 min before the NT (5 μg i.c.v.). In this case also, TRH prevented the hypothermia, but only decreased the catalepsy scores (FIG. 4).

Finally we tested the ability of the opiate antagonist, naloxone, to prevent the effects of NT. Naloxone at a dose of 1 mg/kg (s.c.) was injected 10 min before

NT (2 μg), or 4 mg/kg 30 min before NT (5 μg), or 10 mg/kg 10 min before NT (10 μg). The results of the high dose are shown in FIGURE 5. Naloxone alone did not cause any significant hypothermia or catalepsy. NT alone caused a significant

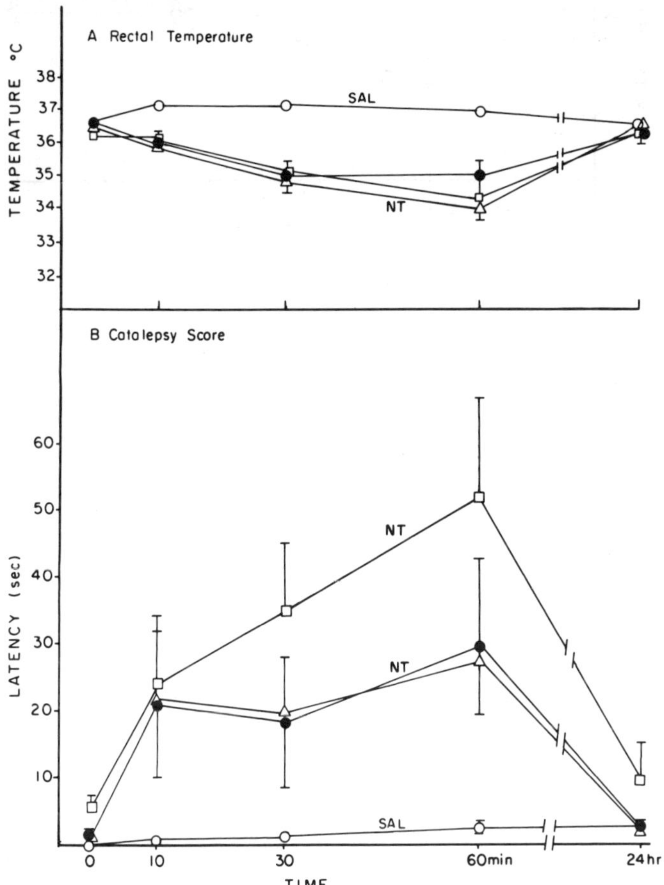

FIGURE 1.   Hypothermic and cataleptic responses to i.c.v. neurotensin. NT was injected i.c.v. at doses of 0 (o), 0.5 (Δ), 1.0 ($\bullet$) and 5.0 ($\square$) μg. Catalepsy and rectal temperature were measured immediately before NT injection and at 10, 30, and 60 min and 24 hr postinjection. A: mean rectal temperatures ($\pm$ SEM); B: mean catalepsy scores ($\pm$ SEM). Rectal temperatures were significantly lowered by all doses of NT at 30 and 60 min ($p < 0.001$, Student's t-test). Catalepsy scores were significantly elevated by 0.5 μg NT at 10, 30, and 60 min and by 5 μg NT at 30 and 60 min ($p < 0.05$, Wilcoxon matched pairs signed rank test). n $\geq$ 12. (Reproduced from Snijders *et al.*[4] with permission.)

hypothermia at 30, 60, and 120 min. Naloxone attenuated this hypothermia so that rectal temperature was not significantly different from control at any time point. At 30 min, body temperature in mice treated with NT plus naloxone was significantly different from those treated with NT alone ($p < .05$). Catalepsy, however,

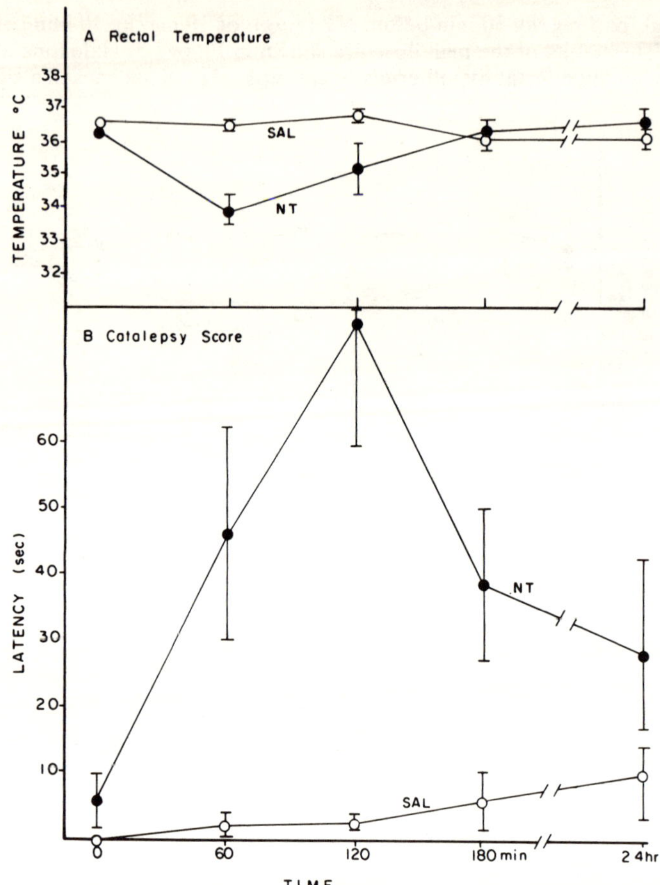

FIGURE 2.   Hypothermic and cataleptic responses to i.c.v. NT. Same as FIGURE 1 except that testing occurred at 0, 1, 2, 3, and 24 hrs after 1 μg NT. Hypothermia due to NT was significant at 60 (p < 0.005, t-test) but not at 120 or 180 min, and NT-induced catalepsy was significant at 60, 120, and 180 min (p < 0.01, Wilcoxon matched pairs signed rank test). n $\geq$ 7. (Reproduced from Snijders *et al.*[4] with permission.)

was not modified by naloxone pretreatment. Catalepsy scores were increased by both NT and NT plus naloxone at all time points and there were no significant differences between the two groups.

DISCUSSION

The present results clearly demonstrate that NT given i.c.v. can induce catalepsy in mice. This finding extends the catalog of behavioral and physiological effects of NT that resemble dopamine receptor antagonists.[1, 2] The only obvious

difference between the present study and an earlier one[3] that failed to find catalepsy following i.c.v. NT was our use of mice as opposed to rats.

The counteraction of the hypothermic effects of NT by TRH confirms previous
findings.[1] The partial antagonism by TRH of NT-induced catalepsy suggests that
the catalepsy and hypothermia do not share a common mechanism, and is consistent with the previously suggested dopamine agonist properties of TRH.

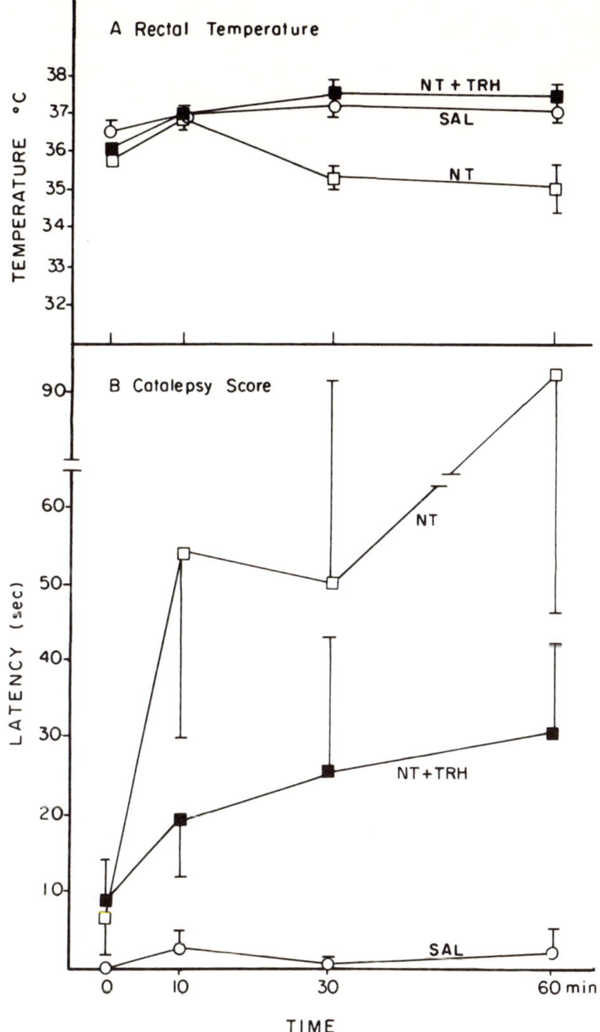

FIGURE 3.   Effect of i.c.v. TRH on NT-induced hypothermia and catalepsy. A and B: as
FIGURE 1, but TRH (5 μg) was injected i.c.v. concomitantly with the NT (5 μg). Hypothermia
was significant (p < 0.05) at 30 and 60 min after NT but not NT + TRH. Catalepsy was
significant after NT (□) or NT + TRH (■) at 10, 30, and 60 min (p < 0.05). n ≥ 7.

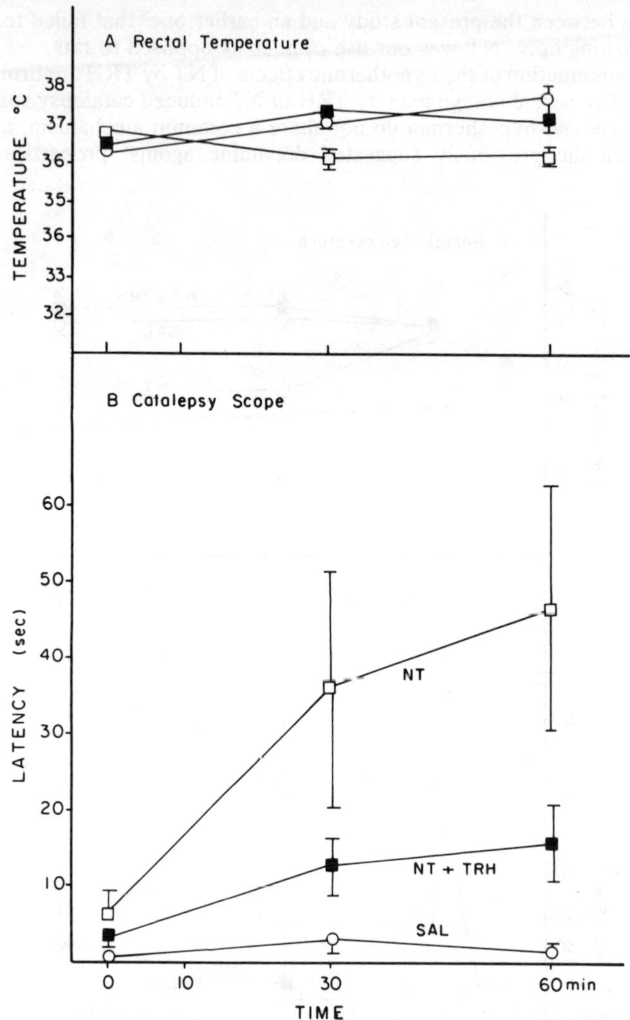

FIGURE 4.   Effect of i.p. TRH on NT-induced hypothermia and catalepsy. A and B: as
FIGURE 1, but TRH (10 mg/kg) was injected i.p. 5 min before i.c.v. NT (5 μg). NT-induced
hypothermia was significant at 60 min (p < 0.05). Catalepsy scores for NT were significantly
different from saline at 30 and 60 min (p < 0.01) and for NT + TRH at 60 min (p < 0.05).
n $\geq$ 7.

Remarkably, Holaday et al.[9] found that TRH (20 μg, i.c.v.) reversed the hypothermic effects of β-endorphin (30 μg, i.c.v.) while only partially reversing the
catalepsy.

Catalepsy has been reported to be caused by dopaminergic antagonists such as
reserpine, chlorpromazine and haloperidol on the one hand, and by a variety of
opiates on the other. A recent careful analysis by de Ryck et al.[10] has shown that

the two forms of catalepsy produced are distinguishable. Our observations suggest that NT-induced catalepsy resembles that induced by dopamine antagonists in bracing reactions to displacement, broad-based support, and weakness. This is

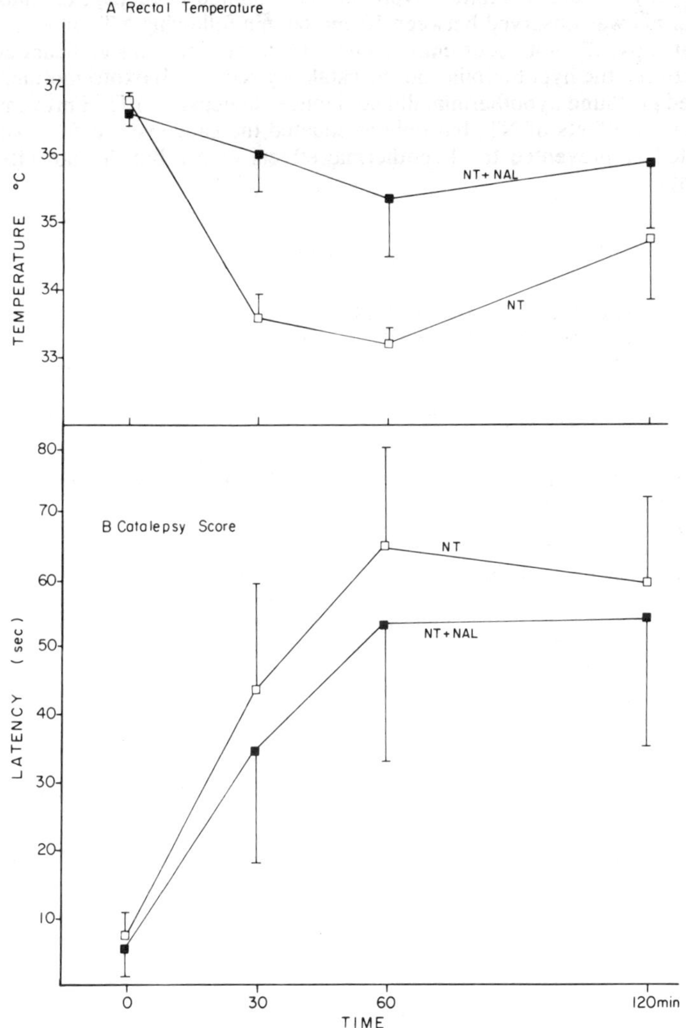

FIGURE 5.   Effect of naloxone on NT-induced hypothermia and catalepsy. Naloxone (10 mg/kg) was injected i.p. 10 min before NT (10 μg/i.c.v.). Hypothermia due to NT alone was significantly different from control at 30 min and 60 min (p < .001) and at 120 min (p < .05). In naloxone-treated animals, body temperature was not significantly different from control at any time point. A comparison of NT alone versus NT plus naloxone indicated a significant difference between the two at 30 min (p < .05). Catalepsy in NT-plus-naloxone-treated animals was similar to that in those treated with NT alone. Both groups showed significant catalepsy at 30, 60, and 120 min (p < .05). n ≥ 9.

supported by the failure of naloxone to block the catalepsy. Thus these data once again favor a neuroleptic-like activity for neurotensin.

Although NT induced pronounced hypothermia as well as catalepsy, we do not believe that the two effects were causally related for the following reasons: (1) Not all mice that exhibited hypothermia exhibited catalepsy; (2) maximum hypothermia was observed between 10 and 60 min following NT, whereas maximum catalepsy did not occur until 120 min; (3) there were no significant correlations between the hypothermia and the catalepsy scores; (4) oxotremorine, which produced profound hypothermia, did not induce catalepsy; (5) TRH prevented the hypothermic effects of NT, but only attenuated the catalepsy; and (6) naloxone attenuated or prevented the hypothermic effects of NT but did not affect the catalepsy.

REFERENCES

1. NEMEROFF, C. B., G. BISSETTE, P. J. MANBERG, A. J. OSBAHR, G. R. BREESE & A. J. PRANGE. 1980. Neurotensin-induced hypothermia: Evidence for an interaction with dopaminergic systems and the hypothalamic-pituitary-thyroid axis. Brain Res. **195:** 69–84.
2. NEMEROFF, C. B. 1980. Neurotensin: Perchance an endogenous neuroleptic? Biol. Psychiatry **15:** 283–301.
3. JOLICOEUR, F. B., A. BARBEAU, F. RIOUX, R. QUIRION & S. ST-PIERRE. 1981. Differential neurobehavioral effects of neurotensin and structural analogues. Peptides **2:** 171–176.
4. SNIJDERS, R., N. R. KRAMARCY, R. W. HURD, C. B. NEMEROFF & A. J. DUNN. 1982. Neurotensin induces catalepsy in mice. Neuropharmacology. **21:** 465–468.
5. MARTIN, G. E., C. R. BACINO & N. L. PAPP. 1980. Hypothermia elicited by the intracerebral microinjection of neurotensin. Peptides **1:** 333–339.
6. DUNN, A. J., S. R. CHILDERS, N. R. KRAMARCY & J. W. VILLIGER. 1981. ACTH-induced grooming involves high-affinity opiate receptors. Behav. Neural Biol. **31:** 105–109.
7. WAND, P., K. KUSCHUNSKY & K.-H. SANTOG. 1973. Morphine-induced muscular rigidity in rats. Eur. J. Pharmacol. **24:** 189–193.
8. BROWN, J. & S. L. HANDLEY. 1980. The development of catalepsy in drug-free mice on repeated testing. Neuropharmacology **19:** 675–678.
9. HOLADAY, J. W., L. F. TSENG, H. H. LOH & C. H. LI. 1978. Thyrotropin-releasing hormone antagonizes β-endorphin-induced hypothermia and catalepsy. Life Sci. **22:** 1537–1544.
10. DE RYCK, M., T. SCHALLERT & P. TEITELBAUM. 1980. Morphine versus haloperidol catalepsy in the rat: A behavioral analysis of postural support mechanisms. Brain Res. **201:** 143–172.

DISCUSSION OF THE PAPER

G. BISSETTE (*University of North Carolina, Chapel Hill*): We agree on the ability of TRH to block neurotensin-induced hypothermia. However, as you implied, we do have some disagreement on the naloxone data.

We found that in rats at 4°C ambient temperature, a 2 mg/kg dose of naloxone is ineffective in blocking neurotensin-induced hypothermia. In mice at an ambient temperature of 24°C, naloxone (5 mg/kg) did not block NT-induced hypothermia.

A. J. Dunn (*University of Florida, Gainesville*): I really do not know how to explain the difference. I can only say that we saw it in three experiments each of which used 10 animals per group. The naloxone produced an attenuation not complete blockade.

A. J. Prange, Jr. (*University of North Carolina, Chapel Hill*): Would you describe in a bit more detail what you mean by catalepsy?

Dunn: One is not aware of the NT-induced catalepsy until one starts testing latencies. We did look at some animals in the grip test, which is the test that van Ree has used with various endorphins. I can say that neurotensin is much more potent in our hands than des-Tyr-γ-endorphin, which gives a very transient catalepsy both by our test and the grip test. I would, certainly, emphasize that you are not seeing the same kind of catalepsy with neurotensin that you see with haloperidol, though it has some similar characteristics.

Prange: Is there any difficulty in rationalizing what we call the muscle-relaxing effect of neurotensin in the Joulou-Courvoisier test with your findings in the catalepsy tests, and the action of TRH in antagonizing both?

Dunn: No, none at all.

HUMAN BRAIN DISTRIBUTION OF NEUROTENSIN IN NORMALS, SCHIZOPHRENICS, AND HUNTINGTON'S CHOREICS*

P. J. Manberg,†,‡ C. B. Nemeroff,‡ L. L. Iversen,§ M. N. Rosser,§
J. S. Kizer,‡ and A. J. Prange, Jr.‡

†*Burroughs Wellcome Co.*
Research Triangle Park, North Carolina 27709

‡*Biological Sciences Research Center*
Departments of Psychiatry and Medicine and the Neurobiology Program
University of North Carolina School of Medicine
Chapel Hill, North Carolina 27514

§*MRC Neurochemical Pharmacology Unit*
Departments of Neurology and Neurosurgery
Addenbrooke's Hospital
Cambridge, England CB2 200

INTRODUCTION

Numerous biochemical and behavioral effects occur following administration of neurotensin (NT). This suggests that NT may play a physiological role in the regulation of several homeostatic mechanisms, including endocrine function, thermoregulation, reaction to painful stimulation, carbohydrate metabolism, and other aspects of mental function. Interactions between NT and dopaminergic systems, described elsewhere in this volume by Nemeroff *et al.*, are especially provocative in light of current theories concerning the cause of schizophrenia. In fact, the observed similarities between NT and antipsychotic drugs have even led to speculation that NT may be an endogenous neuroleptic (Nemeroff[1]). However, such a theory requires that the tridecapeptide or a closely related compound be present in areas of human brain critical to the functions that are disordered in that condition. In this report, we will present data that suggest that NT is distributed within the human brain in a pattern similar to that reported in lower species. Furthermore, studies on cerebrospinal fluid and postmortem brain tissue from patients with a diagnosis of schizophrenia or Huntington's chorea show that NT levels may in fact be altered in these states.

MATERIALS AND METHODS

Radioimmunoassay (RIA) of NT

The RIA employed for NT was developed with a synthetic standard supplied by Dr. J. Rivier (Salk Institute, La Jolla, CA) and Bachem (Torrance, CA). Antiserum was produced in New Zealand white rabbits by an antigenic challenge

*Supported by NIMH MH-34127, MH-33126, MH-33127, MH-2536, MH-00114 and NICHHD HD-03110.

of an NT-bovine serum albumin (BSA) conjugate. Radioiodination (^{125}I) of NT was accomplished with the chloramine T method followed by Sephadex gel-permeation chromatography.

All RIA samples (standards and unknown) were prepared in 200 μl of assay buffer (0.01 M NaPO$_4$, 0.15 M NaCl, 0.01% NaN$_3$, 0.4% dialyzed BSA, pH 7.6 = PBS:BSA). Antiserum was used in a final dilution of 1 : 1600 and incubations were extended for 1–2 days as previously recommended by Carraway.[2]

The site specificity of the antiserum (X-161) was determined by preparing standard curves (5–10,000 pg) for the following NT analogues or neuropeptides: D-Pro^7-NT, D-Pro^{10}-NT, GLN^4-NT, DES AA$^{1-7}$NT, DES AA$^{1-8}$NT, TRH, LHRH, SRIF, TSH, MIF, and oxytocin. NT immunoreactivity for "unknown" samples were calculated from plots of bound counts versus log dose of NT. All samples were assayed in duplicate in at least two dilutions.

TABLE 1

DEMOGRAPHIC CHARACTERISTICS OF PATIENT GROUPS

	Study 1 (FIG. 2)	Study 2 (FIG. 4)		
	Normals	Controls	Schizophrenics	Huntington's choreic
N	19	50	46	24
% male	68	66	59	58
Mean age (range)	67 (28–86)	71 (27–103)	58 (20–86)	54 (26–78)
Time to freezer (hrs; mean ± SEM)	40 ± 4.1	44 ± 3.5	40 ± 5.0	25 ± 4.0

Preparation of Human Brains

Human brain tissue was collected and dissected at the MRC Brain Bank (Cambridge, England) as previously described (Spokes[3,4]). In an initial study of normal subjects, brain samples from 13 males and 6 females who had no history of psychiatric or neurological illness were used. Their mean age was 67 years (range 28–86) and the mean interval between death and autopsy was 40 hours (range 7–94). Demographic characteristics of these patients and those in the subsequent comparison study are shown in TABLE 1.

Frozen brain samples were weighed, homogenized in approximately 20 volumes (w/v) of 1 M HCl, centrifuged at 12,000 × g for 30 minutes, and the clear supernatants lyophilized. Aliquots of brain homogenates were removed and protein concentrations were determined by the Lowry[5] method. The extent of recovery was assessed by adding small amounts (100 or 250 pg) of synthetic standard to aliquots of brain homogenates that were then prepared as described above. With this technique, recovery was found to be 85 ± 6.8% (n = 42). However, the values presented in this report have not been corrected for recovery.

Cerebrospinal fluid obtained from patients providing informed consent was frozen immediately in siliconized glass vials on dry ice before lyophilization.

All samples were reconstituted in buffer (PBS:BSA) and sonicated briefly before assay. Samples were coded and assayed in ignorance of their source.

Chromatographic Conditions

The chromatographic properties of human hypothalamus and nucleus accumbens were studied. Lyophilized supernatants from pooled samples of each tissue were reconstituted in 0.002 M ammonium acetate and initially purified by cation-exchange chromatography on a carboxymethyl cellulose (Whatman 52) column (1.4 × 25 cm) with 3 M acetic acid as the eluent. This initial purification was necessary in order to remove substances that interfered with subsequent high-pressure liquid chromatography (HPLC). After ion-exchange chromatography, samples were redissolved in 50% methanol:H_2O and were injected onto a Whatman Partisil 10 (C_{18} alkyl, 8% reverse-phase) magnum column. An isocratic elution solvent consisting of 50% distilled propanol and 50% aqueous solution containing pyridine (0.02 M pyridine, 0.009 M acetic acid) was used at a flow rate of 3 ml/minute. Duplicate aliquots from fractions that were collected every 18 seconds were dried and then assayed by RIA.

RESULTS

FIGURE 1 presents data on the cross-reactivity of NT analogs in the RIA. Endogenous neuropeptides unrelated to NT did not show cross-reactivity at the concentrations tested. Data from this experiment indicated that the antiserum (X-161) recognizes the midportion (amino acids 5–9) of the NT molecule. Scatchard analysis of NT binding activity showed a K_a (high affinity) of 1.1×10^{10} L/M. Intra-assay variability was 6–7%.

The regional distribution of immunoreactive NT in normal human brain is shown in FIGURE 2. As in other species, highest levels were present in the hypothalamus and certain limbic areas (i.e., nucleus accumbens, amygdala, septum) while relatively low levels were present in the cerebellum and many cortical areas. Low levels were seen in the caudate nucleus, in agreement with similar findings in the rat (Kobayashi *et al.*[6]) and monkey (Kataoka *et al.*[7]), but not the calf (Uhl

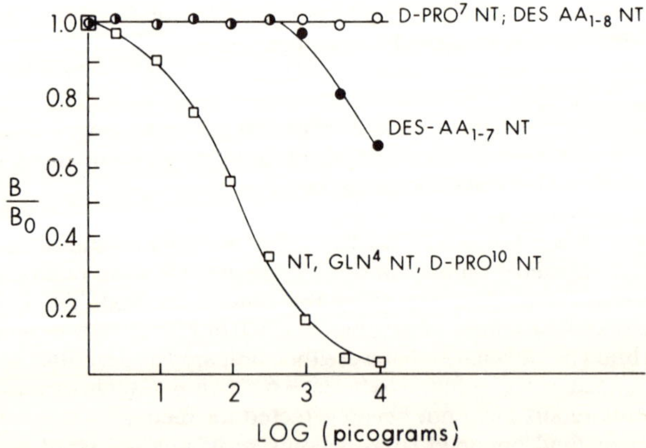

FIGURE 1. Standard curves for neurotensin and related peptides. All peptides were concurrently tested for cross-reactivity in the NT-RIA using antiserum X-161. The following peptides did not cross-react: TRH, LHRH, SRIF, TSH, MIF, and oxytocin.

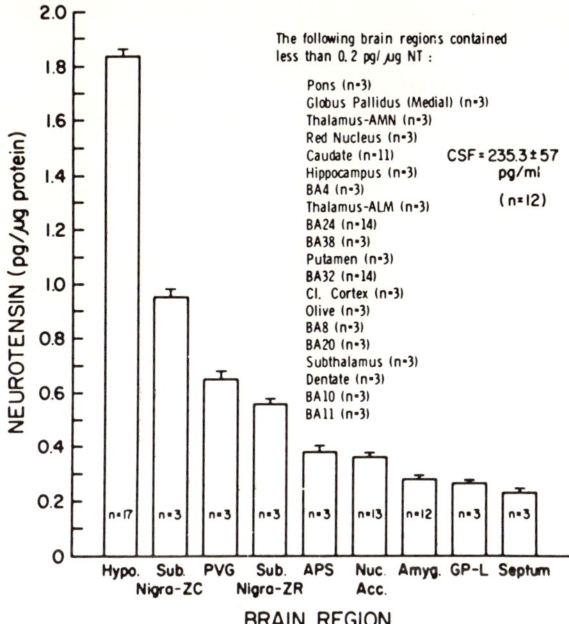

FIGURE 2. Regional distribution of immunoreactive NT in normal human brains and CSF.

and Snyder[8]). Higher levels of NT in the substantia nigra of humans in comparison to levels in other species were also noted.

Data were consistent with the results of Cooper *et al.*,[9] who also studied the distribution of NT in human brain. A rank order comparison of data from these two independent laboratories, using dissimilar antisera, yielded a correlation co-efficient of 0.94 (n = 16 brain areas). This remarkable consistency provides compelling evidence that the immunoreactive compound present in the human is similar if not identical to the synthetic peptide. Additional support is derived from the chromatographic evidence described below.

FIGURE 3 shows the HPLC elution profiles for human hypothalamus and nucleus accumbens. It is evident that the vast majority of immunoreactive material coeluted with synthetic standard after ion exchange and HPLC. A small (10%) secondary peak was observed for the nucleus accumbens, thus indicating that an immunologically related compound, possibly a precursor or a metabolite, may also be present in areas other than the hypothalamus. Nevertheless, because most of the material did coelute with synthetic standard, it is reasonable to assume that direct RIA provides an accurate approximation of actual NT brain levels.

The distribution of NT in brain regions from schizophrenic and Huntington's choreic patients is shown in FIGURE 4. It appears that NT concentrations are elevated in several areas, including the nucleus accumbens and caudate of the choreic's and Brodmann's area (BA 32) of the schizophrenics. If a standard Student's t-test or analysis of variance (ANOVA) is applied to these data, the changes in the caudate and BA 32 reach statistical significance (p < .05). However,

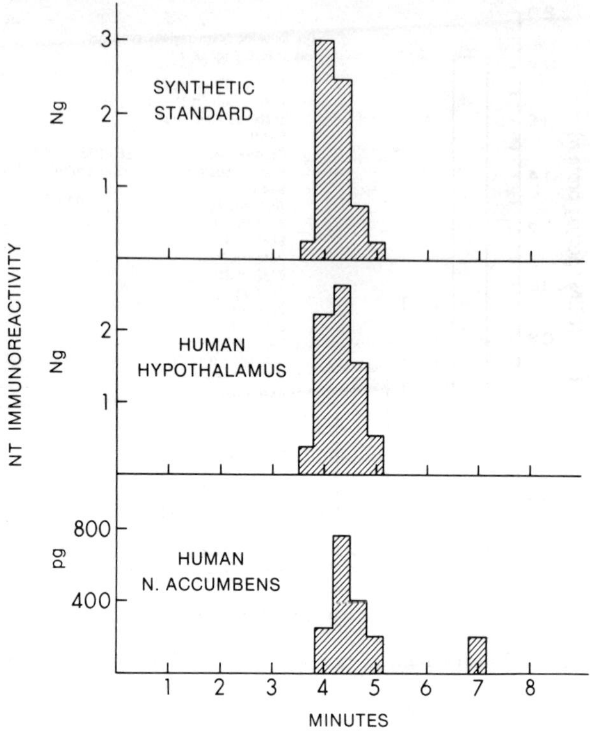

FIGURE 3. High-pressure liquid chromatography elution profiles of immunoreactive NT in human brain.

an alternative presentation, of individual NT values for each brain area, shown in FIGURES 5–11, provides a clearer understanding of the sources of the mean differences between groups. In the schizophrenic patients, concentrations of NT appear to be more variable than in the controls, thus agreeing with the experience of prior investigators of other possible biochemical indices of psychopathology (Bellak[10]). This increased variability is especially evident in the hypothalamus, caudate, BA 24, and BA 32, but not the amygdala. A tendency toward a bimodal distribution of NT in BA 32, the cortical area that shows a significant elevation in schizophrenics, can also be seen.

DISCUSSION

In previous distribution studies of NT, the peptide showed similar relative concentrations in such diverse species as the rat, calf, and monkey. Whichever the species, highest concentrations were usually found in hypothalamus, some limbic areas, periventricular gray, and substantia nigra, with lower amounts found in cortical areas, spinal cord, and striatum. Bovine caudate is a notable exception, Uhl and Snyder[8] finding high levels in this region.

Neurotensin is found in highest concentrations in those regions that subserve the functions affected by administration of the peptide. Thus, NT exerts effects on the regulation of body temperature, neuroendocrine function, locomotor activity, and nociception (Nemeroff *et al.*[11]), and the hypothalamus, nucleus accumbens, amygdala, and substantia nigra are anatomical areas important in one or another of these functions. The congruence of localization and certain functions suggest that such functions may reflect physiological roles. A detailed review of the anatomical sites that are most sensitive to the actions of NT is provided by Kalivas *et al.* in this volume.

In this report, we have presented data that suggest that the human distribution of NT resembles that seen in lower species. The conclusion that NT exists in human brain is reinforced by the similarity between the present results and those of Cooper *et al.*[9] This, along with the chromatographic profiles presented above, provide convincing evidence that the immunoreactive material in human brain is NT. Because the NT regional distribution in man and in other species is similar, it appears likely that the functions of NT have been conserved phylogenetically. The neuroleptic-like effects of NT in animals may thus be relevant to humans.

Present data, we think, demonstrate that NT exists in the brain of humans, as

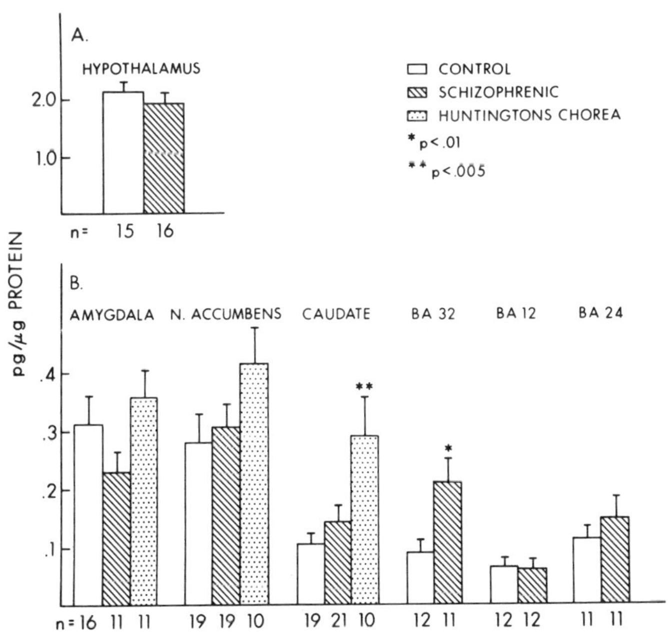

FIGURE 4.   Regional distribution of human immunoreactive NT from controls, schizophrenics, and Huntington's choreics.

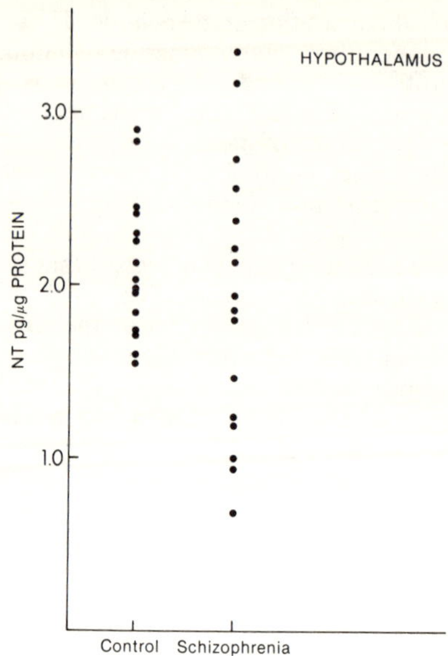

FIGURES 5–11.   Immunoreactive NT concentrations in various human brain regions. Symbols represent values for each individual included in the grouped data presented in FIGURE 4. Open symbols (△) denote Huntington's chorea patients who did *not* receive neuroleptic treatment. All schizophrenics received neuroleptics; controls did not.

it does in the brain of animals, and its regional localization tends strongly to be similar in all species. Furthermore, this localization is such that NT may plausibly be expected to interact with dopamine, a surmise that is supported by the results of the pharmacological experiments already mentioned. A further question is whether present data offer clues as to whether NT plays a role in either of the pathological conditions—schizophrenia or Huntington's chorea—studied by the biochemical examination of brains of patients with these disorders. Variations from normal NT levels were found in both conditions and each condition must be considered briefly. First, it is important to acknowledge the limitations that are imposed by having as data only the levels of NT or, for that matter, the levels of any substance. Turnover rates are surely the best indicator of neuronal activity, but we are presently unable to measure any of the parameters of the metabolism of NT, or of any peptide, that would assess peptide turnover. When we report that the level is, say, elevated, this finding, it must be acknowledged, is compatible with increased, decreased, or unaltered turnover.

A leading biological theory of the pathophysiology of schizophrenia is that in crucial central pathways, notably the mesolimbic system, dopamine activity is excessive in at least a subpopulation of schizophrenic patients. This concept is compatible with the finding of Widerlöv *et al*. (this volume) that a subpopulation of schizophrenics show a reduction of NT in the cerebrospinal fluid. However, present findings of normal levels of NT in the structures of the limbic system contribute little to the acceptance or rejection of this hypothesis for the reasons given above. We found NT to be elevated in one area of the cerebral cortex of the schizophrenics, and mental abnormalities attributable to cortical dysfunction

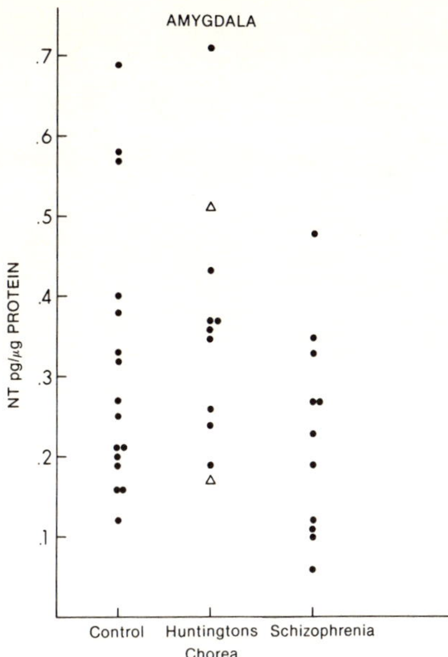

FIGURE 6.   See legend to FIGURE 5.

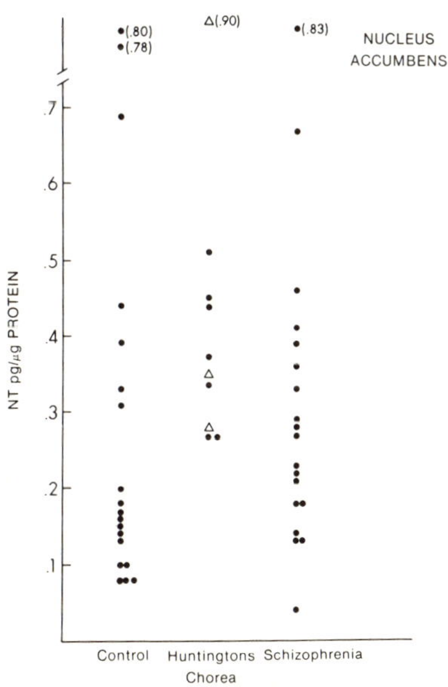

FIGURE 7.   See legend to FIGURE 5.

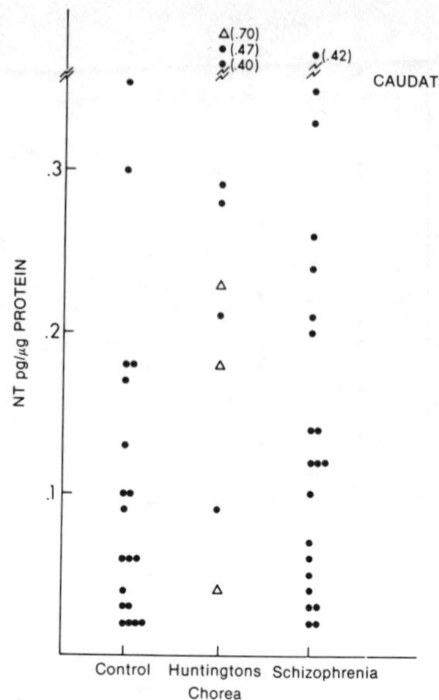

FIGURE 8. See legend to FIGURE 5.

can be noted in some schizophrenic patients. However, the possibility must be entertained that this observation is just a chance association. It is worth noting in passing that measurement of dopamine levels in brain areas from schizophrenic patients have not contributed decisively to a rejection or acceptance of the dopamine hypothesis of schizophrenia. Thus, some investigators have indeed found elevated levels of dopamine in critical areas such as the nucleus accumbens (Bird *et al.*[12]), but others have reported normal levels (Crow *et al.*[13]).

Present findings pertaining to the levels of NT in Huntington's chorea are more tantalizing. This condition involves, among other disabilities, a gross movement disorder and the caudate nucleus is known to modulate voluntary movement. In Huntington's chorea, dopamine has been reported to be elevated in the caudate nucleus (Bird[14]). It is in this context that the elevated levels of NT in the caudate nucleus of patients with Huntington's chorea suggest that NT may indeed play a role in this condition, quite possibly through an interaction with dopamine. This possibility is all the more plausible because of the many demonstrations of NT-DA interactions in both biochemical and behavioral models.

A final question on which present data might come to bear is the question of whether NT levels are responsive to the persistent use of neuroleptic drugs. Govoni *et al.*[15] have reported that repeated daily injections of various neuroleptics into rats produced a significant elevation of NT levels in the nucleus accumbens and caudate, but not in the hypothalamic system, cortex, or amygdala. Widerlöv *et al.* (this volume) have also shown that neuroleptic treatment can elevate NT levels in those schizophrenics who have abnormally low levels of CSF-NT before treatment. This evidence does indicate that NT levels can be altered by neuroleptics. However, the present data on postmortem NT brain levels do not support this conclusion since no relationship between neuroleptic drug treatment and NT

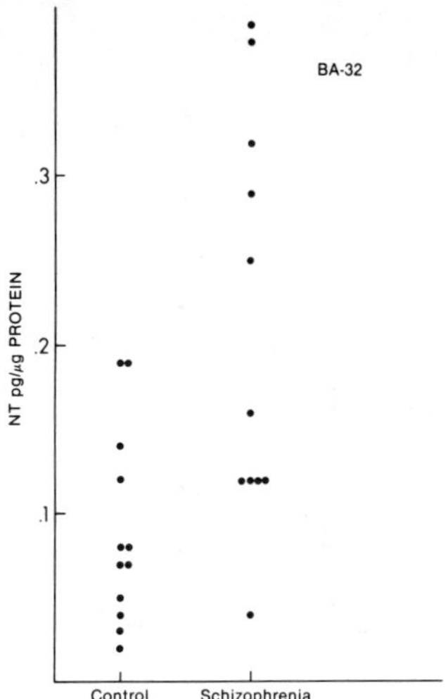

FIGURE 9.   See legend to FIGURE 5.

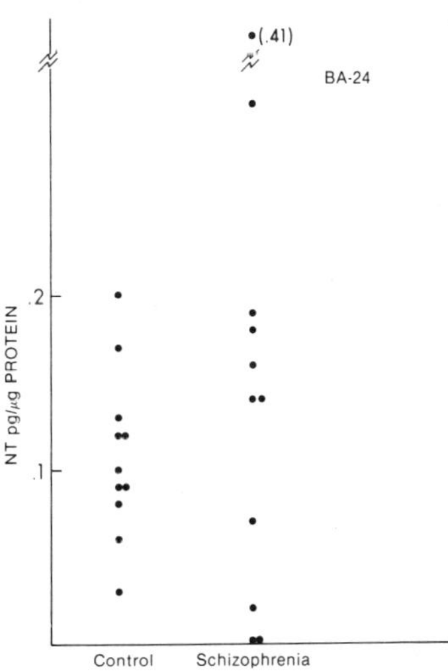

FIGURE 10.   See legend to FIGURE 5.

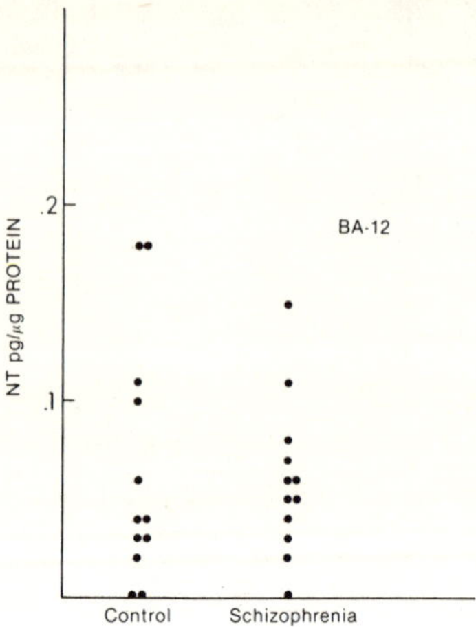

FIGURE 11.   See legend to FIGURE 5.

levels was observed. All of the schizophrenic patients and most of those with Huntington's chorea had received neuroleptic therapy but did not show uniform elevations in NT levels in the nucleus accumbens. While caudate NT levels were elevated in the Huntington's choreic patients, no increase was seen in the schizophrenics. We therefore conclude that the effect of neuroleptic drugs on human NT levels remains as an unresolved issue in this field.

In conclusion, the work described in this report demonstrates the presence of NT in human brain and supplies preliminary evidence that shows that the levels of this peptide may be altered in various diseases. These findings suggest that an elucidation of the physiological actions and behavioral effects of NT in laboratory animals may one day provide greater insight concerning human mental function.

REFERENCES

1. NEMEROFF, C. B., 1980. Neurotensin: Perchance an endogenous neuroleptic? Biol. Psychiatry **15:** 283–302.
2. CARRAWAY, R. E. 1979. Neurotensin and related substances. *In* Methods of Hormone Radioimmunoassay, 2nd ed. B. M. JAFFEE & H. R. BERHMAR, Eds. Academic Press. New York. pp. 139–169.
3. SPOKES, E. G. S. 1979. An analysis of factors influencing measurements of dopamine, noradrenaline, glutamate decarboxylase, and choline acetylase in human post-mortem brain tissue. Brain **102:** 333–346.
4. SPOKES, E. G. S. 1980. Neurochemical alterations in Huntington's chorea. A study of post-mortem brain tissue. Brain **103:** 179–210.
5. LOWRY, O. H. N. H. ROSEBROUGH, A. L. FARR & R. J. RANDALL. 1951. Protein measurement with the folin-phenol reagent. J. Biol. Chem. **193:** 265–275.
6. KOBAYASHI, R. M., M. R. BROWN & W. VALE. 1977. Regional distribution of neurotensin and somatostatin in rat brain Brain Res. **126:** 584–588.
7. KATAOKA, K., N. MIZUNO & L. A. FROEHMAN. 1979. Regional distribution of immunoreactive neurotensin in monkey brain. Brain Res. Bull. **4:** 57–60.
8. UHL, G. R. & S. H. SNYDER. 1977. Regional and subcellular distribution of brain neurotensn.

9. COOPER, P. E., M. H. FERNSTROM, O. P. RORSTAD, S. E. LEEMAN & J. B. MARTIN. 1981.
    The regional distribution of somatostatin, substance p. and neurotensin human brain.
    Brain Res. **218:** 219–232.
10. BELLAK, L., Ed. 1979. Disorders of the Schizophrenic syndrome. Basic Books, New
    York.
11. NEMEROFF, C. B., D. LUTTINGER & A. J. PRANGE, JR. 1980. Neurotensin: Central
    nervous system effects of a neuropeptide. Trends Neurosci. Sept. 1980. pp 212–215.
12. BIRD, E. D., E. G. SPOKES, J. BARNES, A. V. P. MACKAY, L. L. IVERSEN & M. SHEPHERD.
    1977. Increased brain dopamine and reduced glutamicacid decarboxylase and
    choline acetyl transferase activity in schizophrenia and related psychoses. Lancet **2:**
    1157–1158.
13. CROW, T. J., H. F. BAKER, A. J. CROSS, M. H. JOSEPH, R. LOFTHOUSE, A. LONGDEN, F.
    OWEN, G. J. RILEY, V. GLOVER & W. S. KILLPACH. 1979. Monoamine mechanisms in
    chronic schizophrenia: Post-mortem neurochemical findings. Br. J. Psychiatry **134:**
    249–256.
14. BIRD, E. D. 1980. Chemical pathology of Huntington's disease. Ann. Rev. Pharmacol.
    Toxicol. **20:** 535–551.
15. GOVONI, S., J. S. HONG, H.-Y. T. YANG & E. COSTA. 1980. Increase of neurotensin
    content elicited by neuroleptics in nucleus accumbens. J. Pharmacol. Exp. Ther. **215:**
    413–417.

DISCUSSION OF THE PAPER

R. E. CARRAWAY (*University of Massachusetts, Worcester*): I wanted to ask about the use of your antibody, possibly for looking at NT precursors. I noticed on your chromatogram that there was no high molecular weight material, but, since your antibody is directed toward the middle of the molecule, it would be ideal for viewing high molecular weight forms of neurotensin that are extended on either side. What kind of extraction procedure you did use?

P. J. MANBERG (*Burroughs Wellcome Co., Research Triangle Park, NC*): We used 1 M hydrochloric acid homogenization rather than the weak acid and boiling method used by other groups.

CARRAWAY: Have you pursued the precursor?

MANBERG: We have not observed it. When we did our chromatographic studies we first put our samples through a carboxymethylcellulose ion exchange column.

CARRAWAY: That is good reasoning.

MANBERG: We first did the ion exchange and, then, took the fractions off the ion exchange and put it on the HPLC.

CARRAWAY: Did you perform chromatography on any of the brain samples in any of the different patient populations?

MANBERG: We did not have enough material to do that.

D. KAHN (*Columbia University, New York*): I think that the real strength of your CSF study is that it recognized a bimodal distribution of the finding. Clearly talking about schizophrenia as a single entity is like talking about cancer or infection as single entities. Schizophrenia has many clinical representations and undoubtedly many etiological subtypes.

I have two question. Could you talk about the clinical subtypes within the CSF study? Secondly, would it be possible in any way to reanalyze the schizophrenia data in the post-mortem samples based on clinical subtypes or other characteristics of the English patients? There was such broad variance within that group that one wonders if there were subtypes.

MANBERG: The group that had low CSF NT values had significantly higher scores on the psychomotor retardation subscale that was used. The three catatonic patients were all included in the low NT CSF group.

S. R. BLOOM (*Royal Postgraduate Medical School, London*): The data are extremely interesting. You are looking at multiple correlations, and, since you are looking at more than 20 different variables, you may find a significant correlation by chance; it certainly looked a bit that way. Have you performed any statistical correction?

I am not sure that your distribution was normal. The values show a very wide range. It is skewed at both ends. So, I am not sure that the standard statistics apply.

When we studied post-mortem brain samples, we found that while initially we did observe a difference in neurotensin concentrations this was all age-corrected and when this was taken into account there was, then, no difference.

One wonders whether your 103-year-old control subject was completely healthy. I know that in the United States you regard old people as quite able to run affairs. Can you assure me that a 103-year-old was quite normal psychologically?

MANBERG: According to the data print out that was the age. Apparently the subject was healthy for at least 102 years. Again, we get all of our clinical information from the MRC brain bank. We did look for a number of correlations and we did not find any correlation between age and the individual NT values. Could you tell me what NT values you saw changed and in which brain area this occurred? If it was an age function you would probably see it in several brain areas rather than in just one.

BLOOM: If we corrected statistically for the fact that we were looking at nine neuropeptides across 27 brain areas, we found that nothing was statistically significant. Findings that appeared to be significant at first sight were nullified by the correction factor. I think this illustrates a difficulty when you realize that these people have died of pneumonia, or cancer or pancreatitis or whatever and that they were not particularly well psychologically. They also received lots of drugs.

C. B. NEMEROFF (*University of North Carolina, Chapel Hill*): The Cambridge MRC group that provided us with the samples were very careful about identifying the causes of death, such as bronchopneumonia, pancreatitis, etc. After talking to Tim Crow, whom I believe you collaborated with about the age effect, we had our biostatistician examine that issue closely. So, we observed no age correlation whatsoever nor has, I think, anybody else seen this in animal studies.

In addition, we looked at TRH and somatastatin in each of these brain regions to see whether there was a nonspecific effect on peptide concentration. In the region that Dr. Manberg saw the increase in neurotensin in the frontal cortex we saw a significant decrease in TRH. There was not a homogeneous rise or fall in peptide levels with the exception of the caudate nucleus of the Huntington's patients where we saw an increase in TRH, somatostatin, and neurotensin, suggesting, indeed, that tissue shrinkage is probably responsible for that phenomenon.

I would agree with you that the chance occurrence of a random statistically significant finding is, of course, a problem in studies of this kind where one is at the mercy of large numbers and uncontrollable variables.

R. A. HAMMER (*Northwestern University, Evanston, IL*): I appreciate the care with which the chromatographic work that you presented was done and, I want to echo Dr. Carraway's comments about possible subpopulations of neurotensin cross-reactive substances.

It is possible, I suppose, theoretically that those patients with low levels of neurotensin-like immunoreactivity in CSF have that second variant of neurotensin-like material that was running later on HPLC. I think it will be very valuable to keep enough CSF to do similar chromatography both in healthy patients and in schizophrenics in both populations.

G. Uhl (*Johns Hopkins University, Baltimore, MD*): Dr. Kahn has made a plea for the possible separation of schizophrenic subtypes. I would like to inquire if you have looked at the Huntington's chorea data in terms of the more rigid versus the more choreoathetoid forms of the disease. Have you gone back and looked at which of these patients were of the more rigid type?

Manberg: Not yet.

P. Emson (*Cambridge University, England*): We can confirm the neuroleptic effect that, I think, it was Dr. Govoni reported in the rat. If you give rats drugs like fluopenthixol you get a very, very substantial increase in neurotensin content in the striatum and the nucleus accumbens.

Manberg: We have not observed this effect after chronic administration of perphenazine but have with trifluoperazine.

P. Stokes (*Cornell University Medical College, New York*): I think that this was a very nice presentation, and I think that Dr. Nemeroff offered a useful clarification of how the data may be interpreted. I share Dr. Bloom's concern, and I wonder if you might utilize the high positive correlation between your data and that of Cooper *et al.* in terms of increasing the statistical value of your data? We, certainly, are all concerned about interpretation of cross-sectional single determination of levels of substances that we know are undergoing tremendous rates of production and degradation, but, I wonder if you might reinforce your data that way.

THE MANIFOLD ACTIONS OF NEUROTENSIN:
A FIRST SYNTHESIS *

Arthur J. Prange, Jr. and Charles B. Nemeroff

*Biological Sciences Research Center
Department of Psychiatry and the Neurobiology Program
University of North Carolina School of Medicine
Chapel Hill, North Carolina 27514*

Neurotensin (NT) is a substance of what Pearse[1] has called the amine precursor uptake decarboxylase, or APUD, system, that is, a substance that derives from neuroectodermal cells of the primitive neural crest, which is destined to form the nervous system and part of the gastrointestinal tract. Thus NT, like cholecystokinin, bombesin, somatostatin, and many other substances, is found in gut and brain. The amino acid sequences of some peptides can be found embedded in the sequence of one or another larger peptide. For example, the Met-enkephalin sequence is contained within the structure of β-endorphin.[2] However, the sequence of most APUD peptides is not embedded in larger peptides, and in this respect NT is typical of its class. In one chemical respect, NT resembles thyrotropin-releasing hormone (TRH) and luteinizing hormone releasing hormone (LHRH): its NH_2-terminus is a pyroglutamyl moiety, which may lend a measure of protection from enzymatic degradation. The arginine-arginine sequence in positions 8 and 9 render NT a rather basic molecule.

For whatever chemical reasons, NT, like most other peptides,[3] appears to be stringently excluded from penetrating the central nervous system by the blood–brain barrier (BBB). Certain other peptides, such as TRH and LHRH, do appear to cross the barrier, albeit in very small amounts, because after peripheral administration, certain effects that are typical of central administration can be noted.[4,5] The efficiency of the BBB vis-à-vis NT allows a simple classification of the effects of its administration. In TABLE 1, NT's effects on tissues or groups of tissues are listed, electrophysiological effects on single neurons being omitted. Owing to the efficiency of the BBB, peripheral administration of NT appears never to produce central effects. Central administration may produce peripheral effects but apparently only by alterations in neural or humoral outflow from the brain. Whether any effect of administered NT is physiological remains uncertain, but it appears plausible that many are.

Clearly some of the effects of NT are not unique. For example, it shares many properties in common with β-endorphin and bombesin.[6,7] Whether any of its effects are unique is uncertain, because in many systems in which NT is active, other peptides have not been tested. It seems likely that a peptide exerts a spectrum of effects and that the spectrum of any given peptide will overlap to some extent with those of certain others, though potency may vary.[8]

When administered into the brain, NT exerts several effects with considerable potency. Most of these effects have been first observed after the peptide has been introduced into the cerebroventricular system; some have now been noted after direct intracerebral application to discrete brain loci. Neurotensin enhances the

*Supported by NIMH MH-22536, MH-33127, MH-32316, MH-34121 and NICHHD HD-03110.

TABLE 1

EFFECTS OF NEUROTENSIN

I. After Central Administration

  A. On Central Functions

    1. Drug dependent

      a. Enhanced sleep after sedatives
      b. Diminished activity after stimulants
      c. Reduced analgesia in morphine-dependent animals

    2. Not drug dependent

      a. Diminished spontaneous motor activity
      b. Hypothermia
      c. Antinociception
      d. Muscle relaxation
      e. Diminished avoidance behavior in a discrete trial conditioned avoidance paradigm
      f. Diminished rate of electrical self-stimulation
      g. Diminished food consumption
      h. Catalepsy

  B. On Peripheral Function

    1. Endocrine effects

      a. Release of SRIF
      b. Release of vasopressin
      c. Reduced TSH and PRL responses to TRH
      d. Reduced release of TSH, PRL, GH, and LH

    2. Nonendocrine effects

      a. Reduced gastric ulceration after cold-restraint stress
      b. Reduced gastric acid secretion?

II. After Peripheral Administration

  A. On Central Functions (None)

  B. On Peripheral Functions

      a. Hypotension
      b. Vasodilation
      c. Hyperglycemia
      d. Hypoinsulinemia
      e. Hyperglucagonemia
      f. Increased levels of several anterior pituitary hormones
      g. Histamine release

activity of sedatives and antagonizes the activity of stimulants, and effects on drug metabolism do not appear sufficiently to account for these actions. The peptide also lowers body temperature and reduces the response to noxious stimuli (see ref. 6 for review). These effects, we think, promote withdrawal, recuperation, and anabolism and tend to insulate the organism from its environment. This orientation of the organism is essentially what Hess[9] meant by the trophotropic state. We think that NT, probably in conjunction with other peptides, serves to act within the CNS and, through it, to prepare the organism for trophotropism.

Three other observations, reviewed elsewhere,[6] can be viewed as consistent with the notion outlined above. The first is that the actions of NT are most apparent when the organism is subjected to a perturbation in the environment. It does not put animals to sleep; it only prolongs the action of sedatives and antagonizes the action of stimulants. Although it will reduce body temperature in a neutral environment, this effect is greatly amplified in a cold environment. Its antinociceptive effect, of course, cannot be observed unless the animal is given a painful stimulus. The second observation consistent with preparation for trophotropism is the quality of a peripheral effect of CNS administration of NT: reduction in gastric ulceration in rats that have been fasted and then exposed to cold and restraint.[10] The animal appears protected from an effect of generalized stress just as it is protected from pain.

Finally, many of the effects of centrally administered NT are offset by simultaneous administration of TRH, as enumerated in TABLE 2. It should be recalled that TRH has been viewed by Metcalf and Dettmar[11] and by Kruse[12] as an ergotropic substance, that is, a substance that promotes alertness, catabolism, and engagement with the environment. To the extent that this formulation is correct, the dualism between TRH and NT supports the case that the latter is a trophotropic substance.

Is the concept of a trophotropic role for NT in the CNS consistent with its actions in the periphery? Propaedeutically, it should be noted that the two sets of actions need not be harmonious, for apparently, as noted above, CNS NT does not reach the periphery, and in a similar way, peripheral NT probably does not reach the brain in important amounts, though NT injected into the anterior pituitary may do so.[13] When a substance occurs in both the brain and the periphery but each compartment is sequestered from the other, the substance may, in the two areas, be used for functions that tend in unrelated or even opposite directions. Epinephrine, which is stringently excluded by the BBB, provides an illustration. In the brain it stimulates vasodilator centers; in the periphery it modulates vasoconstriction. TABLE 1 lists a variety of effects that occur after the peripheral administration of NT. They address, for the most part, hemodynamics and glucose metabolism. Peripheral NT stimulates prolactin (PRL) release; central NT inhibits it. These contrary effects on a system directly regulated by dopamine (DA) may resonate with the opposed, site-specific effects of NT on DA activity within the brain (see below).

For present purposes, it is necessary to set aside the peripheral actions of NT

TABLE 2

EFFECTS OF NEUROTENSIN THAT ARE ANTAGONIZED
BY THYROTROPIN-RELEASING HORMONE

1. Enhanced sleep after sedatives
2. Hypothermia
3. Antinociception
4. Muscle relaxation
5. Diminished avoidance behavior in a discrete trial
   conditioned avoidance paradigm
6. Diminished food consumption
7. Catalepsy
8. Gastric cytoprotection

and to formulate as far as possible its central actions. Our investigations of the mechanisms by which these effects occur began with pharmacological manipulations. These manipulations were directed to DA systems for two reasons. First, early observations caused us to suspect that NT might function, in part, as a neuroleptic drug, all of which are DA-receptor blockers. Second, an aspect of our early work was the demonstration that NT would reduce the effects of DA releasers such as amphetamine and methylphenidate. Subsequent investigations of possible DA–NT interactions proceeded along both pharmacological and biochemical lines.

In general, whenever the issue has been studied, pharmacological experiments have revealed that impairment of central DA systems leads to enhancement of the actions of administered NT. Conversely, when DA activity is increased acutely, as by amphetamine administration, the effects of NT are generally reduced. It was pharmacological findings of this sort that inspired later biochemical investigations but, it now appears, also confused their interpretation. In a series of studies, we showed that NT, if given centrally in both rats and mice, will block the locomotor-stimulating effects of indirect DA agonists (amphetamine, methylphenidate, cocaine) but not the similar effects produced by DA-receptor agonists (apomorphine, lergotrile). With other observations, these data directed our attention, correctly it appears, to DA metabolism and more specifically, incorrectly it appears, to the presynaptic side of the synapse. We now think we were misled in the following fashion: doses of apomorphine and lergotrile were used that produce not only increased locomotor activity but stereotyped behavior as well; our automatic recording devices will to some extent count stereotypies as locomotion; NT does not block stereotypies. Thus, after apomorphine and lergotrile pretreatment, a false null effect for NT on "locomotor" behavior may have been obtained (see Nemeroff *et al.*, this volume).

Attempts have also been made to study the effects of DA manipulations not on NT effects but on NT itself and to assess the converse relationship, the effects of NT manipulations on DA metabolism. It can be seen from TABLE 3 that more work remains to be done in this area than has yet been accomplished. The effects of manipulations of DA on NT metabolism are unknown except for two studies. Govoni *et al.*[14] reported that neuroleptic administration in rats caused increased levels of NT in brain, and we have recently confirmed these findings (unpublished observations). Our group, with Widerlöv's leadership, found that NT in cerebrospinal fluid is very low in a subgroup of untreated schizophrenic patients.[15] When given neuroleptics, which in these patients always produced clinical improvement, NT levels normalized.

The effects of NT manipulations on DA metabolism have been studied more thoroughly. Although NT antibody has not been used as an independent variable for this purpose, NT administration has been employed extensively. The findings are given in the middle column of TABLE 4. Although TABLES 3 and 4 emphasize the gaps in present knowledge, the broad outline emerges that NT and DA are mutual antagonists. As noted above and in TABLE 2, TRH antagonizes many of the actions of NT. Whether this antagonism is modulated in whole or in part through DA mechanisms is unknown.

Recent pharmacological studies, conducted in our laboratory by Kalivas *et al.*, deserve special emphasis, for they have revealed a remarkable property of NT. After performing site injections of NT bilaterally into many areas of brain and showing that regions in which NT induces hypothermia are, by and large, different from regions in which NT induces antinociception, Kalivas *et al.* (refs. 16, 17, and this volume) built on the earlier work of Ervin *et al.*[18] Ervin and his colleagues

TABLE 3

EFFECTS OF MANIPULATIONS ON CENTRAL DOPAMINE
AND NEUROTENSIN SYSTEMS

Manipulation	Effect on Dopamine	Effect on Neurotensin
Dopamine Inhibition		
Neuroleptics	Receptor blockade with increased turnover	Increased levels
6-hydroxydopamine *	Neuronal destruction	No change
Dopamine Stimulation		
L-DOPA	Increased synthesis	Not done
d-amphetamine	Increased release	Not done
Cocaine	Reuptake blockade	Not done
Neurotensin Inhibition		
Antibody to NT	Not done	Decreased function
Neurotensin Stimulation		
NT central administration (see also TABLE 4)	Increased turnover	Acutely increased total levels

* Destruction of the mesolimbic DA system with 6-hydroxydopamine did not result in any significant reduction in NT concentration in the nucleus accumbens.

reported that NT administered into the nucleus accumbens blocks the locomotor effects of systemically administered amphetamine, though NT in the caudate nucleus does not block the stereotypy-producing effects of the drug. Kalivas and his colleagues [16] adduced the remarkable finding that bilateral injection of NT into the ventral tegmental area (VTA), the site of origin of the mesolimbic DA system, produces *increased* locomotor activity. This stands in sharp contrast to some properties of NT given intraventricularly. It reduces locomotor activity of the untreated animal and locomotor hyperactivity of the animal injected i.p. with amphetamine. The hyperactivity caused by amphetamine is also offset by NT injection into the nucleus accumbens. Still more remarkable was another finding: when given bilaterally into the nucleus accumbens, NT blocks the effects of its own bilateral administration into the VTA [17]; animals behave much as if they have received no treatment at all. This finding goes beyond the principle of site-dependence of a given effect. Not only is the action of NT on locomotor activity site dependent, the quality of its action is site selective. Not only does NT have opposite effects on locomotor activity at different sites, these different actions can cancel each other. Statements of this sort eventually may be applicable to other substances, but to the best of our knowledge there are not yet data to support them.

In the course of further work, it was possible to refine and extend knowledge of biochemical interactions between NT and DA. Widerlöv *et al.*[19] showed that after intracerebroventricular (i.c.v.) administration of NT, turnover of DA is increased in several areas of the brain including nucleus accumbens and striatum, as measured by a variety of standard methods. As summarized in TABLE 4, injection of NT into the nucleus accumbens of the unperturbed animal has no effect on either locomotor activity or DA metabolism. On the other hand, injection into the

TABLE 4

EFFECTS OF NEUROTENSIN INJECTION ON LOCOMOTOR ACTIVITY AND DOPAMINE TURNOVER IN RATS

| | Pretreatment | | | |
| | None or Saline i.p. | | Amphetamine i.p. | Dopamine intra-NA |
Neurotensin Injection Site	Locomotor Activity	Dopamine Turnover	Locomotor Activity	Locomotor Activity
None	—	—	Increased	Increased
Ventricular system	Tend to ↓	↑, NA olfactory tubercle and striatum *	Normalized	Normalized
Nucleus accumbens (NA)	No effect	No effect	Normalized	Normalized
Ventral tegmental area (VTA)	Increased	↑ in NA, not striatum †	—	—
NA and VTA	Normalized	↑ in NA, not striatum †	—	—

*Turnover was assessed by measurement of: (1) DOPAC and HVA levels and (2) DOPA accumulation after DOPA decarboxylase inhibition.
†Turnover was assessed by measurement of DOPAC and HVA levels.

VTA increases locomotor activity and enhances DA turnover, in the nucleus accumbens but not the striatum. When NT is injected into both the VTA and the nucleus accumbens, locomotor activity is reduced, that is, normalized, but DA turnover remains increased in the nucleus accumbens, as after NT injection into the VTA alone. Thus, nucleus accumbens injection of NT blocks the behavioral but not the biochemical consequences of NT injection into the VTA. This strongly suggests that the effect of NT on DA metabolism is not a presynaptic one. Further evidence for this view is summarized in TABLE 4. When DA itself is injected into the nucleus accumbens, animals are rendered hyperactive. When NT also is injected into the same area, activity is normalized. This effect of NT could hardly be modulated at presynaptic sites, for in this preparation, DA release is not an issue; it has been mimicked by DA administration.

Elsewhere in this volume, Kalivas and his colleagues present a diagram that accommodates present knowledge about the interplay of NT and DA in and between the VTA and the nucleus accumbens. Neurotensin is found in high concentration in both the VTA, where DA cell bodies occur in high concentration, and in the nucleus accumbens, to which VTA-DA neurons send axons, thus forming an important component of the mesolimbic system. What is not known is whether the NT neurons in the VTA, like DA neurons in the VTA, project to the nucleus accumbens, or whether alternatively, NT in the nucleus accumbens arises from sites other than the VTA. However, if only a portion of the NT in nucleus accumbens comes from cell bodies in the VTA, then perhaps NT cells, with cell bodies in the VTA, simultaneously exert effects of opposing sign in the mesolimbic DA system. Neurotensin clearly stimulates mesolimbic DA cell bodies to fire; NT from the same population of cells may inhibit the release of or, more likely, the effects of released DA. If a metaphor is allowable, NT simultaneously presses the accelerator and applies the brake. The result of such activity, we think, would be to reduce variance around the rate of propagation of DA-dependent nerve impulses through the VTA–nucleus accumbens system and beyond, whatever the mean rate of transmission may be.

The concept adumbrated above may hold profound implications for understanding the pathophysiology of schizophrenia. In other places we have noted, in descriptive rather then mechanistic terms, the similarities and differences between the actions of administered NT and neuroleptic drugs (ref. 20 and Nemeroff, this volume). TABLE 1 recounts the effects of NT, and many of its central effects are properties of neuroleptics as well. Moreover, still in general terms, DA activity may be disturbed, that is, increased, in at least a subset of schizophrenic patients,[21] and the dualistic counterbalance of DA and NT has been a theme of this review. As mentioned above, CSF NT levels appear grossly diminished in about half of untreated schizophrenic patients. It is in this context that the importance of modulation of DA in the mesolimbic system by NT is striking. A venerable observation about schizophrenic patients is that they are not so much excessive or deficient in one or another function as they are poorly regulated. Schizophrenic patients show striking intersubject variability and, over time, intrasubject variability. This variability may present itself as any of several physiological functions, including pain sensitivity and temperature regulation,[22] or as any of several psychological functions as revealed by formal testing.[23] We now suggest that NT, perhaps in concert with related substances, may reduce this variance by simultaneously stimulating and inhibiting DA pathways in the neuroanatomical system that most plausibly serves the functions at fault in schizophrenia, the mesolimbic system. Such an effect would be a sufficient reason to further consider NT as an endogenous neuroleptic substance.

REFERENCES

1. PEARSE, A. G. E. 1978. *In* Centrally Acting Peptides. J. Hughes, Ed. Macmillan Press. London. pp. 49–57.
2. ROSSIER, J. E., T. M. VARGO, S. MINICK, N. LING, F. E. BLOOM & R. GUILLEMIN. 1977. Proc. Natl. Acad. Sci. USA **74:** 5162–5165.
3. OLDENDORF, W. H. 1981. Peptides 2, Suppl. **2:** 109–111.
4. PRANGE, A. J., JR., C. B. NEMEROFF, P. T. LOOSEN, G. BISSETTE, A. J. OSBAHR III, I. C. WILSON & M. S. LIPTON. 1979. *In* Central Nervous System Effects of Hypothalamic Hormones and Other Peptides. R. Collu, A. Barbeau, J. Ducharme & J. Rochefort, Eds. Raven Press. New York.
5. MOSS, R. C. & S. M. MCCANN. 1973. Science **181:** 177–179.
6. NEMEROFF, C. B., D. LUTTINGER & A. J. PRANGE, JR. 1982. *In* The Handbook of Psychopharmacology. L. L. Iversen, S. D. Iversen & S. H. Snyder, Eds. Plenum Press. New York. Vol. 16: 363–467.
7. NEMEROFF, C. B., D. LUTTINGER & A. J. PRANGE, JR. 1980. Trends Neurosci. **3:** 212–215.
8. NEMEROFF, C. B. & A. J. PRANGE, JR. 1978. Arch. Gen. Psychiatry **35:** 999–1010.
9. HESS, W. R. 1954. Diencephalon: Autonomic and Extrapyramidal Function. Grune and Stratton. New York.
10. NEMEROFF, C. B., D. E. HERNANDEZ, R. C. ORLANDO & A. J. PRANGE, JR. 1982. Am. J. Physiol. **242:** G342–346.
11. METCALF, G. & P. W. DETTMAR. 1981. Lancet **i:** 586–589.
12. KRUSE, H. 1975. J. Pharmacol. (Paris) **6:** 249–268.
13. DORSA, D. M., E. R. DEKLOET, E. MEZEY & D. DEWIED. 1979. Endocrinology **104:** 1663–1666.
14. GOVONI, S., J. HONG, H-Y YANG & E. COSTA. 1980. J. Pharmacol. Exp. Ther. **215:** 413–417.
15. WIDERLÖV, E., L. H. LINDSTROM, G. BESEV, P. J. MANBERG, C. B. NEMEROFF, G. R. BREESE, J. S. KIZER & A. J. PRANGE, JR. 1982. Am. J. Psychiatry. **139:** 1122–1126.
16. KALIVAS, P. W., C. B. NEMEROFF & A. J. PRANGE, JR. 1981. Brain Res. **229:** 525–529.
17. KALIVAS, P. W., C. B. NEMEROFF & A. J. PRANGE, JR. 1982. Eur. J. Pharmacol. **78:** 471–474.
18. ERVIN, G. N., L. S. BIRKEMO, C. B. NEMEROFF & A. J. PRANGE, JR. 1981. Nature **291:** 73–76.
19. WIDERLÖV, E., C. D. KILTS, R. B. MAILMAN, C. B. NEMEROFF, A. J. PRANGE, JR. & G. R. BREESE. J. Pharmacol. Exp. Ther. In press.
20. NEMEROFF, C. B. 1980. Biol. Psychiatry **15:** 283–302.
21. CROW, T. J. 1979. Trends Neurosci. **2:** 52–54.
22. HOSKINS, R. G. 1946. The Biology of Schizophrenia. W. W. Norton & Co., Inc. New York.
23. SHAKOW, D. 1962. Arch. Gen. Psychiatry **6:** 1–17.

NEUROTENSIN AND CHOLESTEROL TRANSPORT *

Ludvik Peric-Golia,[†][‡] Clark F. Gardner,[†] and Milena Peric-Golia[†]

† Veterans Administration Medical Center
Salt Lake City, Utah 84148

‡ Department of Pathology
University of Utah School of Medicine
Salt Lake City, Utah 84132

We observed a significant transient hypercholesterolemia after intravenous injections of low doses of neurotensin (NT) in rats.[1] The degree of hypercholesterolemia was proportional to the dose, reached a maximum 15 min after the injections, and was not affected by adrenalectomy or hypophysectomy.[1] Substance P and bombesin did not affect the plasma cholesterol concentrations, but somatostatin induced a significant hypocholesterolemia.[2] In these and in the following studies, total plasma cholesterol was determined by the method of Weigensberg and McMillan.[3]

In the rats that received 24 hours earlier an intravenous injection of cholesterol-4-^{14}C, NT induced a significant increase in the concentration of the unlabeled, but not of the labeled, plasma cholesterol. This observation suggests that the increase in plasma cholesterol originates from a tissue in which endogenous and exogenous cholesterol had not yet equilibrated, as they had in the plasma.[4]

In guinea pigs, plasma cholesterol levels exceeded the control level by 38% in 5 min (p < 0.001), 47% in 15 min (p < 0.001) and 60% in 35 min (p < 0.001) after intravenous injections of 12.5 pmol/100 g body weight NT. The hypercholesterolemic effect of 125 pmol/100 g body weight NT was moderately more marked and was associated with a severe respiratory distress and tachycardia. Hypercholesterolemia reached 71% above the control level after the injection of 1.25 pmol/100 g body weight NT. This means that in guinea pigs, the hypercholesterolemic response to NT is greatest at a low dose. A decrease in ileal cholesterol concentration coincided with the increase in plasma cholesterol concentration after the intravenous injections of NT. This inverse relationship suggests that NT may stimulate the transport of cholesterol from the small intestine. After a continuous administration of NT, plasma cholesterol concentrations were significantly elevated on the third day but did not exceed the control levels on the fifth day. This indicates that an equilibrium may develop between the endogenous and exogenous NT.

NT is abundant in the small intestine[5] where cholesterol is synthesized, catabolized, and transported.[6] Biosynthesis and catabolism are slow processes, but changes in the transport of cholesterol may be rapid and, therefore, likely to produce the transient hypercholesterolemia after the injections of NT. The inverse relationship between the effects of NT on the plasma and ileal concentrations of cholesterol supports such a possibility.

* Supported by the Veterans Administration.

REFERENCES

1. PERIC-GOLIA, L., C. F. GARDNER & M. PERIC-GOLIA. 1979. The effect of neurotensin on the plasma cholesterol levels in the rat. Eur. J. Pharmacol. **55:** 407.
2. PERIC-GOLIA, L., C. F. GARDNER & M. PERIC-GOLIA. 1980. The effect of some brain/gut peptides on plasma cholesterol levels in the rat. Peptides **1:** 381.
3. WEIGENSBERG, B. J. & G. C. McMILLAN. 1959. Ultraviolet spectrophotometric method for the determination of cholesterol. Amer. J. Clin. Pathol. **31:** 16.
4. HOTTA, S. & I. L. CHAIKOFF. 1955. The role of the liver in the turnover of plasma cholesterol. Arch. Biochem. Biophys. **56:** 28.
5. CARRAWAY, R. & S. E. LEEMAN. 1976. Characterization of radioimmunoassayable neurotensin in the rat. J. Biol. Chem. **25:** 7045.
6. GREEN, P. H. R. & R. M. GLICKMAN. 1981. Intestinal lipoprotein metabolism. J. Lipid Res. **22:** 1153.

ELEVATION OF NEUROTENSIN IN THE CENTRAL REGULATION OF LUTEINIZING HORMONE RELEASE

C. F. Ferris, J. X. Pan, E. A. Singer, N. D. Boyd, and S. E. Leeman

University of Massachusetts Medical Center
Worcester, Massachusetts 02160

Neurons immunoreactive to gonadotropin releasing hormone (GRH) are localized to the medial preoptic septal region of the rat hypothalamus,[1] an area known to be essential for the regulation of ovulation and luteinizing hormone (LH) surge. Consequently, any bioactive substances found in this region of the medial preoptic nucleus are possible candidates as neurotransmitters or neuromodulators affecting GRH release. Since high concentrations of radioimmunoassayable neurotensin (NT) (144 ± 23 fmol/mg protein) are found in this region, this peptide and other bioactive substances were microinjected into this area and circulating levels of LH measured. Experiments were performed on Dial-urethane anesthetized rats three weeks after ovariectomy, since in preliminary experiments, we have found that this anesthetic effectively depresses the pulsatile pattern of plasma LH normally observed in these animals.

Fifty nanoliters of saline containing 20 pmoles of NT was injected into the rostral portion of the medial preoptic nucleus. The precise location of the injection site is given by the following coordinates taken from the stereotaxic atlas of Pellegrino *et al*.:[2] AP = 8.12 mm from vertical zero plane, L = 2.0 mm from the midsagittal suture and angled medially at 8° from the perpendicular, V = 7.0 mm below dura. Thirty minutes after the stereotaxic microinjection, plasma LH levels were increased up to 200 ± 28% above control (p < .01, actual concentrations were 441 ± 42 to 975 ± 128 ng/ml) and remained significantly elevated over the next 30 min (see Fig. 1). A similar stimulatory response was observed following the microinjection of norepinephrine (0.3 nmoles in 50 nl) into the same site (Fig. 1). In contrast, saline, substance P, Leu-enkephalin, and GRH microinjected in the rostral medial preoptic nucleus had no effect (Fig. 1). The small, discrete nature of this NT-sensitive region is demonstrated by the observation that injections of NT 0.4 mm rostral or caudal to AP 8.2 no longer produced an effect on LH levels.

In order to provide additional evidence for a physiological role for NT in this region, the presence of specific binding sites for NT in tissue homogenates of dissected medial preoptic areas was examined using I^{125}-NT. Increasing concentrations of I^{125}-NT (50 Ci/m mol) were incubated in Tris-acetate (50 mM, pH = 7.2) for 30 minutes at 4°C with washed (4×) tissue homogenates of 10 medial preoptic areas (96 mg of total tissue wet weight). 200 μl aliquots of the incubation mixture (containing 62 μg of total protein) were filtered under reduced pressure through GFC glass-fiber filters. A specific component of binding was observed when the nonspecific binding, defined as the binding measured in the presence of 1 μM unlabeled NT, was substracted from the total binding. Scatchard analysis of this specific binding suggested the presence of two binding sites for NT with Kd's of 4.2 nM and 29.0 nM and site concentrations of 0.11 pmoles/mg and 0.68 pmoles/mg, respectively.

Thus, the presence of both NT and specific binding sites of this peptide in the medial preoptic area, together with the observed functional effect of microinjection of small quantities of NT into this region, suggests a possible role for NT in the central regulation of gonadotropin release.

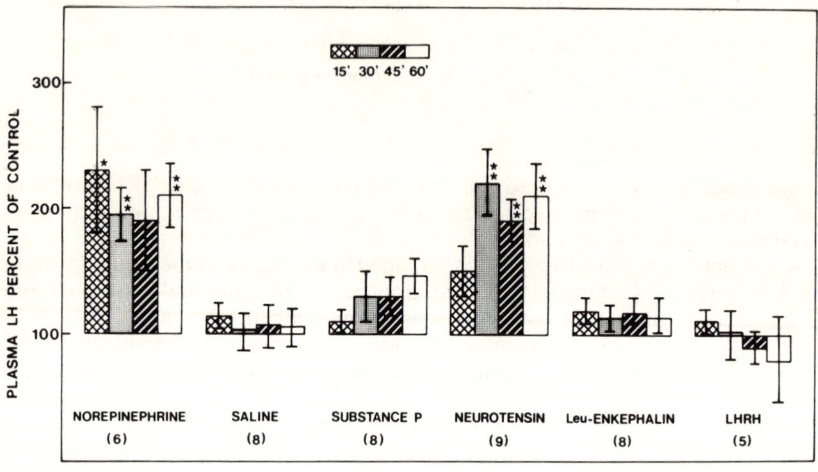

FIGURE 1.   Effect of stereotaxic microinjection (50 nl) into the rostral medial preoptic area (AP 8.2) of norepinephrine (50 ng), substance P (40 ng), neurotensin (40 ng), Leu-enkephalin (50 ng) and LHRH (50 ng) on plasma LH levels. Each bar represents the mean plasma LH concentration measured at 15-minute intervals after microinjection and is expressed as a percent of preinjection control levels obtained one hour before the test microinjection. Statistical comparisons of mean concentrations of LH were performed with a T-test for paired data. The number of animals in each group are shown in parentheses. Vertical bars denote SEM. * p < .05, ** p < .01.

REFERENCES

1. KAWANO, H. & S. OAIKOKU. 1981. Immunohistochemical demonstration of LHRH neurons and their pathways in the rat hypothalamus. Neuroendocrinology 32: 179–186.
2. PELLEGRINO, L. J., A. S. PELLEGRINO & A. J. CUSHMAN. 1979. A Stereotaxic Atlas of the Rat Brain. Plenum Press. New York.

THE STIMULATORY EFFECT OF NEUROTENSIN ON COLONIC MOTILITY IS MEDIATED VIA NERVOUS PATHWAYS

Per M. Hellström and Sune Rosell

Department of Pharmacology
Karolinska Institute
S-104 01 Stockholm 60, Sweden

INTRODUCTION

Previously presented studies support the view that effects of neurotensin *in vivo* are dependent to a great extent on an intact nervous system.[1] Thus, pentagastrin-stimulated gastric acid secretion is inhibited by neurotensin in vagally innervated, but not in denervated, fundic pouches in dogs.[2] Similarly, truncal vagotomy decreases the sensitivity of neurotensin to inhibit gastric emptying and transit of gastrointestinal contents in rats.[3] Moreover, the effect of neurotensin on the activity front of intestinal motility complexes is blocked by hexamethonium and atropine in rats.[4]

In the present study, the ability of neurotensin to stimulate the motility of the proximal colon in the cat[5] has been further examined by application of various blocking agents. The aim was to clarify whether the stimulation of the motility is mediated via nervous pathways or by a direct action on colonic smooth muscle.

METHODS

Colonic Motility

Motility changes in the proximal and distal colon were simultaneously registered. Two water-filled, flaccid balloons were inserted proximally and distally, permitting expulsion of fluid to reservoirs of wide dimension. Colonic motility was measured as changes of the weight of the expelled volume, registered by force displacement transducers.[6]

Administration of Drugs

Neurotensin (Peninsula Inc.) was dissolved in saline with 0.5% bovine serum albumin and aprotinin (Bayer) 100 KIU $\times$ ml^{-1}. Neurotensin was infused i.v. at a rate of 10 pmol $\times$ kg^{-1} $\times$ min^{-1} for 5 min. Acetylcholine 2.7 nmol $\times$ kg^{-1}, isoprenaline 1 nmol $\times$ kg^{-1} and noradrenaline 10 nmol $\times$ kg^{-1} were administered as close i.a. injections. As blocking agents, tetrodotoxin 1–3 μg $\times$ kg^{-1} close i.a., hexamethonium 10 mg $\times$ kg^{-1} i.v., atropine 0.1 mg $\times$ kg^{-1} i.v., guanethidine 3 mg $\times$ kg^{-1} i.v., propranolol 1 mg $\times$ kg^{-1} i.v., and phentolamine 3 mg $\times$ kg^{-1} i.v. were used.

382 Annals New York Academy of Sciences

Nerve Stimulation

Nerve Stimulation

The left vagal nerve and the sympathetic cervical trunk were dissected free. Electrical stimulations were applied to the distal end of the cut left vagal nerve (10 V, 2 msec, 8 Hz) and the proximal end of the cut sympathetic cervical trunk (10 V, 5 msec, 15 Hz).

RESULTS

In one series of experiments (n = 11) tetrodotoxin, hexamethonium, and atropine were consecutively administered. Before the administration of tetrodo-

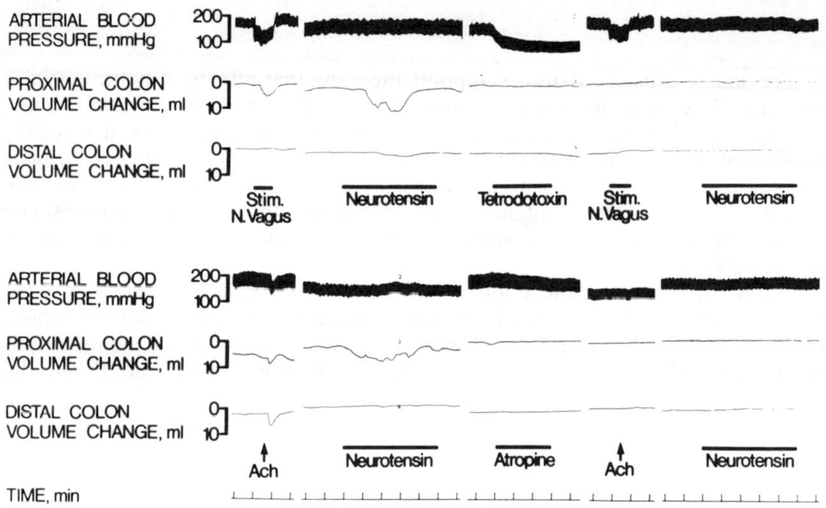

FIGURE 1. The effect of i.v. infusion of neurotensin ($10\ \text{pmol} \times \text{kg}^{-1} \times \text{min}^{-1}$) on arterial blood pressure and proximal and distal colonic motility. Downward deflection of volume change indicates expulsion of fluid from the colon. *Upper panel*: The effects of neurotensin and vagal nerve stimulation (10 V, 2 msec, 8 Hz) are blocked by tetrodotoxin ($3\ \mu\text{g} \times \text{kg}^{-1}$) close i.a. *Lower panel*: The effects of neurotensin and acetylcholine ($2.7\ \text{nmol} \times \text{kg}^{-1}$) close i.a. are blocked by atropine ($0.1\ \text{mg} \times \text{kg}^{-1}$) i.v.

toxin and hexamethonium, control motility and blood pressure responses to vagal nerve stimulation and neurotensin were registered. Tetrodotoxin and hexamethonium powerfully decreased the blood pressure and the spontaneous motor activity in the colon. Twenty minutes after administration of these drugs, neither nerve stimulation nor neurotensin could evoke motility responses in the proximal colon (FIG. 1). However after 1–4 hours, the motility reponses to neurotensin and vagal nerve stimulation reappeared simultaneously. Before administration of atropine, control blood pressure and motility responses to acetylcholine and neurotensin were registered. Atropine decreased the spontaneous motor activity of the colon and the motility response to neurotensin in the proximal colon was abolished. (Fig. 1). However after 4–5 hours, this motility response reappeared simultaneously with the response of acetylcholine.

In another series of experiments (n = 16) guanethidine, propranolol, and phentolamine were consecutively administered. The efficacy of adrenergic blockade with guanethidine was tested by stimulation of the sympathetic cervical trunk. This stimulation did not produce pupillary dilatation after blockade. The efficacy of α- and β-adrenoceptor blockade was tested on arterial blood pressure by administration of noradrenaline and isoprenaline, respectively. The motility response to neurotensin in the colon was markedly increased by all adrenergic blocking agents. Following infusion of neurotensin, contractions were registered in both the proximal and the distal colon, presumably as a consequence of the inhibition of adrenergic input to the colon.

In summary, the effect of neurotensin on colonic motility was inhibited by tetrodotoxin, hexamethonium, and atropine. On the contrary, adrenergic blockade enhanced the response to neurotensin. These findings indicate that neurotensin produces motility responses *in vivo* by acting on cholinergic neurons rather than by a direct smooth muscle stimulation.

References

1. ROSELL, S. & Å. RÖKAEUS. 1981. Clin. Physiol. **1:** 3–20.
2. ANDERSSON, S., S. ROSELL, L. SJÖDIN, D. CHANG & K. FOLKERS. 1980. Scand. J. Gastroenterol. **15:** 253–256.
3. HELLSTRÖM, P. M., G. NYLANDER & S. ROSELL. 1981. Svensk Kirurgi. **39:** 40.
4. AL-SAFFAR, A. 1982. Ann. N.Y. Acad. Sci. **400:** 384–386. This volume.
5. HELLSTRÖM, P. M. & S. ROSELL. 1981. Acta Physiol. Scand. **113:** 147–154.
6. HULTÉN, L. 1969. Acta Physiol. Scand. **78,** suppl. 335.

INFLUENCE OF THE NERVOUS SYSTEM ON THE ACTION OF NEUROTENSIN IN THE SMALL INTESTINE OF RATS

Ahmad Al-Saffar

Department of Pharmacology
Karolinska Institute
S-104 01 Stockholm 60, Sweden

INTRODUCTION

We have reported recently that i.v. infusion of neurotensin (NT) replaces the small intestinal migrating myoelectrical complexes (MMCs), in the small intestine of fasted rats, by irregular spiking activity.[1] The motility pattern induced by neurotensin resembles that seen after ingestion of food. In contrast to the effects of neurotensin, several COOH- and NH_2-fragments of neurotensin do not convert the fasting pattern of the myoelectrical activity, indicating that the effect is produced by the intact neurotensin sequence rather than by small fragments of neurotensin. In the present experiments, the importance of the innervation of the intestine for the effects of NT on the MMCs has been analyzed by means of vagotomy and cholinergic and adrenergic blocking agents.

METHODS

The experiments were performed on 14 male Sprague-Dawley rats (300–400 g). Under pentobarbital anesthesia, four bipolar, insulated, nickel/chrome electrodes, 2 mm apart, were implanted in the muscular wall of the small intestine at 5, 15, 25, and 35 cm distal to the pylorus. Vagotomy and pyloroplasty were performed in seven rats at the time of the implantation of the electrodes. All rats were provided with a jugular vein catheter for drug administration. Following surgery, the animals were allowed to recover for 10 days before experiments. The myoelectrical activity was recorded on a Grass Polygraph, model 7B, with the EEG preamplifier set at a time constant of 0.1 sec. Before an experimental run, the animals were fasted for 48 hr. Three phases of intestinal MMC were identified:[2] (1) phase of "slow waves" or inactivity, (2) phase of "irregular spiking activity" characterized by randomly occurring spiking potentials, and (3) phase of "regular spiking activity" or "the activity front" where spike potentials superimposed all of the slow waves. Synthetic neurotensin (Peninsula) was administered intravenously at a dose of 3.6 and 7 pmol $\times$ kg^{-1} $\times$ min^{-1} for 20 min. Each rat received two or three infusions of NT at two different doses. Atropine, 0.5 mg $\times$ kg^{-1}, hexamethonium, 10 mg $\times$ kg^{-1}, and guanethidine, 3 mg $\times$ kg^{-1}, were administered intravenously in a volume of 0.5 ml for 10 min.

384

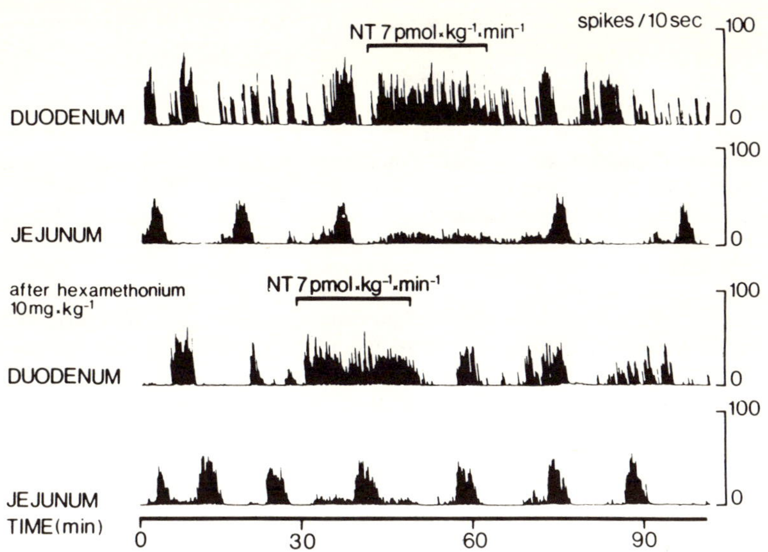

FIGURE 1.   Integrated changes of the intestinal myoelectrical activity at 5 (duodenum) and 25 cm (jejunum) distal to the pylorus. The activity front is represented by maximal amplitude. Neurotensin abolishes the activity front at both levels (*upper tracing*). In the jejunum, pretreatment with hexamethonium prevented the inhibition of the activity front in response to neurotensin (*lower tracing*).

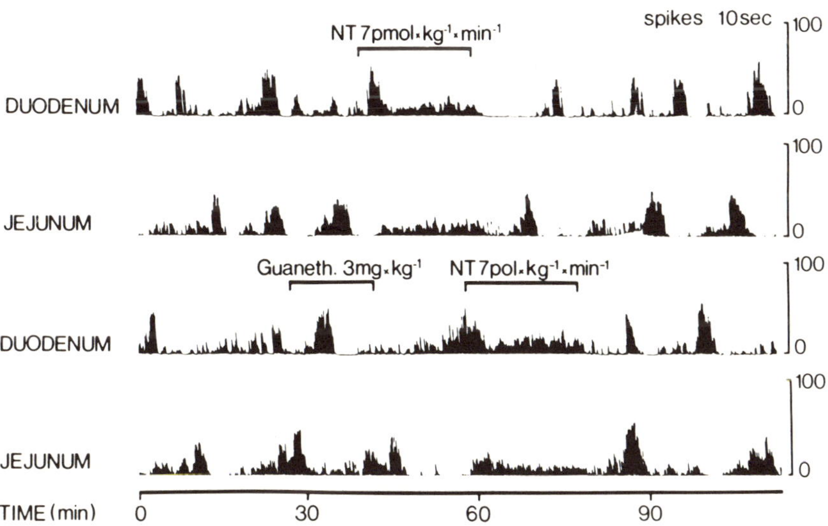

FIGURE 2.   Interruption of MMC by neurotensin before and after administration of guanethidine. For explanation of tracing, see legend of FIGURE 1.

Results and Conclusion

In conscious intact and vagotomized rats, MMCs were recorded from the small intestine. I.v. infusion of neurotensin to intact rats abolished the MMCs and induced a pattern of irregular spiking activity (Figs. 1 & 2). In the duodenum, neurotensin (7 pmol $\times$ kg^{-1} $\times$ min^{-1}) increased the incidence of spike potentials by about 90%. However in the jejunum, the incidence of spike potentials did not change significantly. In vagotomized rats, neurotensin interrupted the MMC in 4 of 7 rats at the lower dose and in 5 of 7 rats at the higher dose. However, increased spiking activity was observed in all experiments. Following atropine or hexamethonium, the duodenum showed a normal response to neurotensin. However in the jejunum, atropine or hexamethonium eliminated the inhibitory action of neurotensin on the activity front (Fig. 1). Guanethidine did not alter the normal response to neurotensin (Fig. 2). These data indicate that the induced irregular spiking activity and the inhibition of the activity front in the duodenum are due to a direct myogenic action of neurotensin or to an indirect noncholinergic, nonadrenergic action, while the inhibition of the activity front in the jejunum seems to involve cholinergic, enteric neurons.

References

1. AL-SAFFAR, A. & S. ROSELL. 1981. Acta Physiol. Scand. **112:** 203–208.
2. BUENO, L., J. FIORAMOTI & Y. RUCKEBUCH. 1975. J. Physiol. (London) **249:** 69–85.

NEUROTENSIN-LIKE IMMUNOREACTIVITY IN PLASMA FROM PATIENTS WITH ENDOCRINE TUMORS OF THE PANCREAS AND GUT

Elvar Theodorsson-Norheim,* Kjell Öberg,† Sune Rosell,*
and Harry Boström†

*Department of Pharmacology
Karolinska Institute
S-104 01 Stockholm 60, Sweden

†Department of Internal Medicine
University of Uppsala
Uppsala, Sweden

Endocrine tumors of the pancreas and gut are known to contain and secrete many humoral substances. In the present study, 34 consecutive patients with previously diagnosed gastrointestinal endocrine tumors were investigated, 18 with malignant pancreatic tumors and 16 with gut tumors. In patients with gut tumors, the primary tumor was located in the ileum (midgut carcinoids) (n = 13), in the foregut (n = 3), in the gastric mucosa (n = 2) and in the respiratory tract (bronchial carcinoid) (n = 1). All patients had regional and hepatic metastases except one (G.O.) who had multiple gastric polyps containing enterochromaffin-like cells. Blood samples were drawn after an overnight fast, collected in chilled tubes containing heparin and Trasylol (400 KIE/ml) and centrifuged. The plasmas were analyzed with antiserum 0-7701, which is NH_2-terminal specific. In 9 healthy subjects after an overnight fast, the mean of p-NTLI was 44.3 ± 5.8 pM (± SEM). The table below shows the clinical syndrome and secretion of peptides in the patients. The p-NTLI levels given are those found in the plasma first analyzed.

The mean p-NTLI values from the other patients with endocrine tumors were 12.7 ± 1.8 pM (SEM). It is of interest to note that in patients with gut tumors (G.O. and L.U.) the p-NTLI was considerably lower than in patients with pancreatic or bronchial tumors. Moreover, peroral ingestion of fat (55 ml Intralipid, 200 mg × ml^{-1}) caused a considerable increase in p-NTLI in patient G.O., whereas in patients B.C. and A.F. the p-NTLI did not change significantly. This is further evidence for the hypothesis that neurotensin-containing cells must be exposed to ingested fat in order to increase p-NTLI. We have been able to monitor the p-NTLI levels for up to 4 years in 4 of the patients. In 3 of them, the p-NTLI levels correlated well with the effect of streptozotocin treatment on tumor size. However, the fourth patient died from an insulinoma syndrome despite the fact that the neurotensin levels were falling after streptozotocin treatment. In one patient (A.F.), infusion of streptozotocin into the arterial supply of the tumor caused an increase in p-NTLI within 45 minutes. This was not the case in patients having endocrine tumors with normal p-NTLI (n = 2). Gel filtration performed on the plasma from 3 of the patients indicated the presence of an NH_2-terminal sequence of neurotensin eluting in a position similar to $(Gln^4)-NT^{1-8}$. There was also a small peak eluting in a position similar to that of synthetic NT^{1-13}. In one plasma sample from patient B.C. taken 4 years ago when he had a Verner Morrison syndrome and high plasma levels of VIP (vasoactive intestinal polypeptide) and

TABLE 1

Patient	Primary Tumor Location	Clinical Syndrome	Secreted Peptides*	p-NTLI (pM)
A. Probable "neurotensinomas"				
A.B.	pancreas	Zollinger-Ellison	gastrin, PP, HCG a and b, neurotensin	310
S.L.	pancreas	Zollinger-Ellison	gastrin, PP, HCG b, neurotensin	103
A.F.	pancreas	Verner Morrison	calcitonin, neurotensin	315
B.C.	pancreas	Verner Morrison	VIP, PP, gastrin, calcitonin, neurotensin	175
K.N.	pancreas	Hypoglycemia	insulin, gastrin HCG b, neurotensin	235
I.D.	pancreas	Glucagonoma	glucagon, insulin, gastrin, PP, calcitonin, neurotensin	921
L.N.	bronchus (foregut carcinoid)	Nil	neurotensin	500
B. Possible "neurotensinomas"				
G.O.	gastric mucosa	Flushing	gastrin, neurotensin	75
L.U.	ileum	Carcinoid	serotinin, neurotensin, HCG a	73

*PP is ?, HCG is human chorionic gonadotropin, and VIP is vasoactive intestinal polypeptide.

neurotensin, only one peak was detected eluting in a position similar to synthetic NT^{1-13}. After streptozotocin treatments, his plasma is now dominated by the small NH_2-terminal sequence of neurotensin.

In our patients, we have not been able to demonstrate with certainty any clinical symptoms related to elevated p-NTLI. This may be due to secretion of biologically inactive sequences of neurotensin. Our data indicate that one must be careful in proposing new clinical syndromes related to one peptide, especially when multiple hormone secretion is demonstrated.

NEUROTENSIN-LIKE IMMUNOREACTIVITY (NTLI) IN ADRENAL MEDULLA: LOCALIZATION, QUANTITATION, CHARACTERIZATION, SUBCELLULAR FRACTIONATION AND RELEASE UPON SPLANCHNIC NERVE ACTIVATION*

Åke Rökaeus,† Gabriel Fried,‡ and Jan M. Lundberg†

Departments of Pharmacology† and Physiology‡
Karolinska Institute
S-104 01 Stockholm 60, Sweden

Neurotensin (NT), a putative gastrointestinal hormone[1] with main localization in endocrine cells of the ileum, has recently also been found in preganglionic sympathetic nerves and in noradrenaline (NA) cells in the adrenal medulla of the cat.[2]

We have examined the presence of NTLI in adrenal glands of several species by immunohistochemistry (IH) using the antiserum "NT[1-11]" and radioimmunoassay (RIA) using the NH_2-terminally directed antiserum "0-7701". The results have been compared with the number of NA-containing cells (defined as DBH-positive and PNMT-negative cells) in the adrenal medulla and the content of NTLI in the ileum and the heart (TABLE 1). Hen, cat, and rat were found to have NTLI-positive cells in the adrenal medulla. In these species NTLI was confined to the NA cells. In contrast, in dog, man, and guinea pig as well as in frog and monkey (not included in the table), no NTLI-positive cells were found. Radioimmunoassay data also revealed high levels of adrenal NTLI in the cat and hen, while in all the other species adrenal NTLI was comparatively low and could not with certainty be distinguished from the content of NTLI in the heart. The reason for the discrepancy between RIA and IH data may depend on differences in the structure of NT between the species. Indeed, it is clear from the hen-tissue NTLI dilution curves (not parallel with those of synthetic bovine NT) that species differences in NT-structure may occur. Gel filtration of acetate buffer (0.5 M, pH 4) extractable NTLI from the adrenal gland of one cat revealed three major peaks as determined with antibody "0-7701." The peaks eluted in order: void volume, before that of synthetic bovine NT and at the position of NT (dominant peak). Subcellular distribution studies of NTLI in homogenates of cat adrenals showed that NTLI in cat adrenals may be stored in a large subcellular organelle, comigrating with (or identical to) chromaffin granules. Electrical stimulation of the left splanchnic nerve (5 msec; 10 V; 2, 6 and 15 Hz) in chloralose-urethane-anesthetized cats (n = 5) increased the output of NTLI and NA from the left adrenal vein in a frequency-dependent manner (FIG. 1). The increased output of NTLI, following a 15 Hz stimulation (n = 2), was reduced by 65% and 74% following pretreatment with hexamethonium (20 and 40 mg/kg, respectively).

In summary, NTLI has been found in the adrenal medulla of hen, cat, and rat. NTLI seems to be localized to the noradrenaline-containing chromaffin cells. In the cat, NTLI seems to be stored in a large subcellular organelle, possibly identical

---

*The present study was supported by grants from the Karolinska Institute and the Swedish Medical Research Council (3518). We thank Dr. M. Brown for the generous supply of antiserum "NT1-11".

TABLE 1

NEUROTENSIN-LIKE IMMUNOREACTIVITY IN ADRENAL GLANDS*

		Hen†	Cat	Rat	Dog	Man	Guinea pig
Heart	NTLI	3.7±0.8 (4)	0.6±0.2 (3)	2.2±0.3 (3)	0.8 (1)	N.D.§	3.1 (2)
Adrenal	NTLI	7.5±1.1 (4)	41±15 (4)	2.4±0.2 (3)	2.5±0.6 (4)	0.5 (2)	1.6±0.4 (3)
Ileum	NTLI	21±2.2 (4)	441±72 (4)	73 (2)	137 (2)	53 (2)	9.1 (2)
NTLI	cells	40–60 (4)	30–50 (6)	10–15 (6)	0 (3)	0 (4)	0 (6)
NA	cells‡	70–80	30–50	5–15	10–40	10–20	0

*The NTLI values (pmol/g tissue weight) are expressed as mean ± S.E.M. The number of NTLI and NA cells are given as % of the total number of chromaffin cells in the adrenal medulla. Number of observations are given in brackets.

† Extractable tissue NTLI does not serial dilute with synthetic bovine NT.

‡ NA cell-values are taken from *Acta Physiol. Scand.* 30: 55–68, 1953, and *The Natural History of the Chromaffin Cell* by R. E. Coupland. Longmans 1965.

§ Not determined.

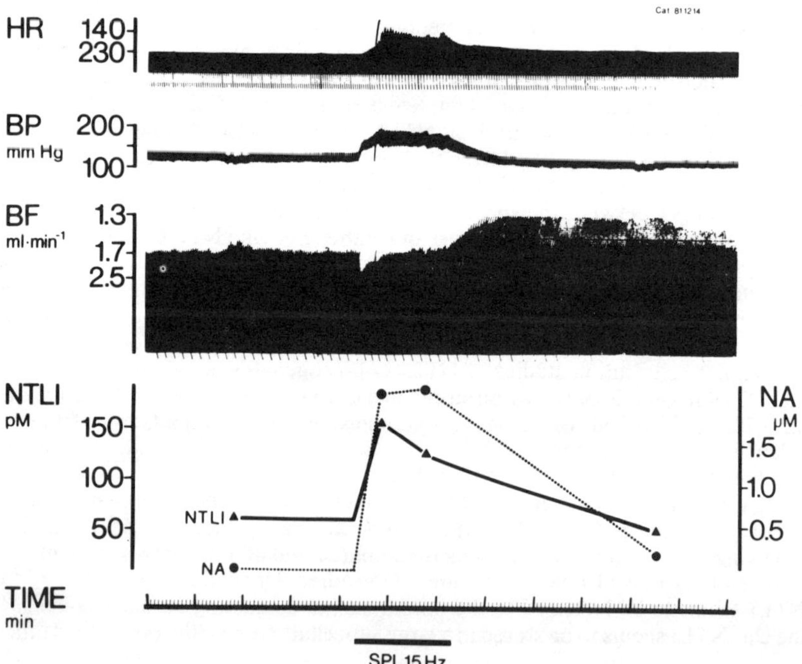

FIGURE 1. Effects of splanchnic nerve stimulation (SPL) at 15 Hz on cat heart rate (HR, beats/min), blood pressure (BP), blood flow (BF), and plasma concentrations of NTLI and NA in the adrenal vein (NTLI ▲ ——▲, NA ●·····●)

to the NA-storing chromaffin granules. Splanchnic nerve stimulation releases NTLI from the cat adrenal gland in a frequency-dependent fashion via activation of nicotinic receptors. NT may thus function as an adrenomedullary hormone.

REFERENCES

1. RÖKAEUS, Å. 1981. Studies on neurotensin as a hormone. Assay and release of neurotensin-like immunoreactivity and actions of neurotensin. Acta Physiol. Scand. Suppl. 501.
2. LUNDBERG, J. M., Å. RÖKAEUS, T. HÖKFELT, S. ROSELL, M. BROWN & M. GOLDSTEIN. 1982. Neurotensin-like immunoreactivity in the preganglionic sympathetic nerves and in the adrenal medulla of the cat. Acta Physiol. Scand. **114:** 153–155.

INTRAVENOUS ADMINISTRATION OF (Gln⁴)-NEUROTENSIN INDUCES VASOCONSTRICTION IN HUMAN ADIPOSE TISSUE

B. Linde,* S. Rosell,† and Å. Rökaeus †

*Department of Clinical Physiology
Huddinge University Hospital
Stockholm, Sweden

†Department of Pharmacology
Karolinska Institute
S-104 01 Stockholm 60, Sweden

Neurotensin is a tridecapeptide that when administered intravenously in the dog and man inhibits gastric acid secretion (Andersson et al.[1] Blackburn et al.[2], changes the intestinal motility from a fasting to a fed-type (Andersson et al.[3]; Thor et al.[4]; Al.Saffar & Rosell[5]), and delays gastric emptying (Blackburn et al., 1980). Neurotensin-like immunoreactivity (NTLI) increases in plasma following the ingestion of fat (Rosell & Rökaeus[6]). Apart from its gastrointestinal actions, neurotensin constricts the vascular bed of canine subcutaneous adipose tissue in doses that do not affect blood pressure, heart rate, or skeletal muscle blood flow (Rosell et al.[7]; Rosell et al.[8]).

The selective vasoconstrictor effect in subcutaneous adipose tissue suggests that neurotensin regulates not only postprandial gastrointestinal function but also the postprandial uptake and storage of substrates in adipose tissue, since these processes are dependent upon substrate supply. In an attempt to evaluate this hypothesis in humans, we have infused (Gln⁴)-neurotensin (18 pmol × kg⁻¹ × min⁻¹ i.v.) during twenty minutes in ten healthy male volunteers. Adipose tissue blood flow was continuously monitored by measuring the local clearance of ⁹⁹ᵐTc-pertechnetate injected s.c. on the abdomen and thigh (Linde & Hjemdahl[9]). Heart rate and blood pressure were measured and calf blood flow studied by venous occlusion plethysmography. NTLI in serum was determined. Infusion of isotonic saline served as control.

During intravenous infusion of (Gln⁴)-neurotensin, the disappearance rate constants of ⁹⁹ᵐTc-pertechnetate from the abdomen decreased by $21.9 \pm 5.9\%$ $(p < 0.01)$. The reduction in clearance persisted throughout the thirty-minute observation period following infusion. The decrease in clearance from the abdomen started after a latent period of five to ten minutes. A significant correlation $(r = 0.87, p < 0.001)$ was found between the percentage decrease in clearance from abdominal fat and the body fat content of the subjects, as determined from the sum of four skinfolds (Fig. 1). Clearance from thigh fat did not change significantly, nor did heart rate, blood pressure, or calf blood flow. During infusion of (Gln⁴)-neurotensin, plasma NTLI increased successively. Ten minutes after starting the infusion, the mean plasma NTLI level had increased from ± 5.1 pM to 302 ± 31 pM. At 20 min, it was 473 ± 51 pM. These concentrations are within the range found after eating a fatty meal.

The results indicate that neurotensin plays a physiological role in the regulation of human adipose tissue blood flow and may be of importance for the postprandial uptake of substrates in adipose tissue in certain regions. We suggest that neurotensin deserves consideration as an endocrine hormone affecting the regional deposition of fat postprandially.

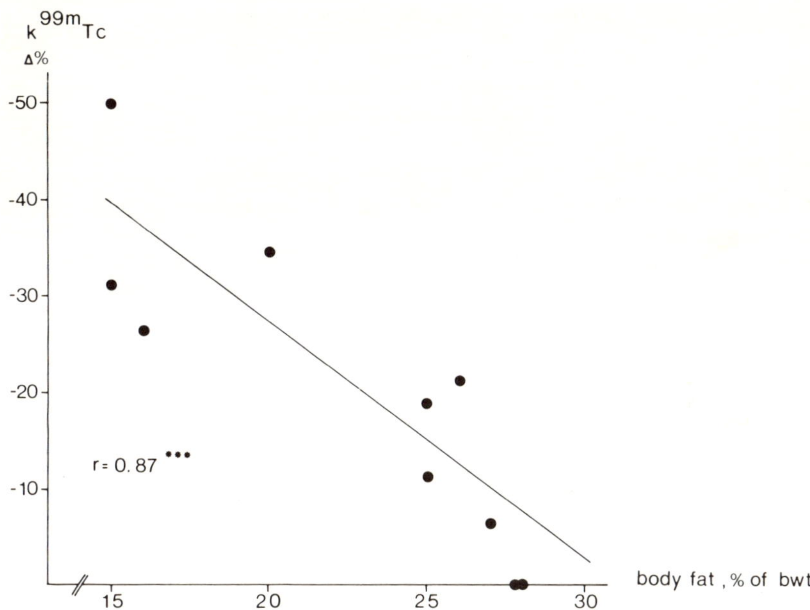

FIGURE 1.   The relation between percentage decrease of disappearance rate constants of ^{99m}Tc-pertechnetate from subcutaneous adipose tissue on the abdomen during infusion of (Gln⁴)-neurotensin (18 pmol × kg⁻¹ × min⁻¹) and body fat content as percentage of body weight in ten healthy male volunteers. (Taken from original article in press in Acta Physiol. Scand.)

REFERENCES

1. ANDERSSON, S., D. CHANG, K. FOLKERS & S. ROSELL. 1976. Life Sci. **19:** 367–370.
2. BLACKBURN, A. M., S. R. BLOOM & R. G. LONG. 1980. Lancet **1:** 987–989.
3. ANDERSSON, S., S. ROSELL, U. HJELMQUIST, D. CHANG & K. FOLKERS. 1977. Acta Physiol. Scand. **100:** 231–235.
4. THOR, K., A. RÖKAEUS, L. KAGER & S. ROSELL. 1980. Acta Physiol. Scand. **110:** 327–328.
5. AL-SAFFAR, A. & S. ROSELL. 1981. Acta Physiol. Scand. **112:** 203–208.
6. ROSELL, S. & Å. RÖKAEUS. 1979. Acta Physiol. Scand. **107:** 263–267.
7. ROSELL, S., E. BURCHER, D. CHANG & K. FOLKERS. 1976. Acta Physiol. Scand. **98:** 484–491.
8. ROSELL, S., Å. RÖKAEUS, D. CHANG & K. FOLKERS. 1978. Acta Physiol. Scand. **102:** 143–147.
9. LINDE, B. & P. HJEMDAHL. 1982. Amer. J. Physiol. **242:** H161-H169.

REGULATION OF MAST CELL HISTAMINE RELEASE
BY NEUROTENSIN

Sandra S. Rossie and Richard J. Miller

Department of Pharmacological and Physiological Sciences
University of Chicago
Chicago, Illinois 60637

Neurotensin (NT) is a 13 amino acid peptide found in neurons of the central nervous system and endocrine cells of the intestinal mucosa. Several observations have suggested that some of NT"s effects observed *in vivo* may be mediated by histamine. We have observed that addition of NT in concentrations as low as 1–10 nM stimulated histamine release from purified rat peritoneal mast cells. Release reached a plateau of 15–20% total cellular histamine with 10^{-7}–10^{-6} NT. Under our experimental conditions, 1 μg/ml compound 48/80 (48/80), a mast cell degranulating agent, released 60–70% total histamine content. Release was optimal at pH 7 and 37°C, and was complete within 2 minutes of peptide addition. The biological activity of NT resided in the COOH-terminus, as the peptide fragment NT8–13 was at least as potent as NT itself. The analogues D-Leu^{13}-NT and D-Trp^{11}-NT stimulated release only at 10^{-5} M. D-Lys^{11}-NT was completely inactive in concentrations up to 10^{-5} M, arguing against the possibility that NT"s ability to stimulate release may merely be due to its positively charged amino acid residues. The potencies of these analogues were similar to their relative activities in other tissues responsive to NT. At 10^{-7} M–10^{-6} M, D-Trp^{11}-NT was able to antagonize the action of NT itself.

Maximal histamine release was obtained at a calcium concentration of 0–1 mM. Manipulations involving calcium utilization in secretion indicate that NT stimulation depends on intracellular stores of calcium. Treatment of cells with ethylenediaminetetraacetic acid in a Ca^{2+}-free medium inhibited histamine release by NT or 48/80. The response was restored by readdition of calcium. Lanthanum, an ion that can block Ca^{2+} channels in some systems, greatly reduced 48/80-stimulated release, but only mildly inhibited that by NT. N,N-(diethylamino) octyl 3,4,5-trimethoxybenzoate, an agent thought to block use of intracellular Ca^{2+}, inhibited NT stimulation to a greater extent than release by 48/80. The voltage-sensitive Ca^{2+} channel antagonists nifedipine and methoxyverapamil had no effect on release by either agent. Quercetin, a naturally occurring flavonoid shown to inhibit IgE-mediated histamine release from mast cells and basophils, also inhibited NT action. The mechanism of quercetin action is unknown. It has been suggested that this drug may act through inhibition of ion transport-associated ATPase activity.

Adenosine 3'5-cyclic monophosphate (cAMP) may be involved in regulation of NT stimulation of mast cells, as manipulations increasing effective cellular cAMP levels reduced release in response to NT. Incubation with the cAMP analogue, 8-bromo-cAMP, inhibited NT-stimulated release in a dose-dependent manner. Combined treatment with prostaglandin E_2, which stimulates cAMP formation, and isobutyl methylxanthine, a phosphodiesterase inhibitor, also inhibited release by NT. 48/80-induced release was reduced to a much lesser extent with either of these treatments.

Treatment with indomethacin, which blocks cyclooxygenase metabolism of arachidonic acid, mildly enhanced NT stimulation, while 8,5,11,14 eicosa-tetraynoic acid, an inhibitor of both cyclooxygenase and lipoxygenase, reduced mast cell response to NT. These results suggest that lipoxygenase metabolites of arachidonic acid may positively modulate NT-mediated histamine release, whereas cyclooxygenase products inhibit release, possibly through generation of cAMP.

Thus, it appears that NT induces histamine release from mast cells under physiologic conditions. NT-mediated release may be an important component of the inflammatory response. Alternatively, the interaction of NT and mast cells suggests a view of mast cells as mediators of hormonal action in circumstances other than inflammation.

NEUROTENSIN ALTERS CUTANEOUS VASCULAR PERMEABILITY AND STIMULATES HISTAMINE RELEASE FROM ISOLATED SKIN*

D. E. Cochrane,† C. Emigh,† G. Levine,† R. E. Carraway,‡
and S. E. Leeman‡

†Tufts University
Medford, Massachusetts 02155

‡University of Massachusetts Medical School
Worcester, Massachusetts 02160

Mast cells of the skin are an important source of histamine (H) and are involved in several common pathological conditions.[1] Their significance to normal regulatory physiology is unknown. Neurotensin (NT) has been shown to bind to specific receptor sites on the isolated mast cell surface,[2] and we have recently shown neurotensin to elicit H release from isolated mast cells, to elevate H levels when given intravenously, and to increase vascular permeability when injected intradermally.[3] Here we show that this change in vascular permeability induced by NT can be prevented by pretreatment of the animal with the specific mast cell H releaser, compound 48/80, that NT can elicit the release of H from isolated skin, and that NT-like material can be generated from extracts of skin.

Rats were pretreated (2 times/day for 3 days) with compound 48/80 (1 mg/kg) intraperitoneally. The animals were then anesthetized, injected intravenously with Evans Blue dye,[3] and 5 min later saline, histamine (200 μg), 48/80 (1 μg), or NT (15 fmole, 1.5 pmole, 150 pmole) injected intradermally. After an additional 5 min, the animals were sacrificed and the abdominal skin observed for blueing around the site of the injection. NT and 48/80 failed to induce any blueing while H elicited a pronounced blueing response. In animals not pretreated with 48/80, NT and 48/80 have previously been shown to elicit prominent blueing responses.[3]

To directly test NT's effect on cutaneous H, we have used slices of skin isolated from rats and continuously perfused with buffered saline. Perfusate samples were collected at regular intervals and assayed for H either fluorometrically or by specific radioenzymatic assay (see ref. 3). NT (10^{-13} to 10^{-6} M) significantly elevated H levels above the baseline, prestimulation value. At a concentration of 10^{-11} M, NT increased H release some 25% above baseline. Pretreating isolated slices of skin with either compound 48/80 (1 μg/ml for 10 min) or disodium cromoglycate (100 μM for 10 min)[3] abolished the response to NT. Pretreating animals with capsaicin (50–150 mg/kg subcutaneously) for 5 days had no significant effect on the ability of NT to release H from skin.

When slices of skin from rat or chicken were extracted by acid/acetone or by hydrochloric acid, neither extract registered in our specific radioimmunoassay (RIA)[4, 5] using both NH_2-terminal- and COOH-terminal-directed antisera. However, extended incubation of the acid extract resulted in the generation of significant amounts of NT-related material. When this extracted material was separated by high-performance liquid chromatography (HPLC), three fractions showed significant activity in our RIA and each of these fractions showed signifi-

*Supported in part by the Tufts Biomedical Research Fund.

cant biological activity as judged by their ability to stimulate H release from isolated mast cells. Each of these HPLC fractions also showed significant binding in our radio-receptor assay[4, 5] using NT receptors from rat brain. Interestingly, each HPLC fraction gave a higher reading in the receptor assay than in the RIA. This suggests that this material shares with NT an ability to interact with NT receptor, but differs slightly from NT in its immunoreactive structure.

We conclude that NT can stimulate cutaneous mast cells to release histamine. We suggest that the skin has the potential to generate NT-related material.

REFERENCES

1. EADY, R. A. J. 1979. The Mast Cell. J. Pepys and A. M. Edwards, Eds. p. 544. Pitman Publishers. London.
2. LAZARUS, L. H., M. H. PERRIN & M. R. BRAEN. 1977. J. Biol. Chem. **252:** 7174–7179.
3. CARRAWAY, R. E., D. E. COCHRANE, J. LANSMAN, S. E. LEEMAN, B. PATERSON & H. WELCH. 1982. J. Physiol. **323:** 403–414.
4. KITABGI, P., R. E. CARRAWAY & S. E. LEEMAN. 1976. J. Biol. Chem. **251:** 7053–7058.
5. KITABGI, P., R. E. CARRAWAY, J. VAN RIETSCHOTEN, C. GRANIER, J. L. MORGAT, A. MENEZ, S. LEEMAN & P. FREYCHET. 1977. Proc. Natl. Acad. Sci. USA **74:** 1846–1850.

NEUROTENSIN, EVIDENCE FOR MULTIPLE RECEPTORS FOR GASTROINTESTINAL MOTILITY ACTION IN DOGS

J. E. T. Fox, Y. Sakai, J. Jury, J. McLean, and E. E. Daniel

Department of Neurosciences
McMaster University
Hamilton, Ontario L8N 3Z5 Canada

We investigated the sites and mechanisms of action of neurotensin on motility of the canine stomach and small intestine *in vivo* in anesthetized animals and the canine gastric corpus *in vitro*. Neurotensin (Sigma) and norepinephrine injected as a bolus into a small artery caused inhibition of field-stimulated (40 V, 5 Hz, 0.5 msec) or spontaneous contractions as recorded by serosal strain gauges (TABLE 1). Clonidine was equipotent with norepinephrine, phenylephrine five times less potent, and isopropylnoradrenaline was ineffective (up to 8×10^{-8} mols). Responses to intra-arterial acetylcholine were much less affected than those induced by field-stimulated release of acetylcholine. In the small intestine, phentolamine and yohimbine increased the ED_{50}'s of neurotensin and norepinephrine, but prazosin and propanolol were less effective (TABLE 1). The responses of the small intestine and the gastric corpus to neurotensin were reduced by guanethidine. Thus, intra-arterial neurotensin inhibits motility by releasing norepinephrine (the receptor may be on adrenergic nerves) and this acts primarily on α_2-adrenergic receptors to inhibit release of acetylcholine in response to field stimulation. At higher doses, a direct action to inhibit smooth muscle responses was observed. Late excitatory responses were rarely seen in either the small intestine or the stomach *in vivo*.

In vitro, gastric corpus circular muscle strips were suspended in organ baths in 37°C Krebs solution and gassed with 95% O_2, and 5% CO_2. Neurotensin either produced inhibition or a biphasic response of inhibition followed by excitation. (ED_{50} inhibition was $6 \times 10^{-6}M$; excitation was $3 \times 10^{-4}M$). The inhibitory response was not altered by effective concentrations of phentolamine, propanolol, atropine, methysergide, mepyramine, cimatadine, tetrodotoxin, or scorpion venom. The excitatory response was reduced by mepyramine, substance P and histamine tachyphylaxis and abolished by prior treatment with 48/80 or Na cromoglycate. Thus *in vitro* receptors are found on smooth muscle eliciting relaxation and on mast cells releasing histamine and other excitatory agents, but the receptor on α-adrenergic nerves was not evident *in vitro*. Thus three receptor sites in the gastrointestinal tract for neurotensin can affect motility. That with the highest affinity on the α-adrenergic nerve is only evident *in vivo*. The neurotensin concentrations necessary to stimulate this receptor are in the range of the circulating levels of neurotensin seen after a fat meal (Rosell 1980). Thus this receptor may be stimulated by "physiological" levels of neurotensin. The smooth muscle and mast cell receptors are more evident *in vitro* than *in vivo*. The reasons for discrepancy are unknown but adrenergic nerve function may not survive in isolated strips. Other peptides, that is, pentagastrin, motilin, and substance P have been shown to release acetylcholine *in vivo* but to lack that action *in vitro;* pentagastrin and substance P also excite smooth muscle directly, an effect less prominent or absent *in vivo*. Thus there are multiple receptor sites for peptides such as neurotensin in the gut; those on nerves seem to have the highest affinity *in vivo*, but those on smooth muscle or on mast cells may be predominant *in vitro*. Extrapolation of conclusions about peptides from one type of study to another is therefore unjustified.

TABLE 1

ED_{50} (MOLES) FOR INHIBITION OF FIELD-STIMULATED CONTRACTIONS *in Vivo*

	Control	Phentol- amine*	Prazosin*	Yohimbine*	Guanethidine†	Propanolol‡
Norepinephrine	2×10^{-9}	1×10^{-8}	5×10^{-9}	4×10^{-8}	2×10^{-9}	2×10^{-8}
Neurotensin	2×10^{-11}	6×10^{-11}	2×10^{-11}	5×10^{-10}	1×10^{-9}	6×10^{-11}

*2 mg/kg i.v.
†10 mg/kg i.v.
‡1 mg/kg i.v.

REFERENCE

ROSELL, S. 1980. Acta Physiol. Scand. **110:** 325.

RELATIONSHIPS BETWEEN NEUROTENSIN AND PANCREATIC SECRETION IN THE DOG

I. Baća, G. E. Feurle, W. Knauf, and R. Carraway*

Surgical Clinic, Medical Polyclinic
University of Heidelberg
6900 Heidelberg 1, Federal Republic of Germany

**Department of Physiology*
University of Massachusetts Medical School
Worcester, Massachusetts 02160

Fat stimulates pancreatic secretion and leads to a rise in the plasma level of immunoreactive neurotensin (iNT). We found that exogenous neurotensin (NT) stimulates pancreatic secretion similar to a fatty meal.[1, 2] Furthermore, it is known that pancreatic secretion increases when pancreatic juice is diverted from the duodenum. In the present study, we examined the effect of pancreatic juice diversion on plasma levels of iNT.

METHODS

Six dogs (18–32 kg) were prepared with chronic pancreatic fistulae by a modification of the method of Herrera *et al.* In the feeding experiment, dogs ate 2 g fat kg^{-1} (Lipofundin^R). Pancreatic secretion was elicited by either fat or by rising sequential doses of NT from 2.5 up to 320 pmol kg^{-1} min^{-1} (each dose was infused for 30 min). Volume, protein, and bicarbonate were determined in the collections of the pancreatic fistula. Venous blood samples were taken in 15-min intervals. In an independent series, venous blood samples were taken when the pancreatic fistula was closed. Levels of iNT were determined in unextracted plasma using a radioimmunoassay kit for NT (Immunonuclear, Stillwater, MN, USA), and also in extracts of plasma using a published procedure employing antiserum HC-8.[3]

RESULTS

Pancreatic bicarbonate response was 1.72 ± 0.33 mmol 60 min^{-1} and pancreatic protein response 0.82 ± 0.072 g 60 min^{-1} (M $\pm$ SEM) after the fatty meal. The integrated plasma iNT response determined in unextracted plasma was 487 ± 87 pmol 1^{-1} 60 min when the pancreatic fistula was closed, and 934 ± 168 pmol 1^{-1} 60 min with pancreatic juice diversion (p < 0.05); in extracted plasma 510 ± 185 pmol 1^{-1} 60 min when the pancreatic fistula was closed, and 1791 ± 504 pmol 1^{-1} 60 min with pancreatic juice diversion (M $\pm$ SEM) (p < 0.05) (Fig. 1).

The lowest dose of NT that caused a significant stimulation of pancreatic secretion was 2.5 pmol kg^{-1} min^{-1}, while a half-maximal response occurred with around 20 pmol kg^{-1} min^{-1} (Fig. 2). The maximal effect of NT on bicarbonate output was about 30% and on volume was about 50% that produced by a maximal dose of secretin.

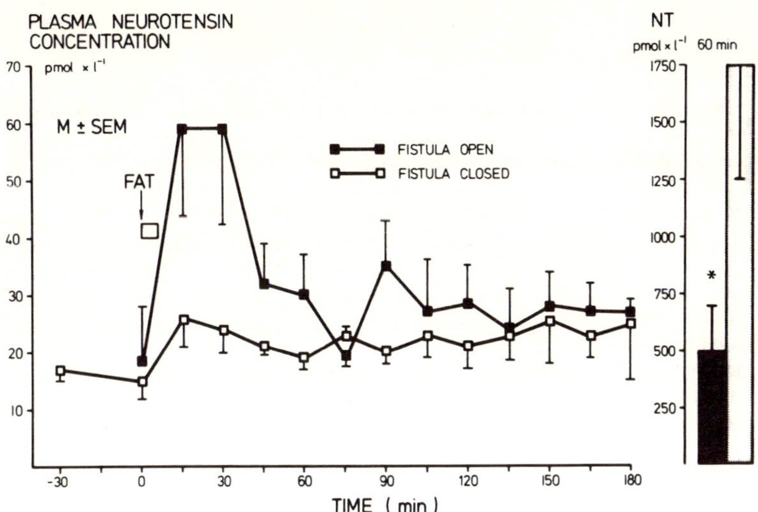

FIGURE 1.   Plasma iNT after a fatty meal (extracted plasma).

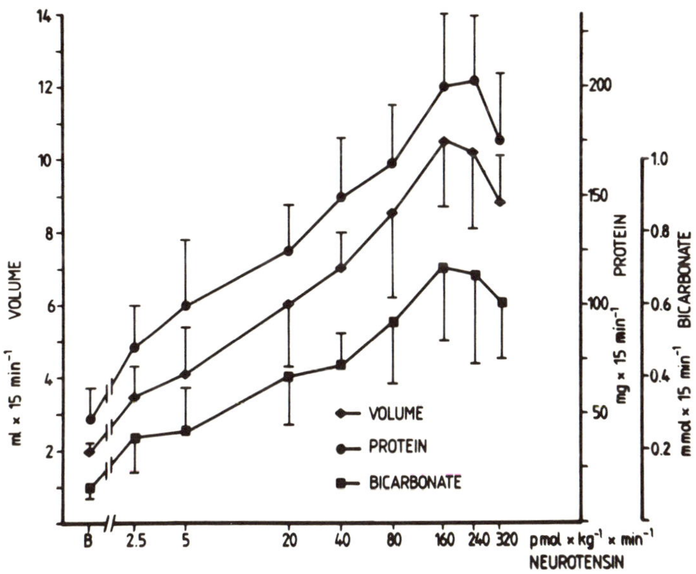

FIGURE 2.   Pancreatic volume, protein, and bicarbonate output in response to NT (n = 5).

CONCLUSION

The present results indicate that the rise in the plasma level of iNT seen after ingestion of fat by dogs is exaggerated when pancreatic secretions are prevented from entering the duodenum. It could be that dietary fat is a stronger stimulus for the release of NT-related substances than are the products of fat digestion by pancreatic enzymes. Another possibility is that the presence of pancreatic juice in the intestinal lumen initiates a feedback mechanism that inhibits the release of NT-like peptides from the gut. It is also not yet clear whether the forms of iNT released into plasma in these two instances are identical. The HC-8 antiserum employed has been shown to react with the COOH-terminal region of NT and not to respond to the major circulating metabolites; however, variants of NT, sharing its COOH-terminal region, have been found to register in this assay.

Since synthetic NT can stimulate pancreatic secretion and pancreatic secretion can in turn affect the plasma levels of NT, it is suggested that endogenous NT or a related peptide might participate in the regulation of the exocrine pancreas.

REFERENCES

1. BACA, I., G. E. FEURLE, U. MITTMANN, A. SCHWAB, W. KNAUF & R. CARRAWAY. 1981. Exocrine pancreatic response and plasma neurotensin concentrations after meal and during infusion of low-dose neurotensin in the dog. Eur. Surg. Res. 13: 36.
2. BACA, I., G. E. FEURLE, A. SCHWAB, U. MITTMANN, W. KNAUF & T. LEHNERT. 1982. Effect of neurotensin on exocrine pancreatic secretion in dogs. Digestion 23: 174.
3. CARRAWAY, R., R. HAMMER & S. E. LEEMAN. 1980. Neurotensin in plasma: Immunochemical and chromatographic character of acid/acetone-soluble material. Endocrinology 107: 400.

TOPOGRAPHY AND FUNCTIONAL IMPLICATIONS OF NEUROTENSIN-CONTAINING STRUCTURES IN THE RAT BRAIN

Lothar Jennes, Walter E. Stumpf, William C. Beckman, Jr., Peter Kalivas, Charles B. Nemeroff, and Arthur J. Prange, Jr.

*Department of Anatomy and Biological Research Center
University of North Carolina
Chapel Hill, North Carolina 27514*

Immunohistochemistry of neurotensin (NT) reveals stained uni-, bi-, or multipolar cell bodies as well as thin and beaded or patchy nerve fibers. The staining is restricted to the cytoplasmic portion of the neuron and is often concentrated in small granules while the cell nucleus remains unstained. Clusters of positive cell bodies occur in the lateral septum and are continuous ventrally with NT-containing cells in the bed nucleus (n.) of the stria terminalis and ventrolaterally with neurons in the caudate putamen. Scattered single or small groups of neurons are seen in the region of the median forebrain bundle throughout its course, in the diagonal band, the preoptic periventricular n. and the n. preopticus medianus. Only a few labeled neurons are seen in the n. preopticus medialis. In the periventricular hypothalamus, NT-positive neurons occur scattered throughout its preoptic, anterior, and central portion. The n. paraventricularis contains, in its parvocellular as well as in its dorsal and medial magnocellular parts, strong accumulations of immunoreactive cell bodies. In the central hypothalamus, the n. ventromedialis, the n. dorsomedialis, and the ventral premammillary n. are virtually free of positive cell bodies, while labeled neurons are seen in the vicinity, including the perifornical region and the lateral hypothalamus. In the n. periventricularis thalami, numerous immunoreactive perikarya are seen. Only a few occur in the caudal ventral thalamus. In the amygdala, most NT cell bodies are observed in the n. centralis and n. medialis. A few scattered neurons occur in the n. basalis pars medialis and in the n. corticalis. The hippocampus does not show NT-positive perikarya. In the midbrain, immunoreactive neurons are scattered throughout the central gray and the formatio reticularis, n. raphe dorsalis and pontis, n. linearis, n. centralis superior, n. parabrachialis, locus ceruleus, and n. ambiguus. Accumulations of positive cell bodies are seen in the n. parabrachialis lateralis, n. interstitialis decussationis tegmenti pontis, n. paranigralis, and in a region medial and ventral to the unlabeled n. tegmenti pontis dorsalis. In the medulla, immunoreactive neurons are seen in different portions of the n. tractus solitarii and the substantia gelatinosa of the n. sensibilis of the trigeminus that are continuous with labeled neurons in the spinal cord dorsal horn as well as the spinal cord dorsal lamina X. A large number of nonneuronal immunoreactive NT-containing cell bodies exist in the anterior pituitary and in the subarachnoid space.

NT-containing fibers are present in all areas in which labeled neurons occur as well as in many other brain regions. They are found in the olfactory bulb in the external plexiform layer and in the n. olfactorius anterior. In the frontal pole, a network of positive fibers is seen in the outer layers with highest concentration in the area close to the sulcus rhinalis in lamina I. This accumulation can be seen throughout the extent of the sulcus. It is connected ventrally with fibers of the piriform cortex, medially with a projection of the claustrum and dorsomedially with a supracallosal projection. This supracallosal projection can be followed mostly in lamina VI from the cingulum through the cingulate cortex toward the medial longitudinal fissure. In the olfactory tubercle, a particular patchy type of immunoreactive nerve fiber occurs in the n. accumbens and in an area close to its ventral and medial surface. A similar type of fiber is seen in the caudate-putamen, the lateral septum, and in the bed n. of the stria terminalis. The tractus of the stria terminalis, however, contains only thin immunoreactive fibers. In the amygdala, the most dense network of NT fibers is seen in the n. centralis. In the diencephalon, the outer lamina of the median eminence shows the highest density of NT fibers. Other networks occur in the area surrounding the thalamic n. parataenialis, the n. rhomboides, the zona incerta, the n. reuniens, and caudally in an area dorsal and medial to the leminiscus medialis, probably including portions of the n. gelatinosus, the n. centrum medianum and the medial n. ventralis thalami as well as in the n. centralis of the corpus geniculatum mediale. In the midbrain and hindbrain, labeled fibers exist in addition to the areas in which labeled neurons occur, in the medial pretectal area, substantia nigra pars compacta, n. pretectalis, caudal collicular n., nuclei reticularis pontis, n. trapezoides and in the n. oralis, sensibilis, and caudalis of the tractus spinalis nervi trigemini. Additional fibers are seen in the area postrema and in different laminae of the spinal cord with highest density in lamina II, followed by lamina I, III, IV and X.

The widespread but select anatomical distribution of NT systems suggests involvement in many brain functions, especially somatosensory and autonomic/endocrine regulation. NT may participate in the endocrine regulation of anterior pituitary secretion either directly after being released into the portal blood system of the median eminence or by affecting brain centers such as raphe nuclei, amygdala, lateral septum or nucleus tractus solitarii, all of which are known to influence, for example, gonadotropin secretion or reproductive behavior. Furthermore, the presence of NT in neurons of the spinal trigeminal tract and in the substantia gelatinosa as well as in related pathways indicates an involvement in the mediation of certain types of pain and in relaying this information to the autonomic system.

IMMUNOCYTOCHEMICAL STUDY OF NEUROTENSIN LOCALIZATION IN THE MONKEY SPINAL CORD*

Marian DiFiglia,† Neil Aronin,‡ § and Susan E. Leeman §

†*Department of Neurology*
Massachusetts General Hospital
Boston, Massachusetts 02114

‡*Departments of Medicine and* §*Physiology*
University of Massachusetts Medical Center
Worcester, Massachusetts 02160

INTRODUCTION

Radioimmunoassayable neurotensin (NT) has been found in the rat spinal cord [1] and immunocytochemical studies at the light microscopic level have shown that immunoreactive neurotensin (iNT) is present almost exclusively in the dorsal horn of the spinal cord both within numerous fibers in laminae I and II [2-4] and in cell bodies located primarily in the region adjoining laminae II and III.[5,6] At the electron microscopic level in the rat, axon terminals that contain iNT have been observed to participate in axodendritic synapses.[3,7,8] Previous studies in our laboratory have examined the distribution and morphologic features of substance P [9] and enkephalin [10] neuronal elements in the monkey spinal cord, where differences have been observed in the pattern of distribution of these peptides among rat, cat, and monkey. In the present study, we investigated the localization of iNT in the monkey spinal cord.

METHODS

The spinal cords of four monkeys (M. fascicularis) were fixed by cardiac perfusion with buffered 4% paraformaldehyde to which 0.2% glutaraldehyde (pH 7.3) had been added. A modification [10] of the peroxidase antiperoxidase method of Sternberger [11] was applied to transverse sections of the spinal cord cut with a Vibratome [R] (Lancer). The processed sections were then mounted on glass slides with 0.2% gelatin for examination at the light microscopic level. For electron microscopy, smaller segments of the spinal cord containing the dorsal horn were dissected, treated with osmium, stained with uranyl acetate, and then embedded in Epon for serial thin sectioning.

Control sections were processed in the presence of NT antiserum preincubated with 20 µg/ml of synthetic NT (Beckman) or in the absence of NT antiserum. The NT antiserum (code : HC8) employed in the immunohistochemistry has been

* This work was supported by NIH Grants NS16367 (M.D.), AM01126-01 (N.A.), BRSG from the University of Massachusetts Medical Center (N.A.), and AM29876-02 (S.E.L.).

† Please address correspondence to: Dr. Marian DiFiglia, Warren 3, Department of Neurology, Massachusetts General Hospital, Boston, MA 02114.

previously described.[12] In brief, the antiserum recognizes the carboxyl-terminus of NT and cross-reacts 2% with NT9–13 and less than 0.01% with NT1–8, NT1–10, NT1–11, NT1–12 and substance P.

RESULTS

At the light microscopic level, most of the iNT was present within numerous small punctate elements, presumed to be axon terminals, which were distributed in the superficial aspect of the dorsal horn. Immunoreactive NT fibers of small diameter were seen infrequently. The most intense staining was visible 250 μm deep to the rim of the dorsal horn (FIG. 1). This region in the monkey includes laminae I and outer II. In many sections, iNT terminals and some fibers were seen to extend deeper into laminae II and III. A less intense staining of iNT was present in small punctate profiles in the regions surrounding the central canal (lamina X). In the dorsal horn, most of the iNT swellings visible at the light microscopic level were approximately 1.0 μm. In the most superficial 100 μm of the posterior horn, large swellings of about 3 μm in size were seen usually in discrete clusters. Although colchicine was not employed, a few cell bodies were labeled in some sections within laminae II and III and ranged from 12–18 μm in size. Control sections either failed to exhibit or showed markedly reduced staining of fibers and terminals.

At the electron microscopic level, most of the iNT in laminae I and II was localized to axon terminals that ranged from 1–3 μm in size and formed primarily axodendritic synapses (FIG. 2).

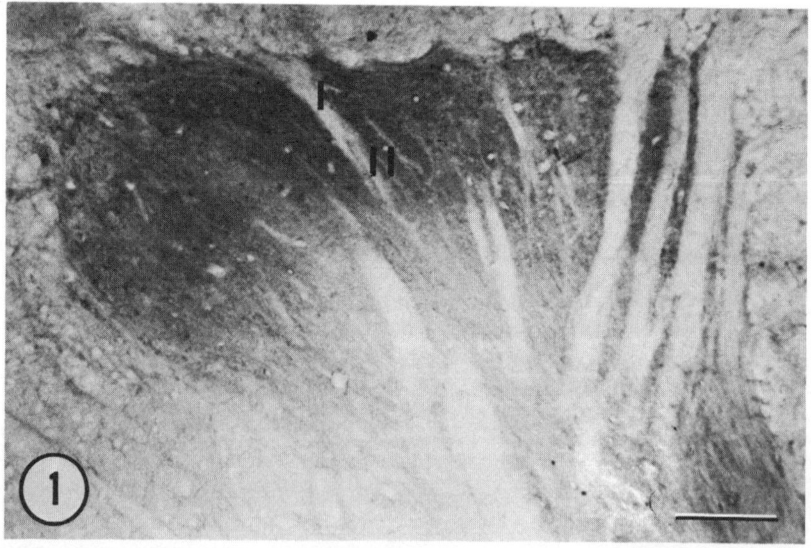

FIGURE 1. Immunoreactive NT in the monkey lumbar dorsal horn. Most of the labeled terminals and fibers were distributed uniformly throughout layers I and outer II. Scale, 200 μm.

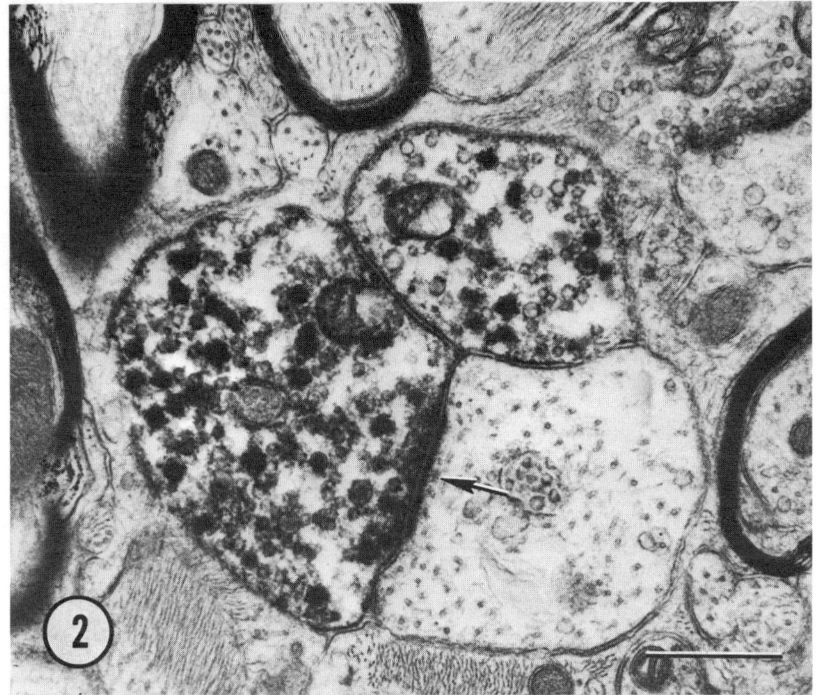

FIGURE 2. NT-positive axon terminals in lamina I. The two labeled profiles contain small clear and large dense core vesicles and one of the terminals forms an asymmetric synapse with a dendrite at arrow. Scale, 0.5 μm.

DISCUSSION

The relatively uniform distribution of iNT axons in the superficial aspect of the monkey spinal cord differs from that reported in the rat where it has been localized to two discrete bands, one appearing in lamina I and another in the deeper aspects of lamina II.[5, 6] The distribution of iNT profiles overlaps with the pattern observed for substance P and enkephalin-containing axons[10] within superficial aspects of the monkey posterior horn. Immunoreactive NT is observed mostly in axon terminals and, unlike the latter two peptides, it is seen in relatively few fibers.

Neurotensin[8] and enkephalin[13] axons in the dorsal horn have been thought to arise principally from intrinsic neurons and to be involved in the modulation of nociceptive inputs. Since the axon terminals containing these peptides primarily form axodendritic synapses, their influence is considered to be largely at post-synaptic sites. The physiologic effects of enkephalin and NT in spinal cord appear to differ. Unlike enkephalin, which has an inhibitory influence on the response of dorsal horn neurons to nociceptive afferent input,[14] NT appears to have both stimulatory and inhibitory electrophysiologic properties that are not easily interpreted (see Henry in this volume). Morphologic properties of iNT axons, such as the different sizes of terminals observed in this study, may be related to the heterogeneity of responses of cells receiving exogeneously applied NT. Further studies at the ultrastructural level are required to examine the synaptic pattern of iNT axon terminals in the monkey.

REFERENCES

1. KOBAYASHI, R. M., M. BROWN & W. VALE. 1977. Regional distribution of neurotensin and somatostatin in rat brain. Brain Res. **126:** 584–588.
2. UHL, G. R., R. R. GOODMAN & S. H. SNYDER. 1979. Neurotensin-containing cell bodies, fibres, and nerve terminals in the brain stem of the rat: Immunohistochemistry mapping. Brain Res. **167:** 77–91.
3. LEEMAN, S. E., M. DIFIGLIA & N. ARONIN. 1980. Neurotensin (NT) and substance P (SP) localization in monkey and rat spinal cord: Light and electron microscopy. Soc. Neurosci. Abstr. **6:** 354.
4. SEYBOLD, V. & R. ELDE. 1980. Immunohistochemical studies of peptidergic neurons in the dorsal horn of the spinal cord. J. Histochem. Cytochem. **28:** 367–370.
5. GIBSON, S. J., J. M. POLAK, S. R. BLOOM & P. D. WALL. 1981. The distribution of nine peptides in rat spinal cord with special emphasis on the substantia gelatinosa and on the area around the central canal (Lamina X). J. Comp. Neurol. **201:** 65–80.
6. HUNT, S. P., J. S. KELLY, P. C. EMSON, J. R. KIMMEL, R. J. MILLER & J.-Y. WU. 1981. An immunohistochemical study of neuronal populations containing neuropeptides on γ-aminobutyrate within the superficial layers of the rat dorsal horn. Neuroscience **6:** 1883–1898.
7. SEYBOLD, V. & B. MALEY. 1981. Ultrastructural localization of neurotensin like immunoreactivity in the dorsal horn of the rat spinal cord. Soc. Neurosci. Abstr. **7:** 59.
8. NINKOVIC, M., S. P. HUNT & J. S. KELLY. 1981. Effect of dorsal rhizotomy on the autoradiographic distribution of opiate and neurotensin receptors and neurotensin-like immunoreactivity within the rat spinal cord. Brain Res. **230:** 111–119.
9. DIFIGLIA, M., N. ARONIN & S. E. LEEMAN. 1982. Light microscopic and ultrastructural localization of immunoreactive substance P in the monkey dorsal horn. Neuroscience **7:** 1127–1139.
10. ARONIN, N., M. DIFIGLIA, A. S. LIOTTA & J. B. MARTIN. 1981. Ultrastructural localization and biochemical features of immunoreactive Leu-enkephalin in monkey dorsal horn. J. Neurosci. **1:** 561–577.
11. STERNBERGER, L. A. 1979. Immunocytochemistry. Wiley. New York.
12. CARRAWAY, R. E. & S. E. LEEMAN. 1976. Radioimmunoassay for neurotensin, a hypothalamic peptide. J. Biol. Chem. **251:** 7035–7044.
13. HÖKFELT, T., A. LJUNGDAHL, L. TERENIUS, R. ELDE & G. NILSON. 1977. Immunohistochemical analysis of peptide pathways possibly related to pain and analgesia. Enkephalin and substance P. Proc. Natl. Acad. Sci. USA **74:** 3081–3085.
14. ZIEGLGANSBERGER, W. & I. F. TULLOCH. 1979. The effects of methionine and leucine-enkephalin on special neurones of the cat. Brain Res. **167:** 53–64.

BIOCHEMISTRY AND PHYSIOLOGY OF NEUROTENSIN-LIKE PEPTIDES IN THE BRAIN AND GUT OF THE MOLLUSC *APLYSIA*

C. H. Price, S. E. Ruane,* and R. E. Carraway*

Department of Biology
Boston University
Boston, Massachusetts 02215

**Department of Physiology*
University of Massachusetts Medical Center
Worcester, Massachusetts 02160

Neurotensin (NT) appears to be a peptide with an ancient lineage since NT-like immunoreactivity (iNT) is present in a wide variety of vertebrates and invertebrates.[1] Certain invertebrate species offer special advantages for studying specific aspects of neurophysiology. Molluscan peptidergic systems are especially useful preparations (due to the large size and identifiability of some neurons) for biochemical and electrophysiological investigations of peptides at the single cell level. This report concerns the biological and chemical properties of iNT found in extracts of *Aplysia* tissues and the effects of mammalian NT on electrical activity in *Aplysia* ganglia *in vitro*.

Tissue samples for radioimmunoassay were dissected from live *Aplysia californica* purchased from a marine supplier and frozen on dry ice. Without thawing, samples were homogenized in 10 volumes of either 0.1 N HCl or 2 N acetic acid and lyophilized. All samples were assayed using antiserum HC-8, which is directed towards the carboxy-terminus of bovine NT. NT-like immunoreactivity was found to be distributed widely in the nervous system (up to 115 fm/ganglion) as well as digestive gland (6 pm/g wet weight) with only small amounts present in other tissues such as skin and body wall muscle (both < 1 pm/g wet weight). There were 200-fold differences in the content of iNT among neurons individually dissected from the abdominal ganglion as single or small groups of cells. Most neurons had < 0.1 fm/cell while a few giant ones had 3–24 fm/cell. The pericardial area, consisting mostly of nerves, contained up to 16.6 pm iNT/g wet weight. Since we estimate the cross-reactivity of *Aplysia* NT-like peptides with the HC-8 antiserum to be < 0.1%, the real levels of peptide in these tissues may be much higher.

Large-scale samples for chromatography were from *Aplysia brasiliana* that were collected off Galveston Island, Texas and extracted in 10 volumes of 0.1 N HCl. Gel filtration was performed using a column of Sephadex G-25 under conditions recommended by the manufacturer with 0.2 N acetic acid. Ion exchange chromatography was performed using a column of sulfylpropyl Sephadex C-25 and eluted with a gradient of pyridine acetate (0.05 M PyAc, pH 3.0 to 0.5 M PyAc, pH 5.5).[2] HPLC was performed using a Waters System and a column of μ-Bondapak C-18 with an acetonitrile gradient. The NT-related substances from *Aplysia* were characterized and partially purified. Extracts of *Aplysia* brain and gut, as well as both those tissues from different *Aplysia* species, gave similar profiles of immunoreactivity at each stage of purification.

Biological activity of *Aplysia* iNT was quantified in an assay based upon an increase in hematocrit in response to intravenous injection of NT into anesthetized (Nembutal; 45 mg/kg) rats as described previously.[3] Partially purified samples of digestive gland from HPLC were resuspended in distilled water and injected into 90–100 g male albino rats (Charles River Breeding Labs) via tail vein. In the rats given intravenous injections of *Aplysia* iNT, hematocrit increased 4.4 ± 0.6% (n = 3, p < .001) over saline controls (37.6 ± 0.61, n = 5) when a dose of 2.4 pmol was injected. Cyanosis was not observed at this dose. Synthetic NT produced cyanosis and a significant increment in hematocrit of 14.7 ± 0.8% (n = 3, p < .001) at a dose of 200 pmol.

Biological activity of synthetic bovine NT ($10^{-9}M$ in seawater) was studied using an *in vitro* preparation of *Aplysia* head ganglia and pedal nerves. The ring of head ganglia was pinned to a recording chamber and perfused with plain seawater and nerve activity monitored by suction electrodes on two main ganglionic nerve trunks. When head ganglia were perfused with NT, there was an increase in the number (5–10-fold) of large amplitude spikes after a delay of 14.3 min. These continued as long as NT was present. Within 3 min of perfusion with seawater to wash out NT, large potentials were reduced and baseline activity returned.

These results argue for the presence of biologically active NT-related substances in the brain and intestinal tissues of *Aplysia*. These substances appear to share COOH-terminal immunodeterminants with bovine NT that is consistent with their shared ability to affect hematocrit. That these may represent "messenger" substances of the nervous system is supported by the demonstrated electrical effects of synthetic NT on isolated *Aplysia* head ganglia. These encouraging findings suggest that it may be possible to utilize some of the advantages of working with an invertebrate system such as *Aplysia* to study the physiological actions of neurotensin.

REFERENCES

1. CARRAWAY, R., S. E. RUANE & H. R. KIM. 1982. Distribution and immunochemical character of neurotensin-like material in representative vertebrates and invertebrates: Apparent conservation of the COOH-terminal region during evolution. Peptides **3:** 115.
2. CARRAWAY, R. & S. E. LEEMAN. 1973. The isolation of a new hypotensive peptide, neurotensin, from bovine hypothalami. J. Biol. Chem. **248:** 6854.
3. CARRAWAY, R. E., L. M. DEMERS & S. E. LEEMAN. 1976. Hyperglycemic effect of neurotensin, a hypothalamic peptide. Endocrinology **99:** 1452.

NEUROTENSIN IN THE PITUITARY GLAND

M. Goedert, S. L. Lightman,* and P. C. Emson

MRC Neurochemical Pharmacology Unit
MRC Centre
Cambridge, England

Neurotensin is a typical brain–gut peptide since it has been isolated from both hypothalamus and small intestine.[1,2]

The presence of neurotensin in the pituitary gland has so far received little attention, although we find high concentrations of neurotensin in the rat anterior lobe and somewhat lower levels in the neurointermediate lobe. Using antisera directed against the amino-and carboxy-terminal ends of the neurotensin molecule in conjunction with high-performance liquid chromatography on reverse phase and with gel chromatography, we have established that most of the neurotensin-like immunoreactivity (NTLI) in rat anterior and posterior pituitary coelutes with synthetic neurotensin. Using indirect immunofluorescence, we visualised NTLI in a population of anterior pituitary cells, although it remains to be seen whether it coexists with any of the known anterior pituitary hormones. The release of NTLI by elevated potassium in a calcium-dependent manner was shown using hemisected rat anterior pituitaries *in vitro.* A possible relationship between anterior pituitary neurotensin and classical endocrine systems was investigated using various hormonal manipulations, such as surgical adrenalectomy or gonadectomy, drug-induced hyperprolactinemia and chemical thyroidectomy using propylthiouracil. Of all these maneuvers, only hypothyroidism produced a large decrease in the levels of neurotensin in the anterior lobe of the pituitary gland. This effect was reversible by substituting L-tri-iodothyronine (TABLE 1). Since hypothyroidism is accompanied by an increased release of thyrotropin-releasing hormone (TRH), we administered TRH to normal rats by means of osmotic minipumps and measured neurotensin in the anterior pituitary gland. As shown in TABLE 1, TRH had the same effect as hypothyroidism. This suggests that the decrease in neurotensin seen after propylthiouracil may be consequent to increased TRH levels. Interestingly, TRH has been shown to be a functional neurotensin antagonist in several systems.[3,4] Besides the rat, we have detected NTLI in guinea pig, cow, pig, and post-mortem human pituitary glands.

The presence of neurotensin in a cell population in the anterior pituitary gland, its release by depolarizing stimuli and the functional relationship between neurotensin and the hypothalamus/pituitary/thyroid axis suggest a possible peripheral hormonal role for neurotensin. Future studies will have to investigate whether pituitary neurotensin contributes to the circulating neurotensin levels, and what could be the peripheral target sites for neurotensin. Preliminary studies indicate that neurotensin binding sites are present in some peripheral endocrine tissues.

---

*Present address: Medical Unit, St Mary's Hospital, Praed Street, London, England.

TABLE 1

EFFECTS OF MANIPULATIONS OF THE HYPOTHALAMUS/ANTERIOR PITUITARY/THYROID
AXIS ON THE NEUROTENSIN CONTENT OF RAT ANTERIOR PITUITARY GLAND

Treatment	Neurotensin Content (pmol/g)
A. Controls	79 ± 4.76
Propylthiouracil	5 ± 0.22 *
Propylthiouracil + L-tri-iodothyronine	71 ± 4.66
B. Controls	76 ± 2.76
Propylthiouracil	4 ± 0.36 *
Propylthiouracil + L-tri-iodothyronine	78 ± 3.96
C. Controls	83 ± 4.45
L-tri-iodothyronine	77 ± 6.88
D. Controls	89 ± 3.66
Thyrotropin-releasing hormone	4 ± 0.41 *

*p < 0.001

REFERENCES

1. CARRAWAY, R. E. & S. E. LEEMAN. 1973. The isolation of a new hypotensive peptide neurotensin from bovine hypothalami. J. Biol. Chem. **248:** 6854–6861.
2. HAMMER, R. A., S. E. LEEMAN, R. CARRAWAY & R. H. WILLIAMS. 1980. Isolation of human intestinal neurotensin. J. Biol. Chem. **255:** 2476–2480.
3. NEMEROFF, C. B., G. BISSETTE, P. J. MANBERG, A. J. OSBAHR, G. R. BREESE & A. J. PRANGE. 1980. Neurotensin-induced hypothermia: Evidence for an interaction with dopaminergic systems and the hypothalamic-pituitary-thyroid axis. Brain Res. **195:** 69–84.
4. OSBAHR, A. J., C. B. NEMEROFF, D. LUTTINGER, G. A. MASON & A. J. PRANGE. 1981. Neurotensin-induced antinociception in mice: Antagonism by thyrotropin-releasing hormone. J. Pharmacol. Exp. Ther. **217:** 645–651.

DEGRADATION OF NEUROTENSIN BY BRAIN SYNAPTIC MEMBRANES *

Frédéric Checler, Patrick Kitabgi, and Jean-Pierre Vincent

Centre de Biochimie du Centre National de la Recherche Scientifique
Université de Nice
Parc Valrose
06034 Nice, Cedex, France

Several lines of evidence indicate that neurotensin may function as a neurotransmitter in the central nervous system. It has been shown for all classical neurotransmitters that there is a mechanism by which the interaction of these substances with their synaptic receptors is terminated. Accordingly, there might exist a degradation process by which neurotensin is inactivated in brain tissues. In this context, we have studied the degradation of [^{3}H]neurotensin by purified rat brain synaptic membranes that have previously been shown to possess specific neurotensin receptors and therefore represent a suitable material to study neurotensin degradation of possible physiological relevance.

We have set up a simple and rapid method that separates intact [^{3}H] neurotensin from its tritiated degradation products on small (500 μl) SP-Sephadex C-25 columns equilibrated and eluted with 0.01 M acetic acid, pH 5.5. In these conditions, neurotensin is retained on the column while the degradation products are eluted. Using this method, both membrane-bound and soluble neurotensin-degrading activities were detected in rat brain synaptic membranes. The soluble activity was released by hypotonic treatment (5 mM TRIS-HCl, pH 7.5) of the membrane preparation whereas the membrane-bound activity remained associated with the membranes even after extensive washing in either hypotonic (5 mM TRIS-HCl, pH 7.5) or hypertonic (1 M KCl) solution. The K_m values for the degradation of neurotensin by membrane-bound and soluble activities were 8 μM and 5 μM respectively, as derived from linear Eadie-Scatchard plots. A similar pattern of activity for various peptidase inhibitors was observed with both membrane-bound and soluble degrading activities. Thiol protective agents and inhibitors of serine- and thiol-endopeptidases had no effect. Puromycin (an aminopeptidase inhibitor) had a weak inhibitory effect. In contrast, chelating agents such as 1,10-phenanthroline, EDTA, and dithiothreitol were all effective inhibitors of neurotensin-degrading activities. Captopril, a specific and highly potent inhibitor of angiotensin-converting enzyme (K_i = 20 nM) and thiorphan, a highly potent inhibitor of enkephalinase (K_i = 5 nM) only partially inhibited the degradation of neurotensin at concentrations 4 to 5 orders of magnitude above those required to inhibit their respective enzymes. The degradation products generated after incubation of unlabeled neurotensin and trace amounts of [^{3}H]neurotensin with membrane-bound and soluble degrading activities were submitted to reverse-phase high-performance liquid chromatography (HPLC). Solvents were acetonitrile (CH$_3$CN) and 0.02 M ammonium acetate, pH 6.4. Immediately following

*This work was supported in part by grant 81.E.0542 from the Délégation Générale à la Recherche Scientifique et Technique (D.G.R.S.T., Paris, France) and by the Fondation pour la Recherche Médicale Française.

injection of degradation products in the HPLC column, a nonlinear gradient (curve 7 of the model 660 programmer, Waters) from 10% to 35% CH_3CN was run for 35 min. The products were detected by their absorbance at 230 nm and their radioactivity content. Six radioactive peaks with retention times of 5, 15, 22, 25, 27, and 34 min were obtained with both membrane-bound and soluble activities. Peak 5 min could only be detected by its radioactivity content because it eluted with contaminants present in brain membranes. The major products that were generated by both activities were peaks 5, 15, and 34 min. Peaks 15 and 34 min were identified by amino-acid analysis and retention times as neurotensin 1–8 and neurotensin, respectively. Peak 5 min eluted with a retention time similar to that of $[^3H]$Tyr. Peak 22, 25, and 27 min were generated in higher relative amounts by the membrane-bound than by the soluble activity. Actually, peak 25 min was very low and peak 27 min was barely detectable when neurotensin was degraded by soluble activity. Peaks 22, 25, and 27 min could not be unambiguously identified by amino-acid analysis. However, since Ile was not detected in hydrolysates of peak 22 min, this product is likely to consist of NH_2-terminal neurotensin sequence(s). Peaks 25 and 27 min differed by their retention times from neurotensin 9–13.

Our data show that membrane-bound and soluble neurotensin-degrading activities from rat brain synaptic membranes have similar properties with respect to K_m values for neurotensin, and specificity towards peptidase inhibitors. This specificity shows that neurotensin is degraded by metalloendopeptidase(s). Characterization of the degradation products indicates that with both membrane-bound and soluble activities, cleavage occurs between residues Arg^8 and Arg^9 of the neurotensin molecule. This yields the octapeptide neurotensin 1–8. The COOH-terminal pentapeptide neurotensin 9–13 probably undergoes other cleavages into smaller peptide(s) lacking the COOH-terminal residue of neurotensin. Other degradation products that are generated in small amounts by the soluble activity and in higher amounts by the membrane-bound activity are currently being analyzed for identification. In conclusion, degradation of neurotensin by rat brain synaptic membranes results in the destruction of the COOH-terminal active region of the peptide. This might represent an inactivation process of physiological relevance for neurotensin at the synaptic level.

ANALYSIS OF [³H]NEUROTENSIN RECEPTORS BY COMPUTERIZED DENSITOMETRY: VISUALIZATION OF CENTRAL AND PERIPHERAL NEUROTENSIN RECEPTORS

Remi Quirion,* T. Andreo Larsen,† Donald Calne,†
Thomas Chase,† Francis Rioux,‡ Serge St-Pierre,‡
Hilleary Evarist,* Agu Pert,* and Candace B. Pert *

*Neurosciences Branch
National Institute of Mental Health
Bethesda, Maryland 20205
†NINCDS Bethesda, Maryland 20205

‡Department of Pharmacology
Faculty of Medicine
University of Sherbrooke
Quebec, Canada

INTRODUCTION

Neurotensin (NT) is known to exert multiple central and peripheral actions such as hypothermia, analgesia, and hypotension (for a recent review, see ref. 1). The presence of specific receptors for NT in the brain has been reported.[2,3] Recently, Young and Kuhar[4] described the autoradiographic distribution of [³H]NT receptors in rat brain using a classical emulsion autoradiography technique. In this paper, we reported the distribution of [³H]NT binding sites in rat brain and kidney. We used the new tritium-sensitive film technique,[5] which allows analysis of autoradiographs by computerized densitometry and subsequent color-coding translation to analyze the distribution of [³H]NT receptors.

METHODS

Slide-mounted tissue sections were prepared as described before.[5] Briefly, slide-mounted brain or kidney sections were incubated in 10 ml of incubation buffer (50 mM Tris-HCl, pH 7.4 at 0°C plus 0.1% BSA and 0.5 mg/ml bacitracin) for 60 min in presence of 4.0 nM [³H]NT (or various concentrations for saturation curves) (New England Nuclear). At the end of the incubation period, slides were transferred sequentially through 4 rinses of cold incubation buffer (2 min in each) and then dried under a stream of cold air. For biochemical experiments, binding of [³H]NT to the tissue slice was quantitated by counting the tissue-laden slide fragment in 10 ml Aquassure scintillation cocktail. Specific binding was calculated as the difference in counts bound in presence and absence of 1 μM NT. For autoradiographic experiments, dried slides were juxtaposed with tritium-sensitive

* Address all correspondence to: Remi Quirion, Ph.D, Neurosciences Branch, Bldg. #10, Room 3N256, National Institute of Mental Health, Bethesda, MD, 20205.

film (Ultrofilm, LKB Instruments) and stored at room temperature for 6 weeks. After this exposure, films were developed and analyzed by computerized densitometry as described before.[5]

RESULTS AND DISCUSSION

Saturation curves and Scatchard analysis of [³H]NT binding in rat brain and kidney slices revealed the existence of a saturable single class of [³H]NT receptors in both tissues (K_D = 5.1 nM in the brain and 8.3 nM in the kidney). Relative potencies of various NT fragments and analogues such as NT8–13, [Phe[11]]-NT and [Trp[11]]-NT suggested that NT receptors in both tissues are very similar to those present in rat stomach strips.[6, 7]

TABLE 1

[³H]NT BINDING SITE DISTRIBUTION IN RAT BRAIN SLICES

External plexiform layer of the olfactory bulb	++++
Accessory olfactory bulb	+++
Anterior cingulate cortex just anterior to the genu of the corpus callosum	+++
Rhinal sulcus	+++
Surrounding the claustrum	+++
Cortex, area 27, lamina I	+++
Lateral septum	+++
Nucleus tractus diagonalis	+++
Nucleus preopticus	+++
Bed nucleus of the stria terminalis	+++
Zona incerta	+++
Habenula	+++
Nucleus tractus olfactorii lateralis	+++
Substantia nigra, zona compacta	+++
Ventral tegmental area	+++
Dentate gyrus	++
Superior colliculus	++
Periaqueductal gray	++
Nucleus tractus solitarii	++
Striatum	++
Cortex	+
Hypothalamus	+

Neurotensin receptor localization in rat kidney slices shows a concentration of receptors in the cortex, a much lower level in the papilla, and practically no binding sites in the medulla. A possible role for these [³H]NT-binding sites in the kidney remains to be demonstrated.

In the brain, the distribution of [³H]NT-binding sites is very discrete and often associated with limbic structures. As shown in TABLE 1, the highest concentration of [³H]NT receptors was observed in the external layer of the olfactory bulb. High levels were found in the anterior cingulate cortex, surrounding the claustrum, in the rhinal sulcus, in the habenula, in the zona incerta, in the substantia nigra, zona compacta, and in the ventral tegmental area. Moderate levels were observed in the striatum, in parts of the amygdala, in the dentate gyrus, in the superior colliculus, in parts of the periaqueductal gray and in the nucleus tractus solitarii. Low levels were observed in most areas of the cortex and the hypothalamus.

As reported recently,[8] we have observed that a 6-hydroxydopamine injection in the substantia nigra, zona compacta induced a marked depletion of [³H]NT receptors in this area without affecting the number of sites in adjacent areas such as the ventral tegmental area. Moreover, we have also observed that such a lesion induced a total loss of [³H]NT-binding sites in the striatum. These results suggest that NT receptors are localized on cell bodies and pre-synaptic terminals of the dopaminergic nigrostriatal pathway thus reinforcing the hypothesis that NT might have potent effects on dopaminergic functions in the brain.

Finally, very preliminary results show an 85% increase in [³H]NT binding in cortical and striatal areas in human schizophrenic brain (n = 2) and a 50% decrease in [³H]NT binding in human parkinsonian brain (n = 4). It is too early to conclude that NT is implicated in the cause of schizophrenia, but our results indicate that future studies evaluating this possibility are certainly worthwhile.

REFERENCES

1. NEMEROFF, C. B., D. LUTTINGER & A. J. PRANGE, JR. 1982. Neurotensin and bombesin. *In* Handbook of Psychopharmacology. L. L. Iversen, S. D. Iversen & S. H. Snyder, Eds. Plenum Press. New York. In press.
2. UHL, G. R., J. P. BENNETT, JR. & S. H. SNYDER. 1977. Neurotensin, a central nervous system peptide: Apparent receptor binding in brain membranes. Brain Res. **130:** 299–313.
3. KITABGI, P., R. CARRAWAY, J. VAN RIETSCHOTEN, C. GRANIER, J. L. MORGAT, A. MENEZ, S. LEEMAN & P. FREYCHET. 1977. Neurotensin: Specific binding to synaptic membranes from rat brain. Proc. Natl. Acad. Sci. USA **74:** 1846–1850.
4. YOUNG, W. S., III & M. J. KUHAR. 1981. Neurotensin receptor localization by light microscopic autoradiography in rat brain. Brain Res. **206:** 273–285.
5. QUIRION, R., R. HAMMER, JR., M. HERKENHAM & C. B. PERT. 1981. Phencyclidine (Angel Dust)/σ opiate receptor: Visualization by tritium-sensitive film. Proc. Natl. Acad. Sci. USA **78:** 5881–5885.
6. QUIRION, R., D. REGOLI, F. RIOUX & S. ST-PIERRE. 1980. The stimulatory effects of neurotensin and related peptides in rat stomach strips and guinea pig atria. Br. J. Pharmacol. **68:** 83–91.
7. QUIRION, R., D. REGOLI, F. RIOUX & S. ST-PIERRE. 1980. Structure–activity with neurotensin: Analysis of positions, 9, 10, 11. Br. J. Pharmacol. **69:** 689–692.
8. PALACIOS, J. M. & M. J. KUHAR. 1981. Neurotensin receptors are located on dopamine-containing neurones in rat midbrain. Nature **294:** 587–589.

SUBNORMAL CSF LEVELS OF NEUROTENSIN IN A SUBGROUP OF SCHIZOPHRENICS AND NORMALIZATION AFTER NEUROLEPTIC TREATMENT

E. Widerlöv,*† L. H. Lindström,* G. Besev,* P. J. Manberg,†
C. B. Nemeroff,† G. R. Breese,† J. S. Kizer,† and
A. J. Prange, Jr.†

*Psychiatric Research Center
Ulleråker Hospital
75017 Uppsala, Sweden

†Biological Sciences Research Center
University of North Carolina School of Medicine
Chapel Hill, North Carolina 27514

Neurotensin (NT) is a tridecapeptide heterogeneously distributed in the central nervous system and the gastrointestinal tract of various mammals, including man. The peptide has been shown to possess many pharmacological properties that are similar to those of neuroleptic drugs: induces hypothermia and muscle relaxation, reduces spontaneous and amphetamine-induced locomotion, and potentiates ethanol and barbiturate sedation.[1] Like neuroleptics, centrally administered NT also increases the synthesis of brain dopamine and elevates the concentration of the principal dopamine metabolites, 3,4-dihydroxyphenylacetic acid (DOPAC) and homovanillic acid (HVA).[2]

In the present study, cerebrospinal fluid (CSF) levels of immunoreactive NT[3] in 12 physically and mentally healthy volunteers (20–42 yr; median 26.5 yr) were compared with the CSF levels of NT from 21 age-matched schizophrenic patients (20–40 yr; median 26 yr). The CSF from most of the patients (n = 15) were drawn during both drug-free and neuroleptic-treated phases. The patients were hospitalized and had suffered from schizophrenia for 3–14 yr (median 5 yr). Diagnoses were assessed according to the Research Diagnostic Criteria. The CSF levels of NT for the healthy volunteers were 240.4 ± 30.5 pg/ml (mean ± SEM; TABLE 1). In contrast to the healthy volunteers, the schizophrenics showed a bimodal distribution of NT during the drug-free phase (p < 0.005). A subgroup of nine patients had markedly lower NT levels (46.7 ± 6.5 pg/ml) compared to the remainder of the schizophrenics (n = 12), who had CSF levels of NT in the same range as the healthy volunteers (249 ± 29.2 pg/ml). Of particular interest was the finding that during neuroleptic treatment, the CSF levels increased in the schizophrenic group with low drug-free values to the same concentrations as for the group with the normal drug-free NT levels (278 ± 37.9 and 282.8 ± 93.6 pg/ml, respectively; TABLE 1).

Observed and reported schizophrenic symptoms were scored according to the Comprehensive Psychopathological Rating Scale (CPRS).[4] The schizophrenics with low NT levels in CSF had significantly higher scores of psychomotor retardation than the schizophrenics with normal drug-free NT levels in CSF (p < 0.02). Also, the three patients with a catatonic form of schizophrenia all belonged to the group with the low drug-free NT levels in CSF. The concept that the pathogenesis of major mental disorders is primarily due to alterations in monoaminergic neu-

TABLE 1

CEREBROSPINAL FLUID LEVELS OF NEUROTENSIN IN HEALTHY VOLUNTEERS
AND SCHIZOPHRENIC PATIENTS *

Group	Number of Subjects	Drug-free	Neuroleptic Treatment
Healthy volunteers	12	240.4 ± 30.5	—
Schizophrenics †	15	156.7 ± 34.6	280.5 ± 51.2
A. with normal drug-free levels (> 100 pg/ml)	8	252.0 ± 40.8	282.8 ± 93.6
B. with low drug-free levels (< 100 pg/ml)	7	47.7 ± 7.83	278.1 ± 37.9 ‡

* Expressed as pg/ml; mean ± SEM.

† Data pertain only to patients from whom CSF samples were obtained in both the drug-free and neuroleptic-treated phases. The division of the schizophrenics into two subgroups is supported by statistical analysis, revealing a bimodal distribution (p < 0.005).

‡ p < 0.0001 versus the drug-free phase (two-tailed test).

ronal activity has gradually been modified in favor of more complex theories. The present data and data from other studies [1,2] support the hypothesis that alterations in CNS NT activity might play a role in the pathogenesis of schizophrenia, presumably by interacting with central dopaminergic pathways.

REFERENCES

1. NEMEROFF, C. B. 1980. Neurotensin: Perchance an endogenous neuroleptic? Biol. Psychiatry 15: 283–302.
2. WIDERLÖV, E., C. D. KILTS, R. B. MAILMAN, C. B. NEMEROFF, A. J. PRANGE, JR. & G. R. BREESE. 1982. Increases in dopamine metabolites in rat brain by neurotensin. J. Pharmacol. Exp. Ther. 222: 1–6.
3. MANBERG, P. J., W. W. YOUNGBLOOD, C. B. NEMEROFF, M. N. ROSSOR, L. L. IVERSEN, A. J. PRANGE, JR. & J. S. KIZER. 1982. Regional distribution of neurotensin in human brain and cerebrospinal fluid. J. Neurochem. 38: 1777–1780.
4. ÅSBERG, M., C. PERRIS, D. SCHALLING & G. SEDVALL. 1978. The CPRS—development and application of a psychiatric rating scale. Acta Psychiat. Scand., Suppl. 271.

THE EFFECT OF NEUROTENSIN ON DOPAMINERGIC NEURONS IN RAT BRAIN*

Avinoam Reches, Robert E. Burke, De-hua Jiang, H. Ryan Wagner,
and Stanley Fahn

Department of Neurology
College of Physicians and Surgeons
Columbia University
New York, New York 10032

Neurotensin (NT) shares several properties with the neuroleptic drugs.[1,2] We have studied the effect of intracerebroventricular (i.c.v.) injections of NT on dopamine (DA) turnover in the nigrostriatal and mesolimbic (olfactory tubercle plus the medial part of the nucleus accumbens) pathways of rat brain.

DA and its metabolites 3,4-dihydroxyphenylacetic acid (DOPAC) and homovanillic acid (HVA) were measured by high-pressure liquid chromatography with electrochemical detection as previously described.[3]

NT (30 μg/10 μl PBS) increases DA turnover in the corpus striatum and mesolimbic pathway after i.c.v. injection. Significant increases in DOPAC levels occur within 15 minutes after injection with peak levels at 60 minutes for both brain regions. This effect of NT was dose dependent. Significant increases in DOPAC levels occurred after 3 μg (the smallest dose tested) with maximal stimulation ($\times$ basal) obtained at 100 μg for the striatum (1.81) and accumbens (2.35). There was no significant difference in the ability of NT to raise DOPAC levels between these two brain regions. At 200 μg NT reduced DOPAC concentrations. HVA levels were similarly affected.

To study possible direct interaction of NT with presynaptic dopaminergic receptors, we injected gammabutyrolactone (GBL), an agent reported to selectively block impulse flow in the dopaminergic neurons.[4] Such blockade of impulse flow results in an immediate increase in the synthesis of dopamine which is probably modified, under these conditions, by presynaptic receptor activity. The accumulation of DOPA after the administration of the DOPA decarboxylase inhibitor NSD-1015 was used to measure striatal tyrosine hydroxylase (TH) activity.[5] Our data suggest that NT (30 μg) has either an inhibitory effect on presynaptic DA receptors or a direct stimulatory effect on TH activity. DOPA accumulation was significantly higher in GBL-treated rats pretreated with NT both in the striatum (5.03 $\pm$ 0.47 vs. 3.47 $\pm$ 0.13 ng/mg tissue, p < 0.025) and mesolimbic pathway (2.14 $\pm$ 0.14 vs. 1.47 $\pm$ 0.08 ng/mg tissue, p < 0.01). These findings are supported by a previous report[6] that describes NT-enhanced DOPA accumulation in various brain regions of NSD-1015 pretreated rats. In the present study, however, using GBL, we were able to demonstrate that this effect of NT is mediated through its action at the presynaptic terminals. Moreover, we were able to demonstrate this effect at a much lower dose of NT. NT and haloperidol

*This work was supported in part by NIH grant NS15959, the Peggy Engl Fellowship awarded to Dr. Reches by the Parkinson's Disease Foundation, and by a Fogarty Public Health International Research Fellows (NIH TW02884) awarded to Dr. Reches. Dr. De-hua Jiang is the recipient of an H. Houston Merritt Fellowship from the Parkinson's Disease Foundation.

differed in their capacity to antagonize the effect of apomorphine on presynaptic DA receptors. NT failed to block apomorphine-induced reductions in DOPA accumulation both in the striatum and the accumbens, while haloperidol completely reversed the apomorphine effect in both regions. The lack of antagonism between NT and apomorphine agrees with the previous observation that NT had no effect on the incidence of apomorphine-induced stereotypies in mice.[7]

Neurotensin also differed from the neuroleptics in its ability to compete for ^{3}H-spiperone binding sites. As previously reported,[2] NT only weakly displaced ^{3}H-spiperone from membrane preparations in either striatum or accumbens. The presence of NT (10 μM) in the binding assay also had no effect on the ability of DA to displace ^{3}H-spiperone. These experiments suggest that NT increases DA turnover in the nigrostriatal and mesolimbic pathways through either inhibition of presynaptic DA receptors or through direct activation of tyrosine hydroxylase.

REFERENCES

1. BISSETTE, G., P. MANBERG, C. B. NEMEROFF & A. J. PRANGE. 1978. Neurotensin, a biologically active peptide. Life Sci. **23:** 2173–2182.
2. NEMEROFF, C. B. 1980. Neurotensin: Perchance an endogenous neuroleptic? Biol. Psychiatry **15:** 283–302.
3. RECHES, A. & S. FAHN. 1982. 3-0-Methyldopa blocks DOPA metabolism in rat corpus striatum. Ann. Neurol. **12:** 267–271.
4. WALTERS, J. R. & R. H. ROTH. 1976. Dopaminergic neurons: An *in vivo* system for measuring drug interaction with presynaptic receptors. Naunyn-Schmiedeberg's Arch. Pharmakol. **296:** 5–14.
5. NOWYCKZ, M. C. & R. H. ROTH. 1978. Dopaminergic neurons: Role of presynaptic receptors in the regulation of transmitter biosynthesis. Prog. Neurol. Psychopharm. **2:** 139–158.
6. GARCIA-SEVILLA, J. A., T. MAGNUSSON, A. CARLSSON, J. LEBAN & K. FOLKERS. 1978. Neurotensin and its amide analogue [G/N^4]-neurotensin: Effects on brain monoamine turnover. Naunyn-Schmiedeberg's Arch. Pharmakol. **305:** 213–218.
7. NEMEROFF, C. B., G. N. ERVIN, P. J. MANBERG, A. J. OSBAHR, S. FELTS, L. S. BIRKEMO & A. J. PRANGE. 1979. Further studies on central nervous system effects of neurotensin on endogenous peptide. Amer. Soc. Neurosci. Abst. **5:** 535.

REWARDING EFFECTS OF NEUROTENSIN IN THE BRAIN

Paul Glimcher, David Margolin, and Bartley G. Hoebel

Department of Psychology
Princeton University
Princeton, New Jersey 08544

The role of neurotensin (NT) in the regulation of reinforcement was investigated using the Conditioned Place Preference (CPP) paradigm to assay the ability of NT to provide positive behavioral reinforcement. Animals preferred to spend time in a place associated with the effects of intracerebral injections of NT. We conclude that NT in the ventral tegmental area (VTA) is rewarding. Rats (n = 11) with bilateral cannulas aimed at the VTA were placed in a rectangular tilt-cage, one side darkened, the other side illuminated. Animals were pretested for three consecutive days, and time spent on each side was recorded. Rats showed a strong preference for the dark side. The following four days, the rats were injected bilaterally with either 5.0 μg/0.5 μl NT or with 0.5 μl saline, and were confined to their nonpreferred side for two hours, to form an association between the drug effects and the bright environment. On the eighth day, all rats received saline injections, were placed in their experimental cages unconfined, and were allowed to demonstrate their place preference.

The CPP paradigm forms an association between an experimental manipulation and a fixed location. The reinforcing quality of a manipulation is demonstrated by its ability to alter place preference. The postconditioning place preferences of individual animals were paired to their respective stable baseline data, collected during three days of pretesting. The data were normalized as percent of total time spent on the nonpreferred side and changes are represented relative to baseline. Rats that received NT showed a significant increase in preference for the side associated with the drug's effect. FIGURE 1 depicts individual data on NT-induced increases in time spent on the formerly nonpreferred side during the initial 15-minute, postconditioning test interval. Note that NT animals universally display a positive change in preference. Throughout the test session, the NT-induced preference for the bright side remained elevated over baseline. FIGURE 2 relates the NT-induced increase in percent time spent on the formerly nonpreferred side during four consecutive 15-minute intervals. The Wilcoxon rank sum test was used to analyze these data. This analysis shows the NT effect to be significant for the first 15 minutes (p < 0.01) when compared to saline controls. The second 15-minute period also demonstrated statistical significance (p < 0.05). Control rats continued to avoid the bright side, spending slightly less time on the nonpreferred side compared to baseline (FIG. 2).

Aghajanian's group (Neurosci. Abstr. **7**: 184.1, 1981) has demonstrated excitation of dopamine (DA) single units in response to iontophoretically applied NT. Additionally, Palacios and Kuhar (Nature **294**: 587–9, 1981) have identified specific NT receptors on the cell bodies of DA neurons. This suggests that some of NT's VTA effects may be mediated by the mesolimbic DA cell group. Kalivas, Nemeroff, and Prange (Neurosci. Abstr. **7**: 13.7, 1981) have shown that VTA injections of NT increased locomotor activity. The midbrain DA systems may be

* Supported by USPHS grant MH-35740.

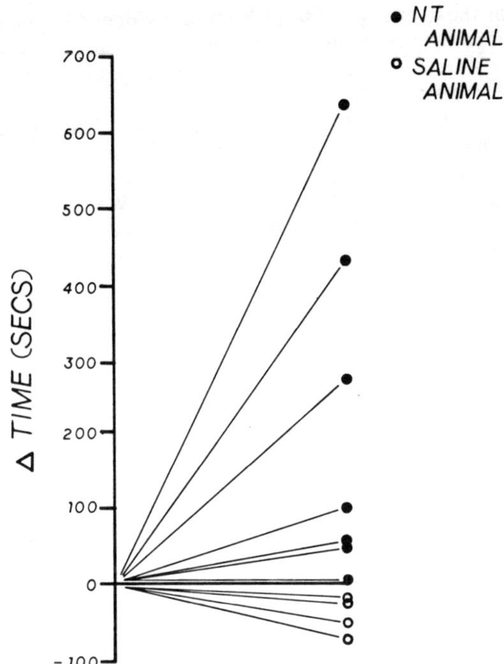

FIGURE 1.   Individual data on neurotensin-induced increase in time spent on formerly nonpreferred side during initial 15-minute interval.

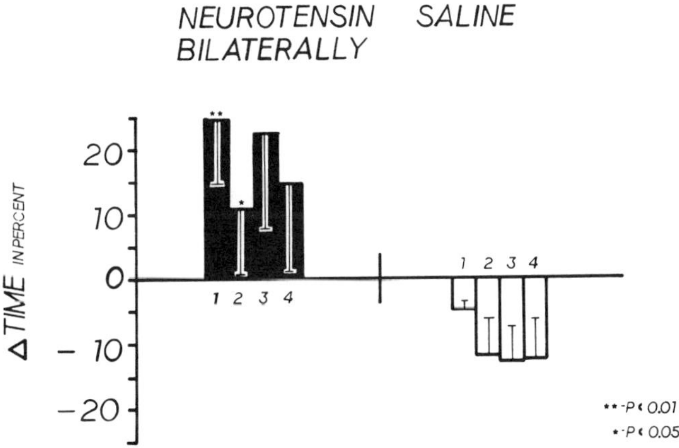

FIGURE 2.   Neurotensin-induced increase in percent time spent on formerly nonpreferred side during four consecutive 15-minute intervals.

a central locus for these effects. A large body of evidence implicates DA fibers in the modulation of reward. In light, as well, of the activational role of DA in motor responses, it seems pertinent to consider the relevance of the ascending DA fibers to the midbrain effects of NT.

These data demonstrate that VTA brain injections of NT produce conditioned place preference in rats. The magnitude of the preference shift is positively correlated with the proximity of the cannulae tips to the VTA (Spearman's rho = 0.741, $p < 0.05$, based on evaluation of histological data). We conclude that NT in the VTA is rewarding. The system whereby NT exerts its rewarding effects, and the physiological relevance of that reward, remains to be described.

NEUROTENSIN INJECTED INTO THE PARAVENTRICULAR HYPOTHALAMUS SUPPRESSES FEEDING IN RATS

B. Glenn Stanley, Nancy Eppel, and Bartley G. Hoebel

Department of Psychology
Princeton University
Princeton, New Jersey 08544

Neurotensin is a tridecapeptide that is released into the blood following a meal (Mashford, *et al.* 1978. Acta. Physiol. Scand. **1044:** 244–246). To investigate its role in feeding behavior, the food intake of 24-hour food-deprived rats (n = 10) was measured following bilateral injection of neurotensin (0, 2.0, 10.0 μg / 0.3 μl / side) into the paraventricular nucleus of the hypothalamus (PVN). As can be seen in FIGURE 1, the 10.0 μg dose of neurotensin produced a significant (p < .001) reduction of deprivation-induced feeding during the 30-minute postinjection test period. To test the behavioral specificity of this effect, the water intake of 24-hour water-deprived rats was measured following bilateral PVN injection of neurotensin (10.0 μg / 0.3 μl / side). The neurotensin injections did not reduce the water intake of these rats during the first 30-minute postinjection interval (12.0 ± 1.0 ml control; 11.7 ± 1.6 ml with neurotensin). This suggests that the neurotensin effect on feeding was specific and was not due to malaise or nonspecific behavioral suppression.

It has been shown that intrahypothalamic norepinephrine (NE) injection elicits feeding, and that the PVN is the area most sensitive to this effect (Leibowitz, 1978. Pharm. Biochem. Behav. **8:** 163–175.). To determine whether the neurotensin effect would generalize to feeding elicited by NE, we gave eight rats bilateral PVN neurotensin (5.0 μg / 0.3 / side) followed by NE (6.4 μg / 0.3 μl / side). As can be seen in FIGURE 2, NE elicted feeding, and this effect was significantly attenuated by neurotensin pretreatment. Thus, PVN neurotensin attenuates feeding elicited by either PVN NE, or by food deprivation.

Neurotensin's feeding suppressive effect could possibly have resulted from leakage into the periphery rather than to an action in the CNS. To test this possibility, we injected neurotensin (0, 0.4, 2.0, 10.0 μg / 0.3 μl) ipsilateral or contralateral to a unilateral PVN NE (12.8 μg / 0.3 μl) injection in eleven rats. As can be seen in FIGURE 3, ipsilateral neurotensin produced a dose-dependent decrease in NE-elicited feeding. The feeding suppressive effect of ipsilateral neurotensin was significantly (p < .05) greater than the effect of contralateral neurotensin. This suggests that the suppression of feeding by neurotensin was largely due to its action in the CNS rather than to leakage into the periphery. These data provide further evidence that neurotensin suppression of feeding was not the result of nonspecific behavioral suppression.

In conclusion, these data suggest that there may be hypothalamic elements that cause a reduction in feeding when activated by neurotensin. Whether receptors are normally activated by neurotensin released from hypothalamic neurons or from the gut remains to be determined.

*Supported by USPHS grant MH-35740.

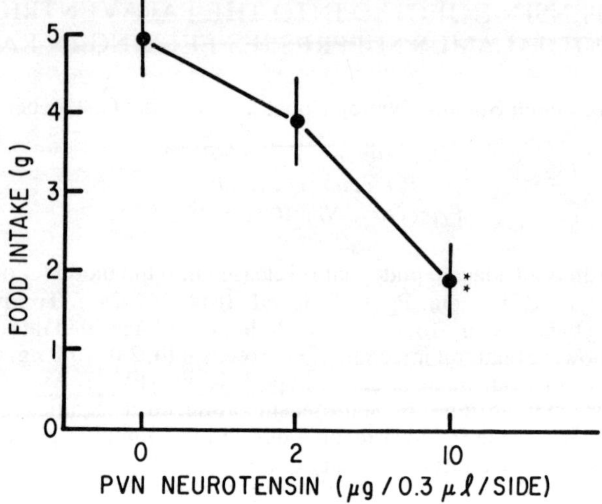

FIGURE 1. Food intake of 24-hour food-deprived rats (n = 10) in the 30-minute period following bilateral PVN neurotensin injection (mean ± SEM; ✶✶ p < .001).

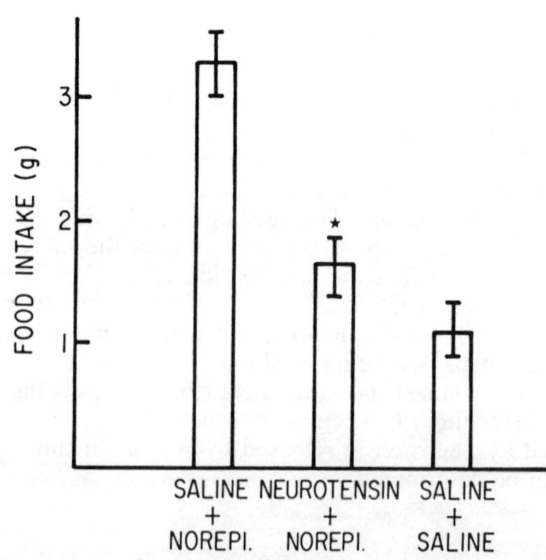

FIGURE 2. Food intake of rats (n = 8) given bilateral PVN injections of saline + NE, neurotensin + NE, or saline ± saline (★ p < .05, saline + NE vs. neurotensin + NE).

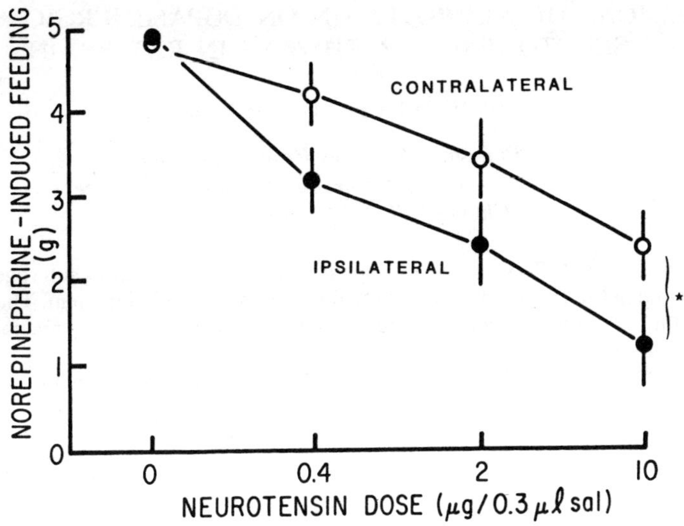

FIGURE 3.   Food intake of rats (n = 11) given PVN neurotensin ipsilateral or contralateral to a PVN NE injection (★ p < .05, ipsilateral vs. contralateral neurotensin).

ACTIONS OF NEUROTENSIN ON DOPAMINERGIC AND SEROTONERGIC PATHWAYS IN RAT BRAIN

Erik Widerlöv and George R. Breese

Biological Sciences Research Center
University of North Carolina School of Medicine
Chapel Hill, North Carolina 27514

Neurotensin (NT) is an endogenous tridecapeptide, which is heterogeneously distributed in the central nervous system in mammals, including man. The peptide has a number of physiological and behavioral effects, many of them resembling the actions of neuroleptic drugs.[1]

The biochemical interaction between NT and some dopaminergic and serotonergic pathways in rat brain (male Sprague-Dawley rats, 150–200 g) have been studied using recently developed high-performance liquid chromatography (HPLC) procedures for simultaneous measurement of dopamine (DA), 3,4-dihydroxyphenylacetic acid (DOPAC), homovanillic acid (HVA), serotonin (5-HT) and 5-hydroxyindole acetic acid (5-HIAA). The accumulation of dihydroxyphenylalanine (DOPA), assessed with HPLC,[3] after DOPA-decarboxylase inhibition with NSD 1015,[4] was increased in olfactory tubercles (p < 0.02) and striatum (p < 0.05) after NT [30 μg intracisternally (i.c.)], compared with saline treatment, suggesting an increased DA synthesis. Behaviorally and physiologically active doses of NT (1–100 μg, i.c.), administered to unanesthetized rats, caused a dose-dependent increase in the concentrations of HVA and DOPAC, the

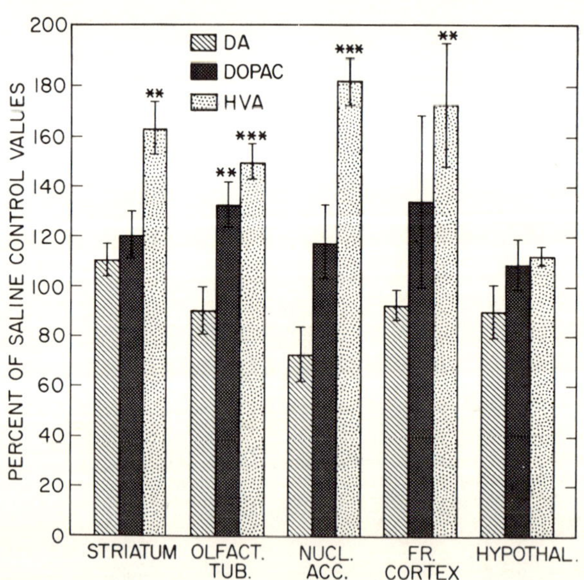

FIGURE 1. Dopamine and its main metabolites in different brain areas at 1 hour after NT 30 μg i.c. The bars represent mean ± SEM of six observations. **p < 0.01; ***p < 0.001 versus saline treatment.

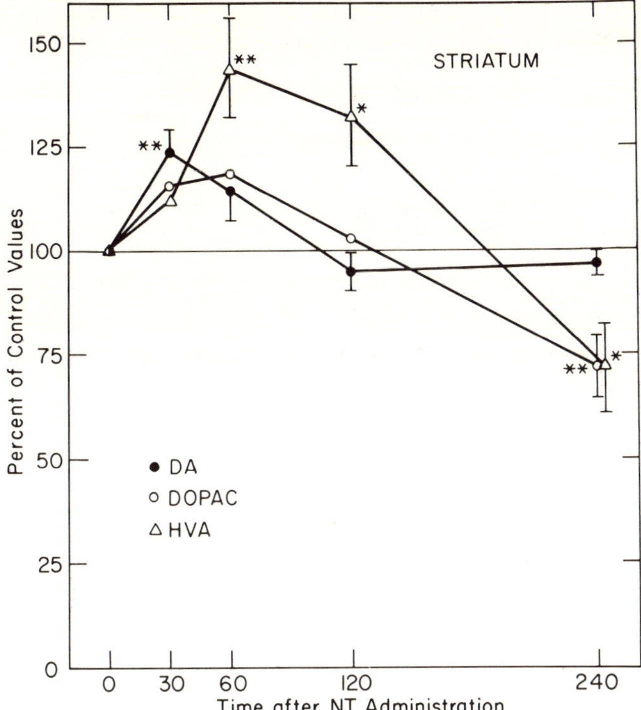

FIGURE 2. Time-course of the effect of NT on dopamine and its main metabolites after NT 30 μg i.c. The symbols represent the mean $\pm$ SEM of 5–6 observations. *p $<$ 0.05; **p $<$ 0.01 versus saline treatment.

primary metabolites of DA, in several brain areas. The relative increase in HVA was higher than that for DOPAC, indicating that NT also might block the reuptake of DA. The action of an intermediate dose of NT, 30 μg, at 1 hr after injection is shown in FIGURE 1. In striatum, an initial increase in DA content after 30 μg NT was followed by an increase and subsequent decrease of DA metabolites (FIG. 2). Similar time courses were observed in olfactory tubercles and nucleus accumbens.

The effects of NT (1–100 μg, i.c.) on the levels of 5-HT and its principal metabolite, 5-HIAA, were studied in different brain areas. After the highest dose of NT, increased levels of 5-HT were observed in hypothalamus and frontal cortex (p $<$ 0.05 analysis of variance followed by Dunnett's t-test) whereas no effect could be seen on the 5-HIAA concentrations.

β-Endorphin, administered in a dose equimolar to 30 μg NT, also increased DA metabolites in certain brain areas. Pretreatment with naloxone completely inhibited the effect of β-endorphin but not that of NT (TABLE 1), suggesting that the two peptides influence DA function by different mechanisms. Considerable data by several investigators have suggested functional links between neuropeptides and dopaminergic mechanisms.[3,5,6] The data provided in the present experiment suggest that NT, too, may have important physiological interactions with central dopaminergic neurons.

Table 1

NALOXONE EFFECTS ON NT- AND β-ENDORPHIN-INDUCED INCREASE
IN STRIATAL DOPAMINE METABOLITES*

i.p. dose	i.c. dose	DOPAC	HVA
Saline	+ saline	100.0 ± 9.2	100.0 ± 8.7
Saline	+ NT (60 μg)	97.5 ± 7.6	$137.7 \pm 6.6\ddagger$
Naloxone (8 mg/kg)	+ NT (60 μg)	103.6 ± 17.7	$138.4 \pm 14.6\dagger$
Saline	+ β-endorphin (20 μg)	141.7 ± 24.0	$138.3 \pm 16.9\dagger$
Naloxone (8 mg/kg)	+ β-endorphin (20 μg)	102.1 ± 9.1	$93.9 \pm 4.6\S$

*Values are expressed as means $\pm$ SEM in percent of the saline-treated controls; n = 6
$\dagger$ p < 0.05
$\ddagger$ p < 0.01 vs. saline + saline group
$\S$ p < 0.05 vs. saline + β-endorphin group

REFERENCES

1. LIPTON, M. A., G. N. ERVIN, L. S. BIRKEMO, C. B. NEMEROFF & A. J. PRANGE, JR. 1979. Neurotensin-neuroleptic similarities: An example of peptide-catecholamine interactions. *In* Catecholamines: Basic and Clinical Frontiers. Pergamon Press. New York. pp. 657–662.
2. KILTS, C. D., G. R. BREESE & R. B. MAILMAN. 1981. Simultaneous quantification of dopamine, 5-hydroxytryptamine and four metabolically related compounds by means of reverse-phase HPLC with electrochemical detection. J. Chromatogr. **225:** 347–357.
3. WIDERLÖV, E., C. D. KILTS, R. B. MAILMAN, C. B. NEMEROFF, A. J. PRANGE, JR. & G. R. BREESE. 1982. Increase in dopamine metabolites in rat brain by neurotensin. J. Pharmacol. Exp. Ther. **222:** 1–6.
4. CARLSSON, A., J. N. DAVIES, W. KEHR, M. LINDQVIST & C. A. ATACK. 1972. Simultaneous measurement of tyrosine and tryptophan hydroxylase activities in brain *in vivo* using an inhibitor of the aromatic amino acid decarboxylase. Naunyn-Schmiedeberg's Arch. Pharmakol. **275:** 153–168.
5. CARLSSON, A., J. A. GARCIA-SEVILLA & T. MAGNUSSON. 1979. Effect of peptides on brain monoamines and on gross behavior. *In* Central Regulation of the Endocrine System. Plenum Press. New York. pp. 223–238.
6. HÖKFELT, T., O. JOHANSSON, A. LJUNGDAHL, J. M. LUNDBERG & M. SCHULTZBERG. 1980. Peptidergic neurones. Nature **284:** 515–521.

THE EFFECTS OF PROSTAGLANDIN E_2 AND PROSTAGLANDIN INHIBITORS ON NEUROTENSIN-INDUCED HYPOTHERMIA IN MICE*

Daniel E. Hernandez, George A. Mason, and Arthur J. Prange, Jr.

*Biological Sciences Research Center
Department of Psychiatry and Neurobiology Program
University of North Carolina School of Medicine
Chapel Hill, North Carolina 27514*

Neurotensin (NT), an endogenous tridecapeptide elicits a profound hypothermia in a variety of mammalian species after intracisternal (i.c.) administration. Conversely, prostaglandin E_2 (PGE_2) produces a significant hyperthermia after central administration. This hypothermic activity of NT is antagonized by central (i.c.) or peripheral administration of thyrotropin-releasing hormone (TRH). For these reasons, we explored the effects of PGE_2 and prostaglandin synthetase inhibitors on NT-induced hypothermia and its antagonism by TRH.

Groups of male Swiss-Webster mice (25–35 g) under light ether anesthesia were injected i.c. (10 μl per injection) with vehicle (0.9% NaCl), NT (0.25, 1, 10 μg), PGE_2 (1–4 μg) or TRH (equimolar to 10 μg of NT). In some experiments, indomethacin (15 mg/kg), acetylsalicylic acid (aspirin, 100 mg/kg) or saline was administered subcutaneously (s.c.) 30 min before i.c. treatment to produce inhibition of prostaglandin synthesis. Indomethacin probably inhibits prostaglandin synthesis in both the brain and periphery whereas aspirin inhibits it only in the periphery. Immediately after the i.c. injections, mice (n = 8/group) were placed in individual cages without bedding at 26°C or 6°C. Colonic temperature was measured at 30-min intervals for two hours by insertion of a lubricated thermocouple probe approximately 1 cm into the rectum. Data were analyzed with the 1-way ANOVA followed by Dunnett's test for multiple comparisons.

Neither in a warm nor a cold environment did aspirin by itself alter body temperature. In a similar way, in both environments, it failed to bias NT-induced hypothermia and failed to bias TRH antagonism of NT-induced hypothermia.

At an ambient temperature of 26°C, PGE_2 (2 or 4 μg) produced significant *hyper*thermia. Both doses of PGE_2 inhibited the hypothermia produced by any of several doses of i.c. NT (0.25–10 μg). However, indomethacin (alone) failed to alter body temperature and failed to affect both NT-induced hypothermia and TRH-induced antagonism of this effect.

At an ambient temperature of 6°C, the action of PGE_2 was not tested. Again, indomethacin (alone) failed to alter body temperature. However, in contrast to its effects in the warm, it now potentiated NT-induced hypothermia and blocked TRH antagonism of this effect.

Because earlier data had shown that NT-induced hypothermia depends upon *central* administration of the peptide, it was not astonishing to find that aspirin, a *peripheral* prostaglandin synthesis inhibitor, had no effect on its actions. It does appear, however, that the hypothermic action of NT is involved with the (central) prostaglandin system. Indomethacin, in the warm, did not alter responses to NT; in the cold it did. This suggests that when the animal's thermoregulation is challenged by a cold environment, the participation of NT becomes at least partly expressed through the central prostaglandin system.

*Supported by NIMH grants MH-32316, MH-34121, MH-22536, MH-33127 and NICHHD HD-03110.

ROLE OF THE AUTONOMIC NERVOUS SYSTEM IN NEUROTENSIN'S CYTOPROTECTIVE EFFECT FOR GASTRIC STRESS ULCERS IN RATS

R. C. Orlando, D. E. Hernandez, C. B. Nemeroff, and A. J. Prange, Jr.

Departments of Psychiatry and Medicine
University of North Carolina School of Medicine
Chapel Hill, North Carolina 27514

We previously reported that intracisternal (i.c.), but not systemic, neurotensin (NT) prevented the development of gastric ulcers in rats subjected to cold-restraint stress (Am. J. Physiol., **242**: G342, 1982). This effect of NT was also shown to be unrelated to changes in gastric pH, body temperature or neuroleptic activity but required an intact prostaglandin synthetic pathway. Since the autonomic nervous system (ANS) modulates many gastric functions including prostaglandin synthesis, we investigated its role in i.c. NT's gastric cytoprotection.

METHODS

Male Sprague-Dawley rats (200–250 gms) were food deprived for 24 hours, then injected, before exposure to 3 hours of cold (4°C) while restrained supine by a wire mesh. After cold-restraint stress, the rats were sacrificed and the gastric mucosa inspected for ulcers.

RESULTS

Alpha- and beta-adrenergic stimulation and cholinergic (muscarinic) inhibition were cytoprotective to a similar degree as i.c. NT. I.c. NT's cytoprotective effect was completely blocked by concomitant beta-adrenergic blockage or cholinergic stimulation. Also, indomethacin blocked the cytoprotective effect of i.c. NT or beta-adrenergic stimulation but not cholinergic inhibition.

CONCLUSION

(1) The ANS can significantly influence the development of gastric ulcers from cold-restraint stress; (2) the ANS acts as an important intermediate in i.c. NT's gastric cytoprotection during cold-restraint stress, and (3) i.c. NT's gastric cytoprotection appears to occur by stimulation of beta receptors in the adrenergic nerve pathways to the stomach.

LIPID STIMULATION OF NEUROTENSIN RELEASE
FROM RAT SMALL INTESTINE

C. F. Ferris, R. E. Carraway, and S. E. Leeman

University of Massachusetts Medical School
Worcester, Massachusetts 02160

Endocrine cells staining immunohistochemically for neurotensin (NT) are present in mucosa of the small intestine (Helmstaedter *et al.*[1]). It has been proposed that gastrointestinal NT is released in response to some intraluminal stimulus and acts as either a hormonal or paracrine agent (Rosell *et al.*[2]; Ferris *et al.*[3]). The present study concerns the levels of NT and NT^{1-8} measured using site-directed RIA after PHLC on extracted hepatic portal and peripheral plasmas from rats during lipid perfusion of the intestine.

In three separate experiments, simultaneous blood samples (2.5 ml) were taken from the superior mesenteric vein and femoral artery of anesthetized rats during control perfusion of the small intestine (0.15 M NaCl, n = 10), or 15 minutes after the onset of lipid perfusion (2.4 mM sodium tauro-deoxycholate, 0.6 mM oleic acid, and 0.3 mM monoolein in 0.15 M NaCl, n = 10). Both the saline and lipid perfusates were warmed to body temperature and delivered at a rate of 3.2 ml/min. The blood collected from each sampling site was pooled and plasma extracted with acid/acetone. Lyophilized samples were dissolved in 2.0 ml of water and injected onto a 3.9 × 300 mm μBondapak C-18 column (Waters Associates, MA) running at 2 ml/min in 0.01 M KH_2PO_4, pH 4.6. Two and one half minutes after sample application, a 20-minute long gradient was run to 15% acetonitrile in 0.01 M KH_2PO_4 using Curve 3. Two minutes thereafter, a 10-minute long gradient was run to 30% acetonitrile using Curve 6. Column fractions were assayed with both amino- and carboxyl-terminal-directed antisera.

There was no significant arteriovenous difference for NT (artery 2 ± 2 fmol/ml − vein 3 ± 2 fmol/ml, FIG. 1) or NT^{1-8} (artery 14 ± 1 fmol/ml − vein 14 ± 2 fmol/ml, FIG. 2) during saline perfusion. During lipid perfusion, arterial levels of NT (2 ± 1 fmol/ml, FIG. 1) and NT^{1-8} (16 ± 1 fmol/ml, FIG. 2) did not differ from the saline group, while levels in the superior mesenteric vein of NT (14 ± 2 fmol/ml, p < .05) and NT^{1-8} (29 ± 4 fmol/ml, p < .05) were significantly elevated.

This study demonstrates that chromatographically identified NT and its metabolite, NT^{1-8} are elevated in the portal circulation but not systemic circulation 15 min after the onset of lipid perfusion; evidence that NT is released from the small intestine and might possibly be degraded in close proximity to its site of release. It is possible that NT^{1-8} is itself released by the intestine, but the fact that it is a breakdown product observed when NT is injected into rats suggests that it is elevated in the portal circulation because of the release and degradation of NT (Carraway, R., this volume). These results are in keeping with either a hormonal or paracrine role for intestinal NT but do suggest that if it acts as a hormone after lipid-stimulated release, the target tissue is likely to be nearby, that is, the heart, liver, or intestine.

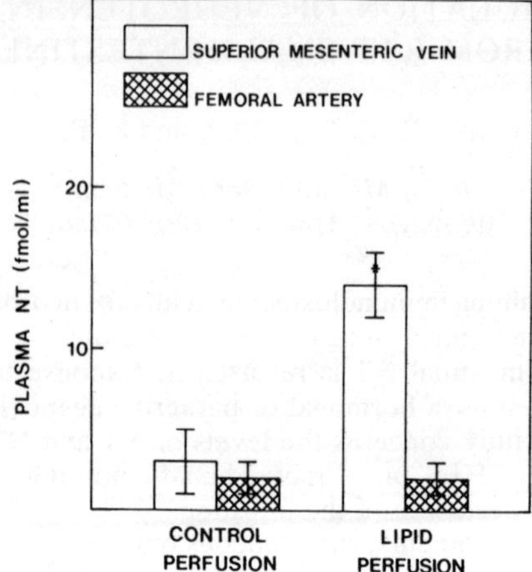

FIGURE 1. Arteriovenous difference in NT concentration in extracted plasma during the perfusion of the small intestine with 0.9% NaCl or a micellar solution of oleic acid. Given are the mean concentration of NT ± SEM (vertical bar) determined by HPLC and RIA (antiserum TG-1) of extracted plasma samples taken simultaneously from the superior mesenteric vein and femoral artery during saline or lipid perfusion (n = 3). *p < 0.05.

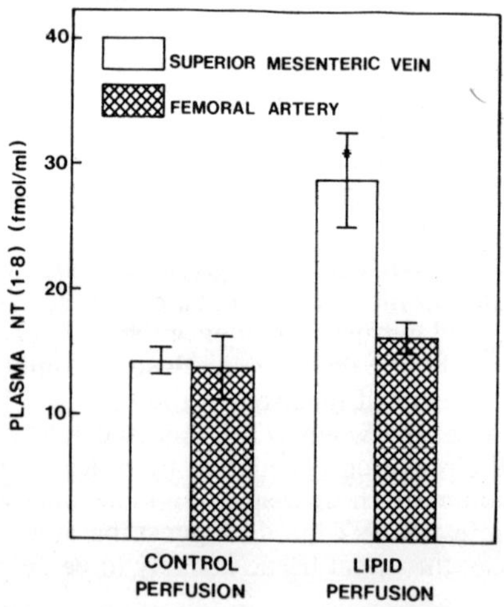

FIGURE 2. Arteriovenous difference in NT[1-8] octapeptide concentration in extracted plasma during perfusion of the small intestine with a control solution of 0.9% NaCl or a micellar solution of oleic acid. Legend is the same as FIGURE 1 except that NT[1-8] was measured. *p < 0.05.

REFERENCES

1. HELMSTAEDTER, V., G. E. FEURLE & W. G. FORSSMANN. 1977. Ultrastructural identification of a new cell type—the N-cell as the source of neurotensin in the gut mucosa. Cell Tissue Res. **184:** 445–451.
2. ROSELL, S., A. ROKAEUS, M. L. MASHFORD, K. THOR, D. CHANG & K. FOLKERS. 1980. Neurotensin as a hormone in man. *In* Neuropeptides and Neural Transmission. C. A. Marsan & W. Z. Traczyk, Eds. Raven Press. New York. pp. 181–189.
3. FERRIS, C. F., R. A. HAMMER & S. E. LEEMAN. 1982. Elevation in plasma neurotensin during lipid perfusion of the rat small intestine. Pettides **2:** 263.

MONOIODO-Trp[11]-NEUROTENSIN, A NEW LIGAND TO STUDY THE INTERACTION OF NEUROTENSIN WITH ITS RECEPTOR*

Jean-Pierre Vincent,† Jean Mazella,† Claudine Poustis,† Frédéric Checler,†
Patrick Kitabgi,† Catherine Labbé,‡ Claude Granier,‡ and
Jurphaas van Rietschoten‡

†Centre de Biochimie du Centre National de la Recherche Scientifique
Université de Nice
Parc Valrose
06034 Nice, Cedex, France

‡Laboratoire de Biochimie et Institut National de la
Santé et de la Recherche Médicale
Faculté de Médecine Secteur Nord
13326 Marseille, Cedex, France

This work deals with the preparation and binding properties to rat brain synaptic membranes of a biologically active derivative of neurotensin that can be labeled at a specific radioactivity of 2,000 Ci/mmol. [^{3}H]neurotensin is the only radioactive ligand of the neurotensin receptor that has been prepared in a pure state. However the specific radioactivity of this tritiated derivative is only 40 to 60 Ci/mmol. Because the number of neurotensin binding sites is small in brain and peripheral tissues, there is a need for a biologically active derivative of neurotensin with a much higher specific radioactivity. Iodination of neurotensin gives a complex mixture of peptides in which both Tyr_3 and Tyr_{11} have incorporated 1 or 2 iodine atoms per residue. Moreover, we have found that iodination of Tyr_{11} in acetyl neurotensin[8-13], a fully active analogue of neurotensin, decreased the biological activity of the hexapeptide by a factor of 20.

Trp[11]-neurotensin, a neurotensin analogue in which Tyr[11] is substituted by a Trp, was synthesized by the Merrifield solid-phase procedure. The biological activity of this analogue was found to be almost identical to that of native neurotensin in an *in vitro* bioassay involving rat ileum. Iodination of Tyr^3 in Trp[11]-neurotensin by the lactoperoxidase method was carried out in the presence of nonradioactive [^{127}I]Na, containing tracer amounts of radioactive [^{125}I]Na. The resulting mixture of diiodo, monoiodo, and unmodified Trp[11]-neurotensin was purified in one step by ion-exchange chromatography at pH 8.6. The mono and diiodo derivatives of Trp[11]-neurotensin were pure as judged by UV spectrophotometry, calculation of the number of iodine atoms per mole of peptide, HPLC, and amino-acid analysis. The activity of monoiodo Trp[11]-neurotensin in the rat ileum bioassay was very similar to that of the unmodified peptide. The use of [^{125}I]Na as the sole source of iodine in the lactoperoxidase-catalyzed iodination reaction allowed to prepare [^{125}I]monoiodo-Trp[11]-neurotensin with a specific radioactivity of 2,000 Ci/mmol. Binding of this analogue to synaptic membranes from rat brain was specific and reversible. The binding isotherm was biphasic and

*This work was supported in part by grant 81.E.0542 from the Délégation Générale à la Recherche Scientifique et Technique (D.G.R.S.T., Paris, France) and by the Fondation pour la Recherche Médicale Française.

436

could be described by postulating the existence of two different classes of independent binding sites. Dissociation constants for the high- and low-affinity binding sites were found to be 0.1 and 4.7 nM, respectively. Corresponding values of binding capacities were 17 and 126 fmol per mg of membrane protein, respectively. While dissociation rate constants are very similar for both types of site, the association rate constant of the high-affinity binding sites is 20-fold greater than that of the low-affinity binding sites. The specificity of a series of neurotensin analogues for both high- and low-affinity binding sites was the same as that previously observed with [^{3}H]neurotensin as labeled ligand. The low-affinity binding sites appear to be similar to the [^{3}H]neurotensin binding sites in terms of both affinity and binding capacity. The high-affinity binding sites that could not be detected with [^{3}H]neurotensin might represent either a new class of neurotensin receptors or a high-affinity state of the [^{3}H]neurotensin receptor.

DEFINING THE ANTINOCISPONSIVE ACTIONS OF NEUROTENSIN

Alan Cowan and Debra E. Gmerek

Department of Pharmacology
Temple University School of Medicine
Philadelphia, Pennsylvania 19140

It has recently been shown that neurotensin (NT), when given intracisternally, can delay the usual avoidance responses of rodents to certain noxious stimuli.[1,2] Naloxone-sensitive sites do not seem to be involved. The antinocisponsive[1] profile of NT is both novel and paradoxical. Thus, NT is active in mouse and rat hot-plate tests[1,3] and mouse tail-flick test[3] but not in the rat tail-flick test.[1] Although it is potent in the mouse (acetic acid) writhing test (0.25 ng is active),[1] we will report here that median effective doses cannot be obtained for NT in (a) the rat hypertonic saline writhing procedure and (b) the rat tail compression test.

We believe that new animal models have to be developed to help evaluate the antinocisponsive properties of peptides such as NT. One approach involves suppressing the wet-dog shakes (WDS) that rats elicit after acute i.p. injection of TRH[4,5] or the dihydrocodeinone, RX 336-M.[6] The rank order of potency of strong analgesics in suppressing shaking correlates well with clinical analgesic data.[6]

The attenuating effects of NT and morphine on TRH-induced shaking are described below. Additionally, we show that (a) tolerance does not readily develop to this suppressant action of NT and (b) naloxone does not precipitate behavioral changes in rats that are receiving NT chronically.

MATERIALS AND METHODS

Male, Sprague-Dawley rats (180–200 g) were each implanted with a 22-guage cannula in the right lateral cerebral ventricle.

Writhing test: Groups of 10 rats were pretreated i.c.v. with NT (Sigma) or morphine and then challenged i.p. 15 min later with 4% (w/v) saline.[7]

Rat tail compression test: The procedure is described in detail by Geller *et al.*[8]

Wet-dog shaking: NT (0.25–2 μg), morphine (0.625–5 μg), or saline (6 μl) was given i.c.v. to individually housed rats who received either TRH (20 mg/kg, i.p.) or saline 15 min later. The number of WDS was counted for 30 min.

Chronic i.c.v. administration: NT (1 μg/hr), morphine (40 μg/hr), or saline (1 μl/hr) was infused into the ventricles of each rat for 7 days from an s.c.-implanted Alzet 2001 osmotic minipump.

RESULTS

An ED_{50} value could not be obtained for NT (10–100 μg) in the rat hypertonic saline test. The ED_{50} for morphine was 2.8 μg.

NT (12.5–100 μg) was not active in the rat tail compression test when measurements were taken at 20, 40, and 80 min. A_{50} values for morphine were 9.7 μg, 4.9 μg, and 3.8 μg.

438

NT was 16 times more potent than morphine (on a molar basis) in attenuating TRH-induced WDS. Naloxone (10 mg/kg, s.c.), given at the same time as either NT (1 μg) or morphine (5 μg), antagonized the suppressant action of only morphine.

Tolerance did not develop, over 168 hr, to the suppressant action of NT. Naloxone (10 mg/kg) did not elicit behavioral changes or a significant weight loss in additional rats that received NT for 170 hr. This dose of naloxone precipitated an abstinence syndrome in the corresponding rats receiving morphine.

DISCUSSION

NT is active in several standard analgesic tests when mice are used[1,3] whereas it is inactive in rats when radiant heat,[1] pressure, or hypertonic saline are the noxious stimuli. Many opioids suppress (in a stereospecific manner) chemically induced WDS in rats[6,9,10]; opioid peptides are also active. In the present work, we have shown that small i.c.v. doses of NT antagonize TRH-induced WDS. Naloxone does not block, and tolerance does not develop to, the suppressant action of NT.

REFERENCES

1. CLINESCHMIDT, B. V., J. C. McGUFFIN & P. B. BUNTING. 1979. Eur. J. Pharmacol. **54:** 129–139.
2. NEMEROFF, C. B., A. J. OSBAHR, P. J. MANBERG, G. N. ERVIN & A. J. PRANGE. 1979. Proc. Natl. Acad. Sci. USA **76:** 5368–5371.
3. OSBAHR, A. J., C. B. NEMEROFF, D. LUTTINGER, G. A. MASON & A. J. PRANGE. 1981. J. Pharmacol. Exp. Ther. **217:** 645–651.
4. COWAN, A., I. R. MACFARLANE & T. WATSON. 1978. Abstr. 2nd World Cong. on Pain **1:** 134.
5. KRUSE, H. & J. MOELLER. 1981. Drug Dev. Res. **1:** 59–66.
6. COWAN, A. 1981. Fed. Proc. **40:** 1497–1501.
7. COLLIER, H. O. J. & C. SCHNEIDER. 1969. Nature **224:** 610–612.
8. GELLER, E. B., L. DURLOFSKY, A. COWAN, C. HARAKAL & M. W. ADLER. 1979. Life Sci. **25:** 139–146.
9. COLLIER, H. O. J., N. J. CUTHBERT & D. L. FRANCIS. 1981. Fed. Proc. **40:** 1513–1518.
10. WEI, E. 1976. J. Pharm. Pharmacol. **28:** 722–724.

PHARMACOLOGICAL EVIDENCE FOR A HETEROGENEITY OF RECEPTORS UNDERLYING VARIOUS CENTRAL AND PERIPHERAL EFFECTS OF NEUROTENSIN

François B. Jolicoeur* and André Barbeau

*Neurobiology Department
Clinical Research Institute
Montréal, Québec, Canada*

and

Remi Quirion, Francis Rioux, and Serge St-Pierre

*Department of Physiology and Pharmacology
University of Sherbrooke
Sherbrooke, Québec, Canada J1H 5N4*

Results from several structure–activity studies have demonstrated previously that the COOH terminal (Arg^9-Pro^{10}-Tyr^{11}-Ile^{12}-Leu^{13}) of neurotensin is primarily responsible for the biological activity of this peptide.[1-3] In our recent *in vitro* studies, the stimulatory effects of neurotensin and several fragments and analogues were examined in rat stomach strips and portal vein as well as in isolated spontaneously beating atria of guinea pigs.[4-6] Results of these studies confirmed the determinant role of the COOH terminal and indicated that, while Arg^9 Pro^{10}, Ile^{12}, and Leu^{13} mainly contribute to the affinity or binding of neurotensin to its receptors, Tyr^{11} appears to be implicated specifically in the activation of neurotensin receptors. In most cases, substitution of Tyr^{11} with certain amino acids yielded compounds with very low potencies or unable to replicate neurotensin's maximum stimulatory effects. These results are summarized in the right portion of TABLE 1 where the relative potencies of some of the fragments and analogues tested are presented.

In parallel to these *in vitro* studies, the neurobehavioral effects of the same fragments and analogues of neurotensin were also examined. All peptides were administered into the left cerebral ventricle of rats and effects on body temperature, locomotor activity and muscular tone were assessed over a 2-hr period after injection. As previously reported,[7-9] neurotensin significantly decreased all three measures in a dose-related fashion. Results obtained with the various fragments indicated that the COOH terminal is also responsible for these central effects of neurotensin. Contrary to neurotensin, some analogues such as {D-Phe^{11}}-NT, {D-Tyr^{11}}-NT, and {D-Trp^{11}}-NT significantly increased locomotor activity. However, similarly to neurotensin, these analogues significantly decreased body temperature and muscle tone of animals. These findings suggest that the putative receptors underlying the influence of neurotensin on locomotor activity may be pharmacologically different from those mediating other central effects of the peptide. These results are summarized in the left portion of TABLE 1 where for each peptide tested, doses producing statistically significant differences on activity, temperature, and muscle tone are given.

Present address: Department of Psychiatry, University of Sherbrooke, Centre Hospitalier Universitaire, Sherbrooke, Que., Canada J1H 5N4.

TABLE 1

CENTRAL AND PERIPHERAL EFFECTS OF NEUROTENSIN

	Central Effects Minimal Effective Doses*			Peripheral Effects Relative Potency†		
PEPTIDE	Motor activity	Hypothermia	Decreased muscle tone	Rat portal vein	Rat stomach strip	Guinea pig atria
NT	1.8 ↓	1.8	1.8	100	100	100
NT[8-13]	7.5 ↓	3.7	15.0	94	69	52
NT[9-13]	15.0 ↓	15.0	30.0	2.4	11	1.3
NT[1-12]	30.0 ↓	30.0	60.0	0.02	0.6	0.09
NT[1-10]	>240.0	>240.0	>240.0	0	0	0
NT[10-13]	>240.0	>240.0	>240.0	0	0	0
{Trp[11]}-NT	0.4 ↓	0.2	0.4	115	130	3.2
{D-Trp[11]}-NT	0.9 ↑	0.2	0.9	0.2	0.7	0.01
{D-Tyr[11]}-NT	3.7 ↑	0.9	3.7	<0.01	0.05	0.01
{D-Phe[11]}-NT	7.5 ↑	0.9	3.7	<0.01	0.02	0.01
{Phe[11]}-NT	60.0 ↓	60.0	60.0	9.4	16	15
{Leu[11]}-NT	>240.0	240.0	>240.0	<0.01	0.03	0.01

*Minimal effective doses for producing significant decreases in body temperature and muscular tone as well as significant changes, increase (↑) or decrease (↓), on locomotor activity are given for each peptide.

†Relative potencies of various peptides tested. Potency of NT was arbitrarily fixed at 100.

When the central effects of the structural analogues are compared to their peripheral effects, some important differences are revealed. For instance, as can be seen in TABLE 1, the analogues {D-Tyr[11]}-NT, {D-Trp[11]}-NT, and {D-Phe[11]}-NT were found to be very potent in their central effects, but relatively weak in their actions on the smooth muscle preparation. Although more work is needed to delineate the exact reasons for these differences, the present results may indicate that the peripheral receptors for neurotensin are distinct from those subserving the central actions of the peptide.

REFERENCES

1. CARRAWAY, R. & S. E. LEEMAN. 1976. Biol. Chem. **251**: 7045.
2. LOOSEN, P. T., C. B. NEMEROFF, G. BISSETTE, G. B. BURNETT, A. J. PRANGE & M. A. LIPTON. 1978. Neuropharmacol. **17**: 109.
3. RIVIER, J. E., L. H. LAZARUS, M. H. PERRIN & M. R. BROWN. 1977. J. Med. Chem. **20**: 1409.
4. QUIRION, R., D. REGOLI, F. RIOUX & S. ST-PIERRE. 1980. Br. J. Pharmacol. **68**: 83.
5. QUIRION R., F. RIOUX, S. ST-PIERRE, F. BELANGER, F. JOLICOEUR & A. BARBEAU. 1981. Neuropeptides. **1**: 253.
6. RIOUX, F., R. QUIRION, D. REGOLI, M. A. LEBLANC & S. ST-PIERRE. 1980. Eur. J. Pharmacol. **66**: 273.
7. JOLICOEUR, F. B., A. BARBEAU, F. RIOUX, R. QUIRION & S. ST-PIERRE. 1981. Peptides **2**: 171.
8. JOLICOEUR, F. B., F. RIOUX, S. ST-PIERRE & A. BARBEAU. 1982. *In* Brain Peptides and Hormones. R. Collu, J. R. Ducharme, A. Barbeau & G. Tolis, Eds. Raven Press. New York. p. 171.
9. NEMEROFF, C. B. 1980. Biol. Psychiat. **15**: 283.

Index of Contributors

(Italicized page numbers refer to comments made in discussion.)